This book is the first to provide both students and researchers in the field of astrophysical jets with a comprehensive and up-to-date account of current research. An important feature of the book is that it integrates studies of jets on all length scales, and combines discussions of both extragalactic and Galactic jets. There are ten chapters, authored by fourteen researchers, each of whom is an expert on their chosen topic, and the book has been edited to provide a cohesive account of this field of study. It will be an important textbook for graduate students, and a valuable reference source for researchers in many areas of extragalactic and Galactic astronomy. It will also be of interest to plasma physicists and space scientists.

The material covered includes, a description and classification of parsec- and kiloparsec-scale extragalactic jets and of Galactic jets; basic jet physics and the derivation of jet parameters; the theories of jet propagation and stability; the processes which accelerate particles and generate magnetic fields; the mechanisms by which jets radiate; hydrodynamic simulations; and the engines which generate and collimate the flows.

Cambridge astrophysics series

Beams and jets in astrophysics

In this series

1 Active Galactic Nuclei
edited by C. Hazard and S. Mitton

2 Globular Clusters
edited by D. A. Hanes and B. F. Madore

3 Low Light-level Detectors in Astronomy
by M. J. Eccles, M. E. Sim and K. P. Tritton

4 Accretion-driven Stellar X-ray Sources
edited by W. H. G. Lewin and E. P. J. van den Heuvel

5 The Solar Granulation
by R. J. Bray, R. E. Loughhead and C. J. Durrant

6 Interacting Binary Stars
edited by J. E. Pringle and R. A. Wade

7 Spectroscopy of Astrophysical Plasmas
by A. Dalgarno and D. Layzer

8 Accretion Power in Astrophysics
by J. Frank, A. R. Ling and D. J. Raine

9 Gamma-ray Astronomy
by P. V. Ramama Murthy and A. W. Wolfendale

10 Quasar Astronomy
by D. W. Weedman

11 X-ray Emission from Clusters of Galaxies
by C. L. Sarazin

12 The Symbiotic Stars
by S. J. Kenyon

13 High Speed Astronomical Photometry
by B. Warner

14 The Physics of Solar Flares
by E. Tandberg-Hanssen and A. G. Emslie

15 X-ray Detectors in Astronomy
by G. W. Fraser

16 Pulsar Astronomy
by A. Lyne and F. Graham-Smith

17 Molecular Collisions in the Interstellar Medium
by D. Flower

18 Plasma Loops in the Solar Corona
by R. J. Bray, L. E. Cram, C. Durrant and R. E. Loughhead

19 Beams and Jets in Astrophysics
edited by P. A. Hughes

BEAMS AND JETS IN ASTROPHYSICS

Edited by

P. A. HUGHES

Department of Astronomy
University of Michigan

CAMBRIDGE UNIVERSITY PRESS

Cambridge

New York Port Chester

Melbourne Sydney

CAMBRIDGE UNIVERSITY PRESS
Cambridge, New York, Melbourne, Madrid, Cape Town, Singapore,
São Paulo, Delhi, Dubai, Tokyo, Mexico City

Cambridge University Press
The Edinburgh Building, Cambridge CB2 8RU, UK

Published in the United States of America by
Cambridge University Press, New York

www.cambridge.org
Information on this title: www.cambridge.org/9780521335768

© Cambridge University Press 1991

This publication is in copyright. Subject to statutory exception
and to the provisions of relevant collective licensing agreements,
no reproduction of any part may take place without the written
permission of Cambridge University Press.

First published 1991

A catalogue record for this publication is available from the British Library

Library of Congress Cataloguing in Publication Data

Beams and jets in astrophysics/edited by P. A. Hughes.
p. cm. – (Cambridge astrophysics series)
Includes index.
ISBN 0 521 34025 X. – ISBN 0 521 33576 0 (paperback)
1. Astrophysical jets. 2. Radio sources (Astronomy) I. Hughes.
P. A. (Philip A.) II. Series.
QB466.J46B43 1991
522′.682 – dc20 2141 CIP

ISBN 978-0-521-34025-0 Hardback
ISBN 978-0-521-33576-8 Paperback

Cambridge University Press has no responsibility for the persistence or
accuracy of URLs for external or third-party internet websites referred to in
this publication, and does not guarantee that any content on such websites is,
or will remain, accurate or appropriate. Information regarding prices, travel
timetables, and other factual information given in this work is correct at
the time of first printing but Cambridge University Press does not guarantee
the accuracy of such information thereafter.

Contents

Journal Abbreviations

Adv. Astr. Ap.	Advances in Astronomy and Astrophysics
Ann. New York Acad. Sci.	Annals of the New York Academy of Sciences
Ann. Rev. Astr. Ap.	Annual Review of Astronomy and Astrophysics
Ann. Rev. Fluid Mech.	Annual Review of Fluid Mechanics
Ap. J.	Astrophysical Journal
Ap. J. Lett.	Astrophysical Journal Letters
Ap. J. Suppl.	Astrophysical Journal Supplement Series
Ap. Lett.	Astrophysical Letters
Ap. Sp. Sci.	Astrophysics and Space Science
Ark. f. fys.	Arkiv för fysik
Astr. Ap.	Astronomy and Astrophysics
Astr. Ap. Suppl.	Astronomy and Astrophysics Supplement Series
Astron. J.	Astronomical Journal
Astron. Zh.	Astronomicheskii Zhurnal
Bull. A. A. S.	Bulletin of the American Astronomical Society
Bull. Astron. Soc. India	Bulletin of the Astronomical Society of India
Can. J. Phys.	Canadian Journal of Physics
Comm. Ap.	Comments on Astrophysics
Compt. Rend. Acad. Sci.	Comptes Rendus de l'Academie des Sciences, U.R.S.S.
Dokl. Akad. Nauk SSSR	Akademiia Nauk SSSR - Doklady (Novaia Seriia)
Geophys. Res. Lett.	Geophysical Research Letters
J. Fluid Mech.	Journal of Fluid Mechanics
J. Geophys.	Journal of Geophysics
J. Geophys. Res.	Journal of Geophysical Research
J. Phys.	Journal of Physics
J. Phys. Soc. Japan	Journal of the Physical Society of Japan
J. Plasma Phys.	Journal of Plasma Physics
J. R. A. S. Canada	Journal of the Royal Astronomical Society of Canada
M. N. R. A. S.	Monthly Notices of the Royal Astronomical Society

Mem. Soc. Astron. Ital.	Memoirs of the Italian Astronomical Society
Phys. Fluids	Physics of Fluids
Phys. Rep.	Physics Reports
Phys. Rev.	Physical Review
Phys. Rev. Lett.	Physical Review Letters
Phys. Scripta	Physica Scripta
Proc. Astron. Soc. Australia	Proceedings of the Astronomical Society of Australia
Proc. Natl. Acad. Sci. U.S.A.	Proceedings of the National Academy of Sciences of U.S.A.
Prog. Theoret. Phys.	Progress of Theoretical Physics
Publ. Astron. Soc. Japan	Publications of the Astronomical Society of Japan
Publ. Lick Obs.	Publications of the Lick Observatory
Rep. Prog. Physics	Reports on Progress in Physics
Rev. Geophys. and Sp. Phys.	Reviews of Geophysics and Space Physics
Rev. Mod. Phys.	Reviews of Modern Physics
Riv. Nuovo Cimento	La Rivista del Nuovo Cimento
Soviet Astr. - AJ	Soviet Astronomy - AJ
Sp. Sci. Rev.	Space Science Reviews
Zeitschr. F. Geophys.	Zeitschrift Für Geophysik

Preface

More than three decades have passed since our present picture of extragalactic radio sources began to unfold. The latter half of that time has witnessed the 'mapping' or 'imaging' of jet-like structures in many of these sources, and the realisation that apparently similar phenomena are associated with many Galactic objects. Numerous books have discussed instrumentation and radiation processes; overviewed the physics underlying both extragalactic and Galactic sources, and their intervening media; and attempted to present a coherent picture of the AGN phenomenon. And yet, although some excellent reviews have appeared, no book has addressed the subject of astrophysical jets in a detailed and comprehensive manner. This volume is an attempt to fill that gap.

What makes such a volume particularly timely, is that we are now digesting the first generation of high-resolution observations of extragalactic jets (MERLIN, VLA and VLBI data), the first generation of numerical simulations (mostly two-dimensional and nonmagnetic), and the first generation of theoretical studies, which have given us a quantitative framework for estimating physical properties and energetics, and for discussing jet formation, propagation, and stability. Now is a time to take stock, as we await the first results of the VLB Array, satellite VLBI, three-dimensional and MHD simulations, and more refined theoretical studies. It also seems timely to compare and contrast the bodies of research on extragalactic and Galactic objects.

In order to achieve a detailed, comprehensive and critical text, it has been necessary to adopt a multi-author approach. We have not tried to disguise the numerous independent voices of this work, and hope that individual chapters will be a valuable reference for the researcher seeking more specialist information. However, we have tried, through the choice, arrangement, and cross referencing of material, to produce a book that may also form a coherent, graduate level introduction to the subject.

The first seven and the tenth chapters of this monograph use SI units (Système International d'Unités). The use of cgs (centimetre gramme second) units in the eighth and ninth chapters reflects the usage in the literature that these two chapters cite. An increasing amount of research, particularly that published in the European journals, uses SI units; while the literature on the electrodynamics of magnetospheres, and on laboratory and space plasma physics (much of it in the Russian literature) continues to utilise cgs units. We have decided that the advantage to the reader of being able to relate the material given here, directly to the large body of work that is cited, outweighs the inelegance of twice straying from the use of SI units.

For the sake of uniformity, we have adopted a Hubble constant of $H_0 = 100$ km s^{-1} Mpc^{-1}. This choice was made on the grounds of convenience and aesthetics, and should not be taken to endorse a particular school of cosmology!

We have attempted to use a uniform style of notation throughout the book. However, the limited number of letters in the Roman and Greek alphabets, together with a desire not to stray too far from convention, means that certain letters and symbols have multiple meaning. A comprehensive list of symbols appears at the end of *each* chapter.

An index of those Galactic and extragalactic objects discussed in the text precedes the general subject index.

It is a pleasure to thank Richard Saunders for many helpful suggestions on the content and arrangement of this monograph, and Anuradha Koratkar for compiling the index.

Astronomy Department
University of Michigan

P. A. Hughes

1

Introduction : Synchrotron and Inverse-Compton Radiation

PHILIP A. HUGHES
Astronomy Department, University of Michigan, Ann Arbor, MI 48109, USA.

LANCE MILLER
Astronomy Department, University of Edinburgh, Edinburgh, EH9 3HJ, UK.

1.1 General introduction

1.1.1 History

In a description of an optical image of M 87 (NGC 4486), Curtis (1918) wrote "a curious straight ray ... connected with the nucleus". By the 1950s the term 'jet' was being used to describe this feature which it seemed plausible to associate with ejection of material from the innermost region of the galaxy (Baade & Minkowski 1954), although the concept of a continuous flow was not then envisaged. Baade (1956) measured the optical polarization of the M 87 jet, supporting the idea that the material was a synchrotron emitting plasma akin to that of the Crab supernova remnant.

Shklovskii (1963), in an attempt to explain the double radio sources and M 87's jet, discussed many ideas that play a role in current theories – accretion of matter in the gravitational potential of a galactic nucleus; the consequent heating of a plasma that breaks out along a preferred axis; the flow of this material into intergalactic space and the re-energization of the electrons within the flow. However, the model still did not encompass the idea of a *continuous* flow, carrying energy in the form of bulk motion. Schmidt (1963) wrote of "a wisp or jet" on the image of the optical counterpart to 3C 273, and by about this time, the term 'jet' was in common usage (*e.g.*, Greenstein & Schmidt 1965; Burbidge, Burbidge & Sandage 1965) – but still without a clear recognition that a continuous flow of matter and energy was involved.

Bolton, Stanley & Slee (1949) made the first tentative identification of a radio source with an extragalactic object (Virgo A – M 87, and Centaurus A – NGC 5128; see also Baade & Minkowski

1954), and so provided an indication of the former's distance. Jennison & Das Gupta (1953), observing Cygnus A, discovered the now widely known 'double-lobed' structure of extragalactic radio sources. Following the suggestion of Shklovskii (1953a,b) that the radio emission might be electron synchrotron radiation, Burbidge (1956) developed the theory for estimating the minimum energy content of these plasma lobes, and so set the stage for theorists to attempt an explanation of the source of this energy. The common theme of early models was the transport of energy, in a largely loss-free form, over scales as large as a Mpc. The ideas included buoyant plasma blobs (Gull & Northover 1973), and plasma blobs (plasmons) confined by self-gravity, inertia (*i.e.*, a density high enough that the expansion at the internal sound speed was negligible over the plasmon lifetime), or ram-pressure due to motion through an ambient medium (De Young & Axford 1967). Plasmons gravitationally bound to supermassive objects, supposedly ejected by a 'gravitational slingshot' process from a parent galaxy, were discussed by Saslaw, Valtonen & Aarseth (1974). Following a suggestion by Rees (1971), Scheuer (1974) and Blandford & Rees (1974) discussed the possibility that energy transport within radio sources is in the form of *beams*; a collimation process (Blandford and Rees suggested that a structure known to hydrodynamicists as a 'de Laval nozzle' might form within the gravitational potential of a parent galaxy) converts much of the internal energy of a hot plasma into energy of bulk motion, and this energy may be recovered at a great distance from its site of generation, where it is randomized, perhaps as the beam interacts with an ambient medium. Within a few years of the birth of the beam model, observers had made the first maps to show jet-like features, (*e.g.*, in 3C 66B, Northover 1973; in 3C 219, Turland 1975b), and associated the radio and optical jet-like structures in M 87 (Turland 1975a).

****

The existence of interstellar molecules has been known for fifty years – the signatures of CN, CH and CH^+ having been observed in stellar spectra in the late thirties (see Bates & Spitzer 1951, and references 1 – 6 therein). However, the discovery of the OH radical and NH_3 (Weinreb *et al.* 1963; Heiles 1968; Cheung *et al.*

1968) marked the start of a period of rapid progress in this field. By the early seventies it was recognised that cold molecular clouds containing remarkably complex molecules form an important phase of the ISM (*e.g.*, Zuckerman & Palmer 1974). Of similar antiquity is the study of T Tauri stars; this started in the early forties, and within a decade it was generally accepted that such objects are stars seen very close to their time of birth (*e.g.*, Herbig 1962). Herbig-Haro objects also, have been known for almost four decades (Herbig 1951; Haro 1952). Litvak *et al.* (1966) and Perkins, Gold & Salpeter (1966) referencing a series of observations made a year earlier, developed the theory of the interstellar OH maser. By the late sixties the existence of an H_2O maser also had been established (*e.g.*, Reid & Moran 1981).

Within the last decade, a remarkable series of observations has tied together this apparently unrelated group of Galactic phenomena. The initial uncertainty over the interpretation of the motions of CO clouds (*e.g.*, Kwan & Scoville 1976; Zuckerman, Kuiper & Rodriguez-Kuiper 1976) soon gave way to a picture of bipolar molecular outflow (*e.g.*, Bally & Lada 1983). Cudworth & Herbig (1979) confirmed the result noted by Luyten in 1963 that the Herbig-Haro objects HH 28 and 29 display large proper motion away from the IR source in Lynds 1551. Observations of systems of maser lines (*e.g.*, Genzel & Downes 1977a,b) and VLBI observations showing the coherent motion of clusters of masing cloudlets (*e.g.*, Genzel *et al.* 1981; Schneps *et al.* 1981) added to the kinematic picture of highly asymmetric outflow – reminiscent of that inferred for the extragalactic radio sources, but associated with young stellar objects. This picture was completed in 1983 and 1984, with the first deep CCD images in $H\alpha$, showing narrow jet-like features associated with T Tauri stars and HH objects (see references 1 – 5 in Mundt 1986). The rapid advances in this field are eclipsed only by those that followed the discovery of the extraordinary moving line-system of SS 433 (Margon *et al.* 1979a,b; Liebert *et al.* 1979) – and which brought old stellar systems into the class of jet-manifesting objects.

Today we have VLA and Merlin images ranging from the barely detected, thread-like jet of Cygnus A, to the plume of emission from M 84; we can follow jets from the Mpc-scale to the subparsec-scale as in 3C 120; we can study the internal structure of jets in the form of knots and filaments as in M 87, and we can watch the interaction of jets with their environment as in NGC 1265. In our own galaxy we see bipolar outflows from young stars, highly collimated flows from old stellar objects (*e.g.*, SS 433) and perhaps, even collimated outflow from the Galactic centre. The observations amassed over the last decade provide remarkable support for the ubiquity of jets that is implicit in the Blandford-Rees-Scheuer model of extragalactic radio sources.

1.1.2 Beams and jets

It is worth making a note on terminology concerning *beams* and *jets*; we always use the term *beam* to refer to figments of the theorist's imagination, and the term *jet* to refer to observed structures. The latter term is now applied indiscriminately in the astronomical literature, and a number of authors have attempted to construct a 'code of practice' for its legitimate usage. For example, Bridle (1984) suggests these three criteria for extragalactic jets:

The structure must be at least four times as long as it is wide (after allowing for instrumental effects).

It should be separable (spatially, or by brightness contrast) from extended structure by high resolution observations.

It should be aligned with a radio core where closest to that core.

Bridle noted some of the problems in applying these criteria to extragalactic sources, and the value of the second and third points when considering Galactic objects is doubtful. Thus here we adopt the view that the term 'jet' should be applied to objects that are substantially elongated (in the spirit of the first criterion above), and for which it is reasonable to suppose that a flow of mass (and energy) is involved. Here it is important never to lose sight of the fact that for many so called 'jets', there is little or no observational evidence for a flow. Such evidence exists almost exclusively for Galactic objects (the moving line systems of SS 433, the proper motion of Herbig-Haro objects, the Doppler shift of OH-masers and the proper motion of H_2O-masers (*e.g.*, IRc2 in Orion), and the Doppler shift of line emission from molecular clouds). In extragalactic sources the scant evidence comprises little more than VLBI observations of proper motion of 'knots' – for which the relationship with a continuous flow is not unambiguously established. Nevertheless, in all these cases, elongated structures exist, where it seems *reasonable* to assume that a collimated flow of matter and energy is involved.

Some so called 'alternatives' to the 'simple' beam picture are really just additional complexities to that picture. For example, the emission from a jet may be from a slowly moving or even stationary sheath, within which a fast core transports energy to extended structure. Observations of the M 87 jet (Hardee, Owen & Cornwell 1988) show limb brightening, although this might just reflect the dense sheath suggested for magnetically collimated flows (see Chapter 8). The flow might be discontinuous; indeed, the many knots of both parsec- and kpc-scale jets, and the patchiness of emission, might reflect alternate quiescent and active periods of the central engine. This is perhaps most easily investigated using numerical hydrodynamics (see Chapter 7) and such studies suggest that substantial variations in the flow speed can induce knots of emission, without invalidating the concept of a collimated flow. Perhaps only when the emitting material is in the form of wholly independent plasmons (as, for example, in the model of ring-like structures in the lobes of Her A (Maiden & Christiansen 1986)) should we give up the concept of a beam.

Other authors have considered that extragalactic 'jets' *do not* delineate a collimated flow of mass and energy from a galactic nucleus. Valtaoja & Valtonen (1984) have modelled the evolution of radio sources (change of power, size and morphology with time) in terms of supermassive objects that are ejected from galactic nuclei, and act as a source of magnetic field and fast particles. In this picture, a 'jet' is the column of field-plus-plasma entrained by interstellar gas during the passage of a supermassive object. Valtonen & Byrd (1980) have been able to reproduce the appearance of the trail source 3C 129, using a similar model – one lobe of emission being associated with a supermassive object falling into the 'parent' galaxy. However, this monograph is based on the premise that jets are manifestations of quasi-continuous outflows of mass and energy, and seeks evidence for this, and an understanding of the physical processes involved.

1.1.3 The selection of material

If we list the major questions that may be answered by observation:

What are jets made of? Are extragalactic jets electron-positron or electron-proton plasmas? Does entrainment change their composition between the parsec- and kiloparsec-scales? What is the relative distribution of neutral and ionized matter in bipolar stellar outflows?

What is the plasma flow speed? Are extragalactic flows relativistic or nonrelativistic? Is the flow speed greatly different from source to source, or from parsec to kiloparsec-scale? What is the spatial relationship between high and low velocity bipolar stellar outflows?

What is the internal pressure, temperature and density (in both relativistic and cold matter)? Are jets confined, and if so, by what? How do they interact with their ambient medium? What causes them to radiate, and how well does the observed pattern of emissivity trace the underlying plasma flow?

What is the strength and geometry of the magnetic field? Does it significantly influence the dynamics of the flow?

and the questions that may be addressed through a better understanding of jets:

What do jets tell us about the central engine from whence they come? What, in particular, can we learn about collimator efficiency, lifetime, symmetry and direction changes? And what do we learn about stellar birth and death? Why are extragalactic jets associated with one class of galaxy but not another; or with one galaxy of a certain type but not another of the same type? Are stellar jets associated with a certain mass of star or a certain environment?

In what way do jets influence their environment? And what can we learn about this environment, in terms of composition, density, temperature, magnetic field? Is there an intergalactic medium?

Is there but one basic phenomenon, in various disguises?

That is, is there a clear link between FR I, FR II, Seyfert, young stellar object and old stellar object jets?

it is striking to note that we are currently unable to give even tentative answers on most of these points. Our philosophy in this monograph has therefore been to present a) a comprehensive description of the observed structures, and b) what we believe to be the physics relevant to an understanding of these structures – a quantitative interpretation of the observations for each of several possible answers to the above questions – that can form the basis for a pursuit of more definite answers to the many outstanding problems. We have put special emphasis on highlighting the assumptions and model dependency involved in the determination of fundamental flow parameters (such as velocity, pressure, and density). And we have attempted to tie all material together to clarify the link between central engine, parsec-scale jet, large-scale jet and lobe; the link between extragalactic and Galactic jets; and the relationship between observations, theory and numerical simulation.

In order to achieve a reasonably thorough treatment within a single volume, we have been forced to omit discussion of a number of important but peripheral subjects:

We have made no attempt to discuss the tools that have been so fundamental in the development of the observational base upon which this volume rests. The reader is warned that, despite recent advances – for example, great improvement in the dynamic range of Long Baseline Interferometry – the limitations of current observing techniques (restricted dynamic range, insensitivity to large scale structure, infrequent and limited frequency range mapping of epoch and frequency dependent structure) must be born in mind when interpreting the data.

In order to present a coherent picture, we have described the sometimes superluminal parsec-scale knots observed in extragalactic sources in terms of propagating structures intrinsic to the flow. We have not discussed the possibility that this motion may be due to gravitational lensing, special particle trajectories (*e.g.*, along large-scale dipole fields), or 'screen effects' wherein emission is excited locally by a propagating wave, but where there is no material flow. It can be noted that some of these models suffer severe difficulties when confronted with observations, and none have found wide favour.

The importance of central engine models for energy generation and collimation of extragalactic jets demands that they be discussed here. However, we have made no attempt to include material on the structure and broadband properties of Active Galactic Nuclei (AGN).

We have said nothing about the cosmological implication of jets. Such a discussion would include observational evidence for a variation of comoving space density, source power and source size with redshift. And it would attempt to interpret these observations, by asking whether there was an enhanced frequency of AGN at high redshift, and perhaps a difference in their power, or in the ambient IGM. We hope that this monograph gives an adequate account of jet physics, into which can be incorporated evolution of the energy source and ambient medium.

For the reader wanting some introductory reading for this volume, or to pursue some of the subject matter that we have not found space to present, we suggest the following:

The article by Bridle & Perley (1984) gives an excellent qualitative and quantitative overview of extragalactic jets, and clearly sets out the principal observed features that need to be explained by theory. The review by Begelman, Blandford & Rees (1984), quite apart from giving a concise but comprehensive review of extragalactic jets, can provide the interested reader with an introduction to the relationship of jets to AGN and their cosmological implications. Miller (1985) and Weedman (1986) cover the subjects of AGN and central engine models.

Lada (1985) reviews the field of outflows associated with young stellar objects, while Schwartz (1983) discusses Herbig-Haro objects in the context of bipolar flows. The review by Margon (1984) covers the first years of study of SS 433.

Within the last decade there have been many conferences and workshops, aspects of which are of undoubted interest to the jet-community. We note the following, only as the most recent and directly relevant: the workshop proceedings edited by Bridle & Eilek (1984) records a lively interaction between observers and theorists; Henriksen (1986) presents the first serious attempt to integrate the study of extragalactic and Galactic (*i.e.*, stellar) jets; Kundt (1987) brings together an interesting selection of papers, covering the subjects of extragalactic and Galactic jets, AGN, central engine models and cosmology; Zensus & Pearson (1987) present recent observational and theoretical results on superluminal radio sources, a class of object whose study depends on advances in VLBI – a subject discussed at I. A. U. Symposium 129 (Reid & Moran 1988).

1.1.4 A synopsis of the monograph

The study of astrophysical jets is an application of radiation physics, plasma physics and (magneto)hydrodynamics, and we present a certain amount of introductory material for each of these disciplines. The relativistic transformation of the radiation field, and MHD shocks are reviewed in Chapter 4; a simple description of laminar and turbulent MHD flows is introduced in Chapters 3 and 5; MHD stability analysis is introduced in Chapter 6; and the plasma physics of waves and shocks is discussed in Chapter 9. We have put special emphasis on the methodology and results of stability analysis and the physics that underlies theories of particle and magnetic field dynamics, in an attempt to make these rather specialized subjects accessible to the whole jet community.

Where appropriate, we have discussed observations in wavebands other than the radio: for example, optical synchrotron, and thermal X-ray observations of extragalactic objects (when instrumental sensitivity permits their detection and when the observations can be related to jets); and optical images, and molecular line observations of Galactic flows. However, nature dictates the devotion of much space to radio observations, because the physical conditions within sources usually favour the emission of radio synchrotron radiation – at least for extragalactic objects. Thus radio synchrotron radiation is a natural diagnostic of the underlying flows and forms the principal interface between theorist, observer, and the flows under study. Because of the fundamental role of the synchrotron and inverse-Compton processes for the understanding of extragalactic jets, the remainder of this chapter is devoted to a discussion of these processes.

The next four chapters are intended to give a comprehensive overview of extragalactic jets. Chapters 2 and 3 lead the reader through the observation and theory of 'large-scale' structures, that include jets on the kpc- and larger-scale, and the lobes of material that cocoon them. Some discussion of parsec-scale jets is given, to clarify the link between these and the larger-scale flows. Terminology is defined, the various types of source are described and illustrated, and a morphological scheme is presented. Systematic trends – particularly the dependence of morphology on source power – are discussed, and the distribution of sources on the 'power-linear size' plane is shown. In Chapter 3 the basic methods, with results, are given for finding source distance, size, power, energy content, magnetic field strength and age. The use of polarization measurements for determining magnetic field direction (and hence probing the underlying flow) and thermal matter content is critically examined. The chapter concludes with a discussion of how the derived parameters, in conjunction with a hydrodynamic description of the flow, are used to determine the sources' energy budget, and explain some of the different morphologies described in Chapter 2. Both chapters emphasize the role of selection effects, assumptions and approximations that colour much of the interpretation.

Chapter 4 concentrates on the parsec- and subparsec-scale objects. Here we are at the limit of resolution for Earth-based telescopes, and much of our knowledge is gained indirectly, or is of a statistical nature. Chapter 4 takes as its premise that apparent superluminal motion is a manifestation of relativistic flow speeds, and critically examines what we have learned and can hope to learn from statistical studies of distributions of apparent velocities and luminosities. It goes on to consider the observational consequences of relativistic collimated flows, both when quiescent, and when subject to perturbation, such as shock waves.

Chapter 5 ties much of the preceding material together within a theoretical framework. It stresses the continuity of central engine/collimator – small-scale flow – large-scale flow – lobe, by considering the mechanisms that may be responsible for collimating flows, bending these same flows, and perhaps leading to the entrainment of ambient material. Within this framework it is asked whether small-scale structures are the 'progenitors' of large-scale jets, with bending leading to the often observed misalignment of these features, and with entrainment of matter being responsible for a slowing from relativistic flow speed on the parsec-scale to subrelativistic flow speed on the kiloparsec-scale.

Chapters 6, 7, 8 and 9 are more specialized discussions that relate in one way or another to the earlier material. Chapter 6 discusses the stability of collimated flows. It sets out the equations of relativistic MHD and demonstrates how an analytic stability analysis may be performed. The material is presented with sufficient explanation, that we hope it can act as an opening into the subject for people hitherto unfamiliar with the field, but also with enough detail that it can act as a good introduction for people wanting to delve deeper into the technicalities. The chapter concludes with an extensive critique of the attempts to explain certain source morphologies in terms of beam instabilities.

Chapter 7 discusses the results of numerical simulations of hydrodynamic flows. It considers the large-scale morphology of extragalactic beams and their backflows (cocoons) and the formation of internal structures such as hotspots and shocks. The relationship between beam speed (subsonic or supersonic) and overall source morphology is considered, and the role of beam-cocoon and cocoon-ambient medium interaction on global morphology is presented. We have made no attempt to address the field of computational technique. Rather, we have presented the results of a family of numerical studies that demonstrate the usefulness of this approach – both in terms of *understanding* different source morphologies, and in terms of determining the hydrodynamic parameters of the flows.

Chapter 8 describes the current models for energy generation and collimation within galactic nuclei. We concentrate on those models favoured by 'current wisdom': those involving accretion structures about a supermassive black hole. Structure and potential role in beam formation are described for thin disc, radiation pressure supported torus, ion pressure supported torus, and thin disc with magnetosphere. There is considerable disparity between the level of development of theories for these objects; but to the extent that this allows, we attempt to assess the strengths and weaknesses of the different theories, describe how beams formed in these different 'environments' might differ, and how the beam may be influenced by the environment of the AGN – for example, the quenching of a counter beam, or the flipping of a beam from one side of an accretion plane to the other.

Chapter 9 describes how particle acceleration to a power-law in energy might be achieved, and how magnetic fields may be maintained, or grow, perhaps with the formation of a large-scale field component. Most of the theories discussed here are rooted in plasma physics – theory, experiment and spacecraft measurements

– whose unfamiliarity to many has prevented them from appreciating advances in this field. The diversity of topics covered means that in one chapter we cannot hope to give a working knowledge of the field. Rather, we have attempted to present enough basic plasma and MHD physics that the reader without a background in these areas can gain a reasonable understanding of current theories of particle and field dynamics, and appreciate the problems and outstanding questions.

Observations of Galactic jets have grown enormously in both quantity and quality over the last decade, so that a single chapter clearly cannot do justice to this field. Ideally, we would have discussed in more detail the Galactic counterparts of the extragalactic objects to which most of this book is devoted. However, as elsewhere in this volume, space limitations have not permitted such an extensive treatment. What we have attempted in Chapter 10 is to give a clear overview of Galactic objects, and the physical processes at work. Where possible, we have stressed the link between Galactic and extragalactic jets. We hope that this can act as a useful interface between the Galactic and the extragalactic literature. We describe the collimated outflows, seen in many different wavebands, that are associated with protostars and young stellar objects, the physical conditions in these flows, and their likely origin. Observations and theory of SS 433 are described, and other candidate 'old stellar objects with jets' are presented. A picture of the complex environment of our Galactic Centre is drawn, and the candidate jet structures (Sgr A, the Northern Galactic Lobe, the Galactic Centre Lobes and the Sofue Jet) are discussed. It appears that our galaxy indeed contains scaled down versions of extragalactic outflows.

1.2 Radiation from extragalactic jets

It can be deduced that most of the photons observed from the nuclei of quasars and active galaxies at infrared, optical and X-ray wavelengths originate on a scale of parsecs, or smaller. Conversely, many of the most luminous radio sources have sizes up to a million times larger. Evidence for a link between the two comes from observations of large-scale radio structures which appear to originate at the nucleus, and of compact radio-emitting nuclei apparently coincident with the nuclei observed at other wavebands.

Direct evidence of outflow from nucleus to extended lobes is provided solely by very high resolution radio observations of those radio nuclei. These show the presence of apparently well-collimated, outward-moving components which link up to the larger-scale structure (see §§1.2.1.5, 2.2.6 and 4.1). Remarkably, the speeds deduced from the observed proper motions are often greater than that of light – the phenomenon of superluminal motion. Thus it is now widely believed that relativistic effects must be taken into account in our interpretation of the nuclear outflow and of the appearance of radio sources.

The inferred presence of relativistic flow speeds means that appropriate transformations must be applied to convert *observed* quantities into the source reference frame, or *computed* quantities into the observer's reference frame. We refer the reader to Chapter 4 of Rybicki & Lightman (1979) for a discussion of the relativistic transformation of the radiation field, and to Appendix C of Begelman, Blandford & Rees (1984) for a summary of the most commonly encountered equations and transformations. The important concepts to bear in mind are:

Because of relativistic *aberration*, radiation emitted at some angle to the flow direction in the source frame is swung round, and seen in the observer's frame at an angle closer to the flow axis;

This effect causes radiation to be *beamed* into the flow direction; in part as a consequence of this, the intensity of radiation is increased for an observer lying near to the flow direction, and is decreased for other observers;

The fact that radiating elements may 'follow' their emitted photons means that an observer may receive, with arbitrarily small time delay, photons emitted by an element of the flow at widely separated times in the source frame – and hence see *apparent superluminal motion.*

The well-known time-dilation effect and the fact that radiating elements may 'follow' their emitted photons can be quantified together through the *Doppler factor*, $\mathcal{D} = \gamma^{-1}(1 - \beta\cos\Psi)^{-1}$, which depends on the flow speed (β), Lorentz factor ($\gamma = 1/\sqrt{1-\beta^2}$) and angle to the observer (Ψ). As there is a similar transformation associated with the cosmological redshift, z, it is often con-

Table 1.1. *Transformation of time (t), frequency (ν) and spectral intensity ($I_\nu(\nu)$) from source (primed) to observer (unprimed) frame; and aberration and apparent transverse motion formulae.*

Quantity	Relation
time	$t = \mathcal{D}^{-1}t'$
frequency	$\nu = \mathcal{D}\nu'$
spectral intensity	$I_\nu(\nu) = \mathcal{D}^3 I'_{\nu'}(\nu')$
aberration of angle Ψ'	$\cos\Psi = (\beta + \cos\Psi')/(1 + \beta\cos\Psi')$
transverse motion	$\beta_{\mathrm{app}} = \beta\sin\Psi/(1 - \beta\cos\Psi)$

venient to replace $\mathcal{D}$ with the function $\mathcal{D}/(1+z)$. The Doppler factor uniquely determines the transformations of such quantities as time, frequency and spectral intensity. These are listed in Table 1.1, together with the aberration and apparent transverse motion formulae. Apparent superluminal motion, and the practical application of the transformation of intensity to realistic flows, are discussed in more depth in Chapter 4.

The radio radiation from the extended radio sources is primarily optically thin synchrotron emission. In the compact nuclear sources opacity is important, and inverse-Compton emission is expected to make a significant contribution, so it is there that the richest behaviour in terms of spectral and polarization properties of the radiation is seen. To set the scene for the coming chapters, we present in the next sections a discussion of these radiation mechanisms, with emphasis on the application of the theory to sub-parsec scale extragalactic jets. For a fuller discussion of many aspects of these processes, the reader is referred to texts by Rybicki & Lightman (1979), Tucker (1975), Pacholczyk (1970), Ginzburg & Syrovatskii (1969) and Chandrasekhar (1950). In discussing synchrotron theory, we have numbered certain constants (*e.g.*, c_5, c_6) following the notation of Pacholczyk (1970), so that the reader needing a more detailed exposition of the subject (in cgs units) can more easily relate the two discussions.

The following discussion of synchrotron radiation is aimed primarily at relating the spectral and polarization properties of the radiation to the physical conditions in the source – and so the source (comoving) values are used throughout. To avoid inelegant and confusing equations, we have omitted primes from all quantities in that subsection, and the reader should bear in mind that all quantities are measured in the source frame. The discussion of inverse-Compton radiation draws heavily on a comparison of *observed* radio and X-ray fluxes – and much of the discussion addresses the issue of determining $\mathcal{D}$ – so there the reader will find expressions involving the observed intensity (I), and the factors of $\mathcal{D}/(1+z)$ that result from having transformed comoving intensity into the observer's frame.

1.2.1 Synchrotron radiation

1.2.1.1 Synchrotron radiation from an electron

The motion of a charged particle in a constant, uniform magnetic field $\mathbf{B}$, comprises a constant velocity component along the field, and a circular motion in a plane perpendicular to the field. As a consequence of the centripetal acceleration of the charge, energy is radiated, and this is termed *cyclotron* radiation if the particle energy is not relativistic, and *synchrotron* radiation in the limit of relativistic particle velocity. v and $v_{\parallel}$ are constants of the motion, so the pitch-angle (defined by $\psi = \cos^{-1}\mu = \cos^{-1}(v_{\parallel}/v)$) also is a constant. The power radiated by a single electron of Lorentz factor, $\gamma\ (= 1/\sqrt{1-v^2/c^2})$, is $P = (2/3)(e^2/4\pi\varepsilon_0)c^{-3}\gamma^4\left(a_{\perp}^2 + \gamma^2 a_{\parallel}^2\right)$, in terms of the components of acceleration perpendicular and parallel to the instantaneous velocity (*e.g.*, Rybicki & Lightman 1979). For an electron gyrating in a magnetic field, $a_{\parallel} = 0$, whilst $a_{\perp} = v_{\perp}eB/\gamma m_e$ (= velocity × gyrofrequency), so

$$\begin{aligned} P_s &= \frac{8\pi}{3}\frac{1}{\mu_0}r_0^2 c\sin^2\psi\gamma^2 B^2 \\ &= 2\sigma_T c\sin^2\psi\gamma^2 u_B, \end{aligned} \tag{1.1}$$

where r_0 is the classical electron radius ($r_0 = e^2/4\pi\epsilon_0 m_e c^2$), σ_T is the Thomson cross section, and u_B the energy density of the magnetic field. The power may be averaged over pitch-angle, in which case the factor $\sin^2\psi$ becomes $2/3$ if the pitch-angle distribution is isotropic.

Equation (1.1) gives the energy radiated over all frequencies and solid angles, but an important consequence of the relativistic particle velocity is that this radiation is beamed into a cone of semi-angle $1/\gamma$ (in radians) about the instantaneous particle direction. If $1/\gamma \ll 1$, an observer sees radiation from an electron only as this cone sweeps across the line-of-sight; thus for the radiation to be 'visible', the angle of the magnetic field to the line-of-sight must be approximately the particle pitch-angle. To receive radiation from particles in a magnetic field that is close to the line-of-sight, the particle pitch-angle must be small, and thus the radiated power will be small ($\sin^2\psi \sim 0$ in equation (1.1)). It follows that a significant flux of radiation can be received only from sources with a significant magnetic field component in the plane of the sky. ($\sin\psi$ may be thought of as giving the projection of B on the sky: $B_\perp = \sin\psi B$.) Radio source fields are often turbulent, not unidirectional, and a 'mean' angle to the line-of-sight may be used.

The characteristic frequency of emitted radiation is the inverse of the pulse width, which is determined by the gyrofrequency of the particle (taking the appropriate pitch-angle, and using the relativistic mass, γm_e) modified by one power of γ to account for the small opening angle of the 'cone' (see above), and two powers of γ (from a '$1-v/c$' term) to allow for the difference between 'emission' and 'arrival' times of the pulse. Thus most radiation is emitted at a circular frequency $\omega \approx \gamma^3 (eB\sin\psi/\gamma m_e) \approx \gamma^2 \Omega_e \sin\psi$. A detailed calculation of the field of a moving charge shows that the spectrum peaks at a frequency

$$\begin{aligned} \omega &= 0.29\,\omega_c \\ \omega_c &= \frac{3}{2}\gamma^2 \Omega_e \sin\psi \qquad (1.2) \end{aligned}$$

and that the spectral power has low and high frequency behaviour

$$\begin{aligned} P_s(\omega) &\propto (\omega/\omega_c)^{\frac{1}{3}} && \omega \ll \omega_c \\ P_s(\omega) &\propto (\omega/\omega_c)^{\frac{1}{2}}\, e^{-\omega/\omega_c} && \omega \gg \omega_c. \qquad (1.3) \end{aligned}$$

This spectrum is shown in Fig. 1.1. ω_c is the median frequency – half the power being radiated above, and half below this value. In some applications it proves adequate to assume that all the power is radiated at a frequency ω_c.

The calculations used to derive equations (1.2) and (1.3) involve

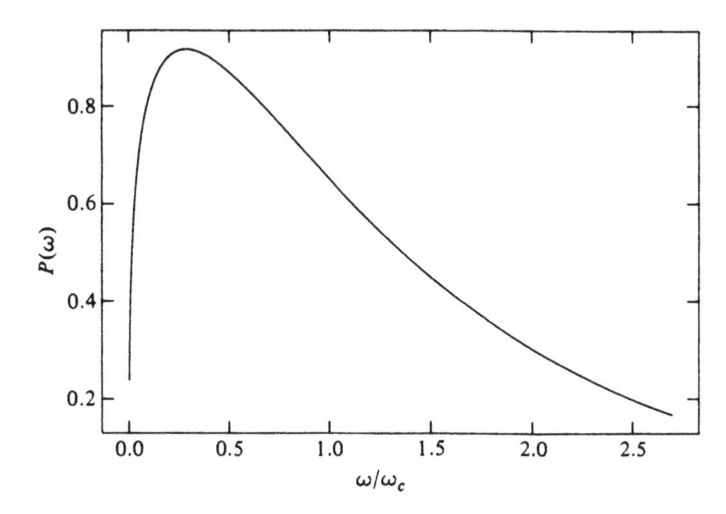

Fig. 1.1. The power radiated by a single electron as a function of angular frequency. This is the function $F(x)$ of Pacholczyk (1970).

finding the power parallel and perpendicular to the direction of the magnetic field projected on the plane of the sky. From these two powers the fractional linear polarization may be computed: it is defined as

$$\pi_L = \frac{P_\perp - P_\parallel}{P_\perp + P_\parallel} \tag{1.4}$$

which is a function of frequency. When integrated over frequency the value of π_L is found to be 0.75 – synchrotron radiation is highly linearly polarized.

1.2.1.2 Synchrotron radiation from an ensemble of electrons

Having discussed the radiation from a single electron, it is now possible to determine the radiation from an ensemble of electrons – the situation of astrophysical interest – by specifying the distribution of particles with respect to energy and pitch-angle, weighting the power for each energy and pitch-angle with this distribution function and integrating over all possible values of these parameters. Scattering by plasma waves (see Chapter 9) should ensure that the particles have an isotropic pitch-angle distribution under most circumstances, and observations imply that density depends on energy as a power law:

$$n(\gamma)\,d\gamma = n_{\gamma 0}\gamma^{-\delta}d\gamma \tag{1.5a}$$

or

$$n\,(E)\,dE = n_{E0}E^{-\delta}dE \tag{1.5b}$$

for energies in the range $E_L < E < E_U$ ($\gamma_L < \gamma < \gamma_U$). Using this distribution, and restricting attention to frequencies $\nu_L \ll \nu \ll \nu_U$, (*i.e.*, $\omega_L \ll \omega \ll \omega_U$, where $\omega_{L,U} \approx \gamma_{L,U}^2 \Omega_e \sin\psi$), the synchrotron emissivity is found to be

$$j_\nu = c_5\,(\alpha)\,n_{\gamma 0}\,(B\sin\theta)^{\alpha+1}\left(\frac{\nu}{2c_{\gamma 1}}\right)^{-\alpha} \tag{1.6a}$$

where

$$c_5\,(\alpha) = \frac{c_3}{4}\frac{\alpha+\frac{5}{3}}{\alpha+1}\Gamma\left(\frac{3\alpha+1}{6}\right)\Gamma\left(\frac{3\alpha+5}{6}\right), \tag{1.6b}$$

and

$$c_3 = \frac{c}{4\pi\varepsilon_0}\frac{\sqrt{3}}{4\pi}\frac{e^3}{m_e c^2}. \tag{1.6c}$$

$c_{\gamma 1} = 3e/4\pi m_e$, $\alpha = (\delta - 1)/2$, and θ is the angle between the magnetic field and the line-of-sight. A useful approximation is that $c_5(\alpha) \approx 8.28 \times 10^{-26}/\,(10\alpha + 1)$ for $0.5 < \alpha < 1.0$. If we replace $n_{\gamma 0}$ in equation (1.6a) with n_{E0}, then $c_{\gamma 1}$ becomes $c_{E1} = 3e/4\pi m_e^3 c^4$. The origin of this power-law is easily seen using the 'monochromatic' approximation that all power is radiated at frequency $\nu \propto \gamma^2$, so that $d\nu \propto \gamma d\gamma$; $j_\nu d\nu \propto P_s n\,(\gamma)\,d\gamma$, and substituting for P_s from equation (1.1), $j_\nu d\nu \propto \nu^{-(\delta-1)/2} d\nu$.

1.2.1.3 Polarization of synchrotron radiation from an ensemble of electrons

Within the frequency range specified, the fractional linear polarization is independent of frequency:

$$\pi_L = \frac{\delta+1}{\delta+7/3} = \frac{\alpha+1}{\alpha+5/3}. \tag{1.7}$$

For many sources $2 < \delta < 3$, implying that the polarization is about 70%.

There is evidence that compact (parsec-scale) sources display a linear polarization of only a few percent (Jones *et al.* 1985), despite the much higher degree anticipated from the homogeneous field theory, due to 'cellular depolarization'. Although a cellular structure has little influence on the total emissivity ($\sin\theta \to \langle\sin\theta\rangle$), it does have a profound effect on the polarization properties of the source.

This view of the magnetic field structure was adopted by Burn (1966), who described the field as a superposition of random and mean components, and by Burch (1979), who described the field in terms of a finite number (N) of reversals within the telescope beam. Use of the Stokes parameters, I_ν, Q_ν, U_ν and V_ν, often facilitates an understanding of polarization properties, (*e.g.*, Jones *et al.* 1985) and we have formulated the following discussion in terms of the Stokes parameters as an introduction to their use. I_ν measures the total energy flux in the electromagnetic waves; Q_ν and U_ν measure the linear polarization ($\pi_L = \sqrt{Q_\nu^2 + U_\nu^2}/I_\nu$, which is equivalent to equation (1.4)); V_ν measures both the magnitude and helicity of the circular polarization. The Stokes parameters are additive for a superposition of independent waves. The previous discussions considered only the magnitude of the linear polarization but, as there is a *plane* of polarization, it is a vector quantity that may be specified by two coordinates in a plane, and Q_ν and U_ν may be thought of as these coordinates. We calculate Q_ν and U_ν in a coordinate system with axes parallel and perpendicular to the projected magnetic field at each point in the source, and for each 'cell', rotate the coordinates into some common reference frame before summing the intensities. If we regard the source as composed of many randomly oriented cells of magnetic field, then Q_ν and U_ν will be the same in each cell *in the local reference frame*, but the random coordinate rotations necessary to express Q_ν and U_ν in a common reference system, scatter the values over the $Q_\nu - U_\nu$ plane, so that when summed, they tend to cancel out. The result is a much reduced degree of polarization; if there are N cells within the telescope beam, $\pi_L \rightarrow \pi_L/\sqrt{N}$. (More precisely, in a large sample of such sources, there will be a distribution of net percentage polarizations with a standard deviation corresponding to reduction by a factor $1/\sqrt{2N}$ from the uniform magnetic field value.)

An important corollary of 'cellular depolarization' is that a shearing or compression of a random field (for example, in a shock or boundary layer), by introducing a preferred direction, can 'repolarize' the emission (Scheuer 1967; Högbom 1979). A compression modifies only those field components perpendicular to its action, and a random field viewed along the axis of compression will maintain its appearance of randomness. However, if the field is com-

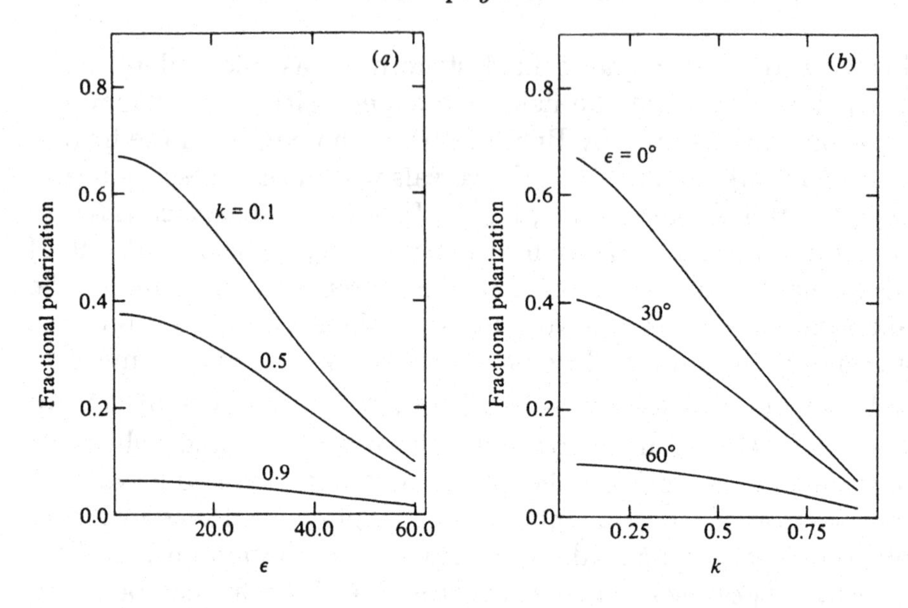

Fig. 1.2. The fractional polarization for synchrotron radiation from a region of compressed, turbulent magnetic field when $\delta = 2$. a) as a function of angle to the plane of compression in the source frame (for three compression factors). b) as a function of compression factor (for three angles to the plane of compression).

pressed into a plane, and viewed in that plane, it will appear indistinguishable from an ordered field and will manifest the approximately 70% polarization associated with an ordered field. (There are field line reversals in the plane of compression, but linear polarization is independent of the *sense* of the magnetic field.) 'Partial polarization' results from either partial compression and/or viewing at some angle to the plane of compression. This description of the field structure is potentially different from the cellular one cited above, in that in the cellular case there is a mean field component, and the fractional polarization can be used as a measure of the fractional magnetic energy in this component, whereas 'partial polarization' due to compression or shearing applies regardless of the degree of tangling of the field – indeed, there need be *no* mean component – and the fractional polarization tells us nothing about the magnitude of a mean field, even if such exists. A detailed analysis of 'partial polarization' may be found in Laing (1980) and Hughes, Aller & Aller (1985). Fig. 1.2 illustrates this effect for various viewing angles to the plane of compression (ϵ) and compressions (defined as unit length being compressed to length k – in a shock, for example, k would be the ratio of upstream to downstream densities, as measured in the upstream and downstream reference frames respectively).

For the reader wanting some 'rule of thumb' for 'partial polarization'

$$\pi_L \approx \frac{\delta+1}{\delta+7/3}\,\frac{(1-k^2)\cos^2\epsilon}{2-(1-k^2)\cos^2\epsilon}. \tag{1.8}$$

As can be seen from Fig. 1.2a, the percentage polarization is insensitive to viewing angle when ϵ is small; almost the maximum possible polarization will be seen when viewing within tens of degrees of the plane of compression. Note that ϵ is measured in the source frame, and that if the source is moving at near light speed with respect to the observer, account must be taken of aberration. For example, an observer looking almost along the axis of a highly relativistic flow will see radiation that – in the source frame – was emitted almost perpendicular to the flow. The polarization properties of a jet are sensitively dependent on its magnetic field configuration (*e.g.*, Laing 1981; Clarke, Norman & Burns 1989) – even in the absence of Faraday effects (§1.2.1.4) – but appear to be determined solely by the distortion of a seed field, being insensitive to the initial 'microscopic' structure of that seed field.

Whatever its origin, low percentage polarization implies a low polarized flux density, and hence that some observations will have a small signal-to-noise ratio. In this situation, the observed polarized flux can be expected to overestimate the actual polarized flux of the source. This is because the observed flux is the magnitude of a vector which is the sum of the intrinsic polarization vector, and a randomly oriented error vector; the probability that the resultant vector has a magnitude larger than that of the intrinsic polarization vector is greater than the probability that it is smaller. This effect becomes more important as the signal-to-noise ratio falls towards unity; a statistical correction may be applied to reduce this bias, the details of which can be found in appendices to papers by Wardle & Kronberg (1974) and Killeen, Bicknell & Ekers (1986).

1.2.1.4 Optical depth and Faraday depth

Thus far we have ignored the effects of the synchrotron plasma on the propagation of the emitted radiation – the above discussion applies to the optically thin domain. The major effects that arise due to the presence of a plasma are self-absorption, rotation of the plane of polarization, and depolarization.

At low frequency the source will be optically thick to synchrotron radiation. As it cannot radiate more efficiently than a 'black-body', we can estimate an upper limit to the intensity from the Rayleigh-Jeans ($h\nu \ll k_B T$) formula, $I_\nu = 2\nu^2 k_B T/c^2$, where $k_B T$ measures the energy per particle of the emitting gas. This energy is just $\gamma m_e c^2$, and we have seen (equation (1.2)) that $\gamma \propto \nu^{1/2}$. Thus we expect $I_\nu \propto \nu^{5/2}$ at low frequency. In a rigorous treatment, a self-absorption coefficient κ_ν is calculated and its influence on the total radiation determined by solving a transfer equation of the form

$$\frac{dI_\nu}{ds} = -\kappa_\nu I_\nu + j_\nu. \tag{1.9}$$

The solution of equation (1.9) for the brightness at the surface of the source is

$$I_\nu = \int_0^L j_\nu \exp\left(-\int_0^s \kappa_\nu d\tilde{s}\right) ds = \frac{j_\nu}{\kappa_\nu}(1 - e^{-\kappa_\nu L}), \tag{1.10}$$

where the last expression assumes that j_ν and κ_ν are independent of s. When the optical depth through the source, $\kappa_\nu L$, is much larger than unity, we have seen that $I_\nu \propto \nu^{5/2}$, so

$$\kappa_\nu \propto j_\nu \nu^{-\frac{5}{2}}. \tag{1.11}$$

The optical depth can be arbitrarily large at any frequency for a given κ_ν if L is large enough. Since κ_ν is a local property of the medium it cannot depend on L and so equation (1.11) is true for all frequencies. Let us define ν_1 as the frequency at which the optical depth becomes unity. For a power-law electron distribution we can write $j_\nu = j_0(\nu/\nu_0)^{-\alpha}$ and therefore $\kappa_\nu L = (\nu/\nu_1)^{-(\alpha+5/2)}$. Then

$$I_\nu = I_0\left(\frac{\nu_1}{\nu_0}\right)^{-\alpha}\left(\frac{\nu}{\nu_1}\right)^{\frac{5}{2}}\left\{1 - \exp\left[-\left(\frac{\nu_1}{\nu}\right)^{\alpha+\frac{5}{2}}\right]\right\}. \tag{1.12}$$

I_0 ($= j_0 L$) is the surface-brightness measured at a frequency ν_0 where the optical depth is negligible. I_ν is plotted in Fig. 1.3a. It may be noted that the transition between high and low frequency spectral forms is very sharp – occupying considerably less than one decade in frequency at the turnover.

It may be shown (*e.g.*, Pacholczyk 1970) that for a homogeneous

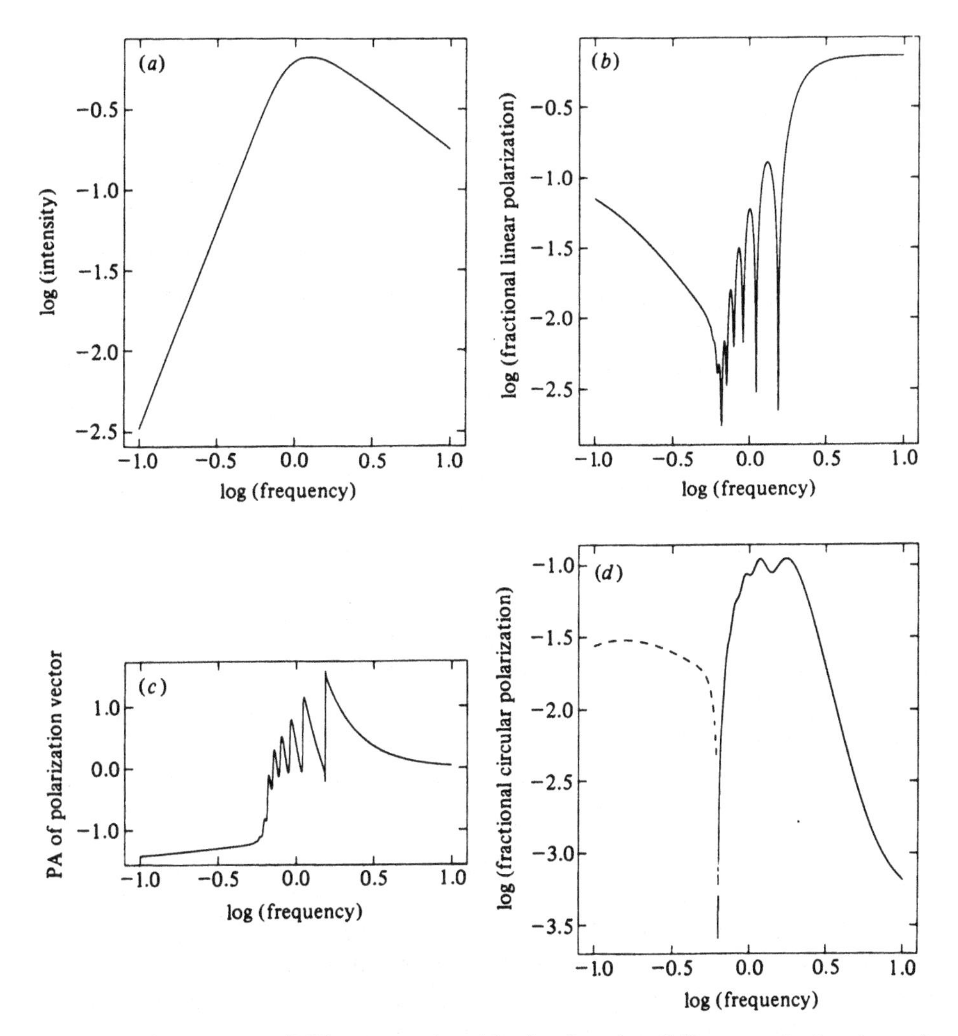

Fig. 1.3. a) The intensity, b) the fractional linear polarization, c) the Position Angle of the linear polarization vector (in radians), and d) the fractional circular polarization (negative values shown dashed) – all as a function of scaled frequency. The curves correspond to the case of characteristic energy near turnover $\gamma_n = 10^3$, low energy cutoff $\gamma_L = 10^2$, energy spectral index $\delta = 2.5$, and the magnetic field at an angle of 45° to the line-of-sight.

source with line-of-sight thickness L,

$$\nu_1 = 2c_{E1}\left(c_6\left(\alpha\right)Ln_{E0}\left(B\sin\theta\right)^{\alpha+3/2}\right)^{\frac{1}{\alpha+5/2}} \tag{1.13a}$$

where

$$c_6(\alpha) = \frac{1}{4\pi\varepsilon_0 c}\frac{\sqrt{3}\pi}{72} e m_e^5 c^{10} \times \left(\frac{6\alpha+13}{3}\right)\Gamma\left(\frac{6\alpha+5}{12}\right)\Gamma\left(\frac{6\alpha+13}{12}\right). \quad (1.13b)$$

Inspection of equation (1.12) shows that ν_1 is the frequency at which the high- and low-frequency limiting forms intersect. Note, however, that this is not the frequency at which the spectrum peaks. The frequency of maximum intensity is given by (Pacholczyk 1970)

$$\nu_{\text{max}} = \tau_{\text{max}}^{-\frac{1}{\alpha+5/2}}\nu_1; \qquad e^{\tau_{\text{max}}} - (\frac{2}{5}\alpha + 1)\tau_{\text{max}} - 1 = 0. \qquad (1.14)$$

For $0.2 \leq \alpha \leq 1.0$, $\tau_{\text{max}} \approx \frac{2}{3}\alpha$; we leave the reader to tabulate precise values of τ_{max}.

In propagating through an electron-proton plasma, the plane of polarization rotates, because a linearly polarized wave may be decomposed into two circularly polarized waves which have different phase velocities. This is the plasma equivalent of the birefringence exhibited by crystals. For a nonrelativistic plasma this angle of *Faraday rotation* depends on the path length traversed, line-of-sight magnetic field, electron density and frequency; it is characterized by the *Faraday depth* a distance s from the observer, given by

$$8.1 \times 10^3 \int_0^s n_{\text{th}}\mathbf{B} \cdot d\tilde{\mathbf{s}} \quad \text{rad m}^{-2} \qquad (1.15)$$

when the density is measured m^{-3}, the magnetic field is in tesla, and distance is measured in parsecs. For a homogeneous plane slab of plasma there is a Faraday rotation of the plane of polarization given by $\text{RM}\lambda^2$ radians (for wavelength in metres). RM is the *Rotation Measure*, which is a polarized emission-weighted mean of the Faraday depth. Thus for a resolved foreground screen this is the Faraday depth (1.15) to its far side, while for a Faraday medium mixed with the emitting plasma, it is one half the Faraday depth to the back of the plasma. Even if the magnetic field is uniform, and there is no external medium to act as a screen, the planes of polarization of radiation arising in different parts of a source undergo a different rotation. Thus, for large Faraday depth there may be a cancellation of the polarized emission arising at different points along a line-of-sight, or from the different lines-of-sight seen by a telescope beam, analogous to that experienced when the field is turbulent. This is known as *Faraday depolarization.* The cellular depolarization discussed in §1.2.1.3 is independent of frequency (except weakly through spectral index gradients within the source) whereas Faraday depolarization has the quite strong frequency dependence just noted. Multifrequency observations are therefore capable, in principle, of distinguishing the two effects; however, high resolution observations of extended structures are often made at a frequency that is too high for Faraday effects to be evident.

The electrons responsible for Faraday rotation may be quite independent of the emitting electrons (as, for example, in the case of a distribution of thermal electrons permeated by a magnetic field with a nonzero line-of-sight component, between the source and the observer; or nonradiating material entrained into the source), or may be those electrons in the low energy part of the spectrum of emitting particles (see below). In the former case, these electrons are often referred to as the 'cold' or 'thermal' particles. Their distribution, and that of the magnetic field, both inside and outside the source, must be known in order to compute the polarization that will be observed, and so no general prescription can be given for this. We draw the reader's attention to a discussion of this issue by Laing (1984). The magnetic field structure, and the origin of rotation and depolarization associated with kpc-scale jets are discussed in Chapter 3. The degree of Faraday rotation scales as $(\gamma m)^{-2}$, so that thermal electrons are more efficient Faraday rotators than mildly relativistic ones, and these in turn are more effective than protons. Furthermore, because of this strong dependence, if the electrons responsible for Faraday effects are relativistic, with power law energy spectrum, it is essential to know the low energy cutoff, γ_L, in order to compute the rotation of the plane of polarization. The details of this are in Appendix C of Jones & O'Dell (1977a) and are discussed further below.

If the plasma is composed entirely of electrons and positrons, the synchrotron emission is polarized, and is subject to the 'cellular depolarization' discussed above. However, the medium does not cause Faraday rotation or depolarization. This is because, in a given region of uniform field, there are as many charges gyrating clockwise as there are charges gyrating counterclockwise, and thus there is no preferential interaction between the particles and either the right-handed or the left-handed circularly polarized wave components.

An extensive analysis of the propagation of synchrotron radiation has been made by Pacholczyk & Swihart (1975) and by Jones & O'Dell (1977a,b). In these papers a complete description of the transfer of polarized radiation is made by writing down a transfer equation for each of the Stokes parameters and finding the particular integral and complementary function solutions for the four, simultaneous, linear differential equations. This leads to analytic expressions for the Stokes parameters if the source is homogeneous. Each transfer equation contains a set of transfer coefficients, derivable from the dielectric tensor of the plasma, which describe the processes of polarized emission, polarized absorption, rotation of the plane of polarization and conversion of linear to circular polarization. Fig. 1.3 shows the results of such calculations performed by the author for a line-of-sight through a uniform plasma with parameters applicable to a compact radio source: Fig. 1.3a shows the total intensity, exhibiting the high and low frequency spectral behaviour discussed above; Fig. 1.3b shows the fractional linear polarization, the oscillations being a manifestation of the depolarization arising from the different Faraday depths of emitting volumes along the line-of-sight; Fig. 1.3c shows the corresponding oscillations in the position angle of the linear polarization; Fig. 1.3d shows the circular polarization, which is higher than would be expected from the simple theory, because of a substantial conversion of linear to circular polarization during propagation through the source. An important point to which we draw attention is that in these calculations the electron spectrum was cut off below a Lorentz factor of 100, and no additional 'cold' matter was included: depolarization and rotation can be major effects due to the contribution to the transfer coefficients of particles with energy near that corresponding to the turnover frequency.

Care must be exercised when asking whether these figures give an accurate picture of the spectrum to be expected for a 'real' source. Observations are made with finite bandwidth, and this will lead to some smearing of spectral structure. The effect is normally small; when we rebinned the curves of Fig. 1.3 to correspond to observing at 1 GHz with a bandwidth of 100 MHz, the change was imperceptible. Integrating the emission over a source with a range

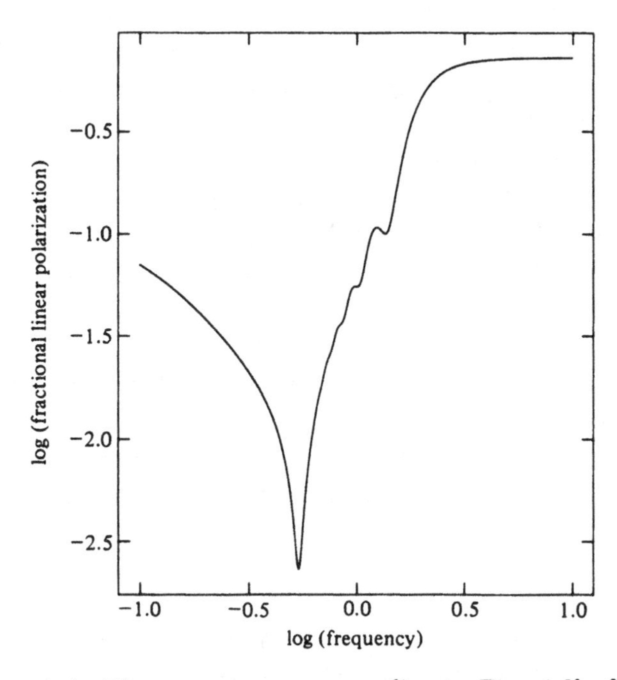

Fig. 1.4. The spectrum corresponding to Fig. 1.3b, for the case of emission from a uniform spherical distribution of synchrotron plasma. The optical depth is different for different lines-of-sight, leading to a 'smoothing' of the spectrum.

of line-of-sight thickness produces much more substantial changes. Fig. 1.4 shows the same as Fig. 1.3b, but integrated over a sphere; it is clear that 'real' sources will not display much of the spectral structure of the simplest models.

1.2.1.5 **Source inhomogeneity**

Lastly we consider the effect of inhomogeneity on the spectral properties of sources. Jones & O'Dell (1977b) have considered the influence of a boundary layer, in which the source magnetic field drops smoothly to zero. This plays little role above the turnover frequency, and although its effects are more important below the turnover frequency, for the most part they are not such as to change radically the character of the radiation spectrum. Exceptions to this are the circular polarization (the sign of which does not necessarily change at the turnover frequency as predicted for a homogeneous source) and the position angle of the linear polarization (which need not exhibit the rotation near turnover frequency, that characterizes the emission from a homogeneous source). However, the intensities fall rapidly with frequency at low frequency, and this part of the spectrum is not readily observed. The sensitivity of the circular polarization to source inhomogeneity, together with the conversion of linear to circular polarization within the source, makes circular polarization as difficult to interpret as it is to measure, and is why we have given it cursory consideration.

The slight inhomogeneity considered by Jones & O'Dell has little influence on the total intensity. A diverging flow constitutes a much more inhomogeneous source and, as noted by Marscher (1980), can lead to a total intensity spectrum that is much flatter near the turnover than is shown in Fig. 1.3a. We have constructed a beam comprising 1110 randomly oriented magnetic cells, superposed on a mean magnetic field containing about 5% of the total magnetic energy density; the field strength is taken to fall as the inverse of the beam radius, and the flow is assumed to be confined and adiabatic. Fig. 1.5a shows the total radiation spectrum near the turnover, computed as for Fig. 1.3, for the beam seen side-on (which is probably applicable to some parsec-scale jets when allowance is made for relativistic aberration).

Note the considerable spectral flattening: the flux is almost constant over one decade in frequency, and changes little over two decades. This contrasts with the corresponding change of two and one half magnitudes for the homogeneous case. Fig. 1.5b shows the profile of the beam at the three frequencies marked on Fig. 1.5a. Note that one 'sees' deeper into the flow at higher frequency, and that at high frequency the emission is localized, so that only the 'throat' of the beam is detected by observation. The 'core' on maps of parsec-scale jets is usually interpreted as being this 'throat'.

It is tempting to believe that the flat spectrum of many compact sources arises from such inhomogeneity. High-resolution mapping shows that these parsec-scale flows are rarely continuous, possessing a number of discrete components (*e.g.*, Fig. 1.6, which shows a core and a one-sided, curved, *knotty* jet at two epochs), and it has been suggested that the spectra of these individual components conspire to produce the flat integrated spectrum (Cotton *et al.* 1980).

However, in that they form segments of a diverging flow, it is not difficult to see that the sum of the spectra is much flatter than the spectrum of any one component. It is these components that,

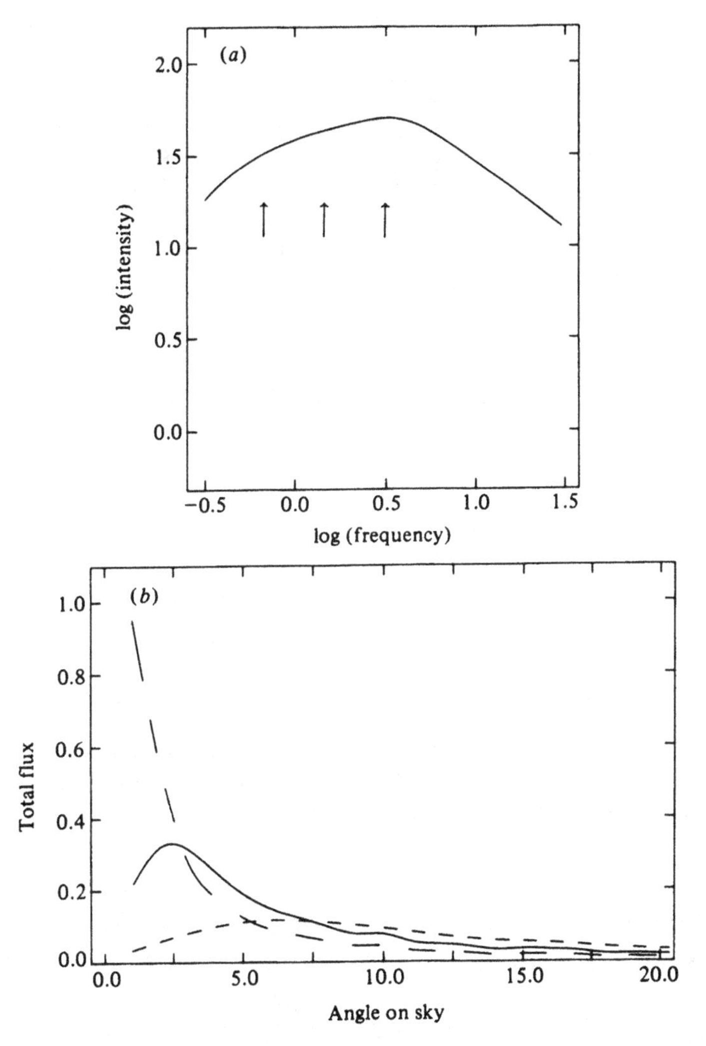

Fig. 1.5. a) The spectrum corresponding to Fig. 1.3a, integrated over a 'side-on' beam of $\sim$ 1000 randomly oriented cells of magnetic field, with magnetic field strength falling with radius as $B \propto r^{-1}$ and radius varying along the beam as $r \propto d^{5/8}$ (which might be appropriate to certain confined flows). Other parameters are as used for Fig. 1.3, applied at some fiducial point in the flow. The scale is the same as in Fig. 1.3 to within a translation, and the marked flattening of the spectrum due to the source inhomogeneity is clear. b) The distribution of brightness for the three frequencies marked on Fig. 1.5a; the long-dashed line is the highest frequency, and the short-dashed line is the lowest frequency.

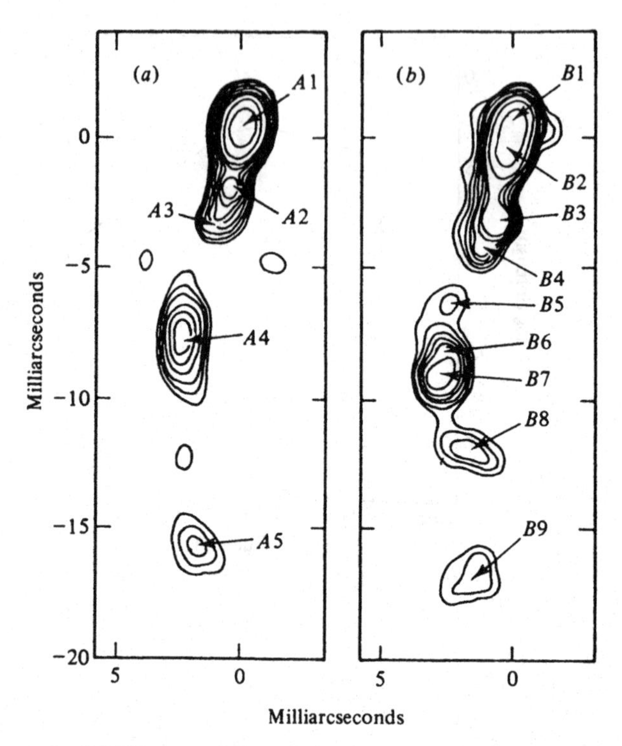

Fig. 1.6. VLBI map of the source 1928+738 at two epochs: a) 1980.72, b) 1982.95, from Eckart *et al.* 1985. The change of structure in this ~ 2 year period is quite evident.

as noted in the introductory comments, often exhibit superluminal motion away from a stationary core (this stationarity being well-established for sources such as 3C 345, *e.g.*, Bartel *et al.* 1986), as seen in Fig. 1.6, and whose link with the larger scale structure reinforces the idea of a transfer of energy and momentum to the outer components. Porcas (1987) lists currently known superluminal sources. These small-scale jets are discussed in §2.2.6, stressing the link between them and the kpc-scale structures, and the fundamental properties of these jets and their knots are summarized in §4.1.

1.2.2 Inverse-Compton radiation

1.2.2.1 Inverse-Compton emission and the Compton catastrophe

The ultra-relativistic electrons in a synchrotron source will also scatter photons by the inverse-Compton mechanism. Provided the photon energy in the rest frame of an electron does not become comparable to the electron's rest mass energy, the scattering can be treated as Thomson scattering in the electron's rest frame and some simple results derived. For electrons of energy $\gamma m_e c^2$ in the observer's frame, scattering photons of frequency ν_s from an isotropic radiation field of energy density $u'_{\rm rad}(\nu_s)$, the mean frequency of the scattered photons will be $\approx \frac{4}{3}\gamma^2\nu_s$ and the power lost by an electron will be $\approx \int \frac{4}{3}\gamma^2 u'_{\rm rad}(\nu_s) c\sigma_T d\nu_s$. These results may be compared directly with the equivalent results for synchrotron emission in a magnetic field with gyro frequency $\Omega_e = eB/m_e$ and energy density $u_B = B^2/2\mu_0$, when the mean frequency of the synchrotron photons is $\frac{4}{3}\gamma^2\Omega_e$ (equation (1.2)) and the power lost by the electron is $\frac{4}{3}\gamma^2 u_B c\sigma_T$ (equation (1.1)). Hence it can be seen that the relative importance of inverse-Compton scattering for the energetics of the electrons is determined by the ratio of the photon to magnetic field energy densities. As the photon energy becomes comparable to the electron rest mass energy in the electron's rest frame (*i.e.*, $\gamma h\nu_s \approx m_e c^2$ for an isotropic photon distribution) the Thomson cross-section should be replaced by the Klein-Nishina cross-section, and the highest photon energies attainable are limited by the energy of the electrons.

It was realized almost immediately after the first redshift determinations of quasars in 1963 that the observed rapid variations in flux density imply very high photon energy densities within quasars, since for variations across a source to be coherent, the light travel time across it should not be shorter than the time scale of the variations. If the emission mechanism is synchrotron then inverse-Compton scattering of the synchrotron photons by the relativistic electrons would almost completely dominate the electrons' energy losses, with most of the radiation being emitted at much shorter wavelengths than the synchrotron emission. Early estimates for the strength of the magnetic field were obtained by assuming that the rapidly-varying optical emission is synchrotron and that the electrons must be sufficiently long-lived to be able to cross the emitting region, so that the whole region can vary coherently (Hoyle, Burbidge & Sargent 1966). The resulting predictions became known as the Compton Catastrophe, as it was thought that successively higher order scatterings would lead to a divergence in the power emitted. In fact, the higher order scatterings do not lead to a divergence, being limited by the electron energy as noted above (Rees 1967), but the power losses by the electrons would still be severe, and for the parameters derived, extremely high X-ray luminosities were predicted.

That method for estimating the field strength is somewhat model-dependent, particularly as electrons can have a greatly increased lifetime in a synchrotron source which is optically thick, as the electrons then absorb as well as emit energy (Rees & Sciama 1966). Rapid variations at radio frequencies, however, also lead to a constraint on the strength of the magnetic field, since there is a maximum surface brightness attainable by an incoherent synchrotron source, dependent on the field strength, which leads to stronger constraints on the field strength at longer wavelengths (cf. equations (1.12) - (1.14)). Rapid variability is often used today to infer the apparent surface brightness of radio sources and often leads to the prediction of extremely large X-ray luminosities. Since the advent of VLBI observations, radio surface brightness can now also be measured directly, and again large inverse-Compton X-ray luminosities are often predicted. However, these large X-ray fluxes are not observed.

1.2.2.2 The maximum brightness temperature

The total power scattered by the inverse-Compton mechanism from an incoherent synchrotron source can be estimated as outlined in the previous section. When combined with a constraint on the energy density of the magnetic field it can be shown that there is a maximum brightness temperature which can be attained, beyond which the inverse-Compton losses become catastrophic.

We have already seen how the ratio of inverse-Compton to synchrotron radiated power is of order

$$\frac{P_{\rm ic}}{P_s} \approx \frac{\int u'_{\rm rad}(\nu)d\nu}{u_B}. \tag{1.16}$$

If the rest-frame photon intensity, I'_ν, is isotropic and has a power-law form of slope $-\alpha$ between ν_L and ν_U (which we identify with the slope and limits of the synchrotron spectrum discussed in §1.2.1) then

$$\frac{P_{\rm ic}}{P_s} \approx \frac{2\pi I'_\nu \nu^\alpha}{u_B c} \frac{\left[\nu^{1-\alpha}\right]_{\nu_L}^{\nu_U}}{1-\alpha} \tag{1.17}$$

for $\alpha \neq 1$. But the maximum *observed* surface brightness at frequency ν attainable by an incoherent synchrotron source is

$$I_\nu \lesssim \left(\frac{8\pi}{3}\frac{m_e^3\nu^5}{eB\sin\theta}\right)^{1/2} c_{56}^{(\alpha)}(\alpha)\left(\frac{\mathcal{D}}{1+z}\right)^{1/2} \tag{1.18}$$

(see §1.2.1.4 and Pacholczyk 1970), where $c_{56}^{(\alpha)}$ is the ratio $c_5^{(\alpha)}/c_6^{(\alpha)}$, the α dependent parts of $c_5(\alpha)$ and $c_6(\alpha)$, equations (1.6b) and (1.13b). Combining this limit with the inverse-Compton estimate transformed to the observer's frame, and so eliminating the dependences on magnetic field strength (through u_B in equation (1.17)), we find for the first-order scattered power

$$\frac{P_{\rm ic}}{P_s} \gtrsim \frac{9}{16\pi\varepsilon_0 c^3}\left(\frac{eI_\nu^2}{m_e^3\nu^5}\right)^2 \frac{I_\nu\nu^\alpha\left[\nu^{1-\alpha}\right]_{\nu_L}^{\nu_U}}{(c_{56}^{(\alpha)}(\alpha))^4(1-\alpha)}\left(\frac{\mathcal{D}}{1+z}\right)^{-6} \tag{1.19}$$

or expressed in terms of the brightness temperature,

$$T = \frac{c^2 I_\nu}{2k_B\nu^2}, \tag{1.20}$$

$$\frac{P_{\rm ic}}{P_s} \gtrsim \frac{18}{m_e c^3}\frac{e^2}{\pi\varepsilon_0}\left(\frac{k_B T}{m_e c^2}\right)^5 \frac{\nu^\alpha}{(c_{56}^{(\alpha)}(\alpha))^4}\frac{\left[\nu^{1-\alpha}\right]_{\nu_L}^{\nu_U}}{(1-\alpha)}\left(\frac{\mathcal{D}}{1+z}\right)^{-6}. \tag{1.21}$$

The inverse-Compton scattered power rises rapidly with brightness temperature. For $P_{\rm ic}/P_s \gtrsim 1$ higher-order scattering leads to catastrophic energy losses for the radiating electrons, limiting the maximum observable brightness temperature to a value such that $P_{\rm ic}/P_s \lesssim 1$. The corresponding maximum brightness temperature has the value

$$T \lesssim 1.6\times 10^{12}(c_{56}^{(\alpha)}(\alpha))^{0.8}\left(\frac{(1-\alpha)}{\nu_U}\left[\frac{\nu_U}{\nu}\right]^\alpha\right)^{0.2}\left(\frac{\mathcal{D}}{1+z}\right)^{1.2} \text{ K} \tag{1.22}$$

for $\alpha < 1$ and $\nu_U \gg \nu_L$, where ν_U is measured in GHz. This value is an extremely weak function of the measured variables and is almost independent of observing frequency. Most radio sources have VLBI-measured brightness temperatures which are no higher than this limit, and Kellermann & Pauliny-Toth (1969) suggested that the energetics of the most compact radio sources were determined by inverse-Compton losses in this way.

It is important to note, however, that it would be possible for a compact radio source to exceed this limit during a flare of radio emission. Such violations would be transient on the time scale of the electron energy losses. Indeed, if the rapid variability of some compact radio sources is indicative of a small emission region then the inferred brightness temperatures are considerably in excess of the 10^{12} K limit during flares (*e.g.*, Marscher *et al.* 1979). Nevertheless, sources radiating near or above the brightness limit should produce a high luminosity of inverse-Compton scattered photons, and since the advent of sensitive X-ray measurements it has become apparent that the predicted luminosities are not observed. That conclusion has greatly influenced the interpretation of superluminal motion (see Chapter 4). The next section indicates how the predicted luminosities are calculated, and discusses some of the problems of applying the theory to the observations.

1.2.2.3 The strength of inverse-Compton emission from a synchrotron source

Detailed calculations, or at least the results of detailed calculations, of the predicted strength of inverse-Compton emission from a synchrotron source have been presented in the literature (*e.g.*, Jones, O'Dell & Stein 1974; Gould 1979; Marscher 1983). This section indicates how these results are obtained, and presents the formulae that are most commonly used in applying the theory to observations.

We start by calculating the intensity of first-order inverse-Compton emission at frequency $\nu_{\rm ic}$ which is produced by the power-law energy distribution of electrons scattering synchrotron photons of frequency ν_s, assuming that both distributions are isotropic, and in the low photon energy limit $\gamma h\nu_s \ll m_e c^2$.

§7.3 of Rybicki & Lightman (1979) presents a clear derivation of the spectrum of monoenergetic photons scattered by monoenergetic electrons, and shows how by integrating over both input photon and electron distributions, a more general inverse-Compton spectrum may be derived (their equations (7.29a,b)). We can easily recast their result to find the inverse-Compton intensity in terms of the synchrotron intensity, by noting that the *spectral* emissivity is $h/4\pi$ times their expression for scattered power, per unit volume, per unit *energy* interval; and that the intensity is this emissivity times the line-of-sight depth of the source, L. We convert from photon energy to photon frequency ($\epsilon = h\nu$), noting that the relation between photon number density, v, photon number intensity, I_N, and energy intensity, I, is

$$v_\epsilon(\epsilon) = \frac{4\pi}{c} I_{N,\epsilon}(\epsilon) = \frac{4\pi}{c}\frac{I_\epsilon(\epsilon)}{\epsilon} = \frac{4\pi}{c}\frac{I_\nu(\nu)}{h\epsilon}. \tag{1.23}$$

Identifying Rybicki & Lightman's electron spectrum constant C with $n_{E0}(m_e c^2)^{-2\alpha}$, and writing $I_\nu(\nu) = [I_\nu \nu^\alpha]\nu^{-\alpha}$ so that the [ ] part is a constant and may be taken outside the integral between ν_L and ν_U, we get

$$I_{\nu,\mathrm{ic}} = \mathcal{F}_1(\alpha)\,\sigma_T n_{E0} L\,(m_e c^2)^{-2\alpha}\, I_{\nu,s}\nu_s^\alpha \ln\left(\frac{\nu_U}{\nu_L}\right)\left(\frac{3\nu_{\mathrm{ic}}}{4}\right)^{-\alpha}, \tag{1.24a}$$

where

$$\mathcal{F}_1(\alpha) = 2^\alpha \left(\frac{3}{2}\right)^{\alpha+1} \frac{(\alpha^2+3\alpha+4)}{(\alpha+2)^2(\alpha+3)(\alpha+1)}. \tag{1.24b}$$

We can also recast Rybicki & Lightman's equation (7.31) to give the intensity of radiation due to inverse-Compton scattering of (black body) microwave photons by the synchrotron emitting electrons:

$$I^{\mathrm{BB}}_{\nu,\,\mathrm{ic}} = \frac{3}{4}\mathcal{F}_2(\alpha)\,\sigma_T n_{E0} L\,(m_e c^2)^{-2\alpha}\,\frac{h}{c^2}\left(\frac{k_B T}{h}\right)^{\alpha+3}\nu_{\mathrm{ic}}^{-\alpha}, \tag{1.24c}$$

where

$$\mathcal{F}_2(\alpha) = 2^{2\alpha+2}\frac{(\alpha^2+3\alpha+4)}{(\alpha+2)^2(\alpha+3)(\alpha+1)}\Gamma(\alpha+3)\,\zeta(\alpha+3). \tag{1.24d}$$

ζ is the Riemann zeta function. These expressions require that the frequency ν_{ic} lies in the range

$$\frac{4}{3}\gamma_L^2\nu_U < \nu_{\mathrm{ic}} < \frac{4}{3}\gamma_U^2\nu_L. \tag{1.25}$$

The striking thing to notice about equations (1.24a,c) is that the inverse-Compton spectra of both scattered synchrotron photons and microwave photons have the same spectral form and slope as that of the synchrotron emission.

The energy density of synchrotron photons is a function of position within the source, except in the limit of infinite synchrotron optical depth, and the distribution will also be anisotropic. To derive the total inverse-Compton emission from a real source we must integrate over the source volume, and hence insert some assumption about the source geometry and the variation of photon energy density within the source. Even in a spherically-symmetric source the photon distribution is anisotropic, although by integrating over the source volume the anisotropy is averaged out. In general, departures from uniformity of the photon energy density will increase the inverse-Compton emission, but appropriate combinations of non-uniformities in the photon energy density and the electron density could lead to reduced inverse-Compton emission (*e.g.*, if the electron density were lower at the centre of our sphere, where the synchrotron photon density is highest).

With these caveats, we can substitute for the number density of electrons at a given energy (n_{E0} in equation (1.24a)), using the expression for the synchrotron intensity ($j_\nu L$, using equation (1.6) for j_ν), to obtain the ratio of inverse-Compton to synchrotron intensities at the respective frequencies as

$$\frac{I_{\nu,\mathrm{ic}}}{I_{\nu,s}} = \frac{16\pi g \mathcal{F}_1(\alpha)}{\sqrt{3}\left(\frac{3}{2\pi}\right)^\alpha c_5^{(\alpha)}(\alpha)} \frac{4\pi\varepsilon_0 c\sigma_T}{e^2} I_{\nu,s}\nu_s^{2\alpha} \ln\left(\frac{\nu_U}{\nu_L}\right)\left(\frac{3\nu_{\mathrm{ic}}}{4}\right)^{-\alpha}$$
$$\times\left(\frac{eB\sin\theta}{m_e}\right)^{-(\alpha+1)}\left(\frac{\mathcal{D}}{1+z}\right)^{-(\alpha+3)} . \qquad (1.26a)$$

Note that the Doppler factor dependence arises from the linear dependence of the right hand side on the synchrotron intensity. We have included a geometrical factor g to allow for possible anisotropy of the photon distribution, the variation of photon energy with position within the source, and inhomogeneities in the electron distribution. Since the same electrons directed towards the observer are responsible for both the synchrotron and inverse-Compton emission, anisotropy of the electron distribution does not itself affect the factor g, but of course any such anisotropy must inevitably result in an anisotropic synchrotron photon distribution. If, as an extreme example, all the synchrotron photons were directed towards the observer, there would be no increase in photon energy in the observer's frame due to inverse-Compton scattering and hence the factor g would be zero at observed frequencies much higher than ν_U. In practice it is difficult to contrive models in which g is suppressed by more than a factor of ten without invoking photon anisotropy. Gould (1979) derives the g factor for a homogeneous, spherical source.

Using the same substitution in equation (1.24c) we can find that

$$\frac{I^{\mathrm{BB}}_{\nu,\,\mathrm{ic}}\nu^{\alpha}_{\mathrm{ic}}}{I_{\nu,\,s}\nu^{\alpha}_{s}} = \frac{\sqrt{3}\,4\pi\mathcal{F}_2(\alpha)}{\left(\frac{3}{2\pi}\right)^{\alpha} c_5^{(\alpha)}(\alpha)} \frac{4\pi\varepsilon_0 h}{e^2 c} \left(\frac{k_B T}{h}\right)^{\alpha+3}$$

$$\times \left(\frac{eB\sin\theta}{m_e}\right)^{-(\alpha+1)} (1+z)^{\alpha+3}. \qquad (1.26b)$$

Notice that no Doppler factor appears because we have calculated the *ratio* of two (supposedly) equally beamed intensities, but that the cosmological factor $(1+z)$ must be retained because of the evolution of the microwave photon energy and number density.

The distinct roles played by equations (1.26a) and (1.26b) are that, given a measure of the ratio of X-ray to radio intensities, the former equation provides a means of estimating the field strength or Doppler factor in a compact source within which the microwave energy density is relatively unimportant, whereas the latter equation provides an estimate of the field strength when inverse-Compton scattering of the synchrotron photons can be ignored – which implies application to the extended lobes of radio sources.

Determination of the Doppler factor requires us to eliminate the dependence on the magnetic field, B. We can deduce B if the intensity of the synchrotron source is measured in the optically-thick part of its spectrum, by recasting equation (1.18) as

$$B = \frac{8\pi}{3}\frac{m_e^3}{e}\frac{\nu_T^5}{I_T^2}\frac{(c_{56}^{(\alpha)}(\alpha))^2}{\sin\theta}\frac{\mathcal{D}}{1+z} \qquad (1.27)$$

where I_T is the intensity at a frequency ν_T where the source is optically thick. This may be substituted into equation (1.26a), to yield

$$\frac{I_{\nu,\mathrm{ic}}}{I_{\nu,s}} = \frac{4\pi\sqrt{3}g}{f(\alpha)}\frac{4\pi\varepsilon_0 c\sigma_T}{e^2} I_{\nu,s}\nu_s^{2\alpha}\nu_{\mathrm{ic}}^{-\alpha}\ln\left(\frac{\nu_U}{\nu_L}\right)$$

$$\times \left(\frac{I_T^2}{3m_e^2\nu_T^5}\right)^{\alpha+1}\left(\frac{\mathcal{D}}{1+z}\right)^{-(4+2\alpha)} \qquad (1.28)$$

where $f(\alpha)$ is a combination of the previous functions of α, which the reader may derive and tabulate.

The intensity in the optically-thick part of the spectrum is not usually measured, as VLB interferometers lack spatial resolution at low frequencies. However, synchrotron spectra of the form considered here have a maximum intensity corresponding to the synchrotron optical depth being of order unity, and if the spectrum is measured near its maximum the field can still be deduced. The intensity at the maximum, relative to the completely optically-thick intensity, depends on both the power-law index α and on the source geometry. A simple slab of emitting material with $\alpha = 1/2$ has

$$\frac{I_{\max}}{\nu_{\max}^{5/2}} \approx 0.29 \frac{I_T}{\nu_T^{5/2}} \tag{1.29}$$

(see §1.2.1.4 and Pacholczyk 1970). A homogeneous sphere has a value lower by about ten percent (Gould 1979). The source geometry plays an important role in estimating the field strength, and we shall return to this point below. The turn-over frequency is usually poorly determined, and the strong dependence on this quantity means that it is less desirable to use this approach than to measure the optically-thick intensity directly. If the true maximum occurs at a frequency below that measured then the intensity of inverse-Compton emission will be greater than that deduced from the equation above.

1.2.2.4 **Low radio-frequency variability**

Rapid variability at low radio frequencies often imposes a stronger constraint on the synchrotron surface brightness than do VLBI measurements at higher frequencies and such data are frequently invoked as evidence of bulk relativistic motion (*e.g.*, Simon *et al.* 1983). There are however severe problems with using the variability information. Firstly, the inverse-Compton problem posed by the rapid variability can be resolved by virtually any of the models that compete with the 'standard' one of bulk relativistic motion, and the conclusions are no longer independent of distance: for example, the light-echo model can result in sufficiently compressed time scales of variability, since the deduced Doppler factor varies as $t_{\rm var}^{2(\alpha+2)/(2-\alpha)}$; and gravitational lensing will amplify the total observed flux density whilst leaving the variability time scale unchanged. Secondly, it now seems likely that radio variability at frequencies $\nu \lesssim 1$ GHz is due to variable refraction by the interstellar medium along the line of sight in our Galaxy (Rickett *et al.* 1984; Cawthorne & Rickett 1985; Rickett 1986) so that the strongest constraints deduced from such data no longer apply.

1.2.2.5 X-ray observations of AGN

The hot gas surrounding extragalactic jets is a source of extended X-ray emission (see §3.3.1), but such emission is also a well-established feature of the nuclei of Active Galaxies. For the entire class of *radio galaxies* there is almost no spectral information capable of constraining the emission mechanism, although the strong correlation between X-ray and nuclear radio luminosities (Fabbiano *et al.* 1984) (together with a weaker correlation between nuclear and extended radio luminosities) leaves no doubt that the X-ray emission is intimately related to the processes of energy generation, and transport within radio jets. More spectral information exists for *quasars* (both radio loud and radio quiet) and *BL Lac* objects, but even here the data are sparse and their interpretation uncertain (Elvis & Lawrence 1985). One possible interpretation of the range of observed X-ray spectral slopes is that in sources with $\alpha_x \sim 0.7$, the X-ray emission is inverse-Compton scattered IR photons, in sources with $\alpha_x \sim 1.2$, the X-ray emission is a continuation from lower frequency of the (energy loss steepened) synchrotron spectrum, while in sources with $\alpha_x \sim 2.2$ the X-ray spectrum is modified by the high frequency tail of a spectral 'big bump' associated with the inner region of an optically thick accretion disc (cf. Chapter 8). Given the uncertainty in interpretation, and the fact that the X-ray emission may have a different origin in different sources, the conservative approach to the use of combined radio and X-ray observations is to assume that even an X-ray detection yields only an upper limit to the flux of inverse-Compton scattered synchrotron photons.

1.2.2.6 Consequences of the inverse-Compton limits

Equation (1.28) is commonly inverted to deduce limits on the Doppler factor, $\mathcal{D}$, given upper limits on the intensity of the inverse-Compton X-rays:

$$\frac{\mathcal{D}}{1+z} \gtrsim \left[4\pi\sqrt{3}\frac{g}{f(\alpha)}\frac{4\pi\varepsilon_0 c\sigma_T}{e^2}\right.$$
$$\left.\times\ \frac{S_{\nu,s} I_{\nu,s} I_{\nu,T}^{2(\alpha+1)}}{S_{\nu,\mathrm{ic}}\left(3m_e^2\nu_T^5\right)^{\alpha+1}}\frac{\nu_s^{2\alpha}}{\nu_{\mathrm{ic}}^{\alpha}}\ln\left(\frac{\nu_U}{\nu_L}\right)\right]^{1/(4+2\alpha)} \tag{1.30}$$

where S_ν is the observed flux density. Note that, apart from the (1+z) redshift factor, the predictions are independent of distance to the source. This equation still contains the assumption that the surface brightness can be measured in both the optically-thin and -thick portions of the spectrum. In practice this is only rarely possible (*e.g.*, Unwin *et al.* 1983) and instead the frequency and intensity of the spectral maximum are used in conjunction with equation (1.29). Such a procedure has to be used with caution, however, and can only be justified where there is clear spectral evidence that the spectrum maximum is not at a higher frequency. If the variability time scale, rather than VLBI observations, is used to give an estimate of source size, it is essential to deduce the surface brightness at the same frequency at which the variability is measured and to use the more conservative approach of deducing the field strength by assuming that the source is optically thick at that frequency, as in the absence of VLBI observations there can be no evidence that the source component in question contains only a limited range of optical depths (*e.g.*, Rees & Sciama 1965). Even with high-resolution maps made at various frequencies one can never eliminate the possibility that the observed component actually contains a wider range of synchrotron optical depths than that assumed in deriving the inverse-Compton limit. That situation could easily arise if the component contained a range of magnetic field strengths. Increasing the range of optical depths broadens the maximum in the spectrum and reduces the maximum intensity, for a given optically-thin intensity, resulting in too low a value for the magnetic field strength. This effect thus leads to a value for the predicted inverse-Compton emission which is systematically too large. The deduced Doppler factor varies as $I_{\rm max}^{(\alpha+1)/(\alpha+2)}\nu_{\rm max}^{-5(\alpha+1)/2(\alpha+2)}$, so an effective change in $\nu_{\rm max}$ of a factor 2 changes $\mathcal{D}$ by a factor 3 for constant $I_{\rm max}$.

Other uncertainties arise from the derivations of the synchrotron intensity and of the optically-thin spectral index. The radio flux density of a component can be measured accurately, but the surface brightness (equivalent to the radiation intensity outside the source) is usually deduced from an apparent source size after allowance has been made for the VLBI synthesised beam profile. Naturally, the highest inferred surface brightness values arise from observations where components are only barely resolved or are unresolved, and the corresponding surface brightness estimates are correspondingly uncertain. Systematic errors can arise if the surface brightness is deduced from the apparent full width at half maximum intensity: a barely-resolved homogeneous emitting sphere has a diameter which is a factor ~ 1.8 larger than its full width at half maximum intensity (Marscher 1977), thereby changing the deduced Doppler factor by a factor 2.6 if $\alpha = 0.75$ (Marscher & Broderick 1985). Also, care has to be taken to ensure that the observed X-ray frequency lies within the limits set by ν_L and ν_U and the deduced field strength (see above).

The derived Doppler factor is insensitive to the value of the spectral index for $\alpha \geq 0.5$, and is smaller for steeper indices. To see this, we can collect terms with the same α-dependence together to obtain

$$\frac{\mathcal{D}}{1+z} \gtrsim \left\{ A \left[\frac{I_T^2 \nu_s^2}{3 m_e^2 \nu_T^5 \nu_{\rm ic}} \right]^{\alpha} \right\}^{1/(4+2\alpha)} \tag{1.31}$$

$$\gtrsim \left\{ A \left[\frac{4}{3} \left(\frac{k_B T}{m_e c^2} \right)^2 \frac{\nu_s}{\nu_{\rm ic}} \right]^{\alpha} \right\}^{1/(4+2\alpha)} \tag{1.32}$$

where the other quantities have been incorporated into the variable A. At $\mathrm{T} = 10^{12}$K the α-dependent term in the brackets has a value $\sim 4 \times 10^{-4}$, so for $\mathcal{D} \gtrsim 1$ changing α to a larger value leads to a smaller value of $\mathcal{D}$. The α-dependence is strongest for higher brightness temperatures, and for such sources the steepest spectral indices consistent with the data should be chosen.

To date this method has lead to the deduction of minimum Doppler factors greater than unity for the moving knots in a few superluminal sources, such as NRAO 140 (Marscher & Broderick 1985), 3C 273 and 3C 345 (Unwin *et al.* 1983). Given the uncertainties in the derivations outlined above, these conclusions should only be considered as indicative of bulk relativistic motion, but nevertheless it is evidence which is independent of distance and independent of the apparent observed motions of the superluminal components. Future high-resolution observations at lower frequencies should help strengthen these results.

But it must be emphasised that if the photon distribution is anisotropic in the source rest frame then the photon energy density is much lower and hence the inverse-Compton problem can be resolved (*e.g.*, Woltjer 1966). An anisotropic photon distribution probably requires that the electron distribution likewise be anisotropic. As yet there is no evidence that these distributions are isotropic, and such an escape clause will probably always exist.

1.3 Extended radio sources

Some of the discussion of §1.2 may be applied to extended sources, and here we comment on this and indicate the results that are obtained. It is current wisdom that the radiation from extended structures is *not* relativistically beamed (cf. Chapter 4), so we may set $\mathcal{D} = 1$ in all formulae. Noting that $I_T \propto \nu_T^{5/2}$ and $I_0 \propto \nu_0^{-\alpha}$ (see §1.2.1.4) we have that

$$\frac{\nu_T^5}{I_T^2} = \frac{\nu_1^{2\alpha+5}}{\left(I_0\nu_0^{\alpha}\right)^2} < \frac{\nu_0^{2\alpha+5}}{\left(I_0\nu_0^{\alpha}\right)^2}. \tag{1.33}$$

Substitution of equation (1.33) in (1.27) provides an expression for an *upper limit* to B that depends only on measurements made in the optically thin part of the spectrum, and which, being determined solely by the spectrum, is independent of H_0. The optimum limit comes from taking ν_0 as close to ν_1 as possible, which requires that the latter be determined accurately – particularly as ν_0 appears to a very high power in equation (1.33). This is best done by fitting a model to the observed spectrum. However, it must be noted that for extended sources the turnover frequency occurs in the tens of MHz band, and the poor instrumental resolution, poor resolution due to interstellar scattering, ionospheric effects, and contribution to spectral steepening from free-free absorption within the Galaxy, make this technique of limited value.

Equation (1.26b) provides a *lower limit* to the magnetic field strength if X-ray observations set an upper limit to the intensity of inverse-Compton scattered microwave photons (convincing detections being rare or nonexistent). This is because an upper limit to the intensity sets an upper limit to the number of scattering electrons, and hence a lower limit to the magnetic field required to explain the radio emission. As with the previous method, this result is independent of H_0. In applying this technique, the extent of the region under study should be borne in mind. For example, if the X-ray observations are made at an energy of ~ 2 keV, scattering of 300 GHz microwave seed photons must have been by electrons with $\gamma \sim 1000$. The dominant radio emission from such electrons in the magnetic fields estimated by equipartition arguments (cf.

Chapter 3) is at $\sim$ 100 MHz, at which frequency radio emission is dominated by that from the diffuse bridges of sources (cf. Chapter 2). It is thus appropriate to compare radio and X-ray fluxes from a region encompassing such a bridge. The derived fields are typically lower than, but within a factor of ten of, the equipartition value (*e.g.*, 0.05 – 0.5 nT for a sample of 3CR sources studied by Miller *et al.* (1985)).

These two methods of placing limits on the magnetic field strength remain valid if the source is inhomogeneous; in such a case the upper limit should be interpreted as an upper limit on the field in those more self-absorbed regions that dominate the optically thin emission at higher frequency, while the lower limit derived from X-ray observations is an average field, weighted with the distribution of microwave photon scattering electrons.

1.4 Discussion

We have presented the radiation physics that is essential for an understanding of extragalactic jets, particularly those flows on the sub-parsec scale, wherein the effects of opacity, and scattering of synchrotron photons in the inverse-Compton process are important.

Our knowledge of the role of these processes in determining the spectral and spatial appearance of sources is incomplete. However, further advances can be expected – particularly when a more accurate description of the underlying flow pattern is available. Hydrodynamic simulations of jets began a decade ago, and it is to be hoped that such schemes will soon provide a more quantitative description of *propagating* structures, such as shocks, and will be formulated to accommodate relativistic bulk motion and equation of state. Studies continue to be made of the radiative properties of jets. For example, Jones (1988) has explored the appearance of compact synchrotron sources in all four Stokes parameters, and such work will form the basis for the interpretation of the next generation of radio observations. Calculations of the anticipated spectrum of radiation from material close to the central engine are essential for the interpretation of X-ray and γ-ray observations. The data are normally too sparse for such models to be well-constrained, but some recent work (*e.g.*, Melia & Königl 1989) relates the spectrum to observable bulk properties of jets; this approach promises to provide much better constrained models of the high-frequency emission.

Space VLBI such as RADIOASTRON (*e.g.*, Kardashev & Slysh 1988) will provide tens of microarcsecond resolution, and so will help to define the shape of propagating components which have hitherto been barely, if at all, resolved. Although without the resolution achieved by the inclusion of an orbiter baseline, the VLB Array (*e.g.*, Romney 1988) will provide a dedicated instrument, with optimum sampling of the aperture plane and good dynamic range, and receivers initially covering a range of frequencies up to 23 GHz. VLB polarimetry is already proving to be a powerful tool (*e.g.*, Wardle & Roberts 1988). This now provides maps in linearly polarized flux, which often give contrast between core and propagating component better than seen on total flux maps, and so is valuable for studying these moving structures, while the mapping of changes in degree and direction of polarized emission along the jet is already aiding in the interpretation of these sources. ROSAT and the next generation of X-ray satellites will provide data to constrain the parameters in a greater sample of sources, aid in the classification of X-ray spectra, and hopefully give insights into the nature of the X-ray generating process(es).

This promise of advances on both the theoretical and observational fronts means that in the coming years we can expect to achieve a much better understanding of these parsec-scale structures that form a critical link between the central engine and the large-scale jets and lobes of extragalactic sources.

It is a pleasure to thank Paddy Leahy for major contributions to the material in this chapter, and Tony Allen, Hugh Aller, Margo Aller, Brian Babler, John Baldwin, Tim Cawthorne, Jean Eilek, Richard Saunders, Richard Sears and Dave Silva for numerous comments and suggestions.

Symbols used in Chapter 1

Symbol	Meaning
$a_{\parallel}$, $a_{\perp}$	components of acceleration

A	constant in expression for $\mathcal{D}/(1+z)$
$\mathbf{B}$, B	magnetic field
$B_\perp$	component of B projected on the sky
c	speed of light
c_3, c_5	constants in expression for j_ν
c_6	constant in expression for ν_1
$c_5^{(\alpha)}$	α-dependent part of c_5
$c_6^{(\alpha)}$	α-dependent part of c_6
$c_{56}^{(\alpha)}$	ratio of $c_5^{(\alpha)}$ to $c_6^{(\alpha)}$
$c_{\gamma 1}$	constant in expression for j_ν
c_{E1}	constant in expression for j_ν
C	spectral constant from Rybicki & Lightman (1979)
d	distance along jet
$\mathcal{D}$	Doppler factor
e	charge of an electron
E	particle energy
$E_{L,U}$	lower, upper limits to particle energy spectrum
$f(\alpha)$	function of α in expression for $\mathcal{D}$
$F(x)$	function describing radiated power as a function of frequency (Pacholczyk 1970)
$\mathcal{F}_{1,2}(\alpha)$	functions of α in expressions for inverse-Compton intensity spectrum (Rybicki & Lightman 1979)
g	fudge factor in expression for inverse-Compton intensity
h	Planck constant
I, I_ν	intensity, spectral intensity, first Stokes parameter
I', I'_ν	intensity, spectral intensity in source frame
I'_0	frequency independent spectral constant in source frame
I_ϵ	energy spectral intensity
I_N	number intensity
I_0	$I(\tau \ll 1)$
I_T $(I_{\nu,T})$	$I(\tau \gg 1)$
$I_{\max}$	peak intensity of spectrum
$I_{\nu,s}$	synchrotron spectral intensity
$I_{\nu,\mathrm{ic}}$	inverse-Compton spectral intensity
$I_{\nu,\mathrm{ic}}^{\mathrm{BB}}$	inverse-Compton spectral intensity due to scattering microwave photons
j_ν	spectral emissivity

j_0	$j_\nu(\tau \ll 1)$
k	compression factor
k_B	Boltzmann constant
L	line-of-sight depth through a source
m_e	electron mass
$n(E)$, $n(\gamma)$	number of relativistic particles of given energy
n_{E0}, $n_{\gamma 0}$	constants in the power-law forms for $n(E)$ and $n(\gamma)$
n_{th}	number density of thermal particles
N	number of turbulent cells within telescope beam
P, $(P_s$, $P_{\mathrm{ic}})$	radiated power (synchrotron, inverse-Compton)
$P_\parallel$, $P_\perp$	components of radiated power in orthogonal directions
Q_ν	second Stokes parameter
r	radius of jet
r_0	classical electron radius
RM	rotation measure
s	line-of-sight distance through source
$S_{\nu,s}$	observed synchrotron flux density
$S_{\nu,\mathrm{ic}}$	observed inverse-Compton flux density
t_{var}	variability time scale
T	brightness temperature
u_B	magnetic energy density
u_{rad}	radiation energy density
U_ν	third Stokes parameter
v, $v_\parallel$, $v_\perp$	velocity, components of velocity along and perpendicular to magnetic field
v	photon number density
V_ν	fourth Stokes parameter
z	redshift
α	frequency spectral index ($S \propto \nu^{-\alpha}$)
α_x	frequency spectral index in X-ray band
β	v/c
β_{app}	apparent transverse β
γ	Lorentz factor of flow or particle
γ_L, γ_U, γ_n	lower and upper cutoff Lorentz factors for power law energy spectrum, characteristic Lorentz factor for this spectrum
Γ	gamma function
δ	energy spectral index ($n \propto E^{-\delta}$)

ϵ	angle between observer and plane of compression
ϵ	photon energy
ε_0	permittivity of free space
ζ	Riemann zeta-function
θ	angle between magnetic field and line-of-sight
κ_ν	absorption coefficient
λ	wavelength
μ	cosine of particle pitch angle
μ_0	permeability of free space
ν	frequency (linear)
ν_L, ν_U	lower and upper limits to frequency spectrum
ν_0	$\nu(\tau \ll 1)$
ν_1	$\nu(\tau = 1)$; intersection frequency
ν_T	$\nu(\tau \gg 1)$
$\nu_{\max}$	frequency at which spectrum peaks
ν_s	frequency of synchrotron photons
$\nu_{\rm ic}$	frequency of inverse-Compton photons
π_L	fractional linear polarization
σ_T	Thomson cross-section
τ	optical depth
$\tau_{\max}$	τ at peak of spectrum
ψ	particle pitch angle
Ψ	angle between bulk flow and observer
Ψ'	emission angle in source frame
ω	frequency (circular)
ω_c	critical frequency in synchrotron spectrum
Ω	solid angle
Ω_e	electron gyrofrequency

References

Baade, W., 1956, *Ap J.*, **123**, 550.

Baade, W. & Minkowski, R., 1954, *Ap. J.*, **119**, 215.

Bally, J. & Lada, C. J., 1983, *Ap. J.*, **265**, 824.

Bartel, N., Herring, T. A., Ratner, M. I., Shapiro, I. I. & Corey, B. E., 1986, *Nature*, **319**, 733.

Bates, D. R. & Spitzer, L., 1951, *Ap. J.*, **113**, 441.

Begelman, M. C., Blandford, R. D. & Rees, M. J., 1984, *Rev. Mod. Phys.*, **56**, 255.

Blandford, R. D. & Rees, M. J., 1974, *M. N. R. A. S.*, **169**, 395.

Bolton, J. G., Stanley, G. J. & Slee, O. B., 1949, *Nature*, **164**, 101.

Bridle, A. H., 1984, In *Physics of Energy Transport in Extragalactic Radio Sources*, eds. Bridle, A. H. & Eilek, J. A. (NRAO: Greenbank, WV), p. 1.

Bridle, A. H. & Eilek, J. A. (eds.), 1984, *Physics of Energy Transport in Extragalactic Radio Sources* (NRAO: Greenbank, WV).

Bridle, A. H. & Perley, R. A., 1984, *Ann. Rev. Astr. Ap.*, **22**, 319.

Burbidge, G. R., 1956, *Ap. J.*, **124**, 416.

Burbidge, G. R., Burbidge, E. M. & Sandage, A. R., 1965, In *Quasistellar Sources and Gravitational Collapse*, eds. Robinson, I., Schild, A. & Schucking, E. L. (University of Chicago Press), p. 337.

Burch, S. F., 1979, *M. N. R. A. S.*, **186**, 519.

Burn, B. J., 1966, *M. N. R. A. S.*, **133**, 67.

Cawthorne, T. V. & Rickett, B. J., 1985, *Nature*, **315**, 40.

Chandrasekhar, S., 1950, *Radiative Transfer* (Clarendon Press: Oxford).

Cheung, A. C., Rank, D. M., Townes, C. H., Thornton, D. D. & Welch, W. J., 1968, *Phys. Rev. Lett.*, **21**, 1701.

Clarke, D. A., Norman, M. L. & Burns, J. O., 1989, *Ap. J.*, **342**, 700.

Cotton, W. D., Wittels, J. J., Shapiro, I. I., Marcaide, J., Owen, F. N., Spangler, S. R., Rius, A., Angulo, C., Clark, T. A. & Knight, A. C., 1980, *Ap. J. Lett.*, **238**, L123.

Cudworth, K. M. & Herbig, G. H., 1979, *Astron. J.*, **84**, 548.

Curtis, H. D., 1918, *Publ. Lick Obs.*, **13**, 9 (p. 31).

De Young, D. S. & Axford, W. I., 1967, *Nature*, **216**, 129.

Eckart, A., Witzel, A., Biermann, P., Pearson, T. J., Readhead, A. C. S. & Johnston, K. J., 1985, *Ap. J. Lett.*, **296**, L23.

Elvis, M. & Lawrence, A., 1985, In *Astrophysics of Active Galaxies and Quasi-Stellar Objects*, ed. Miller, J. S. (University Science Books: Mill Valley, CA), p. 289.

Fabbiano, G., Miller, L., Trinchieri, G., Longair, M. & Elvis, M., 1984, *Ap. J.*, **277**, 115.

Genzel, R. & Downes, D., 1977a, *Astr. Ap.*, **61**, 117.

Genzel, R. & Downes, D., 1977b, *Astr. Ap. Suppl.*, **30**, 145.

Genzel, R., Reid, M. J., Moran, J. M. & Downes, D., 1981, *Ap. J.*, **244**, 884.

Ginzburg, V. L. & Syrovatskii, S. I., 1969, *Ann. Rev. Astr. Ap.*, **7**, 375.

Gould, R. J., 1979, *Astr. Ap.*, **76**, 306.

Greenstein, J. L. & Schmidt, M., 1965, In *Quasistellar Sources and Gravitational Collapse*, eds. Robinson, I., Schild, A. & Schucking, E. L. (University of Chicago Press), p. 175.

Gull, S. F. & Northover, K. J. E., 1973, *Nature*, **224**, 80.

Hardee, P. E., Owen, F. N. & Cornwell, T. J., 1988, In *Active Galactic Nuclei: Proceedings of the Georgia State University Conference*, eds. Miller, H. R. & Wiita, P. J. (Springer-Verlag: Berlin), p. 347.

Haro, G., 1952, *Ap. J.*, **115**, 572.

Heiles, C. E., 1968, *Ap. J.*, **151**, 119.

Henriksen, R. N. (ed.), 1986, *Proceedings of the Conference on Jets from Stars and Galaxies*, *Can. J. Phys.*, **64**, 351.

Herbig, G. H., 1951, *Ap. J.*, **113**, 697.

Herbig, G. H., 1962, *Adv. Astr. Ap.*, **1**, 47.

Högbom, J. A., 1979, *Astr. Ap. Suppl.*, **36**, 173.
Hoyle, F., Burbidge, G. R. & Sargent, W. L. W., 1966, *Nature*, **209**, 751.
Hughes, P. A., Aller, H. D. & Aller M. F., 1985, *Ap. J.*, **298**, 301.
Jennison, R. C. & Das Gupta, M. K., 1953, *Nature*, **172**, 996.
Jones, T. W., 1988, *Ap. J.*, **332**, 678.
Jones, T. W. & O'Dell, S. L., 1977a, *Ap. J.*, **214**, 522.
Jones, T. W. & O'Dell, S. L., 1977b, *Ap. J.*, **215**, 236.
Jones, T. W., O'Dell, S. L. & Stein, W. A., 1974, *Ap. J.*, **188**, 353.
Jones, T. W., Rudnick, L., Aller, H. D., Aller, M. F., Hodge, P. E. & Fiedler, R. L., 1985, *Ap. J.*, **290**, 627.
Kardashev, N. S. & Slysh, V. I., 1988, In *The Impact of VLBI on Astrophysics and Geophysics*, eds. M. J. Reid & J. M. Moran (Kluwer Academic Publishers: Dordrecht, Netherlands), p. 433.
Kellermann, K. I. & Pauliny-Toth, I. K. K., 1969, *Ap. J. Lett.*, **155**, L71.
Killeen, N. E. B., Bicknell, G. V. & Ekers, R. D., 1986, *Ap. J.*, **302**, 306.
Kundt, W. (ed.), 1987, *Astrophysical Jets and their Engines* (D. Reidel: Dordrecht, Netherlands).
Kwan, J. & Scoville, N., 1976, *Ap. J. Lett.*, **210**, L39.
Lada, C. J., 1985, *Ann. Rev. Astr. Ap.*, **23**, 267.
Laing, R. A., 1980, *M. N. R. A. S.*, **193**, 439.
Laing, R. A., 1981, *Ap. J.*, **248**, 87.
Laing, R. A., 1984, In *Physics of Energy Transport in Extragalactic Radio Sources*, eds. Bridle, A. H. & Eilek, J. A. (NRAO: Greenbank, WV), p. 90.
Liebert, J., Angel, J. R. P., Hege, E. K., Martin, P. G. & Blair, W. P., 1979, *Nature*, **279**, 384.
Litvak, M. M., McWhorter, A. L., Meeks, M. L. & Zeiger, H. J., 1966, *Phys. Rev. Lett.*, **17**, 821.
Maiden, M. E. & Christiansen, W. A., 1986, *Can. J. Phys.*, **64**, 490.
Margon, B., 1984, *Ann. Rev. Astr. Ap.*, **22**, 507.
Margon, B., Ford, H. C., Katz, J. I., Kwitter, K. B., Ulrich, R. K., Stone, R. P. S. & Klemola, A., 1979a, *Ap. J. Lett.*, **230**, L41.
Margon, B., Ford, H. C., Grandi, S. A. & Stone, R. P. S., 1979b, *Ap. J. Lett.*, **233**, L63.
Marscher, A. P., 1977, *Ap. J.*, **216**, 244.
Marscher, A. P., 1980, *Ap. J.*, **235**, 386.
Marscher, A. P., 1983, *Ap. J.*, **264**, 296.
Marscher, A. P. & Broderick, J. J., 1985, *Ap. J.*, **290**, 735.
Marscher, A. P., Marshall, F. E., Mushotzky, R. F., Dent, W. A., Balonek, T. J. & Hartman, M. F., 1979, *Ap. J.*, **233**, 498.
Melia, F. & Königl, A., 1989, *Ap. J.*, **340**, 162.
Miller, J. S. (ed.), 1985, *Astrophysics of Active Galaxies and Quasi-Stellar Objects* (University Science Books: Mill Valley, CA).
Miller, L., Longair, M. S., Fabbiano, G., Trinchieri, G. & Elvis, M., 1985, *M. N. R. A. S.*, **215**, 799.
Mundt, R., 1986, *Can. J. Phys.*, **64**, 407.

Northover, K. J. E., 1973, *M. N. R. A. S.*, **165**, 369.
Pacholczyk, A. G., 1970, *Radio Astrophysics* (W. H. Freeman and Co.: San Francisco).
Pacholczyk, A. G. & Swihart, T. L., 1975, *Ap. J.*, **196**, 125.
Perkins, F., Gold, T. & Salpeter, E. E., 1966, *Ap. J.*, **145**, 361.
Porcas, R. W., 1987, In *Superluminal Radio Sources*, eds. Zensus, J. A. & Pearson, T. J. (Cambridge University Press: Cambridge), p. 12.
Rees, M. J., 1967, *M. N. R. A. S.*, **137**, 429.
Rees, M. J., 1971, *Nature*, **229**, 312.
Rees, M. J. & Sciama, D. W., 1965, *Nature*, **208**, 371.
Rees, M. J. & Sciama, D. W., 1966, *Nature*, **211**, 805.
Reid, M. J. & Moran, J. M., 1981, *Ann. Rev. Astr. Ap.*, **19**, 231.
Reid, M. J. & Moran, J. M. (eds.), 1988, *The Impact of VLBI on Astrophysics and Geophysics* (Kluwer Academic Publishers: Dordrecht, Netherlands).
Rickett, B. J., 1986, *Ap. J.*, **307**, 564.
Rickett, B. J., Coles, W. A. & Bourgois, G., 1984, *Astr. Ap.*, **134**, 390.
Romney, J. D., 1988, In *The Impact of VLBI on Astrophysics and Geophysics*, eds. M. J. Reid & J. M. Moran (Kluwer Academic Publishers: Dordrecht, Netherlands), p. 461.
Rybicki, G. B. & Lightman, A. P., 1979, *Radiative Processes in Astrophysics* (John Wiley and Sons: New York).
Saslaw, W. C., Valtonen, M. & Aarseth, S., 1974, *Ap. J.*, **190**, 253.
Scheuer, P. A. G., 1967, In *Plasma Astrophysics, Proceedings of the International School of Physics "Enrico Fermi" Course 39*, ed. Sturrock, P. A. (Academic: New York), p. 262.
Scheuer, P. A. G., 1974, *M. N. R. A. S.*, **166**, 513.
Schmidt, M., 1963, *Nature*, **197**, 1040.
Schneps, M. H., Lane, A. P., Downes, D., Moran, J. M., Genzel, R. & Reid, M. J., 1981, *Ap. J.*, **249**, 124.
Schwartz, R. D., 1983, *Ann. Rev. Astr. Ap.*, **21**, 209.
Shklovskii, I. S., 1953a, *Astron. Zh.*, **30**, 15.
Shklovskii, I. S., 1953b, *Dokl. Akad. Nauk SSSR*, **90**, 983.
Shklovskii, I. S., 1963, *Soviet Astr. - AJ*, **6**, 465.
Simon, R. S., Readhead, A. C. S., Moffet, A. T., Wilkinson, P. N., Allen, B. & Burke, B. F., 1983, *Nature*, **302**, 487.
Tucker, W. H., 1975, *Radiation Processes in Astrophysics* (The MIT Press: Cambridge, MA).
Turland, B. D., 1975a, *M. N. R. A. S.*, **170**, 281.
Turland, B. D., 1975b, *M. N. R. A. S.*, **172**, 181.
Unwin, S. C., Cohen, M. H., Pearson, T. J., Seielstad, G. A., Simon, R. S., Linfield, R. P. & Walker, R. C., 1983, *Ap. J.*, **271**, 536.
Valtaoja, E. & Valtonen, M. J., 1984, *Astr. Ap.*, **130**, 373.
Valtonen, M. J. & Byrd, G. G., 1980, *Ap. J.*, **240**, 442.
Wardle, J. F. C. & Kronberg, P. P., 1974, *Ap. J.*, **194**, 249.
Wardle, J. F. C. & Roberts, D. H., 1988, In *The Impact of VLBI on Astrophysics and*

Geophysics, eds. M. J. Reid & J. M. Moran (Kluwer Academic Publishers: Dordrecht, Netherlands), p. 143.

Weedman, D. W., 1986, *Quasar Astronomy* (Cambridge University Press: Cambridge).

Weinreb, S., Barrett, A. H., Meeks, M. L. & Henry, J. C., 1963, *Nature*, **200**, 829.

Woltjer, L., 1966, *Ap. J.*, **146**, 597.

Zensus, J. A. & Pearson, T. J. (eds.), 1987, *Superluminal Radio Sources* (Cambridge University Press: Cambridge).

Zuckerman, B., Kuiper, T. B. H. & Rodriguez-Kuiper, E. N., 1976, *Ap. J. Lett.*, **209**, L137.

Zuckerman, B. & Palmer, P., 1974, *Ann. Rev. Astr. Ap.*, **12**, 279.

2

Observations of Large Scale Extragalactic Jets

T. W. B. MUXLOW AND S. T. GARRINGTON
University of Manchester, Nuffield Radio Astronomy Laboratories, Jodrell Bank, Macclesfield, Cheshire, SK11 9DL, UK.

2.1 Introduction

The interpretation of the diverse forms of observed radio source structure has always been problematical since this normally involves the use of some form of classification scheme. With the benefit of hindsight, it is clear that this exercise has not always proved to be a total success. Every astronomical object is the product of a unique set of physical circumstances which must, at some level, ultimately preclude the imposition of a generalised classification scheme covering many objects. It remains, however, a necessary basic stage in the process of scientific investigation. Any classification scheme is based upon gross structural features derived from observation. Such observations are of an inhomogeneous set of objects and are limited by sensitivity and imaging techniques. Schemes are therefore subject to strong selection effects and their subdivisions are arbitrary. A scheme can, however, prove useful provided the subdivisions broadly map out differing segments in the parameter space of the physical conditions of radio sources. The problem is, of course, that it is those very same physical conditions that are as yet unknown and that one is attempting to investigate. Thus, any current classification scheme is dominated by the characteristics of the telescopes available to observers at the time, and incorporates the 'conventional wisdom' derived from the interpretation of previous work. Such circumstances are profoundly inelegant but probably unavoidable.

At present, the 'conventional wisdom' derives from observations made in the mid 1970s with imaging arrays such as the Cambridge 5 km telescope which paradoxically, we now realise were not optimised for the detection of jets in powerful extragalactic radio sources. The standard beam models as presented in Scheuer (1974) and Norman *et al.* (1982), infer the presence of a beam which transports the relativistic plasma and energy generated in the active core of a radio source from that core to the outer extended radio structure. With the advent of the Very Large Array (VLA) and MERLIN telescopes in the early 1980s many more jets were discovered in powerful extragalactic sources. These are now unambiguously identified with the 'beams' of the standard beam models. It is thus not possible to describe jets in extragalactic radio sources without reference to the large-scale extended structure within which they are imbedded since the two are inextricably linked.

The telescope which at present dominates the existing database of observations relating to large-scale extragalactic jets is the VLA run by N. R. A. O. and sited in New Mexico. This is complemented by the Jodrell Bank MERLIN array for smaller sources where higher resolution is required and by the Westerbork array in Holland for very large extended sources. The term 'large-scale' restricts this description to jets with sizes $\geq$ 1 kpc. These have been mapped extensively by the VLA and MERLIN at angular resolutions $\geq$ 0.15 arcseconds. Very-Long-Baseline-Interferometry (VLBI) observations of small-scale nuclear jets on the milliarcsecond scale size are discussed in more detail in Chapter 4.

2.2 Source classification

2.2.1 A glossary of descriptive terms

Maps produced by connected-element interferometers such as the Cambridge 5 km in the 1970s gave rise to a simple division in structural classification. There were *extended, steep spectrum* objects and *compact, flat spectrum* objects. The former were roughly collinear doubles with two *lobes* of radio emission extending out to several hundreds of kiloparsecs on either side of the optical identification. The latter were, at the levels of sensitivity and angular resolution then available, unresolved points coincident with the nucleus of the host galaxy. Much work was done in classifying the *extended* objects; the *compact* sources being treated as a separate class of object studied by separate groups of workers with VLBI arrays. With the development of imaging techniques and the advent of more sensitive and higher resolution interferometers, it is now clear that many of the *compact, flat spectrum* sources contain, at some level, some of the elements of extended steep spectrum emission present in *extended, steep spectrum* sources. The *compact* and *extended* source classification has thus now been modified to read *core-dominated* and *lobe-dominated* sources. Since it now appears that all radio galaxies and radio-loud quasars with luminosities at 1 GHz $\geq 10^{25}$ W Hz^{-1} possess flat spectrum *cores* and steep spectrum *jets* and *lobes* when mapped with the required dynamic range, angular resolution, and sensitivity, the question now under consideration is what set of physical conditions and observer viewing angles determine the *relative prominence* of the structural elements seen in powerful radio sources.

For clarity and completeness, we now outline a simple set of descriptive structural elements in general usage by radio astronomers to characterise the different parts of radio source structures. The spectral indices quoted (α) are in the conventional form $S \propto \nu^{-\alpha}$ where S is flux density and ν is frequency.

1. **Cores**
   Radio *cores* are compact flat spectrum components associated with the power source in the nucleus of the radio galaxy or quasar. They are normally unresolved at angular resolutions ≥ 0.1 arcseconds. When studied by VLBI instruments, they are often resolved and show milliarcsecond length, multicomponent jets. The subcomponents often have different turn-over frequencies (cf. §1.2.1) so that their superposition produces a core spectrum which is fairly flat at cm wavelengths (Cotton *et al.* 1980). More correctly, the *core* is the stationary component, which may be the optically thick base of the jet, or the site of the first recollimation shock, before which the jet plasma might have been of too low a surface brightness to be detected. Alternatively, it might be emission from the central object itself; emission from the central object could emanate from synchrotron plasma imbedded in the broad line clouds or from the environment of an accretion disc. Radio imaging is still several orders of magnitude away from the resolutions required to select between such models.
2. **Jets**
   Radio *jets* are linear features linking the *cores* to the outer

extended *lobe* structures. They may be visible over all or only part of their inferred path, one or two-sided, smooth or knotty, centre-brightened or edge-brightened. The term has been the subject of much misuse by workers and we here adopt the criteria of Bridle & Perley (1984) as set out in §1.1.2. Such jets typically have spectral indices around 0.6 and are identified with the transmission beams in the standard models. The radiation detected represents energy loss and leakage from such beams.

3. **Hotspots**

   The bright components located at the outer extremities of the lobes of powerful extended sources are known as *hotspots*. Their linear sizes are generally less than about 1 kpc or so; VLBI observations of 3C 205 (Lonsdale 1989) reveal a compact ridge in the southern hotspot with a width of about 20 pc. Hotspots have spectral indices in the range 0.5 to 1.0, generally somewhat flatter than the lobes in which they are embedded. The fractional polarization at $\lambda 6$ cm is typically 10% and the inferred magnetic field direction is usually perpendicular to the source axis. Hotspots are naturally interpreted as the 'working surface' where the beam meets the ambient medium, creating a strong shock which converts a significant fraction of the beam kinetic energy into relativistic particles. The detailed dynamics and structure of this working surface have been studied using numerical simulations which are discussed in Chapter 7.

   Recent high resolution observations (*e.g.*, Laing 1989) show that in many sources the lobes contain a single bright, very compact primary hotspot, generally to one side of the lobe and set back from its leading edge along with a secondary hotspot with a more complex morphology. Such multiple hotspots have been modelled by Williams & Gull (1985), who propose that the primary hotspot is the site of current beam impact and the diffuse secondary is a 'splatter-spot' formed by the deflected beam.

4. **Lobes**

   Lobes are a general term to describe the extended regions of radio emitting plasma exhausted from the beams. Well used descriptions of *lobe* morphology include the following:

Table 2.1 *Extragalactic Radio Sources*

Population Type	A	B
Emission Type	'Disk Emission'	Standard Beam Model
Power Source	10^4 Massive Stars	Galactic Nucleus
Energy Source	Nuclear Fusion	Gravity
Extreme Example of Type	Starburst Galaxy	Radio Quasar
Number in Radio Surveys	0.5%	99.5%
Median Luminosity (1 GHz)	10^{22} W Hz^{-1}	10^{27} W Hz^{-1}
Space Density at Median Luminosity	10^6 Gpc^{-3}	10 Gpc^{-3}
Typical Distance	10 Mpc	1 Gpc
Maximum Luminosity (1 GHz)	10^{24} W Hz^{-1}	10^{30} W Hz^{-1}

a. **Plumes**
Plumes are a descriptive term often ascribed to the extended regions of radio emission associated with relatively low luminosity radio galaxies (see §2.2.4.1) where the outer boundary is ill-defined and sensitivity dependent since the surface brightness decreases with increasing distance from the galaxy.

b. **Tails**
Tails are lobe structures which are believed to be deflected back by interaction with an external medium resulting in an overall radio structure which is significantly non-linear.

c. **Bridges**
Bridges are the inner lobe regions of high luminosity radio sources which join the two outer halves of the source structure. They contain the oldest and most radiatively aged plasma in the source.

d. **Haloes**
Haloes are low surface brightness amorphous lobe structures containing old radiatively aged plasma. Such structures probably result from severe disruption of the normal linear beam dynamics.

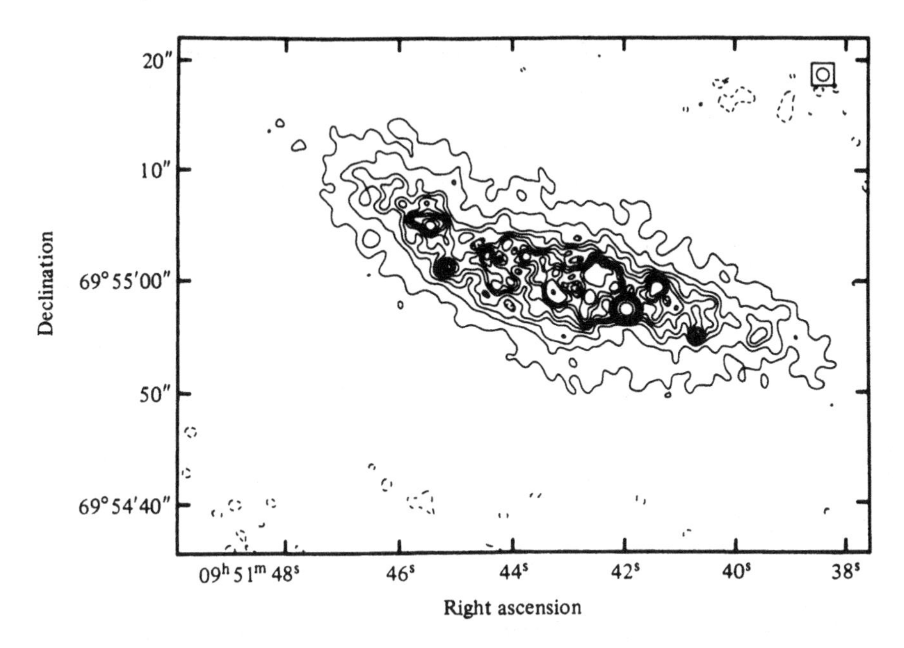

Fig. 2.1. VLA map of M 82 at 1413 MHz. Resolution: 1.1 arcsec. Peak: 264 mJy/beam. Contours: -2, -1, 1, 2, 3, 4, 5, 6, 7, 8, 9, 10, 16, 20, 25, 50%. Reproduced with permission from Kronberg, Biermann & Schwab (1985).

2.2.2 The optical identifications of extragalactic radio sources

Among the extragalactic sources of radio radiation there are two main populations whose properties are outlined in Table 2.1. The term 'extragalactic radio source' is generally applied only to population B objects in Table 2.1 since they are several orders of magnitude more powerful than population A objects and hence totally dominate flux density limited radio samples. It is, however, interesting to note that population A objects by far outnumber population B objects in the universe since they have far higher space densities even though they are very rarely found in source surveys. Their radio emission is generated by stellar activity throughout the galactic disc. An example radio map for one such object 3C 231 (M 82) is shown in Fig. 2.1. Since radio jets are not found in such objects we infer that there is no currently active central power source and associated beam system to transport the energy to the

Table 2.2 *Extragalactic Radio Sources in Large Galaxies*

Population Type	Spirals	Ellipticals
Angular Momentum	Dominates Dynamics	Negligible
10 kpc scale disc	Yes	No
Current Star Formation	Yes	No
$T \leq 10^5$K Gas	10^{-3} Mass of Stars	$\leq 10^{-6}$ Mass of Stars
$T \geq 10^5$K Gas	?	0.4 Mass of Stars
	If an Active Radio Source	
NLR Only Visible	Seyfert 2	Radio Galaxy
NLR + BLR/CONT	Seyfert 1	N-Galaxy/ BL Radio Galaxy
BLR/CONT Dominates	?	Radio-Loud Quasar
Maximum Luminosity (1 GHz)	$10^{25.5}$ W Hz^{-1}	10^{29} W Hz^{-1}
Maximum Size of Source	10 kpc	3 Mpc
Radio Structure	Distorted	Usually Linear

active regions of radio emission. Population A objects are usually associated with spiral/irregular type galaxies.

Population B objects are usually associated with luminous ($M_B <$ -20) galaxies. However, spiral and elliptical galaxies appear to give rise to two distinct classes of population B object. These are presented in Table 2.2 which summarizes their basic differences and optical classification based on the relative prominence of the narrow-line region (NLR), broad-line region (BLR), and optical/UV continuum (CONT).

We now present in more detail the *lobe-dominated* radio structures found in luminous spiral and elliptical galaxies which contain jets.

2.2.3 The structures of radio Seyfert galaxies

S-shaped kiloparsec scale radio structure suggestive of disrupted jets have been found in many Seyfert galaxies (Wilson 1983). It is usual to ascribe the difference in radio properties between Seyfert galaxies and their more powerful elliptical counterparts to *(a)* the lower power output of the central nuclear 'powerhouse' and *(b)* the difficulties encountered by the plasma beams in propagating through the dense rotating interstellar medium found in spiral discs. Some Seyferts however, do not contain obvious continuous jet structures. These have been modelled in terms of expanding radio emitting plasmons photoionized by UV continuum from the nucleus (Pedlar, Dyson & Unger 1985). Where jets seem unambiguously to be present, they are up to 1 or 2 kpc long and run continuously into extended structures of a few kpc in length. An example is shown in Fig. 2.2 where the nuclear region of NGC 4388 has been mapped with the VLA. The 30 arcsecond length structure corresponds to a size of about 2 kpc with the self-absorbed core lying to the northern end of the bright inner jet. The jet in this object is thus two-sided, asymmetrical in length, and provided one is content to describe the extended northern feature as a jet, about 2 kpc in overall extant. It therefore just falls within our definition of 'large-scale' extragalactic jets.

NGC 4388 has been studied by Hummel and his co-workers in a continuing investigation of the polarization characteristics of Seyfert structures. At 4.8 GHz they find that the jets are already completely depolarized, lending weight to the interpretation that much thermal material has been entrained into the jets by interaction with the surrounding interstellar medium.

Further evidence for interaction between the jets and surrounding medium comes from studies of the connection between the radio radiation and the forbidden line optical emission. A broad correlation is found in Seyferts between the radio and forbidden line luminosities (Wilson & Willis 1980). Furthermore, the radio luminosity also appears to be correlated with the width of the optical [OIII] line. Wilson (1982) has suggested that this may be due to acceleration of line emitting clouds which drift into the jet. A refinement of this explanation using a combination of radial acceleration and orbital motion has been used by Booler, Pedlar & Davies (1982) to account for the 35° position angle rotation between the radio structure and extended forbidden line region in NGC 4151.

The typical luminosity of the radio structures lie in the range $P_{1\mathrm{GHz}} = 10^{21} - 10^{25}$ W Hz^{-1}. When present, the jet structure tends to be two sided. The disrupted nature of these objects can, however, give rise to major asymmetries.

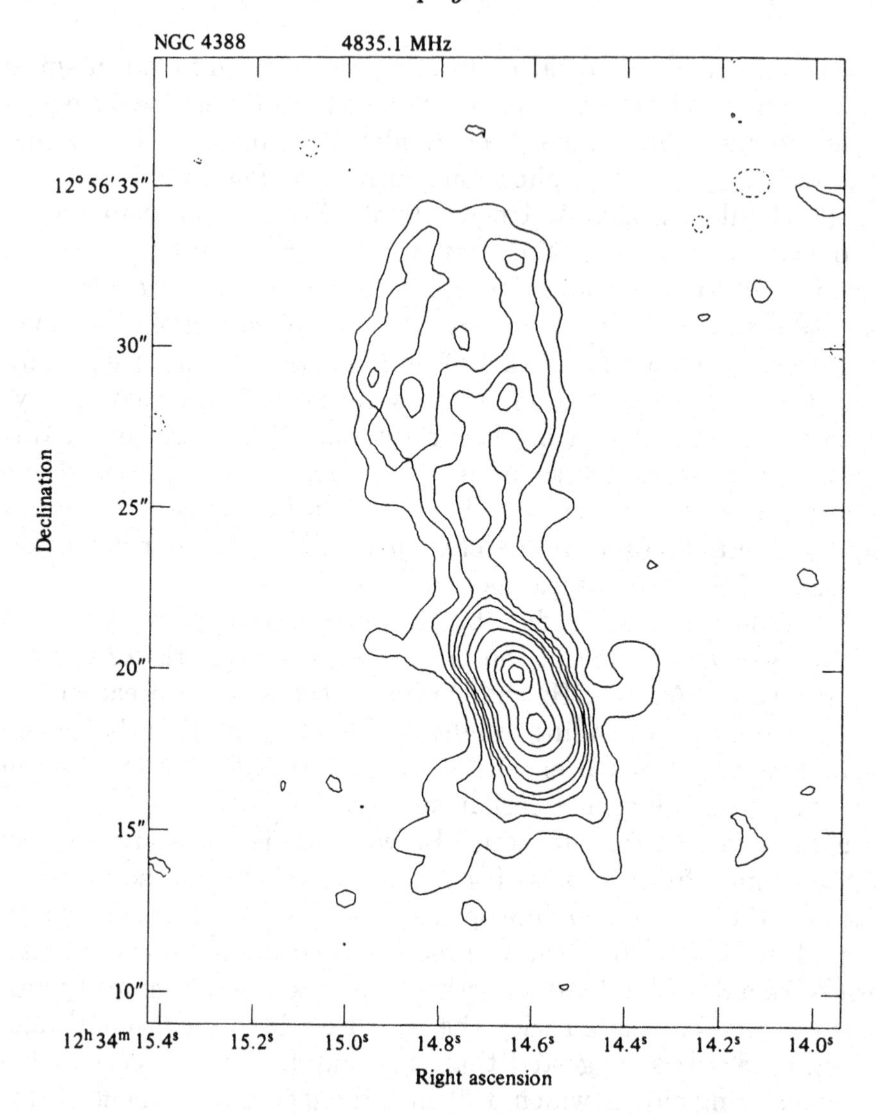

Fig. 2.2. VLA map of NGC 4388 at 4835 MHz. Resolution: 1.2 arcsec. Peak: 5.4 mJy/beam. Contours: -20, 20, 40, 80, 100, 150, 200, 400, 600, 800, 1000 μJy/beam. Map provided by E. Hummel.

2.2.4 Lobe-dominated radio galaxies

The properties of radio sources, as has been already noted, change with luminosity. This is especially true of radio galaxies whose structures seem to undergo an abrupt transition around $P_{178\mathrm{MHz}} = 5 \times 10^{25}$ W Hz^{-1}. This was first noted by Fanaroff &

Riley (1974) who introduced a simple classification scheme around this critical luminosity. Sources below this critical luminosity are known as FR I type objects and those above as FR II type objects. The structural differences as outlined below are broadly interpreted as a major transition in the jet fluid flow, probably from turbulent transonic (FR I) to laminar supersonic (relativistic?) flow (FR II). As shown in Table 2.2, lobe-dominated radio galaxies are usually ellipticals. There are, however, significant differences between the host galaxies of FR I and FR II radio sources (Owen & Laing 1989; Prestage & Peacock 1988; Lilly & Prestage 1987). The host galaxies of FR II sources are usually normal giant ellipticals and are not usually in rich clusters. FR I sources, on the other hand, tend to be brighter and larger, and are generally described, following Matthews, Morgan, & Schmidt (1964), as D or cD. Heckman *et al.* (1986) find that the host galaxies of FR II sources are more likely to appear 'disturbed', perhaps due to interaction with companion galaxies. Owen & Laing (1989) argue that FR II structures are essentially transient phenomena in normal ellipticals whereas FR I structures probably have longer lifetimes since their overall sizes are similar although their implied jet velocities are lower. Recent spectral ageing studies of a number of FR II sources indicate ages in the range 2 – 20 million years (Alexander 1987; Leahy, Muxlow & Stephens 1989).

2.2.4.1 **Fanaroff-Riley type I objects**

These lower luminosity radio galaxies tend to have prominent smooth continuous two-sided jets running into large-scale lobe structures (plumes) which are edge-darkened and whose steepest radio spectra (by implication the most radiatively-aged material) lie in the outermost extended regions furthest from the host galaxy. The term *edge-darkened* simply means that the ratio of the separation of the peaks of radio lobe emission to the total source size, as defined by the outermost detected parts of the lobes, is significantly less than unity (technically the Fanaroff-Riley definition requires this ratio to be less than 0.5 for a Type I object). Most straight jets are one-sided (by $>$4:1 intensity ratio) close to the core but become two sided after a few kpc. The one-sided region lasts typically $<$10% of the total length of the jet and if one of the jets is brighter on the large-scale it tends to be the one with the one-sided base. The jets often contribute over 10% of the total power of the extended structure and where resolved transversely have large opening angles. The opening angle can be estimated by the quantity $d\phi/d\theta$ where ϕ is the deconvolved FWHM jet width and θ is the angle from the radio core. Such studies reveal that the jets in FR I sources often have varying opening angles along their lengths. Typically, $d\phi/d\theta \leq 0.1$ for the first kiloparsec. 'Flaring' may then occur between 1 and 10 kpc with $d\phi/d\theta$ in the range 0.25 to 0.6. Beyond this, recollimation may occur with re-expansion beyond 100 kpc from the core. The jet/plume transition is sometimes associated with a sudden widening and brightening of the radio structure. There is often a distinct gap between the core and the start of the radio jets. Initially the magnetic field is predominantly along the jet ($\mathbf{B}_{\parallel}$), becoming predominantly perpendicular to the jet ($\mathbf{B}_{\perp}$) further out. At 5 GHz the typical degree of polarization is in the range 10 – 40%. Some jets have a $\mathbf{B}_{\perp-\parallel}$ configuration with the field perpendicular to the jet axis near the centre but becoming parallel to the axis near one or both of its edges. This configuration is often found where the jets bend with the outer edge being more strongly polarized suggesting that the $\mathbf{B}_{\parallel}$ is amplified in this region by interaction with the ambient medium.

A typical FR I type structure is illustrated in Fig. 2.3 which shows a VLA image of the 13th magnitude cDE4 ($z = 0.0181$, $P_{1\mathrm{GHz}} \sim 10^{25}$ W Hz^{-1}) radio galaxy 3C 449. Note the smooth jets extending to $\sim$ 1 arcminute (20 kpc) to either side of the host galaxy. The gap between the core and the bright regions of the jets is well illustrated in this galaxy. The jet smoothness can be quantified by taking the ratio between the brightness in any knots present (S_k) and the inter-knot regions (S_i). 3C 449 has a S_k/S_i value of <5 which is typical of FR I type sources. The jets are highly polarized at 5 GHz, averaging $\sim$ 30% along their lengths with $\mathbf{B}_{\perp}$ throughout.

Figs. 2.4 and 2.5 show the inner jet regions of the 11.6 magnitude E1 ($z = 0.0129$, $P_{1\mathrm{GHz}} \sim 10^{25}$ W Hz^{-1}) galaxy IC 4296 (PKS 1333-337) as mapped by Killeen, Bicknell & Ekers (1986). Fig. 2.4 shows the change of polarization position-angle, indicating the transition from $\mathbf{B}_{\parallel}$ to $\mathbf{B}_{\perp}$, together with the 'flaring' of the jet where $d\phi/d\theta$ changes from <0.1 to $\sim$0.45 some 2.6 kpc from the core. Fig. 2.5 shows the onset of the $\mathbf{B}_{\perp-\parallel}$ polarization configuration further

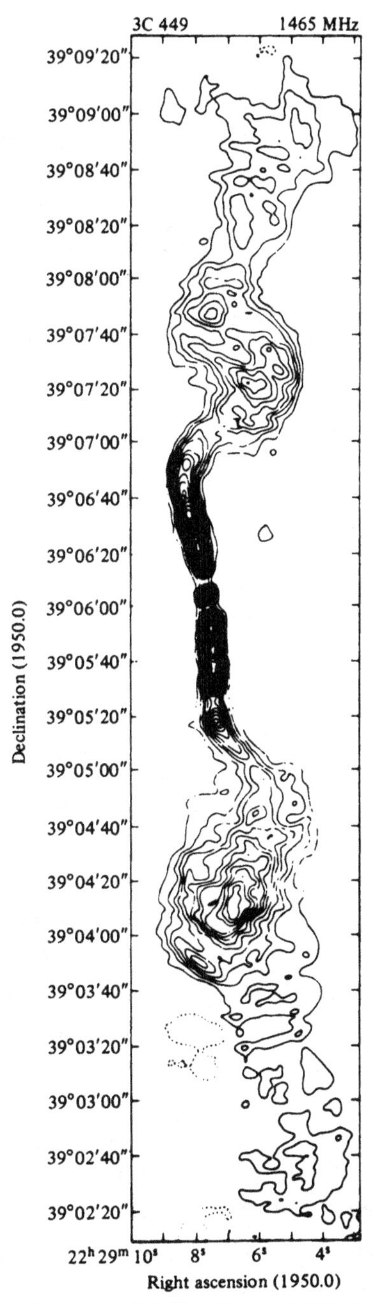

Fig. 2.3. VLA map of 3C 449 at 1465 MHz. Resolution: 4.8 × 3.4 in PA −47°. Peak: 22.2 mJy/beam. Contours: -5, 5, 10, 15 ... 100%. Reproduced with permission from Perley, Willis & Scott (1979). Reprinted by permission from *Nature*, **281**, 437. Copyright © 1979 Macmillan Magazines Limited.

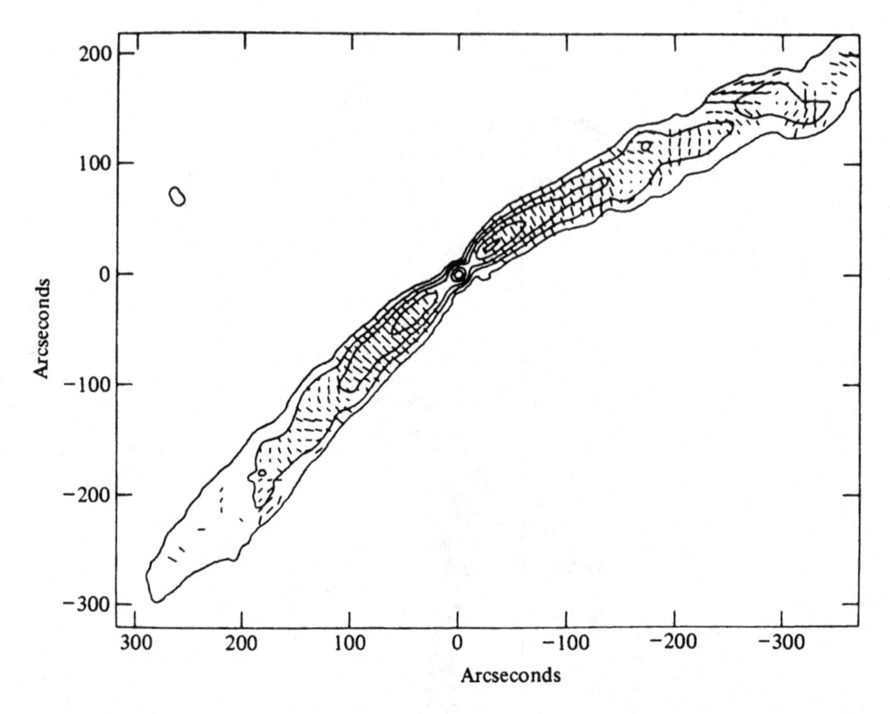

Fig. 2.4. VLA map of IC 4296 at 1465 MHz. Resolution: 10.4 arcsec. Peak: 163 mJy/beam. Contours: -1, 5, 15, 40, 70%. Vectors represent projected magnetic field with lengths proportional to fractional polarization (1 arcsec = 3.8%). Reproduced with permission from Killeen, Bicknell & Ekers (1986).

down the jet beyond 120 arcseconds (20.5 kpc) from the core where transverse oscillations and bends start to appear. Fig. 2.6 shows VLA maps of the weak nearby ($z = 0.0031$, $P_{1\mathrm{GHz}} \sim 10^{24}$ W Hz^{-1}) radio galaxy 3C 272.1 (M 84). This is included to give an example of a one-sided jet close to the core as is seen in the bright base of the northern jet in M 84.

Although two-sided jet symmetry is a general feature of FR I type structures, it should be noted that some sources show significant side to side asymmetry. Two well studied examples are NGC 315 and NGC 6251 (Willis *et al.* 1981; Perley, Bridle & Willis 1984). In NGC 315 the counterjet is detected with similar levels of surface brightness but is much more broken up with S_k/S_i values more typical of FR II jets. In NGC 6251 the jet/counterjet brightness ratio is ~50:1 (Willis *et al.* 1982).

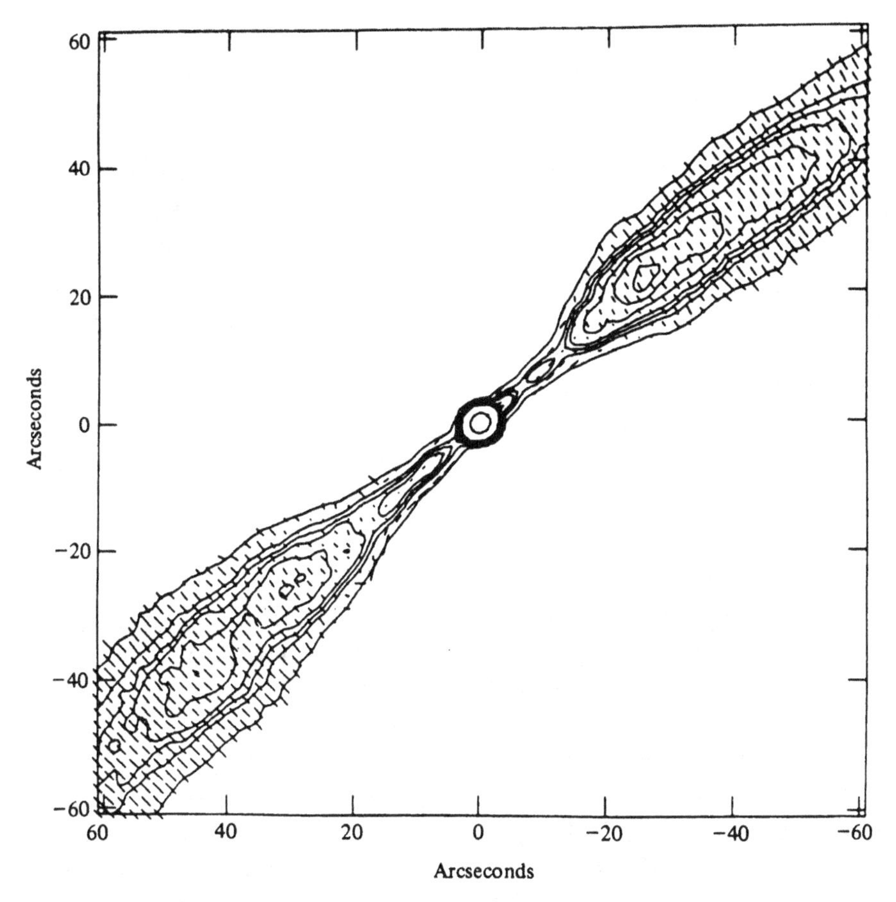

Fig. 2.5. VLA map of IC 4296 at 1465 MHz. Resolution: 3.2 arcsec. Peak: 153 mJy/beam. Contours: -1, 3, 4, 5, 7, 9, 11, 50%. Vectors represent projected magnetic field with lengths proportional to fractional polarization (1 arcsec = 22.1%). Reproduced with permission from Killeen, Bicknell & Ekers (1986).

2.2.4.2 Fanaroff-Riley type II objects

Radio galaxies above the transition luminosity of $P_{178\mathrm{MHz}} = 5\times 10^{25}$ W Hz^{-1} tend to have large-scale structures which are edge-brightened (peak brightness separation to total source size ratio close to unity, cf. edge-darkened in the previous section) with bright outer hotspots. The steepest radio spectra are found in the inner extended regions of the lobes or bridges nearest the optical identification. Although the absolute luminosities of the jets and cores in FR II objects are in general higher than in FR I galaxies, the

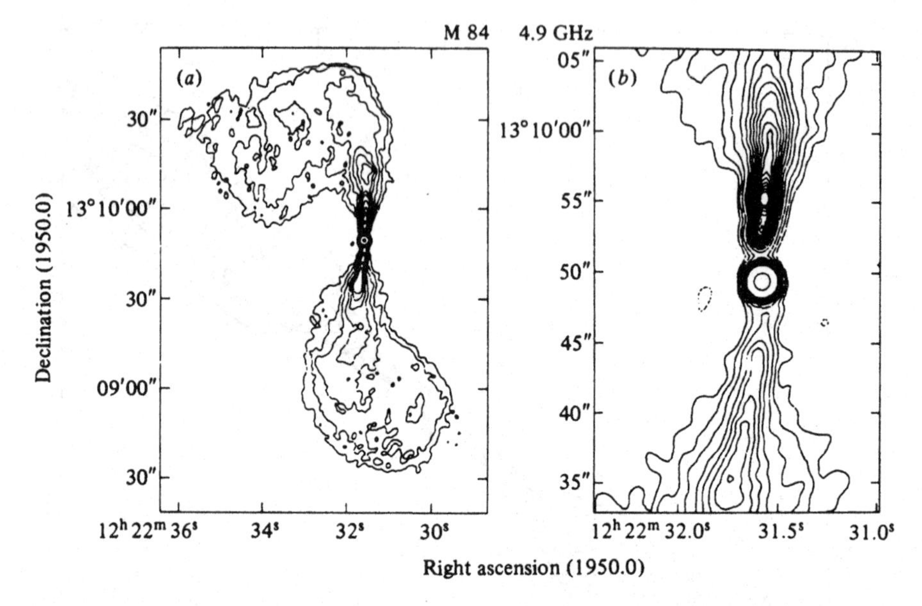

Fig. 2.6. VLA map of M 84 at 4885 MHz. The right panel shows detail of the inner jets. Reproduced with permission from Bridle & Perley (1984); data from Laing & Bridle (1987).

proportionately greater increase in lobe and hotspot luminosities results in the jets and cores being far less prominent in the radio structures than for FR I type objects. This is illustrated in Fig. 2.7 where typical 3C radio source core, jet, hotspot, and lobe luminosities are plotted against total source luminosity. The cores and jets in these structures usually contain <10% of the total source luminosity which results in the non-detection of some cores in existing maps. The extended jets are even more difficult to detect and are found in only a few images of very high dynamic range. Jets are thus unambiguously found in <10% of these luminous radio galaxies compared with about 80% of weak FR I galaxies.

When detected, the jets in FR II sources differ significantly from those in FR I structures. They are usually one-sided with a jet/counterjet ratio >4:1 (Bridle & Perley 1984). The situation is further complicated since the jets are in general not smooth but dominated by bright knots with S_k/S_i values of >4. The opening angles of jets in FR II structures are lower than those for FR I sources. When resolved transversely, FR II jet values for $d\phi/d\theta$ are usually <0.1 with mean values typically around 0.05. The great distance

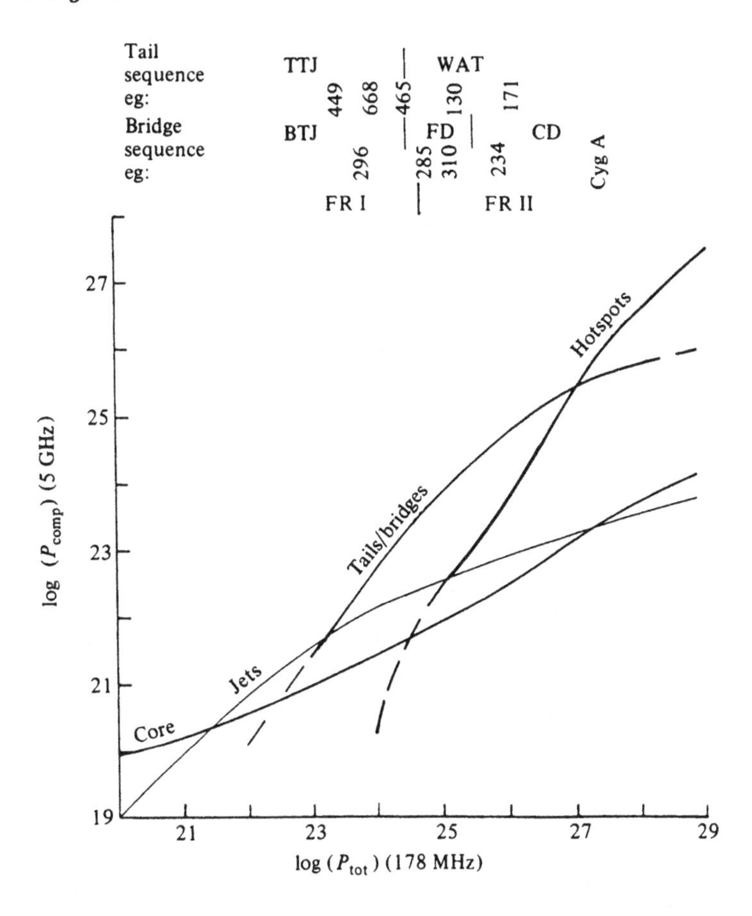

Fig. 2.7. Comparison of the radio power emitted by various components as a function of total source luminosity for typical 3C sources. Reproduced with permission from Leahy (1985).

of many of these objects together with the small jet opening angles results in many jets being only marginally resolved. Although variations in opening angle have been detected, systematic trends have not been established. The magnetic field configuration is usually $\mathbf{B}_{\parallel}$ throughout except for the bright knots where $\mathbf{B}_{\perp}$ dominates.

A typical FR II type jet is illustrated in Fig. 2.8 where the VLA 5 GHz image of part of the radio structure of the 14th magnitude cD galaxy 3C 405 (Cygnus A) ($z = 0.057$, $P_{1\mathrm{GHz}} \sim 10^{29}$ W Hz^{-1}) is shown.

A few sources show a rather more 'relaxed' structure with fatter lobes and little or no evidence for active cores, jets, or hotspots.

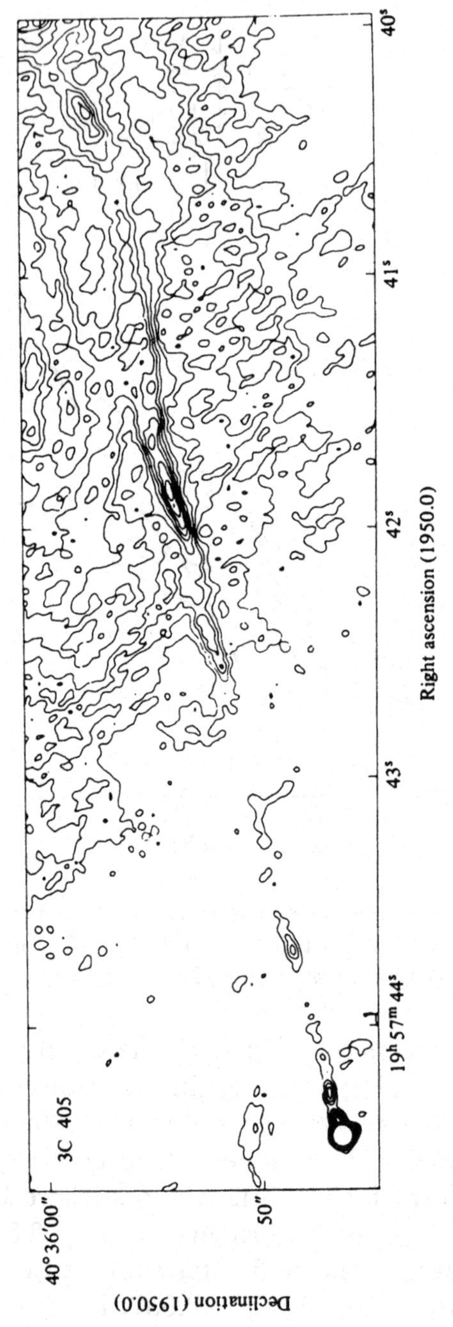

Fig. 2.8. VLA map of the jet in Cygnus A at 4885 MHz. Resolution: 0.4 arcsec. Peak: 3260 mJy/beam. Contours: 0.05, 0.1, 0.2, 0.3, 0.4, 0.5, 0.6, 0.7, 0.8, 0.9, 1.0, 1.1%. Reproduced with permission from Perley, Dreher & Cowan 1984.

These **Fat Doubles** are amongst the weakest FR II structures, lying just above the FR I – FR II break luminosity. They tend to have radio spectra which are steeper than other FR II objects. These may be dying radio sources where the injection of fresh particles and energy from the central power source has ceased.

2.2.4.3 Unusual lobe-dominated radio galaxy structures

1. **Narrow-Angle-Tail Sources**
   Also known as 'Twin-Tail' and 'Head-Tail' sources; objects in this class are, in general, distorted FR I type structures with bent two-sided jets running into an extended tail. Such objects are found in clusters of galaxies where the parent galaxy has a large proper motion with respect to the cluster and the radio structure is thus bent back by ram pressure from the hot external medium found in clusters. Although two-sided and roughly symmetrical, these jets when observed at arcsecond resolution do not appear smooth and tend to have S_k/S_i values intermediate between typical FR I and FR II structures. The magnetic field configuration in the jets is $\mathbf{B}_{\parallel}$ dominated, unlike other two-sided FR I jets, which lends further evidence in support of the interaction hypothesis since it implies that this may be an extreme example of the $\mathbf{B}_{\perp-\parallel}$ configuration found in FR I structures where the jets are deflected.
   The most well-known example of this class of object is NGC 1265 where the 12th magnitude ED3 galaxy ($z = 0.0255$, $P_{1\text{GHz}} \sim 10^{25}$ W Hz^{-1}) has an observed radial velocity of 2300 km s^{-1} relative to the mean value for the Perseus cluster. Fig. 2.9 shows VLA maps of the head region of this radio galaxy together with the low surface-brightness tail, extending some 8 arcminutes to the north. Fractional polarization values along the jet are in the range 10 – 25% with $\mathbf{B}_{\parallel}$ throughout and tend to be higher along the outer edges, over the inner two-thirds of the jets.
2. **Wide-Angle-Tail Sources**
   This morphological class of object was first classified by Owen & Rudnick (1976). They are C-shaped structures associated with the optically dominant galaxies in rich clus-

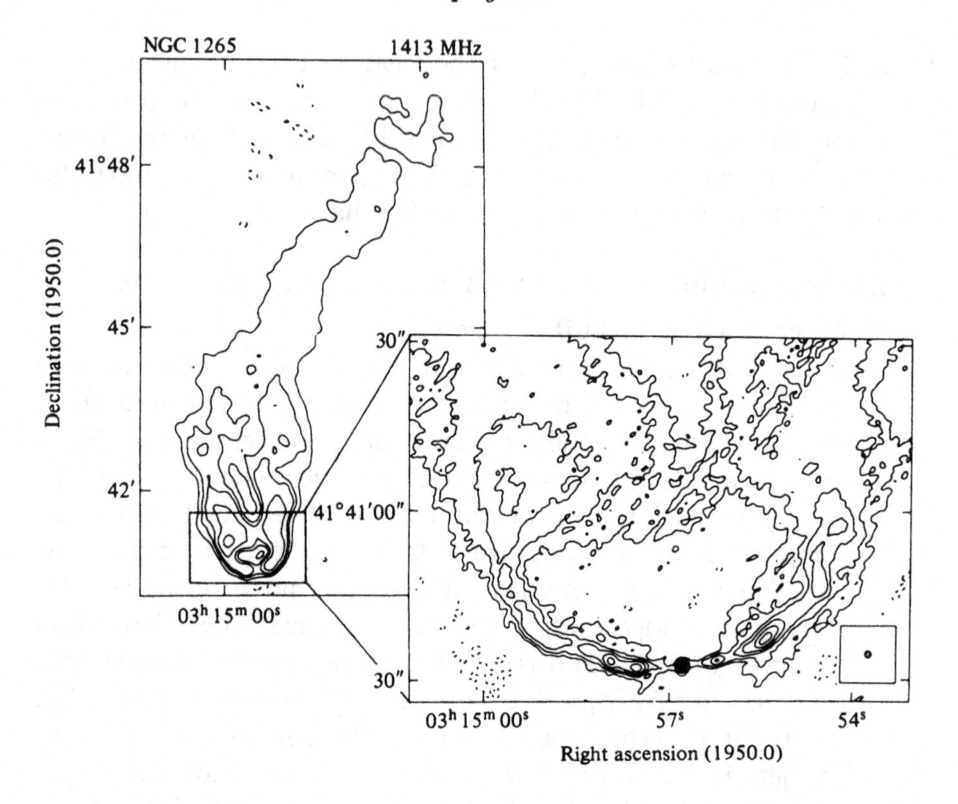

Fig. 2.9. VLA maps of NGC 1265 at λ21 cm. Left panel: Resolution: 12.9 × 11.4 arcsec, in PA −4°. Contours: -10, 10, 30, 50, 100, 200 mJy/beam. Right panel: Resolution: 1.2 arcsec. Contours: -0.3, 0.3, 0.7, 1.5, 3.0, 6.0, 10.0 mJy/beam. Reproduced with permission from O'Dea & Owen (1986).

ters and have total radio luminosities intermediate between Narrow-Angle-Tail sources and classical double FR II structures. Their outer structures are characterised by disrupted FR I type plumes or tails with inner hotspots linked to the central component by jets. There is usually a distinct change in position angle in the radio structure in the vicinity of the hotspot where the jets are suddenly disrupted. The detailed cause of this disruption is not well understood although various workers have suggested the action of internal shocks for the smaller ($<$200 kpc) structures and collisions with cool clouds for the larger sources.

Since there are only of order 10 such sources that have

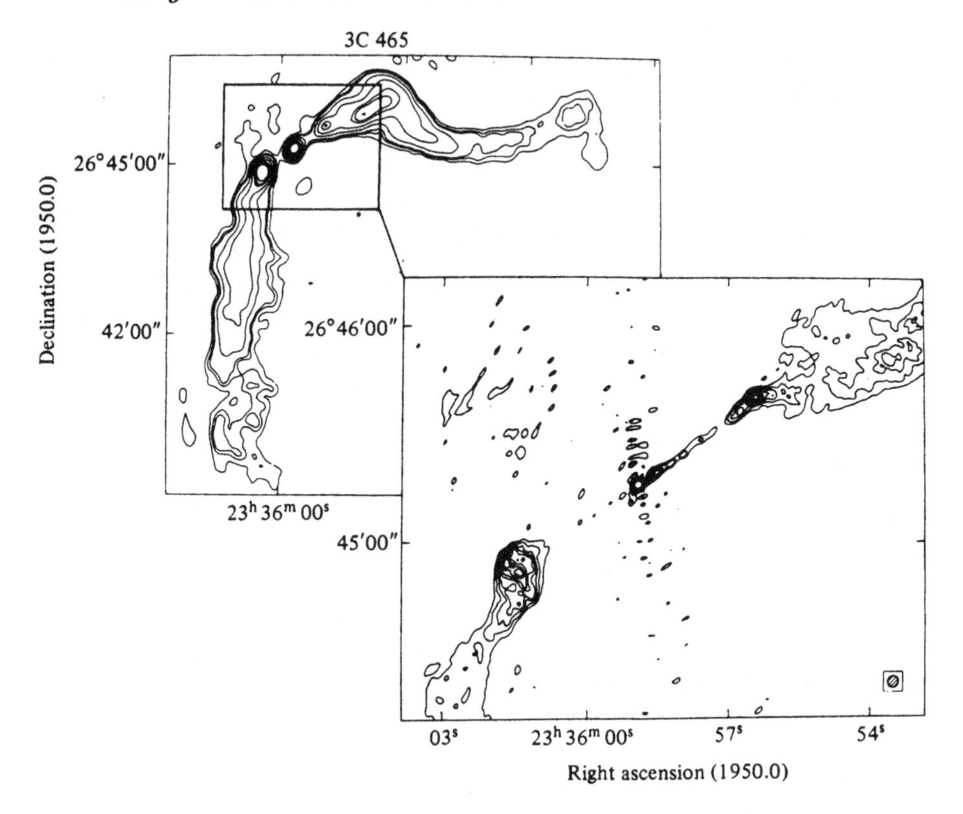

Fig. 2.10. VLA maps of 3C 465 at 1480 MHz. Upper panel: Resolution: 9.5 × 1.2 arcsec. Contours: -5, 5, 10, 20, 40, 80, 100, 150 mJy/beam. Lower panel (inset): Resolution: 2.0 arcsec. Contours: -1.8, 1.8, 3, 5, 7, 10, 13, 17, 20, 80 mJy/beam. Reproduced with permission from Eilek *et al.* (1984).

been studied in detail the inner jet properties are not statistically well established. They do, however, appear to resemble the jets in FR II type structures with jet/counterjet ratios >4:1, S_k/S_i values of >4, and a dominant $\mathbf{B}_{\parallel}$ magnetic field configuration (Saikia & Salter 1988). Fig. 2.10 provides an example of this structural type where the 1.4 GHz VLA map of the inner portion of 3C 465 is shown.

3. **Steep Spectrum Core Sources**
   Also referred to as 'Core-Halo' type sources; this class of object is dominated by a kiloparsec scale steep spectrum component associated with the nucleus of the parent object. This is surrounded by and imbedded within a diffuse low

surface brightness halo component. 'Core' in this case refers to the complete nuclear component rather than the compact flat spectrum feature within it. The statistical properties of this class of object are not well understood but the best studied example of this structural type is the nearby ($z = 0.0043$) radio galaxy M 87 (Virgo A, 3C 274). This source contains an inner 1 arcminute ($\sim$ 5 kpc) double component within an extended halo which at 408 MHz has an extension of 16 $\times$ 12 arcminutes ($\sim$ 80 $\times$ 60 kpc) (Cameron 1971). Joining the core to the north preceding inner component is a jet of length 25 arcseconds ($\sim$ 2 kpc). Over the first half of its length the jet is knotty and expands with an almost constant opening angle with $d\phi/d\theta = 0.07$. At this point the jet appears to undergo a strong oblique shock after which the flow becomes disrupted and turbulent. At 15 GHz the jet has typical fractional polarization values of 20% with $\mathbf{B}_{\parallel}$ throughout except at the bright knot associated with the oblique shock where $\mathbf{B}_{\perp}$ dominates.

2.2.5 Lobe-dominated radio-loud quasars

In flux density limited samples of radio sources, there is a strong selection effect toward high luminosity structures surrounding quasars since at high redshift it is often only these that are detectable. The structures we thus find in this category are typically above the Fanaroff-Riley transition luminosity of $P_{178\mathrm{MHz}} = 5 \times 10^{25}$ W Hz^{-1} and are broadly similar in character to FR II galaxies of equivalent luminosity (§2.2.4.2). There are however some important differences. In the quasar structures the jets are again one-sided but have luminosities that are higher than in galaxies. Thus, bright one-sided radio jets are relatively common in this morphological class, being detected in 40 – 70% of cases (cf. <10% for FR II radio galaxies) (Owen & Puschell 1984; Wardle & Potash 1982). With improvements in imaging sensitivity and dynamic range, we may expect jets to be detected in all radio loud quasars. Indeed, in recent deep VLA observations of a sample of 12 3CR lobe-dominated quasars Hough *et al.* (1988) detected jets in all cases.

In quasars, the core luminosity tends to be higher than in radio galaxies of similar luminosity. The structures are thus char-

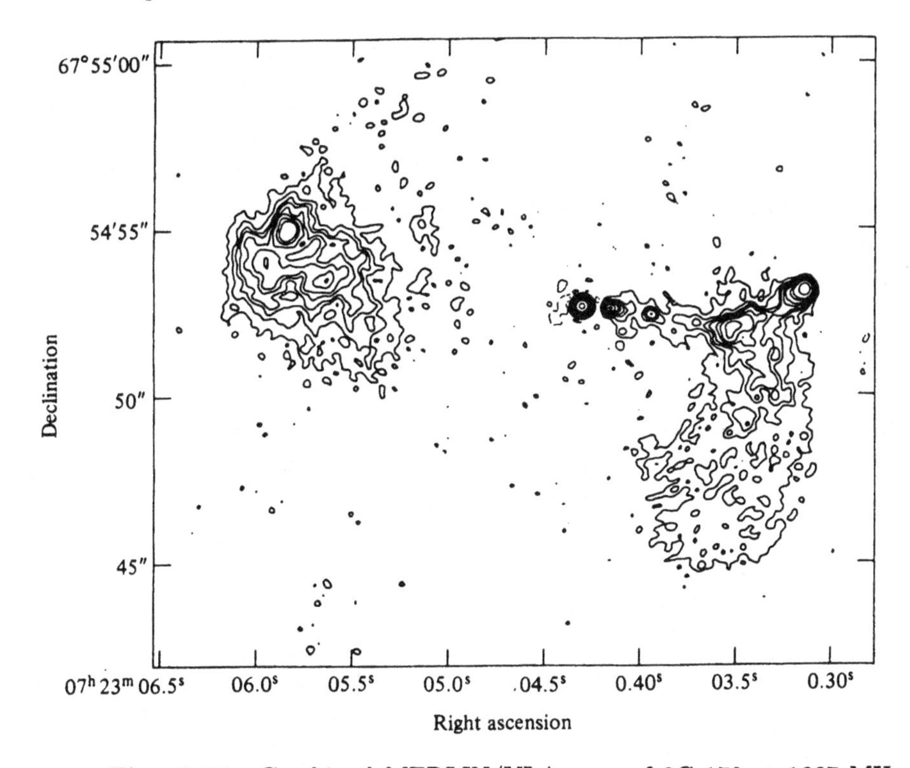

Fig. 2.11. Combined MERLIN/VLA map of 3C 179 at 1667 MHz. Resolution: 0.28 arcsec. Peak: 364 mJy/beam. Contours: -0.3, 0.3, 0.6, 0.9, 1.2, 1.8, 3.6, 7.2, 14.4 ... mJy/beam. Reproduced with permission from Shone, Porcas & Zensus (1985). Reprinted by permission from *Nature*, **314**, 603. Copyright © 1985 Macmillan Magazines Limited.

acterised by prominent cores and strong one-sided jets. In a few cases (*e.g.*, 1857+566, Saikia *et al.* 1983), the bright jet emanates from a relatively weak core. The overall double-lobed structure also shows differences to those found in radio galaxies. In quasars, the lobe structures show a greater degree of bending or distortion than is found in radio galaxies (Leahy, Muxlow & Stephens 1989). In many cases there are significant deviations from the 'classical' linear double-lobed structures found in FR II galaxies.

The jet properties are similar to FR II galaxies. They are one-sided with jet/counterjet ratios > 4:1. When mapped at high resolution they show small opening-angles with values of $d\phi/d\theta$ <0.1 and are knotty rather than smooth with S_k/Si values > 4. The magnetic field configuration is usually $\mathbf{B}_{\|}$ throughout. An example of one such quasar structure is given in Fig. 2.11 where the source

3C 179 ($z = 0.846$) has been mapped with a combination of MERLIN and the VLA at 1.67 GHz (Shone, Porcas & Zensus 1985). This structure exhibits many of the features described above including a bright core, a one-sided knotty jet running into the more compact hotspot and a lobe which is offset with respect to the axis defined by the core, jet and hotspots.

2.2.6 Core-dominated sources

As the name suggests, objects in this class are characterised by a very luminous core which tends to dominate the overall source flux density. In addition, this is usually associated with a bright one-sided radio jet. The very high core surface brightness makes such objects ideal candidates for VLBI studies; and indeed a large (and inhomogeneous) set of such sources has been mapped at very high angular resolution. These studies have revealed that in general, there is a continuity between the small and large-scale structures with a milliarcsecond scale one-sided jet running into the outer arcsecond scale jet. The sense of the one-sided asymmetry is always preserved and those few sources where the VLBI jet appeared to point away from the arcsecond scale jet have now, with more detailed mapping, been shown to simply exhibit very large bend angles close to the core (*e.g.*, 3C 395, Saikia, Muxlow & Junor 1989). Significant bending of the jet, especially near the core, is a common feature in core-dominated sources.

Projection effects will play an important rôle in determining the apparent bends we see in such jets: a slightly bent beam may appear strongly curved if it aligns closely with the line of sight. The probability distribution of apparent bend angles have been calculated by Readhead *et al.* (1983); we note that if the angle of the beam to the line of sight is less than the intrinsic bend angle of the beam then apparent bend angles of up to 180° are possible. The tendency for core dominated sources to have large apparent bends is thus consistent with the jets in these sources being at small angles to the line of sight as expected in quasar 'unified schemes' (see §2.4.3).

Many core-dominated sources show some form of activity indicative of relativistic motion in their core regions. This can take the form of *superluminal* motion of knots in the VLBI jet with implied velocities above the speed of light, or short variability time scales indicating physical dimensions as small as light months, which in turn implies energy densities above the Compton limit – where the radio emission would be quenched by the inverse-Compton scattering of the radio photons off the relativistic electrons (see §1.2).

All the established superluminal motions have been outward. Any suggested contractions (*e.g.*, 4C 39.25, Shaffer *et al.* 1987) have been shown to be anomalous and a consequence of a changing complex structure and stationary components. In the context of unification schemes (see §2.4.3), which attempt to broadly explain core-dominated structures in terms of source alignment and relativistic beaming of the radio emission, we would expect that virtually all core-dominated sources will eventually be found to show superluminal activity. It is clearly important to follow such superluminal motion into the arcsecond scale structure: whilst this is observationally difficult, work in this area is currently under way (see below). Where detected, spectral gradients along these jets are small; with the spectra steepening away from the core, consistent with synchrotron depletion of the higher-energy electrons in the outer parts of the jet.

In addition to the core and jet, a low surface brightness extended component is often found on the opposite side to the jet. In some sources, this whole triple structure is then embedded in a very low surface brightness amorphous halo. The jet properties are broadly similar to those found in lobe-dominated quasars (see previous section), but are of higher luminosity and show a greater degree of bending. Although the general properties of this morphological class have been outlined above, there is a great variation from source to source. We thus give four examples in order to illustrate the range of core-dominated structures found.

Fig. 2.12 shows a MERLIN image of the bright quasar 3C 273 ($z = 0.158$). Superluminal motion is well established in the core (Whitney *et al.* 1971; Cohen *et al.* 1971) and the jet bends by $\sim 20°$ within the first 10 milliarcseconds before extending continuously for 22 arcseconds (40.7 kpc) from the core to beyond the limit of the optical jet, terminating in a bright head. The ridge-line of the jet shows a ripple whose wavelength decreases by a factor of 6 along its length. To the south of the jet there is a region of extended radio emission. No emission is detected on the opposite side to the jet and the brightness ratio of the head of the jet to any counter-jet is

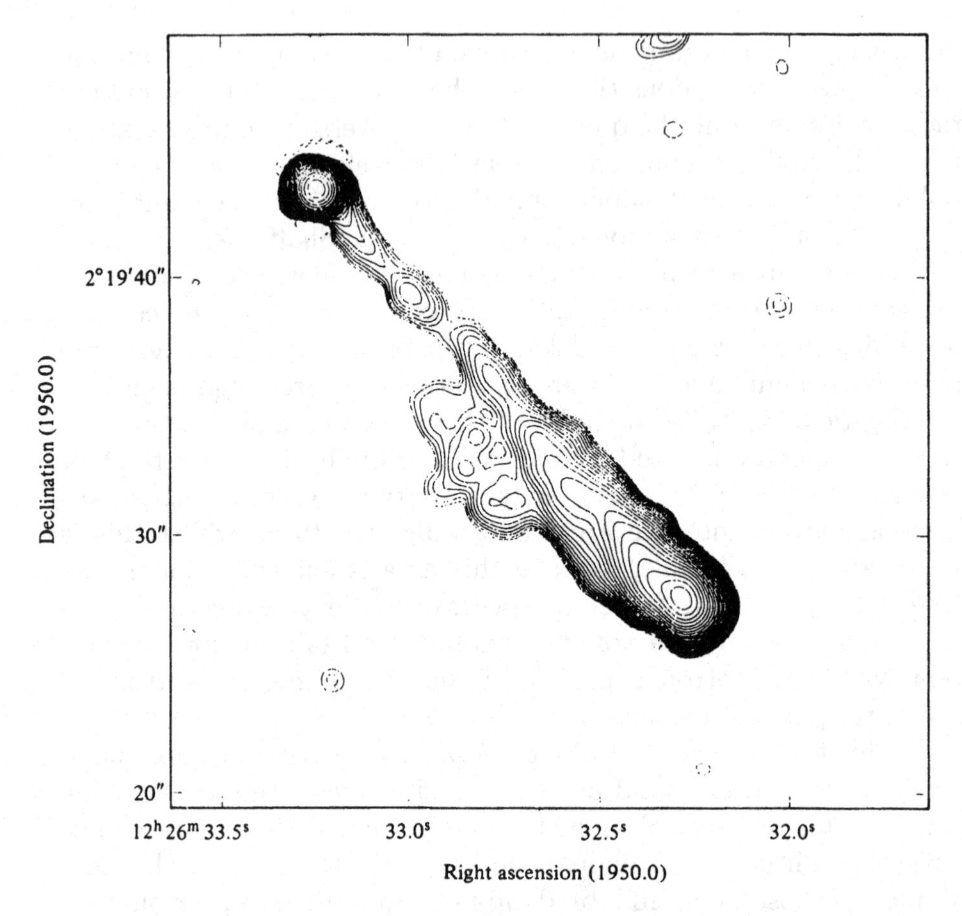

Fig. 2.12. MERLIN map of 3C 273 at 408 MHz. Resolution: 1.0 arcsec. Peak: 1350 mJy/beam. Logarithmic contours at intervals of $\sqrt{2}$ are shown above a bottom contour of 6.7 mJy/beam. Reproduced from Davis, Muxlow & Conway (1985). Reprinted by permission from *Nature*, **318**, 343. Copyright © 1985 Macmillan Magazines Limited.

>5500:1. In this respect 3C 273 is highly atypical; no other object has such a recorded sidedness ratio.

The MERLIN 1.67 GHz image of the nearby ($z = 0.033$) Seyfert galaxy 3C 120 is shown in Fig. 2.13. This is just the inner part of a larger arcminute scale double structure aligned roughly north-south. Between this 1989 epoch and an earlier 1980 epoch image superluminal motion has been detected out to 0.2 arcseconds (100 pc) from the core (Muxlow & Wilkinson 1989). Evidence for similar superluminal motion in the position of the knot 4 arcseconds (2

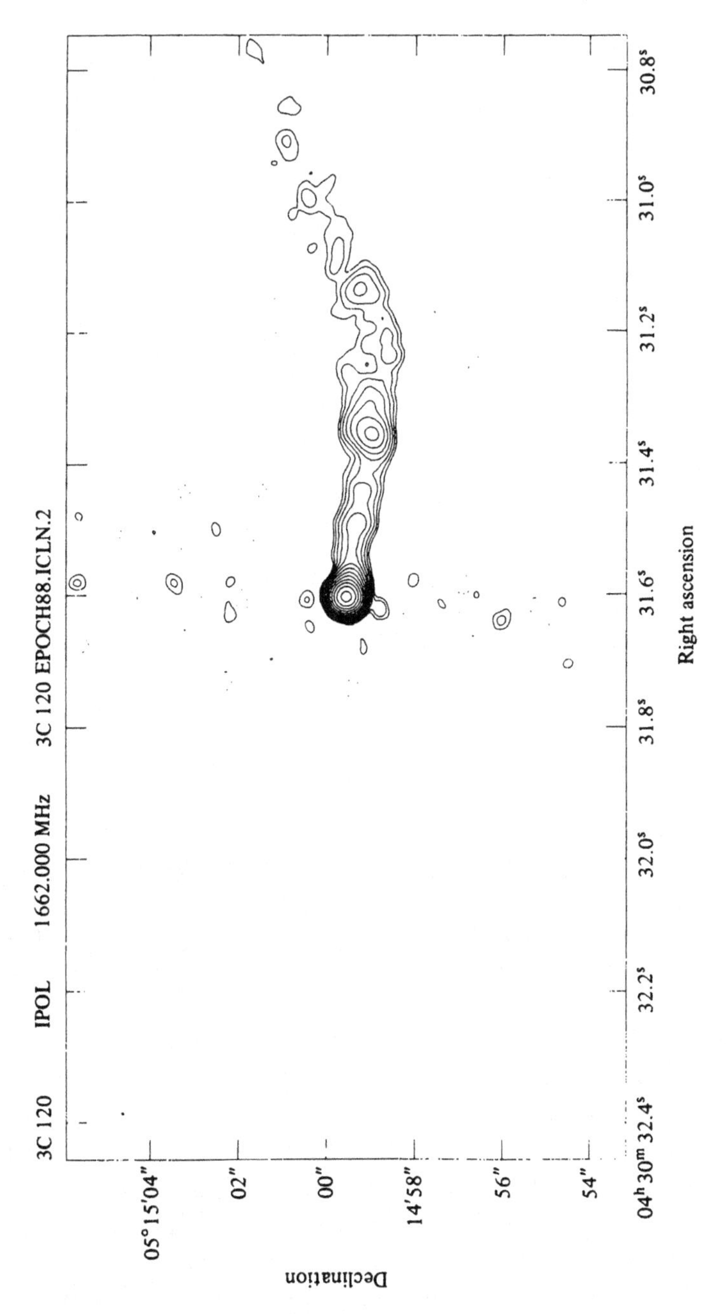

Fig. 2.13. MERLIN map of 3C 120 at 1662 MHz. Resolution: 0.35 arcsec. Peak: 2779 mJy/beam. Contours: -2, -1, 1, 2, 4, 8 ... mJy/beam. Muxlow & Wilkinson (1989).

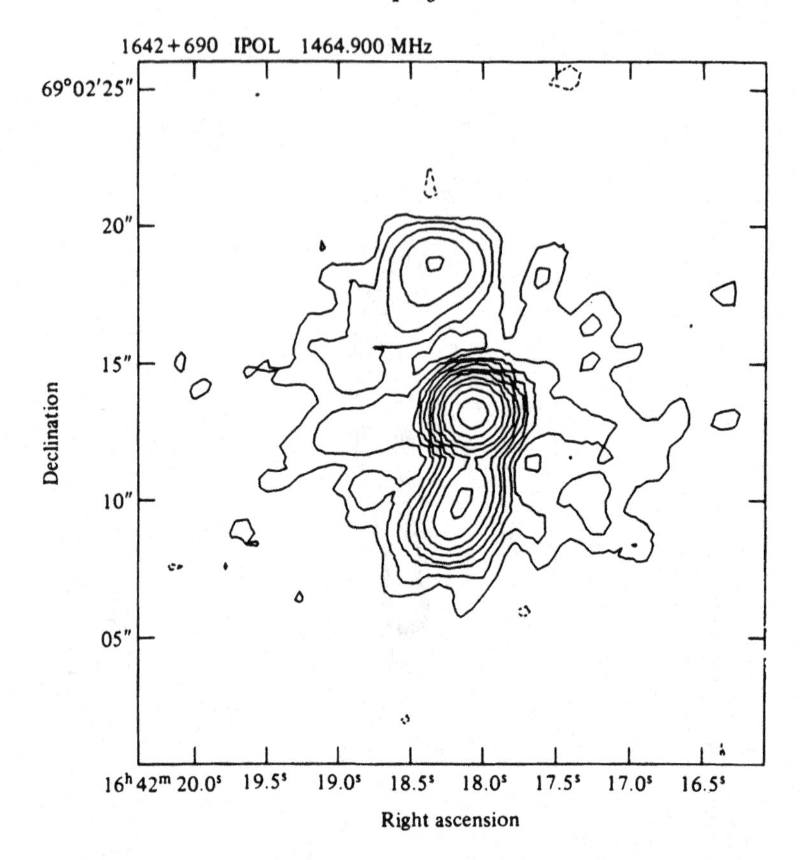

Fig. 2.14. VLA map of 1642+690 at 1656 MHz. Resolution: 1.2 arcsec. Peak: 9991 mJy/beam. Contours: -0.6, 0.6, 1.2, 2.4, 4.8 ... mJy/beam. Reproduced with permission from Browne (1987a); data from Murphy (1988).

kpc) down the jet has been presented by Walker, Walker & Benson (1988). This is the first preliminary detection of superluminal motion on the kiloparsec scale. Fig. 2.14 shows a VLA image of the quasar 1642+690 (z = 0.751). This is included since it shows all the structural features present in this morphological class: a bright core with a curving one-sided jet together with an extended counter component imbedded in an amorphous halo.

Fig. 2.15 shows the extremes of curvature which can be found in a few examples of this class. The object is the quasar 3C 418 (z = 1.698) and the image is from a combined MERLIN/EVN dataset at 1.67 GHz (Muxlow, Neff & Lind 1989). The total bending of the jet integrated along its length is nearly one complete turn. The

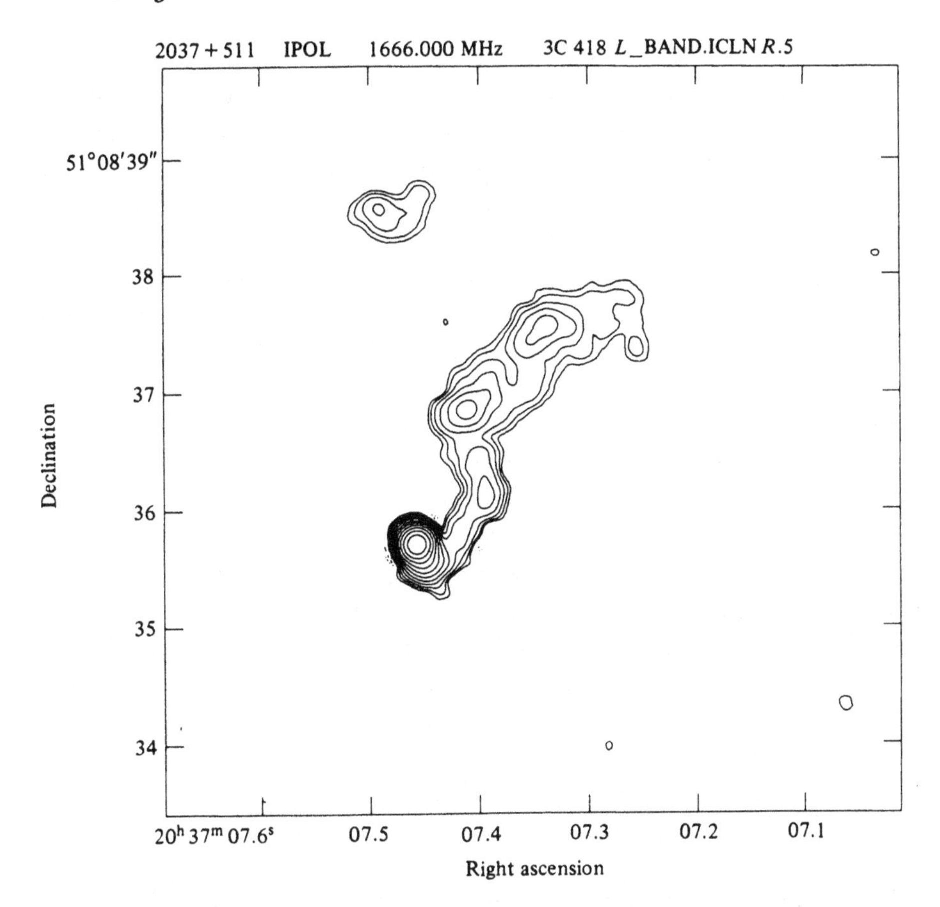

Fig. 2.15. MERLIN/EVN map of 3C 418 at 1666 MHz. Resolution: 0.16 arcsec. Peak: 4115 mJy/beam. Contours: -2, 2, 4, 8, 16 ... mJy/beam. Muxlow *et al.* (1989).

jet brightens where it bends, and the spectral steepening along its length is interrupted at the bend points by regions of flatter spectral index. This suggests that the jet bending is likely to be a real deflection from interaction with the surrounding medium rather than an apparent bend produced by the precession of the jet ejection angle at the core. No convincing case of proven jet precession has yet been found in this class of object.

2.2.7 Compact steep-spectrum sources

This morphological class originated with observations with instruments such as the Cambridge 5 km telescope. Some sources, whilst retaining a conventional steep radio spectral index around 0.7, remained unresolved at resolutions about 2 arcseconds. Observations with the VLA, MERLIN, and VLBI arrays have now successfully mapped many of these objects. More correctly, this class is now seen to comprise objects whose overall radio structures are of galactic or subgalactic dimensions. This implies linear sizes of less than 50 kpc, and in a number of cases less than 10 kpc. These high resolution studies have revealed a wide variety of structures. These sometimes resemble the doubles seen in larger sources whilst others appear as core-jet structures. Many cannot be clearly classified at all, appearing as extremely distorted complex structures. Fanti *et al.* (1985) have shown that there is a clear division between quasar and galaxy radio structures in this class: the quasars are generally core-jet or complex whilst the galaxies are doubles, although not necessarily simple doubles. The relationship of this morphological class to the radio source population in general is unclear. Some workers (*e.g.*, Phillips & Mutel 1980) have argued that the double structures represent the young progenitors of classical double radio sources. Others reason that these sources are a separate class of intrinsically small objects where the jets encounter great difficulty in escaping from the inner regions of the host galaxy or quasar (Fanti & Fanti 1986). In some sources, projection effects may also be significant.

The properties of the jets found in the handful of quasar core-jet structures which have been mapped are not well established, except that in many cases they are extremely disrupted. The distortions found are such that the term 'jet' may, in some cases, be inappropriate since collimated outflow from the core along such structures may have broken down. One of the best examples of a jet in this class of object is illustrated in Fig. 2.16, which shows a VLBI map of the quasar 3C 48 ($z = 0.367$) at 6 milliarcsecond resolution. A one-sided jet is seen emanating from the core. After 55 milliarcseconds (170 pc) the jet runs into a bright knot after which it flares with a value for $d\phi/d\theta$ approaching 0.7. After about a further 150 pc the angle of the flare decreases with the $d\phi/d\theta$ value dropping to around 0.05. This regime continues for about 175 milliarcseconds (525 pc) before flaring again with $d\phi/d\theta$ again rising to values around 0.7. The total length of the jet is around 0.44 arcseconds

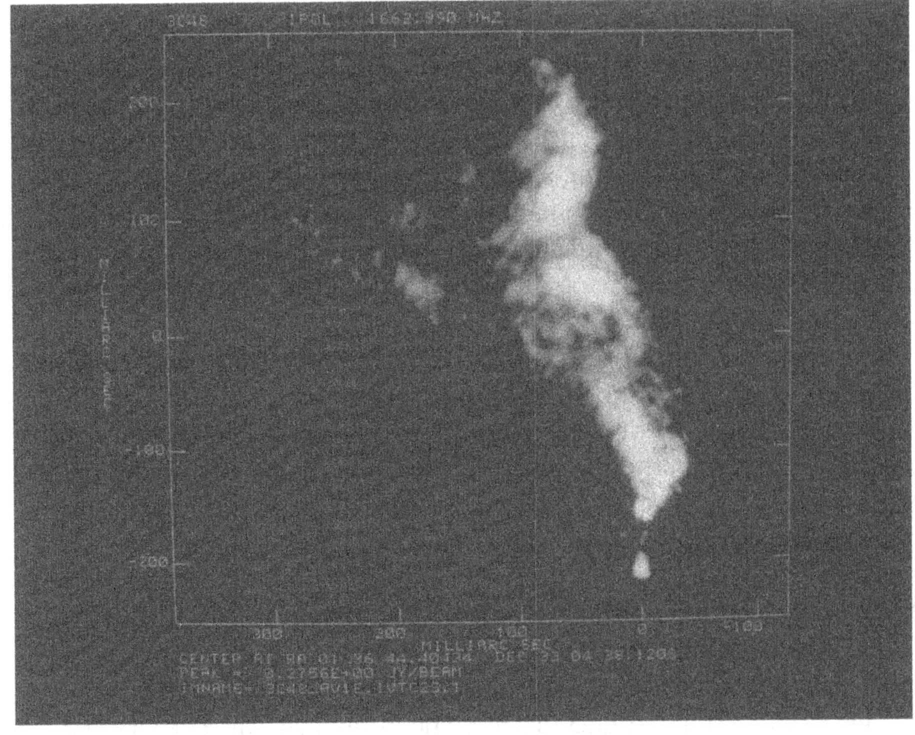

Fig. 2.16. VLBI map of 3C 48 at 1660 MHz. Resolution 6 milliarcsec. Map provided by P. Wilkinson and A. Tzioumis.

(1.3 kpc) and is entirely one-sided throughout. In general where jets do exist in this class of object, they appear to be one-sided.

2.3 Jet emission at other wavelengths

2.3.1 Optical and infrared continuum

Optical continuum emission from radio jets is detected in only a handful of objects because the optical flux density is very low. These objects are therefore selected to be local or very luminous. Emission coincident with bright knots in the radio jet have been unambiguously found in 3C 31, 3C 66B, M 87, 3C 273, and 3C 277.3; more tentative detections exist for a few other objects. For the low luminosity radio jets (*e.g.*, M 87) the radio-infrared-optical spectral index is generally 0.7 ± 0.1 (Butcher, van Breugel & Miley 1980). In M 87, Smith *et al.* (1983) find a good fit to a power law between the radio and the optical B band with only minor variations ($< \pm 0.02$) in index from knot to knot. The spectrum steepens in the optical with a measured *VBU* spectral index of 1.7 (Visvanathan & Pickles 1981). For 3C 273, Röser & Meisenheimer (1986) find a steeper value of 0.96 for the radio-optical index of the knot at the end of the jet. In this case the optical and radio jet profiles are different indicating a spectral steepening along the jet. The failure to detect optical continuum emission with present sensitivities from other strong quasar radio jets indicates that this steeper value of radio-optical spectral index may be the more typical value for higher luminosity objects.

The optical continuum is up to 20% linearly polarized in M 87 (Schmidt *et al.* 1978) and 14% polarized in 3C 277.3 (Miley 1983). In 3C 273 Röser & Meisenheimer find the optical knot to be 13% polarized (the same value as at 5 GHz) with $\mathbf{B}_{\parallel}$ (again in agreement with the radio). It is thus clear that the optical continuum is synchrotron radiation from the same regions as the radio jets.

2.3.2 Optical and infrared emission lines

Line-emitting regions are found to lie adjacent to jets near the outer edges of bends and close to, but just outside bright radio jet knots. This suggests that the line-emitting gas is pushed outwards by the radio jets and that occasionally this gas has sufficient inertia to deflect them. This is clearly not emission from the jet itself but may originate in clouds in normal galactic rotation which have been accelerated, heated, and ionized by the action of the jet. In most cases the synchrotron pressure (derived from the usual minimum energy conditions) within the jet roughly matches the thermal pressure of the line-emitting gas with typical densities of $10^8 - 10^9$ m^{-3} and temperatures around 20,000 K. Observed line-widths are in the range 300 – 500 km sec^{-1} and bulk velocities within a few hundred km sec^{-1} of the optical nucleus. As noted in §2.2.3, the [OIII] line-width is found to be correlated with the integrated luminosity on the kiloparsec scale size. In addition the broader lines are sometimes seen to lie close to the brightest radio regions. Both shock heating and photoionisation may be present; the detailed balance of the various excitation mechanisms may vary from source to source.

2.3.3 X-rays

High resolution imaging from the *Einstein* X-ray Observatory has detected emission from just two radio jets (M 87 and Centaurus A). In M 87 the individual knots cannot be resolved but the measured optical-X-ray spectral index of 1.7 (Smith *et al.* 1983) is in good agreement with the value in the optical band suggesting a synchrotron origin for the X-rays. The structural similarity between the radio and X-ray knots in the jet of Centaurus A again strongly suggests synchrotron as the emission mechanism (Feigelson *et al.* 1981). In 3C 273 there is a marginal detection of the jet in a deep *Einstein* HRI exposure (Feigelson 1980; Willingale 1981). This result is heavily dependent upon point source subtraction since the detection is at a level of 1/400 that of the nucleus. If confirmed, this gives an optical-X-ray spectral index of ~ 1.3 which when taken with the radio-optical value of 0.96 is qualitatively similar to M 87.

2.4 Systematic trends

We have described a morphological classification of extragalactic radio sources and their jets, arranged essentially in order of increasing luminosity. We have already seen that in many respects the different classes differ in degree rather than kind. As a complementary approach, this section explores the systematic trends among four of the key observational parameters, namely source luminosity and size and the fractional luminosity in the core and jets.

2.4.1 The P-D diagram

We start with two fundamental properties of radio sources: their radio power, P, and linear size, D. While there is an analogy between the P-D plot for radio sources and the Hertzsprung-Russell diagram for stars, the P-D diagram is much harder to interpret: largely because, unlike main sequence stars, radio sources are not equilibrium systems. The derivation of these two parameters from observed flux density and angular size is discussed in more detail in the next chapter. The most complete flux density-limited sample of sources for which these parameters have been accurately determined is the revised 3C sample of Laing *et al.* (1983), hereafter the LRL sample. Because this sample has been selected at low frequency (178 MHz) it contains predominantly lobe-dominated sources, mainly radio galaxies, with steep spectra. An important feature of any flux density-limited sample is that luminosity and redshift are strongly correlated: only the most powerful sources are detected at large distances.

Fig. 2.17 plots the linear sizes and luminosities for the LRL sample. There is a trend, evident in this diagram, for the more powerful sources to have smaller linear sizes. It is hard, however, to separate this correlation from one with redshift (*e.g.*, Masson 1980; Macklin 1982).

Interpretation of the P-D diagram is complicated by a combination of physical and observational selection effects. Limited surface brightness sensitivity may exclude weak extended sources, while the minimum linear size which can be resolved decreases at higher redshift and hence higher power. In fact, for the LRL sample, these effects are not very important because of the variety of telescopes that have been used.

Other physical selection effects determine the distribution of source types in the P-D diagram and may be largely responsible for the overall decrease of linear size with increasing luminosity. Core-dominated sources occupy a narrow band at high P. The apparent high power and small linear size of these sources has been attributed to relativistic beaming of the small-scale jets. Sources at small angles to the line of sight will appear brighter due to Doppler beaming and smaller due to projection effects and the fact that the extended emission may be hard to detect in the presence of the very bright core. Nearly all of the remaining compact sources with $D < 10$ kpc have high luminosity ($P_{178\mathrm{MHz}} > 10^{26}$ W Hz^{-1}) and are known as compact steep spectrum (CSS) sources. They are thought to represent a class of sources distinct from the larger classical doubles (see §2.2.7).

The largest sources with $D > 1$ Mpc, the so-called giant radio galaxies, occupy a narrow range in luminosity just above the transition between FR I and FR II sources. The lack of higher power giant radio galaxies may be a selection effect: the extended lobes of more powerful sources may be significantly reduced by synchrotron ageing and inverse-Compton losses so that the source appears as two unrelated hotspots and would not be classified as a classical double radio source (Baldwin 1982). Whether such losses are sufficient to account for the upper bound to D is not yet clear – an alternative cause might be that there is a maximum total energy

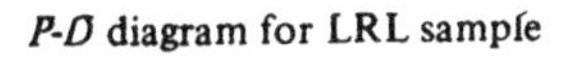

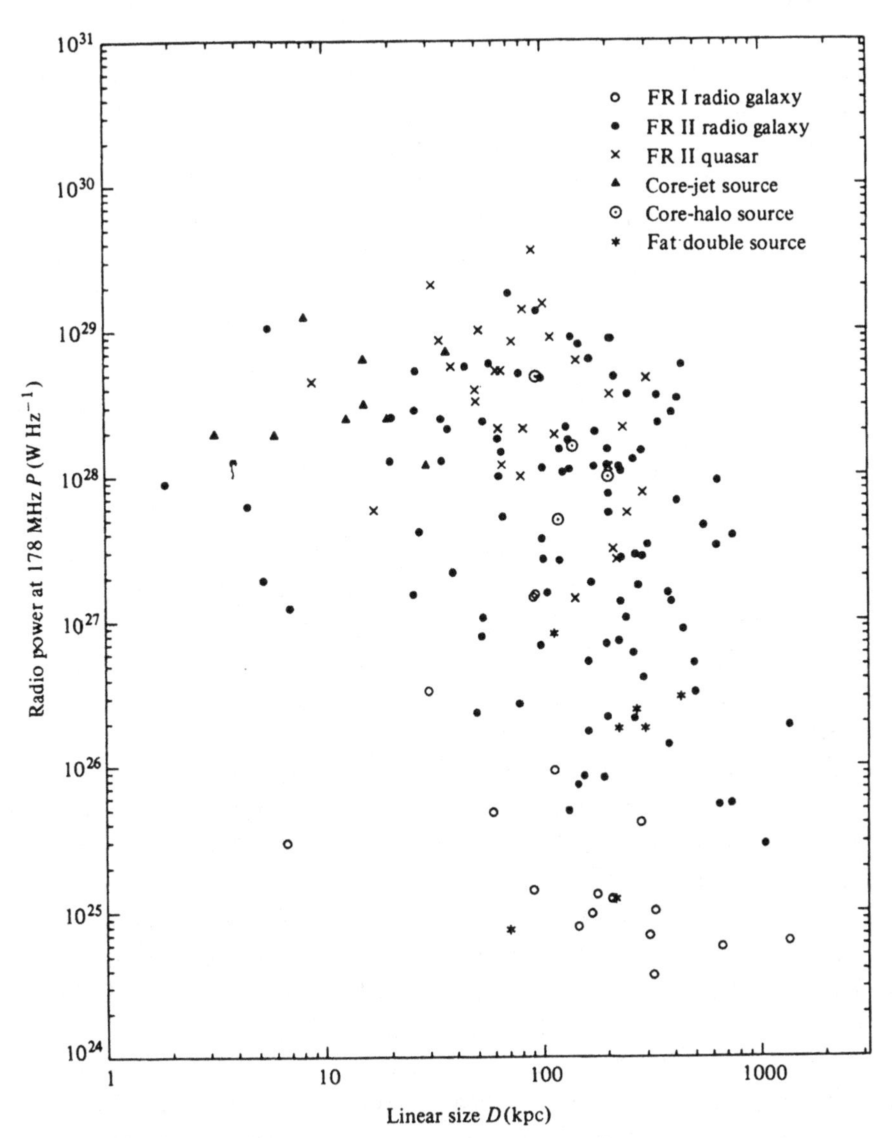

Fig. 2.17. Luminosity – Linear size diagram for the LRL sample. Symbols denote different source types. Data provided by J. P. Leahy; sizes and luminosities derived assuming $H_0 = 100$ km s^{-1} Mpc^{-1} and $q_0 = 0$.

for a radio source, thus defining an upper limit to the linear size distribution which is a decreasing function of P (Allington-Smith 1984).

Samples with lower flux limits such as the Leiden-Berkeley Deep Survey (Oort *et al.* 1987) and the B2 sample (Ekers *et al.* 1981; de Ruiter *et al.* 1989) which contain a higher proportion of lower luminosity sources, show that for $P_{178\mathrm{MHz}} < 10^{26}$ W Hz^{-1} linear size increases with radio power roughly as $D \propto P^{0.5}$. There is a saturation at higher powers, corresponding to a maximum source size of around 1 Mpc.

The distribution of sources in the P-D diagram has been used to place some constraints on their evolutionary tracks (Baldwin 1982). We may assume that source lifetimes are considerably less than the age of the universe and that sources die relatively quickly. Relic radio sources are indeed rare: less than 4% of radio galaxies lack evidence of ongoing activity (Giovannini *et al.* 1988; Cordey 1986). The fact that the numbers of sources in successive decades in D increase by less than factors of 10 implies that the expansion velocity must increase with D if the radio power remains fairly constant. Baldwin (1982) has modified the basic twin beam model of Scheuer (1974) for sources expanding into an atmosphere where the density falls as (radius)$^{-\delta}$, so that the expansion velocity increases as $v \propto D^{\delta/2-1}$. He finds that $\delta = 1.9$, increasing to 2.9 for $D > 300$ kpc gives a satisfactory fit to the numbers of sources as a function of D. Sources would then evolve as $P \propto D^{-0.4}$ for $D < 300$ kpc and $P \propto D^{-1.2}$ for $D > 300$ kpc.

Gopal-Krishna & Wiita (1987, 1988) have extended this model to consider what happens when the source expands through the interface of the gaseous halo and the hot intergalactic medium, as inferred from the X-ray background (*e.g.*, Guilbert & Fabian 1986). As the beam crosses this interface, the head of the beam expands and the velocity of advance increases. When this velocity becomes subsonic relative to the IGM, the radio emission quickly fades; although for high power beams, inverse-Compton losses dominate the decay. This sudden decay for low power beams would then account for the observed flattening of the radio luminosity function for $P_{1\mathrm{GHz}} < 10^{24}$ W Hz^{-1}.

2.4.2 Prominence and collimation of radio jets

We have already noted that the jets of low-luminosity FR I

sources are more prominent than the jets of higher luminosity FR II sources. Amongst FR I sources the fraction of emission from the jet generally decreases with increasing total luminosity (Parma *et al.* 1987). De Ruiter *et al.* (1989) have noted that since higher power jets are longer (see above) and have smaller opening angles, the jet volume scales as $P^{0.5}$, which implies that the jet emissivity is independent of the source power, at least for low power sources.

In powerful FR II radio galaxies the jets rarely contribute more than a few percent of the total luminosity. However, even the weakest FR II jets are more luminous in absolute terms than the majority of FR I jets. Therefore although the ratio of jet luminosity to total luminosity decreases, there is still an overall correlation between jet luminosity and total luminosity. There is also a correlation between jet luminosity and core luminosity, both for FR I sources (Parma *et al.* 1987) and FR II sources (Burns *et al.* 1984). It is less clear, however, whether the fraction of emission from the jets and cores relative to the total emission are correlated in radio galaxies.

The collimation of the jets in FR I sources varies considerably from source to source. In about 50% of sources the opening angle is constant along the jet and more powerful sources have smaller opening angles. In other FR I sources the opening angle varies along the jet and there is usually a pattern of expansion followed by collimation and then expansion again. Bicknell *et al.* (1989) have fitted the turbulent jet model of Bicknell (1986) to a sample of 23 low luminosity jets from the collection of Parma *et al.* (1987) (see Chapter 3). In the majority of cases the model fits indicate jets with Mach numbers of 1 to 3, but about one-third of the fits admit solutions with Mach numbers up to 10. These high Mach number jets are lighter, more collimated and occur in more powerful sources than the lower Mach number jets. Such jets may represent the transition to the narrow jets of FR II sources which are generally unresolved with opening angles of less than a few degrees.

In radio quasars, jets are detected much more frequently. Of the 28 classical double quasars in the LRL sample, jets have been detected in all but two and are all one-sided. The most prominent jets occur in the quasars with complex or one-sided structures (Bridle 1988) and prominent cores (Saikia 1984a). Bridle (1988) finds that there is a statistical correlation between the relative prominence of jets and cores in quasars, although Lonsdale (1986) finds that there is no such correlation in a sample of high redshift quasars. There is also a trend for more prominent jets to occur in sources of smaller linear size (Bridle 1988), although it is interesting to note that Wardle & Potash (1984) found jets in each of the eight largest quasars of a complete sample of 4C sources.

2.4.3 Trends with core prominence

Flat-spectrum radio cores are seen in all quasars and in about 80% of radio galaxies. The degree of core prominence is parameterised by R, the ratio of core flux to extended flux, usually at a given emitted frequency of 5 GHz.

In radio galaxies the core luminosity is correlated with the total luminosity (Giovannini *et al.* 1988; de Ruiter *et al.* 1989), but the core fraction R falls with increasing total luminosity as $R \propto P^{-0.5}$ from 0.04 at $P_{1\text{GHz}} = 10^{23}$ W Hz^{-1} to 0.003 at $P_{1\text{GHz}} = 10^{27}$ W Hz^{-1}. This relation holds for both FR I and FR II radio galaxies.

In quasars there is a greater variation in core fraction with R values ranging from 0.001 to 10, which are not correlated with the extended luminosity. The wide variation in core emission may be elegantly explained by relativistic beaming. The strongest evidence that the parsec scale jets seen in radio cores have bulk relativistic velocities comes from VLBI observations of superluminal motion (see Chapters 1 and 4). The radio emission from such jets is discussed in detail in Chapter 4; here we merely note that since the degree of relativistic beaming increases rapidly as the angle of the jet to the line of sight decreases, the degree of core prominence may be used as an indication of the angle of the source axis to the line of sight. This assumes that the emission from the lobes is isotropic and that the nuclear jets and the large-scale jets are reasonably aligned (VLBI observations show that the intrinsic misalignments are typically only 10° (Rusk & Rusk 1986)). The idea that core and lobe-dominated quasars may only differ in orientation (Orr & Browne 1982) has become known as the 'unified scheme'.

Several studies have examined correlations between the large-scale properties of classical quasars and their R values in order to check whether they are consistent with R being used as a measure of source orientation (Kapahi & Saikia 1982; Saikia 1984a; Hough &

Readhead 1988; Murphy 1988). Their findings may be summarised as follows:

1. There is no strong correlation of R with redshift or extended luminosity. This is consistent with the idea that R depends merely on orientation.
2. Linear size anticorrelates with R as expected due to projection effects. This is unlikely to be a secondary effect due to any correlations of linear size with redshift or luminosity, given the null correlations mentioned above.
3. Sources with high R appear more misaligned than those with low R, as expected if projection effects amplify small intrinsic departures from collinearity. The intrinsic misalignments are estimated to be around 10°.
4. Sources with bright one-sided jets have higher R values. This supports the idea that the one-sided jets are also relativistically beamed.
5. The fraction of sources which show extended emission only on one side of the core is higher among sources with high R.

These trends are consistent with the unified scheme of Orr & Browne (1982) in which the core-dominated sources are simply at a smaller angle to the line of sight. There are, however, several problems with this interpretation. Several studies (*e.g.*, de Bruyn & Schilizzi 1986) have noted that the projected linear sizes of core-dominated sources, and in particular sources which show superluminal motion in the parsec scale jets, are still too large for them to be at small angles to the line of sight. Browne (1987a) has suggested a physical correlation between core Lorentz factor and linear size so that selecting core-dominated sources would introduce a bias towards large linear size.

There are also problems in accounting for the observed differences in the optical properties of lobe and core-dominated quasars as simply the results of orientation effects. Lobe-dominated quasars show broader permitted emission lines and weaker FeII emission lines (Miley & Miller 1979) and stronger [OIII] emission in the nebulosity surrounding the quasar nucleus (Boroson & Oke 1984). The differences in nuclear emission lines may arise if the broad lines are produced in a rotating disc perpendicular to the jet axis. The differences in the surrounding nebulosity are harder to explain by orientation effects and further work is needed to establish the strength of these differences and take account of the correlation between [OIII] emission and extended radio luminosity (Wills & Browne 1986). These differences in optical properties have motivated alternative models in which the beams of core-dominated quasars dissipate much of their energy in the broad line region rather than form extended radio sources (Norman & Miley 1984).

There is some evidence that core-dominated sources may inhabit regions of space with lower galaxy density than lobe-dominated sources (Prestage & Peacock 1988). Again, further work is required to test this effect in samples of matched extended radio luminosity, given the trend for more powerful sources to occur in regions of higher density (Yates *et al.* 1986). If confirmed, this would argue very strongly against orientation-based unification schemes for core and lobe-dominated quasars.

Finally, several authors have suggested a broader unified scheme in which radio galaxies may simply be quasars seen at large angles to the line of sight (Barthel 1989; Scheuer 1987). Barthel (1989) has shown that in a complete sample of 3C sources between redshifts of 0.5 and 1.0 the median linear size of radio galaxies is twice that of quasars, suggesting a critical angle of 45° for determining whether a source is seen as a quasar or a radio galaxy. The broad optical emission lines characteristic of quasar nuclei may be obscured by a torus of dust between the narrow and broad line regions. Relativistic beaming in the core may boost the continuum emission in quasar nuclei above that of galaxies. It is not yet clear whether all the differences in optical properties of radio galaxies and quasars can be ascribed to differences in orientation (Browne 1987b; Hutchings 1987).

2.4.4 Asymmetries in jets and double radio sources

The original twin beam model was motivated by the overall symmetry of classical double radio sources. When high resolution radio maps revealed the presence of jets in powerful radio galaxies and quasars, thus providing a striking confirmation of the beam model, they immediately raised a new problem: the jets are almost invariably seen on one side only. The parsec scale jets detected by VLBI observations are also nearly all one-sided. High resolution observations have also shown considerable asymmetries in the twin jets of lower luminosity FR I sources.

While it is generally accepted that the VLBI scale jets are asymmetric due to Doppler beaming, given the relativistic velocities implied by the observed superluminal proper motions, it is less clear whether these velocities persist out to kiloparsec scales to account for the one-sided kiloparsec scale jets. Alternatively, the asymmetries could be intrinsic but then one must find an alternative explanation for the superluminal velocities, unless the jets decelerate between parsec and kiloparsec scales. These questions are discussed in detail in the following chapters: here we examine the observed asymmetries in jets and how they correlate with asymmetries in the large-scale properties in the lobes.

2.4.4.1 **Asymmetries in FR I jets**

In general the asymmetry in flux density between the jets of low luminosity twin-jet sources increases with luminosity (Parma *et al.* 1987) up to $P_{178\mathrm{MHz}} = 10^{25}$ W Hz^{-1} beyond which jets appear to be completely one-sided. The observed asymmetry depends strongly on the linear resolution: most jets appear to be one-sided near the core but more symmetric on larger scales. Most jets show gaps of a few kpc between the core and a 'flaring point', often the brightest part of the whole jet, and in general the brighter jet has a brighter flaring point which is closer to the core, typically by a factor of two.

The observed dust lanes in some host galaxies of twin jet sources can be used to determine the orientations of the jets, assuming that the jets are perpendicular to the dust lanes (as they appear in projection). Laing (1984a) finds that in seven such sources there is no tendency for the approaching jet to be the brighter of the two, suggesting that the brightness asymmetry is not due to mild relativistic beaming. However, the assumption of perpendicularity may be invalid given the general lack of correlation between jet directions and stellar rotation axes or optical major axes found in other samples of radio galaxies (see §3.3).

There are a few very asymmetric low-luminosity jets such as NGC 315 and NGC 6251 where the large-scale jets are one-sided but these sources do show weaker counter-jets near the core. Wide-angle tailed sources have generally symmetric tails which flare out from narrow jets connecting them to the core. These jets are usually quite asymmetric (*e.g.,* 3C 465). The distinction between jets and tails is less clear in narrow-angle tailed sources and these tend to be more symmetric.

2.4.4.2 Asymmetries in FR II radio galaxies and quasars

The jets of the vast majority of FR II radio galaxies and quasars appear to be completely one-sided. Cygnus A shows a faint main jet and a weaker counter-jet (Perley, Dreher & Cowan 1984) both of which extend much of the way into the lobes. In 3C 219 (Bridle *et al.* 1986), on the other hand, the counter-jet consists of an elongated knot 10 kpc from the core. Concerted efforts are now under way to detect the kpc scale counter-jets in more FR II sources (Bridle 1988). Glimpses of counter-jets are seen in several quasars with jet to counter-jet intensity ratios up to several hundreds.

Several studies have investigated large-scale asymmetries in powerful double sources; their findings are summarised below.

The two lobes of classical double sources, both radio galaxies and quasars, are in fact rather symmetric in flux density and separation from the core, typically differing by less than a factor two (Macklin 1981). There has been some debate concerning the distribution of the ratio of the separations of the two components from the core. Rudnick & Edgar (1984) and Neff *et al.* (1989) find a deficit of symmetric sources and argue that this provides evidence that the two lobes are formed by alternating ejection from the nucleus. Other samples, however, find no such deficit (Ensman & Ulvestad 1984; Saikia 1984b; Padrielli *et al.* 1988).

The hotspots of quasars show a greater asymmetry: the ratio of peak brightness is typically 4 for a sample of quasars drawn from the Jodrell Bank 966 MHz survey (Shone 1985) and tends to increase as sources are observed at higher resolution. In sources with jets there is a strong tendency for the jet to point to the brighter, or only, primary hotspot (Laing 1989). Double hotspots appear to be more common on the jet side with the jet feeding the more compact of the two hotspots within that lobe (Laing 1989; Lonsdale 1989). There is also a tendency for the jet side to have a flatter spectrum than the opposite side (Reid 1987; Garrington *et al.* 1989). This effect may be partially explained by the fact that hotspots generally have flatter spectra than their associated lobes, and the jetted lobe has a more prominent hotspot.

The two components of powerful double sources often show considerable asymmetries in their polarization characteristics (Conway & Strom 1985). In sources with one-sided jets, the jet side almost invariably depolarizes less with increasing wavelength than the opposite side (Laing 1988; Garrington *et al.* 1988).

All these asymmetries are consistent with, but not conclusive evidence of, mild relativistic beaming of the jets and associated hotspots. The depolarization asymmetry has also been used to give strong support to the idea that the one-sided jets are beamed (Laing 1988). If the depolarization is due to differential Faraday rotation (see Chapter 3) in a halo of ionised gas and tangled magnetic field surrounding the radio source, then the side with the approaching (beamed) jet would be seen through a shorter path length of this depolarizing medium and would depolarize less than the receding lobe. However, there is observational evidence that the cool gas, as revealed by optical emission lines, surrounding radio galaxies is often very asymmetric. In particular, the optical emission appears stronger on the side which is closer to the nucleus (Pedelty *et al.* 1989). This asymmetry in the environment would cause asymmetric depolarization and could give rise to asymmetries in the large-scale jets.

2.5 Concluding remarks: towards a unified picture

We have discussed a wide range of radio sources which cover several decades in radio luminosity and size. A morphological sequence of source types in order of luminosity can be defined and is determined chiefly by the relative dominance of the cores and jets. Nearly all the structures we see may be interpreted in terms of the standard model of twin plasma beams interacting with the ambient gas to form the extended regions of radio emission.

The appearance of a source seems to be governed by three main factors: the power of the jets, the environment and the angle of the jets to the line of sight. The observed morphology of low-power jets, in particular, the trend for the spreading rate to decrease with source luminosity, have been used to argue that these beams are trans-sonic and turbulent, dissipating much of their energy as radio emission and forming the plumes of FR I radio galaxies. Detailed

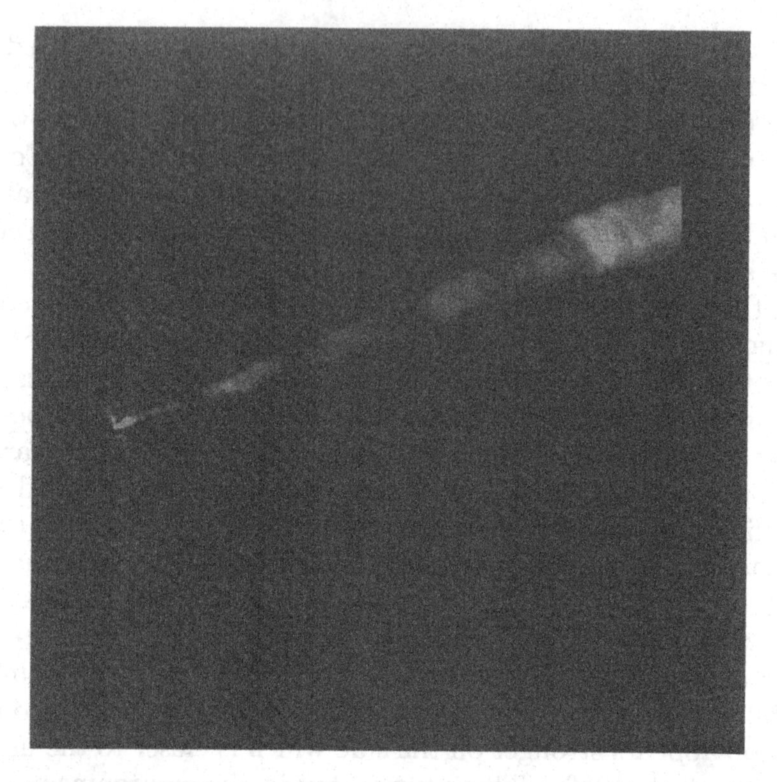

Fig. 2.18. VLA map of the inner part of the jet of M 87 at 15 GHz. Resolution 0.1 arcsec. Reproduced with permission from Owen *et al.* (1989).

models of such jets have been developed by Bicknell and are discussed in §3.5.4. More powerful jets in FR II sources appear to be supersonic, forming strong shocks (which we observe as hotspots) where they interact with the surrounding gas. Hydrodynamic simulations of supersonic jets (described in Chapter 7) reproduce many of the observed features of FR II sources. The observed differences between the host galaxies and cluster environments of FR I and FR II sources suggests that the environment also plays an important rôle in determining source morphology.

Many of the other differences between sources have been ascribed to differences in orientation, via relativistic beaming and projection effects: the relative prominence of the jets and cores and the asymmetries in jets and hotspots and perhaps the differences between radio galaxies and quasars may all depend on orientation.

So far most of our information on jets has come from studies of the large-scale properties of radio sources. Only now are detailed images of some of the brightest jets becoming available. Fig. 2.18 shows a recent VLA image of the one-sided jet in M 87 (Owen *et al.* 1989). The complex pattern of filaments along the jet suggests that most of the emission that we see from jets arises in a surface layer. This is even inferred for the nuclear jet in this source since it shows significant limb brightening (Reid *et al.* 1989). While it seems that observations of the jet material which carries most of the momentum will continue to be elusive, such images provide the most striking confirmation of the beam model.

Symbols used in Chapter 2

Symbol	Meaning
$B_{\parallel}$, $B_{\perp}$	components of magnetic field
D	linear extent of source
M_B	B-band magnitude
P	power (luminosity) of source
$P_{1\mathrm{GHz}}$	spectral power at 1 GHz
$P_{178\mathrm{MHz}}$	spectral power at 178 MHz
R	ratio of core flux to extended flux
S	flux density
S_i	flux density of interknot jet
S_k	flux density of knot
T	temperature
v	expansion velocity of source
z	cosmological redshift
α	frequency spectral index ($S \propto \nu^{-\alpha}$)
δ	density-radius index ($\rho \propto r^{-\delta}$) for ambient medium
θ	angle on sky, measured from core
ν	frequency
ϕ	FWHM across jet

References

Alexander, P., 1987, *M. N. R. A. S.*, **225**, 27.

Allington-Smith, J. R., 1984, *M. N. R. A. S.*, **210**, 611.

Baldwin, J. A., 1982, In *Extragalactic Radio Sources, I. A. U. Symposium 97*, eds. Heeschen, D. S. & Wade, C. M. (Reidel: Dordrecht, Netherlands), p. 21.

Barthel, P. D., 1989, *Ap. J.*, **336**, 606.

Bicknell, G. V., 1986, *Ap. J.*, **305**, 109.

Bicknell, G. V., de Ruiter, H. R., Fanti, R., Morganti, R. & Parma, P., 1989, *Preprint*, European Southern Observatory.

Booler, R. V., Pedlar, A. & Davies, R. D., 1982, *M. N. R. A. S.*, **199**, 229.

Boroson, T. A. & Oke, J. B. 1984, *Ap. J.*, **281**, 535.

Bridle, A. H., 1988, In *Active Galactic Nuclei*, eds. Miller, H. R. & Wiita, P. J. (Springer: Berlin), p. 329.

Bridle, A. H. & Perley, R. A., 1984, *Ann. Rev. Astr. Ap.*, **22**, 319.

Bridle, A. H., Perley, R. A. & Henriksen, R. N., 1986, *Astron. J.*, **92**, 534.

Browne, I. W. A., 1987a, In *Superluminal Radio Sources*, eds. Zensus, J. A. & Pearson, T. J. (Cambridge University Press: Cambridge), p. 111.

Browne, I. W. A., 1987b, In *Superluminal Radio Sources*, eds. Zensus, J. A. & Pearson, T. J. (Cambridge University Press: Cambridge), p. 129.

Burns, J. O., Basart, J. P., De Young, D. S. & Ghiglia, D. C., 1984, *Ap. J.*, **283**, 515.

Butcher, H. R., van Breugel, W. J. M. & Miley, G. K., 1980, *Ap. J.*, **235**, 749.

Cameron, M. J., 1971, *M. N. R. A. S.*, **152**, 439.

Cohen, M. H., Cannon, W., Purcell, G. H., Shaffer, D. B., Broderick, J. J., Kellermann, K. I. & Jauncey, D. L., 1971, *Ap. J.*, **170**, 207.

Conway, R. G. & Strom, R. G., 1985, *Astr. Ap.*, **146**, 392.

Cordey, R. A., 1986, *M. N. R. A. S.*, **219**, 575.

Cotton, W. D., Wittels, J. J., Shapiro, I. I., Marcaide, J., Owen, F. N., Spangler, S. R., Rius, A., Angulo, C., Clark, T. A. & Knight, C. A., 1980, *Ap. J.*, **238**, L123.

Davis, R. J., Muxlow, T. W. B. & Conway, R. G., 1985, *Nature*, **318**, 343.

de Bruyn, A. G. & Schilizzi, R. T., 1986, In *Quasars, I. A. U. Symposium 119*, eds. Swarup, G. & Kapahi, V. K. (Reidel: Dordrecht, Netherlands), p. 203.

de Ruiter, H. R., Parma, P., Fanti, C. & Fanti, R., 1989, *Preprint*, Istituto di Radioastronomia, Bologna.

Eilek, J. A., Burns, J. O., O'Dea, C. P. & Owen, F. N., 1984, *Ap. J.*, **278**, 37.

Ekers, R. D., Fanti, R., Lari, C. & Parma, P., 1981, *Astr. Ap.*, **101**, 194.

Ensman, L. M. & Ulvestad, J. S., 1984, *Astron. J.*, **89**, 1275.

Fanaroff, B. L. & Riley, J. M., 1974, *M. N. R. A. S.*, **167**, 31p.

Fanti, C. & Fanti, R., 1986, In *Superluminal Radio Sources*, eds. Zensus, J. A. & Pearson, T. J. (Cambridge University Press: Cambridge), p. 174.

Fanti, C., Fanti, R., Parma, P., Schilizzi, R. T. & van Breugel, W. J. M., 1985, *Astr. Ap.*, **143**, 292.

Feigelson, E. D., 1980, PhD Thesis, Harvard University.

Feigelson, E. D., Schreier, E. J., Delvaille, J. P., Giacconi, R., Gridlay, J. E. & Lightman, A. P., 1981, *Ap. J.*, **251**, 31.

Feretti, L., Giovannini, G., Gregorini, L., Parma, P. & Zamorani, G., 1984, *Astr. Ap.*, **139**, 55.

Garrington, S. T., Conway, R. G. & Leahy, J. P. 1989, in preparation.

Garrington, S. T., Leahy, J. P., Conway, R. G. & Laing, R. A., 1988, *Nature,* **331**, 147.
Giovannini, G., Feretti, L., Gregorini, L. & Parma, P., 1988, *Astr. Ap.,* **199**, 73.
Gopal-Krishna & Wiita, P. J. 1987, *M. N. R. A. S.,* **226**, 531.
Gopal-Krishna & Wiita, P. J. 1988, *Nature,* **333**, 49.
Guilbert, P. W. & Fabian, A. C. 1986, *M. N. R. A. S.,* **220**, 439.
Heckman, T. M., Smith, E. P., Baum, S. A., van Breugel, W. J. M., Miley, G. K., Illingworth, G. D., Bothun, G. D. & Balick, B., 1986, *Ap. J.,* **311**, 526.
Hough, D. H., Bridle, A. H., Burns, J. O. & Laing, R. A., 1988, *Bull. A. A. S.,* **20**, 734.
Hough, D. H. & Readhead, A. C. S., 1988, Caltech preprint no. 21.
Hutchings, J. B., 1987, *Ap. J.,* **320**, 122.
Kapahi, V. K. & Saikia, D. J., 1982, *J. Ap. Astr.,* **3**, 465.
Killeen, N. E. B., Bicknell, G. V. & Ekers, R. D., 1986, *Ap. J.,* **302**, 306.
Kronberg, P. P., Biermann, P. & Schwab, R. F., 1985, *Ap. J.,* **291**, 693.
Laing, R. A., 1984a, In *Physics of Energy Transport in Extragalactic Radio Sources*, eds. Bridle, A. H. & Eilek, J. (NRAO: Greenbank, WV), p. 119.
Laing, R. A., 1984b, In *Physics of Energy Transport in Extragalactic Radio Sources*, eds. Bridle, A. H. & Eilek, J. (NRAO: Greenbank, WV), p. 128.
Laing, R. A., 1988, *Nature,* **331**, 149.
Laing, R. A., 1989, In *Hotspots in Radio Galaxies*, eds. Meisenheimer, K. & Röser, H.-J. (Springer: Berlin), p. 27.
Laing, R. A. & Bridle, A. H., 1987, *M. N. R. A. S.,* **228**, 557.
Laing, R. A., Riley, J. M. & Longair, M. S., 1983, *M. N. R. A. S.,* **204**, 151.
Leahy, J. P., 1985, PhD Thesis, University of Cambridge.
Leahy, J. P., Muxlow, T. W. B. & Stephens, P. W., 1989, *M. N. R. A. S.,* **239**, 401.
Lilly, S. J. & Prestage, R. M., 1987, *M. N. R. A. S.,* **225**, 531.
Lonsdale, C. J., 1986, *Can. J. Phys.,* **64**, 445.
Lonsdale, C. J., 1989, In *Hotspots in Radio Galaxies*, eds. Meisenheimer, K. & Röser, H.-J. (Springer: Berlin), p. 45.
Macklin, J. T., 1981, *M. N. R. A. S.,* **196**, 967.
Macklin, J. T., 1982, *M. N. R. A. S.,* **199**, 1119.
Masson, C. R., 1980, *Ap. J.,* **242**, 8.
Matthews, T. A., Morgan, W. W. & Schmidt, M., 1964, *Ap. J.,* **140**, 35.
Miley, G. K., 1983, In *Astrophysical Jets*, eds. Ferrari, A. & Pacholczyk, A. G. (Reidel: Dordrecht, Netherlands), p. 99.
Miley, G. K. & Miller, J. S., 1979, *Ap. J.,* **228**, L55.
Murphy, D. W., 1988, PhD Thesis, University of Manchester.
Muxlow, T. W. B., Neff, S. G. & Lind, K. R., 1989, *M. N. R. A. S.,* To be published.
Muxlow, T. W. B. & Wilkinson, P. N., 1989, *Nature,* To be published.
Neff, S. G., Hutchings, J. B. & Gower, A. C., 1989, *Astron. J.,* **97**, 1291.
Norman, C. & Miley, G. K., 1984, *Astr. Ap.,* **141**, 85.
Norman, M. L., Smarr, L., Winkler, K.-H. & Smith, M. D., 1982, *Astr. Ap.,* **113**, 285.
O'Dea, C. P. & Owen, F. N., 1986, *Ap. J.,* **301**, 841.

Oort, M. J. A., Katgert, P., Steeman, F. W. M. & Windhorst, R. A., 1987, *Astr. Ap.*, **179**, 41.

Orr, M. J. L. & Browne, I. W. A., 1982, *M. N. R. A. S.*, **200**, 1067.

Owen, F. N., Hardee, P. E. & Cornwell, T. J., 1989, *Ap. J.*, **340**, 698.

Owen, F. N. & Laing, R. A., 1989, *M. N. R. A. S.*, **238**, 357.

Owen, F. N. & Puschell, J. J., 1984, *Astron. J.*, **89**, 932.

Owen, F. N. & Rudnick, L., 1976, *Ap. J.*, **205**, L1.

Padrielli, L., Rogora, A. & de Ruiter, H. R., 1988, *Astr. Ap.*, **196**, 49.

Parma, P., Fanti, C., Fanti, R., Morganti, R. & de Ruiter, H. R., 1987, *Astr. Ap.*, **181**, 244.

Pedelty, J. A., Rudnick, L., McCarthy, P. J. & Spinrad, H., 1989, *Astron. J.*, **97**, 647.

Pedlar, A., Dyson, J. E. & Unger, S. W., 1985, *M. N. R. A. S.*, **214**, 463.

Perley, R. A., Bridle, A. H. & Willis, A. G., 1984, *Ap. J. Suppl.*, **54**, 291.

Perley, R. A., Dreher, J. W. & Cowan, J. J., 1984, *Ap. J.*, **285**, L35.

Perley, R. A., Willis, A. G. & Scott, J. S., 1979, *Nature*, **281**, 437.

Phillips, R. B. & Mutel, R. L., 1980, *Ap. J.*, **236**, 89.

Prestage, R. M. & Peacock, J. A., 1988, *M. N. R. A. S.*, **230**, 131.

Readhead, A. C. S., Hough, D. A., Ewing, M. S., Walker, R. C. & Romney, J. D., 1983, *Ap. J.*, **265**, 107.

Reid, A., 1987, PhD Thesis, University of Manchester.

Reid, M. J., Biretta, J. A., Junor, W., Muxlow, T. W. B. & Spencer, R. E., 1989, *Ap. J.*, **336**, 112.

Röser, H.-J. & Meisenheimer, K., 1986, *Astr. Ap.*, **154**, 15.

Rudnick. L. & Edgar, B. K., 1984, *Ap. J.*, **279**, 74.

Rusk, R. & Rusk, A. C. M., 1986, *Can. J. Phys.*, **64**, 440.

Saikia, D. J., 1984a, *M. N. R. A. S.*, **208**, 231.

Saikia, D. J., 1984b, *M. N. R. A. S.*, **209**, 525.

Saikia, D. J., Muxlow, T. W. B. & Junor, W., 1989, *M. N. R. A. S.*, In Press.

Saikia, D. J. & Salter, C. J., 1988, *Ann. Rev. Astr. Ap.*, **26**, 93.

Saikia, D. J., Shastri, P., Cornwell, T. J. & Banhatti, D. G., 1983, *M. N. R. A. S.*, **203**, 53p.

Scheuer, P. A. G., 1974, *M. N. R. A. S.*, **166**, 513.

Scheuer, P. A. G., 1987, In *Superluminal Radio Sources*, eds. Zensus, J. A. & Pearson, T. J. (Cambridge University Press: Cambridge), p. 104.

Schmidt, G. D., Peterson, B. A. & Beaver, E. A., 1978, *Ap. J.*, **220**, L31.

Shaffer, D. B., Marscher, A. P., Marcaide, J. & Romney, J. D., 1987, *Ap. J.*, **314**, L1.

Shone, D. L., 1985, PhD Thesis, University of Manchester.

Shone, D. L., Porcas, R. W. & Zensus, J. A., 1985, *Nature*, **314**, 603.

Smith, R. M., Bicknell, G. V., Hyland, A. R. & Jones, T. J., 1983, *Ap. J.*, **266**, 69.

Visvanathan, N. & Pickles, A. J., 1981, *Proc. Astron. Soc. Australia*, **4**, 177.

Walker, R. C., Walker, M. A. & Benson, J. M., 1988, *Ap. J.*, **335**, 668.

Wardle, J. F. C. & Potash, R. I., 1982, In *Extragalactic Radio Sources, I. A. U. Symposium 97*, eds. Heeschen, D. S. & Wade, C. M. (Reidel: Dordrecht, Netherlands), p. 129.

Wardle, J. F. C. & Potash, R. I., 1984, In *Physics of Energy Transport in Extragalactic Radio Sources*, eds. Bridle, A. H. & Eilek, J. (NRAO: Greenbank, WV), p. 30.

Whitney, A. R., Shapiro, I. I., Rogers, A. E. E., Robertson, D. S., Knight, C. A., Clark, T. A., Goldstein, R. M., Marandino, G. E. & Vandenberg, N. R., 1971, *Science,* **173**, 225.

Wilkinson, P. N., Tzioumis, A. K., Simon, R. S., Benson, J. M. & Walker, R. C., 1989, *Nature,* To be published.

Williams, A. G. & Gull, S. F. 1985, *Nature,* **313**, 34.

Willingale, R., 1981, *M. N. R. A. S.,* **194**, 359.

Willis, A. G., Strom, R. G., Bridle, A. H. & Fomalont, E. B., 1981, *Astr. Ap.,* **95**, 250.

Willis, A. G., Strom, R. G., Perley, R. A. & Bridle, A. H., 1982, In *Extragalactic Radio Sources, I. A. U. Symposium 97*, eds. Heeschen, D. S. & Wade, C. M. (Reidel: Dordrecht, Netherlands), p. 141.

Wills, B. J. & Browne, I. W. A., 1986, *Ap. J.,* **302**, 56.

Wilson, A. S., 1982, In *Extragalactic Radio Sources, I. A. U. Symposium 97*, eds. Heeschen, D. S. & Wade, C. M. (Reidel: Dordrecht, Netherlands), p. 179.

Wilson, A. S., 1983, In *Highlights in Astronomy 6*, ed. West, R. M. (Reidel: Dordrecht, Netherlands), p. 467.

Wilson, A. S. & Willis, A. G., 1980, *Ap. J.,* **240**, 429.

Yates, M. G., Miller, L. & Peacock, J. A., 1986, *M. N. R. A. S.,* **221**, 311.

3

Interpretation of Large Scale Extragalactic Jets

J. P. LEAHY
N. R. A. O., P. O. Box O, Socorro, NM 87801, USA.

3.1 Introduction

3.1.1 What are extragalactic radio sources?

We believe that extragalactic radio sources are *interactions* between large-scale jets and the hot, diffuse gas that surrounds elliptical* galaxies. The interactions not only cause the hotspots, bridges, tails etc., but also the radio emission from the jets themselves. Radio sources should be thought of as processes rather than objects: the overall radio structure must change substantially on the shortest possible dynamical time scale, the sound-crossing time. Another way to put this is that, at least in the FR II sources, there is no steady-state description. This is the main reason why radio sources are much harder to understand than, say, main-sequence stars. On the other hand, the lack of equilibrium means that the structure of radio sources reflects their past history, so that in principle it should be much easier to deduce the life-cycle of radio sources than that of stars. At present, the most successful theoretical models concentrate on regions for which a local steady-state description is likely to be appropriate, notably the bases of jets, far from any end-effects, and in the co-moving frame of the hotspots, where the jets terminate.

It is worth emphasizing that in the standard model of FR II sources, and to a lesser extent FR Is, we only see half the story in the radio. As we shall see, it is usually assumed that synchrotron radiation is only emitted by plasma which has entered the system via the jet. Equally important is the shock-heated external medium surrounding the jet plasma; in principle this can be studied in X-

* Analogous interactions occur in spiral (Seyfert) galaxies, but their radio emission is much less luminous. See §3.5.6.

rays, although as yet there are no reliable detections. We can hope that the AXAF X-ray satellite (due for launch in the late 1990s) will revolutionize the field even more profoundly than the VLA has done in this decade.

The rest of this chapter is organized as follows. §3.2 reviews the standard arguments which are used to estimate the physical conditions in the sources. §3.3 reviews data on the source environments. §3.4 presents a simple approach to jet physics, and discusses how observations of the structure of jets can constrain the parameters of the system. §3.5 considers the main types of dynamical models that have been applied to various classes of sources. §3.6 reviews some major open questions and considers some of the ways we might go about answering them. But first, to set the following sections in context, I will review some of the basic astrophysics which underlies most of the models.

3.1.2 **What are radio sources made of?**

3.1.2.1 **fluids...**

The space-filling medium surrounding radio sources is a fully ionised, virtually collisionless plasma. Table 3.1 gives the characteristic parameters for the well-studied Coma cluster, in which several radio sources are embedded. Of course most radio sources are not in rich clusters of galaxies like Coma; but the densities and temperatures in Coma are probably within a couple of orders of magnitude of the environments of most kiloparsec-scale radio sources. The formulae assume for simplicity that the gas is pure hydrogen; inclusion of a cosmic abundance of helium would change some parameters by $\sim 10\%$, generally less than the associated uncertainties.

We know far less about the physics of the *synchrotron plasma* which makes up the radio-emitting regions, in particular radio jets, although it is a safe bet that it, too, is fully ionised and collisionless. Even the composition is a matter for debate: while a hydrogen plasma is usually assumed by default, a number of authors have argued on various grounds that an electron-positron plasma is the major constituent. Evidently synchrotron plasma differs from its surroundings in the density of relativistic electrons (cosmic rays) and/or the intensity of the magnetic field. It is usually assumed that both of these components are much enhanced, but some models invoking diffusion of electrons into the ambient medium require

Table 3.1. *Plasma parameters for the centre of the Coma cluster.*

Parameters determined from observations:				
Electron number density [a,b]	$n_{\rm th}$	—	4×10^3	m^{-3}
Temperature [b]	T	—	1×10^8	K
Magnetic field [a,c,d]	B	—	0.1	nT
Derived parameters:				
Thermal pressure	$P_{\rm th}$	$2n_{\rm th}k_BT$	1×10^{-11}	Pa
Magnetic energy density	u_B	$(B^2/2\mu_0)$	4×10^{-15}	$\mathrm{J\,m^{-3}}$
Plasma parameter	β_P	P/u_B	4×10^3	
Adiabatic index	Γ	—	5/3	
Thermal sound speed	c_s	$\sqrt{\frac{2\Gamma k_BT}{m_p}}$	2×10^6	$\mathrm{m\,s^{-1}}$
Alfvén speed	v_A	$\sqrt{\frac{2u_B}{n_{\rm th}m_p}}$	3×10^4	$\mathrm{m\,s^{-1}}$
rms electron speed	w_e	$\sqrt{\frac{3k_BT}{m_e}}$	7×10^7	$\mathrm{m\,s^{-1}}$
Plasma frequency	ω_p	$\sqrt{\frac{n_{\rm th}e^2}{\varepsilon_0 m_e}}$	3×10^3	$\mathrm{rad\,s^{-1}}$
Debye screening length	λ_D	$w_e/(\sqrt{3}\omega_p)$	1×10^4	m
Coulomb logarithm [e]	L_c	$\ln\frac{2\lambda_D}{(\hbar/m_ew_e)}$	37	
Effective collision frequency [f]	ν_0	$\frac{n_{\rm th}e^4}{4\pi\varepsilon_0^2m_e^2w_e^3}L_c$	4×10^{-13}	Hz
Electron mean free path	Λ_e	w_e/ν_0	6	kpc
Larmor frequency	Ω_e	eB/m_e	20	$\mathrm{rad\,s^{-1}}$
Larmor radius	r_{ge}	$\sqrt{\frac{2}{3}}w_e/\Omega_e$	4×10^6	m
Measures of macroscopic scale size D:				
Cluster core radius [b]			150	kpc
GE galaxy effective radius [g]			10	kpc
radius of typical jet			2	kpc

[a] $H_0 = 100\mathrm{km\,s^{-1}\,Mpc^{-1}}$ has been assumed.
[b] From Hughes, Gorenstein & Fabricant (1988).
[c] From Kim *et al.* (1986).
[d] Strictly B is the magnetic flux density, but since the quantity H is never used it is easier to refer to B as the magnetic field.
[e] This is Spitzer's approximation, valid for $T > 10^5$K.
[f] Collision time for electrons colliding with protons.
[g] From Owen & Laing (1989).

that the magnetic field is comparable inside and outside the radio-emitting regions. In addition the synchrotron plasma could contain relativistic protons or positrons, and non-relativistic ("thermal") protons and electrons. Charge balance requires positive particles, but they need not be relativistic. It is worth noting that there is room for a dominant contribution to the energy density from mildly-relativistic electrons whose synchrotron radiation occurs at unobservably low frequencies. Note also that the *mass* density might be dominated by a thermal component which contributes negligibly to the energy density.

Most models of radio sources treat both the external medium and the synchrotron plasma as continuous, electrically-neutral media, and here I briefly outline the justification for this approximation. Familiarity with basic plasma physics is assumed (see Spitzer (1962) for an excellent overview).

Neutrality is a very good approximation since the Debye screening length λ_D is always many orders of magnitude smaller than the jet radius. Even so, small-scale charge separation in strong turbulence, shock waves or plasma boundary layers may have important implications for particle acceleration (cf. §9.5).

At first sight the continuum (fluid) approximation seems quite incorrect since, even in the external medium, the collisional mean free path is somewhat larger than the jet radius. Such collisionless plasmas can nevertheless be treated as continuous fluids in some respects, provided that the magnetic field is sufficiently strong. The importance of the magnetic field in plasmas is parameterised by the ratio β_P of the particle pressure to the magnetic energy density. This is very poorly known (§§3.2.8, 3.2.9), and it is still possible to find treatments in the literature which assume that the field is zero in the X-ray gas. However the effective mean free path will be determined by magnetic effects as long as $\beta_P < 10^{30}$, and many arguments suggest that β_P must be far lower than this. In the synchrotron plasma we shall see that β_P is within a few orders of magnitude of unity.

Magnetic fields will suppress diffusion perpendicular to the field lines provided that the Larmor radius $r_{ge} \ll D$. They are also believed to suppress diffusion *along* the field lines. This is because particles streaming along field lines excite MHD and plasma waves which then scatter the particles by resonant interactions. Some details of the scattering process are described in §9.2.2; see also Melrose (1980). Resonant scattering also dominates the transport of particles across field lines, as is well known from studies of Solar System plasmas. Generally the length scale for wave-particle scattering is within a few orders of magnitude of the proton Larmor radius, and so is negligible on the macroscopic scale. This of course is just the condition for the fluid approximation to hold.

3.1.2.2 ...and fields

Evidently magnetic fields are an essential component of the dilute plasmas under consideration. The basic fact about such fields is that they are 'frozen-in' to the plasma under most circumstances. The usual proof of this (§9.6.1) is based on a scalar conductivity, $\mathbf{j} = \sigma\mathbf{E}$. While appropriate for weakly-ionized plasmas (*i.e.*, those for which electrons collide more often with atoms than with ions), this equation is totally wrong for fully-ionised plasmas, irrespective of the collisional rate. Generally the currents and electric fields in such plasmas are not causally connected. However, as one might expect, flux freezing is a valid approximation in fully-ionised plasmas; see Parker (1979), p. 43 for a derivation in this case.*

The basic equations for the magnetic and electric fields and the currents are then:

$$\frac{\partial \mathbf{B}}{\partial t} = \nabla \times (\mathbf{v} \times \mathbf{B}), \tag{3.1}$$

$$\mathbf{E} = -\mathbf{v} \times \mathbf{B}, \tag{3.2}$$

$$\mu_0 \mathbf{j} = \nabla \times \mathbf{B}. \tag{3.3}$$

The electric field and the currents can be derived from the magnetic and velocity fields and do not need to be taken into account explicitly.

Some apparent paradoxes appear when one tries to translate from the MHD language of frozen-in fields in a fluid to the plasma physics language of currents and particles. For instance, the field exerts a force on the fluid through the $\mathbf{j} \times \mathbf{B}$ force, which implies that only the current perpendicular to the field has a dynamical effect. How can such currents exist when transport across the field

* Parker implies that the conductivity is scalar for fully-ionised collisional plasmas, but this is not so (*e.g.*, Spitzer 1962). However his derivation for collisionless plasmas also applies to the former case.

lines is forbidden? *Pace* Parker (1979), the answer is not the drift currents associated with variation in the magnetic field. Actually these are $O(\beta_P)$ times *larger* than the bulk currents and are oppositely directed*. They are largely cancelled by purely macroscopic currents associated with gradients in the particle density and temperature. In the presence of such gradients, the currents associated with the gyration of the particles around field lines do not cancel out at each point. (This is just the reason why 'pressure' (*i.e.*, nT) gradients affect a fluid in which particle-particle collisions are negligible). Thus in high-β_P plasmas the bulk currents are a small residual effect. Alfvén & Fälthammer (1963) give a very clear discussion of the microscopic-macroscopic connection in such plasmas.

For simplicity bulk magnetic forces are usually neglected in the external medium and the values of β_P suggested by observations are consistent with this. In the synchrotron plasma, mostly for historical reasons (see §3.2.8), it is usually assumed that relativistic particles and magnetic fields have near-equal energy densities. If the non-relativistic components have negligible pressure, which is likely in the hotspots at least because of the extremely high relativistic pressures, then β_P is around unity. Magnetic forces should then be taken into account.

The anisotropy of magnetic forces makes life difficult for theorists, but some simplification is possible if it is assumed that the magnetic field is highly tangled on a small scale. This assumption crops up in several contexts in the rest of this chapter, so it is useful to summarise here the properties of various degrees of field tangling. We consider four cases: a uniform field; a stretched field in which the field lines are aligned but their direction reverses on a small scale (for instance a collection of long thin field loops all pointing in the same direction); a compressed field in which the field lines are nearly confined within parallel planes (for instance a collection of field loops with their planes parallel); and finally a field with no large-scale order of any kind. It does not matter whether individual field lines are closed into small loops or execute random walks, although the latter is statistically necessary unless the magnetic field generation process produces closed field lines. For

* Strictly, $\mathbf{j}_{\rm drift} \cdot \mathbf{j}_{\rm bulk} \leq 0$.

Table 3.2. *Properties of tangled magnetic fields.*

Type:	Uniform	Stretched	Compressed	Tangled
f	—	$\gg$	$\ll$	1
$\mathbf{\Psi}_B$	$u_B(1,1,-1)$	$u_B(1,1,-1)$	$u_B(f^2,f^2,1)$	$u_B(\frac{1}{3},\frac{1}{3},\frac{1}{3})$
$\frac{\pi_L}{\pi_{L,max}}$	1	1	$\frac{\sin^2\theta}{1+\cos^2\theta}$	$\frac{d}{3W+d}$
$\frac{\phi_F}{C_F n_{th} B}$	$W\cot\theta$	$\sqrt{dW}\cot\theta$	$\sqrt{\frac{f dW \tan\theta}{2}}$	$\sqrt{\frac{dW}{3\sin\theta}}$
$g(\theta,\alpha)$ [a]	$a(\alpha)(\frac{\sin\theta}{2})^{1+\alpha}$	$a(\alpha)(\frac{\sin\theta}{2})^{1+\alpha}$	$\frac{6-3\sin^2\theta}{4}$	1

[a] $a(\alpha) = 1/B(\frac{3+\alpha}{2},\frac{3+\alpha}{2})$, where B is the Beta function. Note that $a(1) = 6$.

definiteness, consider an isotropically tangled field with correlation length d, which is stretched or compressed by a factor f parallel to the z axis, so $\langle B_z^2\rangle/f^2 = \langle B_x^2\rangle = \langle B_y^2\rangle$. Table 3.2 lists for each of these the diagonal components of the effective magnetic stress tensor (the rest are zero in this coordinate system); the factor by which the synchrotron polarization is reduced for an angle θ between the observer and the z axis; the Faraday depth ϕ_F (see §1.2.1.4) given a width W perpendicular to the z axis; and the monochromatic anisotropy factor for synchrotron emission $g(\theta,\alpha)$ (see §3.2.3). The polarization and anisotropy parameters are rather complicated for compressed fields with arbitrary spectral index; I quote the simple expressions for $\alpha = 1$ which are indicative of the general behaviour (§1.2.1.3 discusses the polarization in more detail). The bolometric anisotropy factor is given by $g(\theta,1)$. The formula for the Faraday rotation in a compressed field applies when $\tan\theta < 1/f$.

For "Grand design" field structures, *i.e.*, large-scale but non-uniform models such as proposed by Chan & Henriksen (1980), one can either calculate these quantities directly by averaging over the region of interest or note that typically they behave like the semi-random fields with $d \sim W$.

It is worth noting explicitly that for the fully-random field, the effective magnetic stress tensor is isotropic. The dynamical effect of the magnetic field reduces to an effective pressure $\frac{1}{3}u_B$, or one third of the text-book magnetic pressure. The reason for the difference is that in addition to the standard pressure there is the magnetic tension term, equivalent to a negative pressure along the field lines equal to twice the isotropic pressure. In a tangled field we average the effect of the tension term over all directions, giving a net isotropic constriction of $\frac{2}{3}u_B$, to which we add the isotropic term to get the result in Table 3.2. This can be derived more rigorously by integrating the magnetic force

$$\mathbf{j} \times \mathbf{B} = -\nabla \cdot \mathbf{\Psi}_B = -\nabla(B^2/2\mu_0) + \mathbf{B} \cdot \nabla\mathbf{B}/\mu_o \qquad (3.4)$$

over a control volume larger than the field tangling size in a region where there is a gradient in B^2. A final way to remember the result is to note that a static B-field can be considered as a superposition of virtual photons. Hence a random B-field has the same relation between pressure and energy density as a photon gas.

It should also be evident from Table 3.2 that magnetic stresses in compressible flows are never isotropic in general, since even if the field starts out fully-tangled it will be compressed or sheared. Hence purely fluid-dynamical models can never give more than qualitative insight into the physics of radio jets.

Another plausible possibility is that some regions of the jets and lobes are magnetically-dominated, *i.e.*, $\beta_P \ll 1$. In this case the field will try to evolve to a force-free configuration, with $\mathbf{j} \times \mathbf{B} = 0$. Parker (1979) has emphasised that a global force-free configuration is not possible. Any region containing enhanced magnetic field will always try to increase its total volume, *i.e.*, there is always a net magnetic pressure when all directions are averaged. Thus a force-free structure can occur only locally within a larger magnetized region. Force-free fields are considered further in §9.6.3.

A final caveat is that conventional MHD assumes that all fluid velocities are negligible compared to the speed of light. The model for FR II radio sources outlined in Chapter 7 suggests that the jets are relativistic and the jet fluid may have a relativistic sound speed $c/\sqrt{3}$. The relativistic MHD equations are presented in §6.4.2; as noted there, electric fields cannot usually be ignored in this limit. As yet no attempts have been made to model relativistic MHD jets.

3.1.3 **Adiabatic expansion**

The motion of a fluid element of synchrotron plasma will often be adiabatic, or more correctly isentropic, meaning that there

Table 3.3. *Effects of adiabatic expansion.*

Quantity		$B_{\parallel}$	$B_{\perp}$	B_{tangled}
Magnetic field	B	$\mathcal{R}_{\perp}^{-2}$	$\mathcal{R}_{\perp}^{-1}\mathcal{R}_{\parallel}^{-1}$	$\mathcal{R}^{-2}$
Critical frequency	ν_c	$\mathcal{R}_{\perp}^{-10/3}\mathcal{R}_{\parallel}^{-2/3}$	$\mathcal{R}_{\perp}^{-7/3}\mathcal{R}_{\parallel}^{-5/3}$	$\mathcal{R}^{-4}$
Luminosity	L_{ν}	$\mathcal{R}_{\perp}^{-(2+\frac{10}{3}\alpha)}\mathcal{R}_{\parallel}^{-2\alpha/3}$	$\mathcal{R}_{\perp}^{-(1+\frac{7}{3}\alpha)}\mathcal{R}_{\parallel}^{-(1+\frac{5}{3}\alpha)}$	$\mathcal{R}^{-(2+4\alpha)}$
Emission coefficient	j_{ν}	$\mathcal{R}_{\perp}^{-(4+\frac{10}{3}\alpha)}\mathcal{R}_{\parallel}^{-(1+\frac{2}{3}\alpha)}$	$\mathcal{R}_{\perp}^{-(3+\frac{7}{3}\alpha)}\mathcal{R}_{\parallel}^{-(2+\frac{5}{3}\alpha)}$	$\mathcal{R}^{-(5+4\alpha)}$
Energy integrals	$\mathcal{I}_1$	$\mathcal{R}_{\perp}^{\frac{10\alpha-5}{3}}\mathcal{R}_{\parallel}^{\frac{2\alpha-1}{3}}$	$\mathcal{R}_{\perp}^{\frac{14\alpha-7}{6}}\mathcal{R}_{\parallel}^{\frac{10\alpha-5}{6}}$	$\mathcal{R}^{4\alpha-2}$
(see §3.2.7)	$\mathcal{I}_2$	$\mathcal{R}^{2\alpha-1}$	$\mathcal{R}^{2\alpha-1}$	$\mathcal{R}^{2\alpha-1}$

is no heat exchange with neighbouring elements or production of heat by internal stress. The approximation will only fail at strong shocks and to some degree in turbulent flow (cf. §3.5.4). For reference I summarise here the way adiabatic expansion or compression changes the internal properties of the fluid.

An element with streamline cross-section A and length l expands from state 1 to state 2: we define expansion factors $\mathcal{R}_{\perp} = \sqrt{A_2/A_1}$ and $\mathcal{R}_{\parallel} = l_2/l_1$. If the flow is steady, $\mathcal{R}_{\parallel} = v_2/v_1$. The mean expansion factor $\mathcal{R} = (\mathcal{R}_{\perp}^2\mathcal{R}_{\parallel})^{1/3} \propto (Av)^{1/3}$. Evidently the number density $n \propto \mathcal{R}^{-3}$; the pressure is given by the normal adiabatic relations: for the relativistic electrons $P_{\text{re}} \propto \mathcal{R}^{-4}$. These two results imply that the individual particle energy $\propto \mathcal{R}^{-1}$, corresponding to the de Broglie wavelength increasing as $\mathcal{R}$. If $\mathcal{R}_{\perp} \neq \mathcal{R}_{\parallel}$ the parallel and perpendicular field components behave differently, the results being derivable from flux-freezing. (This affects several other relations.) One field component always declines more slowly than the particle energy, so a prolonged anisotropic expansion will result in a magnetically-dominated (low β_P) plasma, unless particle and field energies can be exchanged by some dissipative process. If the flow is turbulent the field will not depart far from isotropy, so β_P might be expected to remain constant; however turbulence may either enhance or destroy the magnetic field (§9.6.1) and the adiabatic relations will not be accurate, both because of turbulent dissipation and because the local values of $\mathcal{R}_\perp$ and $\mathcal{R}_\parallel$ will be very different from their average over the region. Table 3.3 gives the adiabatic relations for some important parameters.

The very strong variation of emissivity with $\mathcal{R}$ makes this a sensitive indicator of expansion, although we shall see that actually deriving j_ν is rather uncertain. If we are not too badly fooled by variations in filling factor, the emissivity in FR II sources changes by factors of $10^3 - 10^7$ between hotspots and the fainter parts of the bridges, while in FR I sources the emissivity contrast between the bases of jets and the outer tails is $100 - 10^4$. These imply $\mathcal{R} \lesssim 10$; larger values would render the source completely invisible.

3.2 Interpretation of the radio data

In this section I present the methods used to estimate physical properties of radio sources directly from the synchrotron emission. I have tried to make clear the limitations of these methods and in particular to emphasize that some depend on assumptions which have never been verified, and may be quite inaccurate. I have also noted some ways in which high-quality X-ray data may one day improve the situation. In addition to the intrinsic problems discussed here, one should always be aware that the 'typical' values of physical parameters will be biassed in more or less benign ways by observational selection effects, as discussed in §§2.4.1 & 4.2.2.

3.2.1 Effective distance $R_{\rm eff}$

Distances to nearly all extragalactic radio sources are derived via the redshift z and Hubble's law. Conveniently, powerful sources are associated with strong optical line emission which is usually detectable at cosmological distances (§3.3.2). In less powerful sources, including FR I radio galaxies and BL Lac objects, emission lines are faint, but for the nearer examples of both classes one can detect stellar absorption lines in the host galaxy. Radio galaxies have a relatively small range in absolute magnitude, at least for $z < 1$, so their redshifts can be roughly estimated from the apparent magnitude if a spectrum is not available (*e.g.*, Laing, Riley & Longair 1983).

Table 3.4 gives formulae for effective distance $R_{\rm eff}$, in Friedman cosmologies (the notation is that of Gunn 1978); and also the projected linear size D, monochromatic radio power P_ν, and surface

Table 3.4. *Cosmological formulae.*

$R_{\rm eff}$	$= cZ_q/H_0$	
Z_q	$= (q_0 z + (q_0 - 1)(\sqrt{1+2q_0 z} - 1))/q_0^2(1+z)$	General
	$= z(1+\frac{1}{2}z)/(1+z)$	$q_0 = 0$
	$= 2(1 - 1/\sqrt{1+z})$	$q_0 = \frac{1}{2}$
	$= z/(1+z)$	$q_0 = 1$
P_ν	$= R_{\rm eff}^2(1+z)S_{(\nu/(1+z))}$	
	$= 8.557 \times 10^{25} h^{-2} Z_q^2 (1+z)^{1+\alpha}(S_\nu/\mathrm{Jy})\,\mathrm{W\,Hz^{-1}\,sr^{-1}}$	
D	$= R_{\rm eff}\Theta/(1+z)$	
	$= 14.53 h^{-1}(Z_q/1+z)(\Theta/\mathrm{arcsec})\mathrm{kpc}$	
I_ν	$= P_\nu/D^2 = (1+z)^{(3+\alpha)}(S_\nu/\Omega)$	

brightness I_ν, as functions of angular size Θ, solid angle Ω, flux density S_ν, 'effective redshift' Z_q, and redshift z.

Although redshifts can be measured with high precision, the derived distances suffer from the factor of two uncertainty in Hubble's constant. Often only *relative* distances need be accurate, in which case the precise value of H_0 does not matter, as long as we are consistent. Since various values are assumed in the literature, I will include the parameter $h = H_0/(100\,\mathrm{km\,s^{-1}\,Mpc^{-1}})$ when discussing derived quantities.

In addition to their cosmological recession, galaxies have so-called 'peculiar' velocities, *i.e.*, real motions in space, of order 500 km $\mathrm{s^{-1}}$, which is a significant source of distance error at redshifts of less than 0.01. The isotropy of the Hubble flow is only established to about 15% accuracy (Sandage 1987) so this is a conservative upper limit to errors in relative distance at intermediate redshifts. At large redshifts ($z > 0.5$), different cosmological models can give significantly different values for parameters such as luminosity and linear size. At these redshifts the q_0 dependence cannot be treated as simply an arbitrary second-order coefficient in a Taylor expansion for the distance in terms of z. Instead, the usual assumption of an isotropic Friedman model with zero cosmological constant has a major effect on the calculated distances. Evidence such as the linearity of the magnitude-redshift diagram for radio galaxies suggests that the Friedman models are not grossly wrong out to a redshift of unity (*e.g.*, Longair 1987).

3.2.2 Linear size D

Given a redshift and an assumed cosmology, the measured *largest angular size* (LAS) can be converted into a projected linear size, D. The only important uncertainty here is that the 'angular size' is rather poorly defined. Formally the LAS is the maximum separation between any two points in the source. It can increase if more sensitive observations reveal faint emission further away from the centre than was previously apparent. This has often occurred for FR I sources, many of which seem to fade into the noise. Optical astronomers avoid the problem by quoting either isophotal or metric sizes, respectively based on fixed surface-brightness or on some accessible parameter of the brightness distribution. Unfortunately no such systematic convention is used in the radio literature.

For FR II sources there is no such problem. The outer edges of the lobes are quite sharp (not just near the hotspots, but all around), so the LAS can be accurately determined. Instead of the LAS, one can measure the separation of the hotspots. Because the latter are sometimes set back from the leading edge the two measures can differ by up to 20%; *caveat emptor* since it is not always clear which was used. The hotspot separation has the advantage that the ratio of size to separation for hotspots is much smaller than for the diffuse component of the lobes, so that projection effects in samples can be modelled more accurately with the simple assumption $D = D_0 \sin i$, which ignores the finite width of the objects. On the other hand, since the hotspots represent short-time scale phenomena relative to the lobes, the use of hotspot separations will introduce additional scatter into the D distribution. Also, the sizes derived are not strictly comparable with those for FR I sources. Finally, many sources have multiple hotspots in each lobe, and the choice of the brightest may depend on resolution. Usually these quibbles at the 10% level are irrelevant because of the large range of D in the population.

3.2.3 Radio power P_ν

The measured flux density S_ν of a radio source can be readily converted into a specific radio power via the formulae given in Table 3.4, the second of which includes a first-order K-correction assuming a constant spectral index. This is the spectrum-dependent correction required because we usually observe at a fixed frequency ν, and wish to obtain the power at that frequency, in the frame of the galaxy, despite the fact that the radiation we observe was emitted at frequency $(1+z)\nu$.

P_ν is the radio power emitted in the direction of the observer. More interesting is the power emitted over all directions, *i.e.*, the specific luminosity L_ν. We saw in §1.2.1.3 that synchrotron radiation is anisotropic, so the conversion from P_ν to L_ν is not trivial. Typical sources contain many regions with different magnetic field orientations, so for the integrated power the anisotropy is largely averaged out (this is confirmed by their low integrated polarization). In high resolution maps the degree of polarization at each point is typically 20% and tends to rise to values near the theoretical maximum around the edge of the source and in faint, steep-spectrum regions; so isotropy is a poor approximation. Table 3.2 lists the angle-dependent factor $g(\theta, \alpha)$ for simple tangled-field models, where

$$g(\theta', \alpha)L'_\nu = 4\pi P'_\nu(\theta'). \tag{3.5}$$

The primes in equation (3.5) indicate that the formula applies in the rest frame of the emitting material. A further potentially major source of anisotropy is relativistic beaming, which is likely to be very important for core-dominated sources and may be significant in many other circumstances. We therefore need to relate quantities in the emitting frame to those in the frame of the host galaxy, which will be denoted by unprimed symbols.* There is an ambiguity in transforming luminosity, since we have to choose which luminosity in the emitting frame is the 'equivalent' of what we observe. In general, again including a K-correction term,

$$P_\nu(\theta) = \mathcal{D}^{(m+\alpha)} P'_\nu(\theta'). \tag{3.6}$$

Here the Doppler factor $\mathcal{D} = [\gamma(1-\beta\cos i)]^{-1}$, where i is the angle between the line-of-sight and the velocity. For a volume fixed in space through which emitting material moves, $m = 2$; while for

* A third frame involved is that of the observer on Earth, but the transformations from the observer to the galaxy frame are implicit in the formulae of Table 3.4.

an identifiable blob of plasma, $m = 3$. These relations are derived and discussed in §4.3.1. For steady-state flow, $m = 2$ is appropriate. Doppler beaming makes even approaching components fainter, unless one is inside the beaming cone, *i.e.*, $i \lesssim 1/\gamma$.

We shall see in §3.2.5 that Doppler beaming anisotropy does not significantly affect the total power of extended sources. On the other hand it might well be important for individual components, in particular the jets and primary hotspots in FR IIs (*e.g.*, Laing 1989).

A final note is that optically-thick structures radiate anisotropically because the observed flux density depends on the projected surface area. Fortunately large-scale emission is always optically thin, while in VLBI jets the effect is small compared to Doppler beaming.

In summary, total luminosities of extended sources can be derived fairly accurately, but in well-resolved regions the accuracy is only a factor of 2 or so because of intrinsic anisotropy, and may be much further off if flow speeds are relativistic. Of course, both these effects can be taken into account when comparing specific models with the observations.

3.2.4 Emission coefficient j_ν

For optically thin emission, the observed surface-brightness is related to the emissivity via:

$$S_\nu/\Omega = (1+z)^3 \int\int\int j_{(1+z)\nu} b(\alpha,\delta)dl\, d\alpha\, d\delta. \tag{3.7}$$

Here $b(\alpha, \delta)$ is the telescope response function (the 'beam') and (α, δ) are angles on the sky. (j_ν and I_ν refer to the *galaxy* frame; I denote the observed surface-brightness by S_ν/Ω.) The inversion of equation (3.7) to derive j_ν is one of the main uncertainties in radio source physics.

Since most radio components (especially jets) are rather elongated, it is usual to assume cylindrical symmetry to estimate the line-of-sight depth through the source. The average emission coefficient of a slice of a cylinder of width W with projected length D along the axis is $\langle j_\nu \rangle = P_\nu/V = 4P_\nu/(\pi D W^2 \csc i)$, where i is now the angle of the axis to the line-of-sight. Note that the derived value for j_ν is proportional to H_0.

Many radio components show clear signs of limb-brightening or of being immersed in a fainter halo; in these cases the assumption of constant emissivity is rather poor. One can do better by trying to determine the run of j_ν with radius, given I_ν as a function of projected distance (y) from the symmetry axis:

$$\langle j_\nu(r)\rangle = -\frac{1}{\pi}\int_r^\infty \frac{\partial I_\nu/\partial y}{\sqrt{y^2 - r^2}}dy. \tag{3.8}$$

This inversion (the Abel integral) is rather unstable in practice, but one can use some regularization technique (*e.g.*, maximum entropy) to stabilize the solution.

Cylindrical symmetry implies that the source should appear mirror symmetric around the supposed axis, and departures from this ideal set a limit to the applicability of the model. In general straight jets (*e.g.*, Perley, Bridle & Willis 1984; Killeen, Bicknell & Ekers 1986) are symmetric to within $\sim 30\%$ except near bright knots, while curved jets (O'Dea & Owen 1986) are markedly asymmetric. Radio bridges (Leahy & Williams 1984) vary a great deal in cross-sectional profile, but are seldom axisymmetric to within 20%.

Even for symmetrical sources, the brightness distribution does not distinguish circular from elliptical cross-sections, and indeed Smith (1984) has suggested that radio jets are actually ribbon-like. The distribution of projected opening angles constrains the average ratio of minor to major axis b/a: for FR I jets the absence of very small opening angles (Parma *et al.* 1987) implies $\langle b/a\rangle \gtrsim 0.3$. If b/a varies along the component, the area is not proportional to the square of the projected width.

How will these uncertainties affect typical values of j_ν? For a randomly oriented sample, $\langle g(\theta,\alpha)\rangle = 1$ in the absence of Doppler beaming, and $\langle \csc i\rangle = \pi/2$. On average, Doppler effects will reduce the apparent emissivity, as will an elliptical cross-section $b/a < 1$ (*i.e.*, usually the line of sight will be shorter than the projected width). Of course strong radiation anisotropy may result in an orientation-biassed sample.

The most serious problem is the possibility of small-scale variations in emissivity. As a gesture towards this, the emitting regions are assumed to occupy only a fraction Φ (the *filling factor*) of the source, thus $\langle j_\nu\rangle = \Phi j_\nu$. Since j_ν is very roughly proportional to u^2, $\Phi \approx \langle u\rangle^2/\langle u^2\rangle$. High-quality images of radio sources often show fine-scale substructure (Fig. 3.1) which is only limited by the reso-

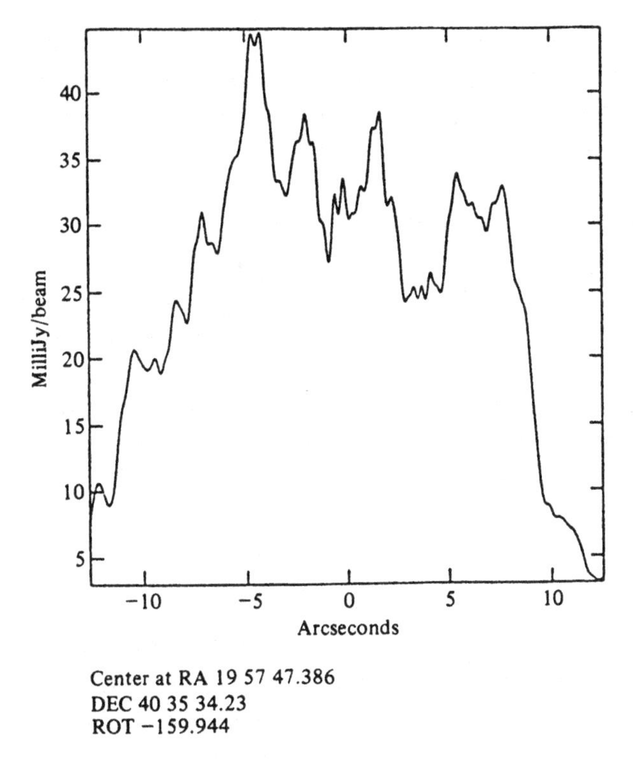

Fig. 3.1. Intensity profile across the East lobe of Cygnus A (Carilli 1989). Reproduced by permission of Chris Carilli.

lution. Perley, Dreher & Cowan (1984) estimate that in the lobes of Cyg A, $\Phi \lesssim 0.1$. Unfortunately there seem to be no observational lower limits to Φ: even if the emission appears to be completely resolved there may be structure on a much finer scale. For instance in a typical galaxy stars have $\Phi \sim 10^{-25}$, and yet the light seems quite smooth at arcsecond resolution. Furthermore, if the emission occurs in sheets, the fraction of lines-of-sight intersecting emitting regions (the *covering factor*) can remain near unity even when the filling factor is small.

From a theoretical point of view, even the inhomogeneity that we observe directly is hard to explain (*e.g.*, Hines, Eilek & Owen 1989). In gas dynamics a random field of pressure fluctuations could have amplitude of order the mean pressure at most, which only gives $\Phi \sim 0.5$. Smaller filling factors require a confinement mechanism, for instance magnetic flux tubes (*e.g.*, Parker 1979) or a radio-quiet fluid mingled with the synchrotron plasma. A variant of the latter is the case of a variable ratio of particle to field energy density, in which case (other things being equal) the emission will be dominated by the regions where the ratio is closest to unity (cf. §3.2.8).

3.2.5 Age (τ) and expansion speed

An obvious lower limit for the age of a radio source is the light-travel time from the nucleus to the projected distance of the outermost component. If the source was really expanding relativistically, it should look very asymmetrical, with any component on the other side of the nucleus both much fainter than the bright component and much closer in. The general symmetry of the large-scale structure in radio sources can therefore be used to argue that the overall expansion speed cannot be relativistic, and hence that the lower limit to the age can be written as:

$$\tau \geq \beta_{\max} cD/2. \tag{3.9}$$

Here $\beta_{\max}$ is the upper limit to β derived from symmetry arguments, and is typically of order 0.2 for extended radio sources, implying ages larger than a few million years. The limits derived are inversely proportional to H_0 through the dependence on D. Obviously this is only an upper limit on the *mean* expansion speed for a class of sources – it is always possible that any given source may be expanding faster.

Fig. 3.2 shows the simple geometry considered. Owing to the light-travel time across the source, the nearer component is seen at a later time and if both components travel outwards at the same speed, the separation ratio $Q = (1+\beta\cos i)/(1-\beta\cos i)$. From the ratio of the forwards and backwards Doppler beaming factors, the flux density ratio $M = [(1+\beta\cos i)/(1-\beta\cos i)]^{(3+\alpha)}$. As before, the relation for steady jets differs in that the exponent should be $(2+\alpha)$. If a sample is randomly distributed in orientation (§4.2.2 discusses the selection of such a sample) then $p(Q)dQ = p(M)dM = p(i)di = (1/4\pi)\sin i di$. The probability distributions for Q and M can be derived by change of variable and compared with the predictions of an assumed velocity distribution.

Both the brightness asymmetry and the separation asymmetry distributions have been used to constrain the expansion speeds of FR II sources. The two methods actually refer to different speeds:

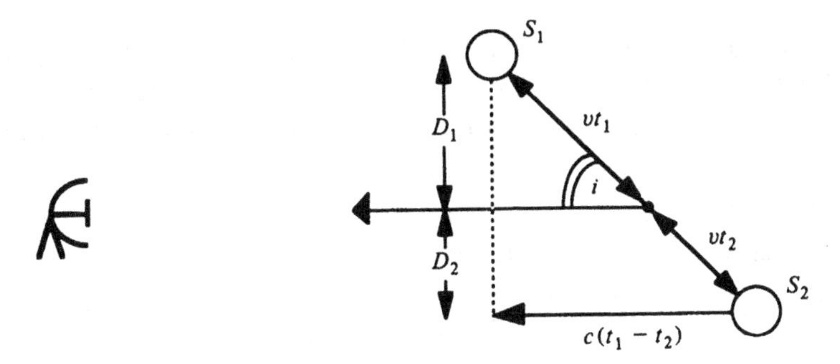

Fig. 3.2. Geometry for a simple relativistically-expanding double source. Both components move away from the centre at speed v. The components have fluxes S_1, S_2, with $M = S_1/S_2$, and projected separations from the core D_1, D_2, with $Q = D_1/D_2$. After Longair & Riley (1979).

the brightness asymmetry constrains the flow speed of the emitting material, while separation ratios constrain the *pattern speed*, *i.e.*, the speed at which the apparent structure propagates. In modern theories these two may be quite different (§§3.5.6 & 4.3.3). Mackay (1973) used the brightness asymmetry of the lobes in FR IIs to show than the β of the bulk of the emitting material is less than 0.08 and hence the Doppler corrections to the source luminosity are unimportant, as noted above. The method can be applied to the brightness ratios in jets (*e.g.*, in FR I sources) to put a limit on the jet speed. A large sample of twin-jet sources has only recently become available (Parma *et al.* 1987): a careful analysis has not yet been made but the general impression of symmetry (except in the central kiloparsec or so) suggests $\beta_{\rm max} \sim 0.3$. O'Dea (1985) finds $\beta_{\rm max} \sim 0.2$ for a sample of twin-tail sources in clusters.

From the point of view of setting a limit to the source age, the pattern speed (which is usually expected to be smaller than the flow speed in the jets but larger than the flow speed in the bridges) provides a stronger constraint. The separation ratio (Q) distribution of FR II sources has a long tail to high asymmetries, in contrast to the distribution expected from a constant β, which has a sharp cutoff (Longair & Riley 1979). A global fit gave $\beta_{\rm max} \sim 0.3$ (Macklin 1981) but this limit is too strong for the most asymmetric sources and too weak for the rest. Banhatti (1980) attempted to "deconvolve" the observed Q distribution to give the velocity distribution, and found a function which peaked at $\beta \sim 0.15$ with an inter-quartile range of 0.1 – 0.4. Saikia (1984) discussed Q in FR II quasars and found that there was no tendency for the jet side to be the longer, which is expected if the jet asymmetry is due to Doppler beaming, and hence most of the asymmetry in component separation must be due to other effects. He argues that selection effects acting on a population of distorted sources might cause this result, in which case strong limits cannot be derived. McCarthy (1989) finds that 3C sources become more asymmetric at $z \gtrsim 0.8$, implying the possibility of higher expansion speeds in the most distant and powerful sources. However he also finds that there is almost always more line-emitting gas on the shorter side, which strongly suggests that the observed asymmetries are mostly due to the external medium and hence that expansion speeds are well below the upper limits derived here.

There are two other routes often used to constrain the age of sources, both rather model dependent. The first, spectral ageing, is discussed next. The second, ram pressure balance in FR II sources, is considered in §3.5.6 and §7.2.3.

3.2.6 Spectral ageing

As mentioned in §2.2.1, the fainter regions of radio sources characteristically have steeper high-frequency spectra, and this is conventionally interpreted in terms of synchrotron/Compton radiation loss combined with expansion as material flows away from sites of active particle acceleration.

The effects of radiative losses on the particle energy spectrum are discussed in §9.2.4.1 (see also Pacholczyk 1970). Adiabatic expansion alters the critical frequency of all particles by a constant factor, and so changes the frequency and amplitude scaling of the spectrum but not its functional form. This form depends on the amount of tangling in the magnetic field, the effectiveness of pitch-angle scattering, and on the way particles are injected. In general the particle spectrum has a characteristic energy, γ_T, which, for the simple case of a tangled field, effective scattering, and single-burst injection (the 'JP' model, Jaffe & Perola 1973), corresponds to the energy of an electron which started with $\gamma = \infty$. The radio spectrum is characterised by the so-called *break frequency*, ν_T, which is the critical frequency corresponding to γ_T.

The break frequency depends on the initial magnetic field and on the subsequent expansion history; if the field is constant in time (implying no expansion or compression) then:

$$\nu_T = c_{\gamma 1} B \gamma_T^2 = \frac{(9/4)c_7 B}{(B^2 + B_{\rm MB}^2)^2 t^2}, \tag{3.10a}$$

where

$$c_7 = 27\pi m_e^5 c^2/\mu_0^2 e^7 = 1.12 \times 10^3 \text{ nT}^3\,\text{Myr}^2\,\text{GHz}, \tag{3.10b}$$

$$B_{\rm MB} = \sqrt{2\mu_0 u_{\rm MB}} = 0.318(1+z)^2 \text{ nT}. \tag{3.10c}$$

$B_{\rm MB}$ is the equivalent magnetic field to the microwave background radiation. Kardashev (1962) and Alexander (1987) consider the more general case of radiation in a smooth expansion. The break frequency roughly marks the point at which the spectrum steepens from the low-frequency power law to the high-frequency regime dominated by radiation loss. However, the transition is very gradual and substantial steepening occurs well below ν_T; thus to determine ν_T it is necessary to fit model spectra to the data; the resulting values are therefore somewhat model-dependent.

Fig. 3.3 shows the radio spectra of some limiting cases, and Fig. 3.4 gives the negative of the gradient of the spectrum, *i.e.*, the local spectral index. Usually, tangled-field models are used, since the observed polarization is generally well below the theoretical maximum; however steep-spectrum regions are often very highly polarized and should really be fitted with uniform-field models. In particular, the characteristic steepening from α_L to $\alpha_H = (4/3)\alpha_L + 1$ in the widely used Kardashev-Pacholczyk (KP) model depends on the tangled-field assumption and could not be observed in such regions. In practice, details of field tangling and pitch-angle scattering have little effect on the derived break frequencies, as long as ν_T is defined consistently between models (*e.g.*, by using equation (3.10a) for all models, as here).

The biggest differences are between the 'single burst' models which assume that a power law is injected at a particular time (*e.g.*, the moment at which a fluid element passed through a strong shock, say in the hotspot), and the 'continuous injection' (CI) models which assume that there is a constant supply of electrons with a power law spectrum. The latter will also apply to the net spectrum of a region containing both the acceleration site and a downstream

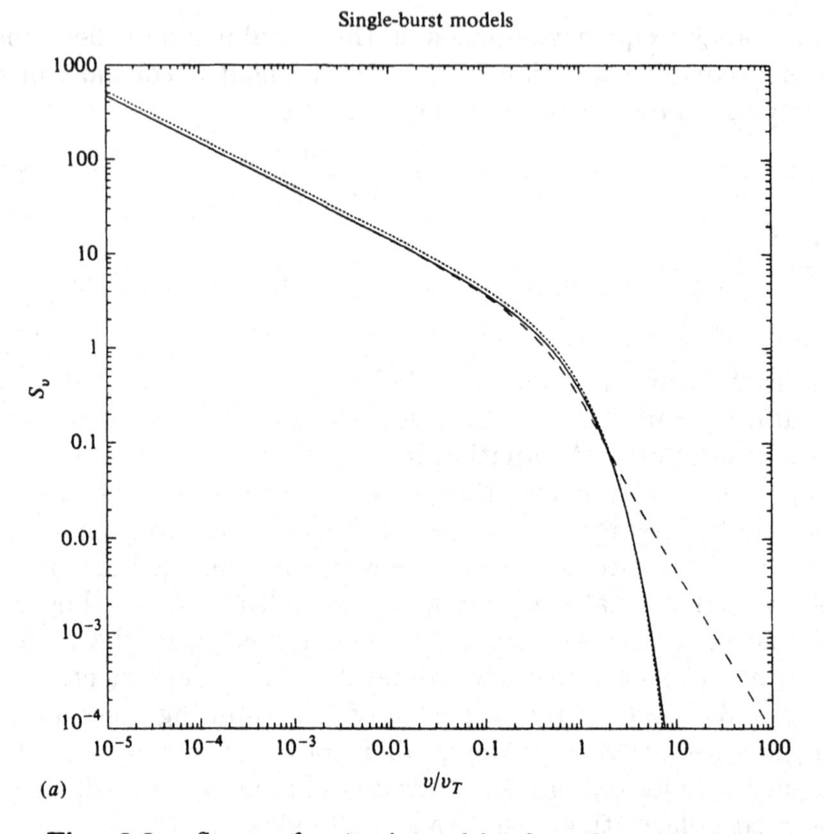

Fig. 3.3a. Spectra for simple models of synchrotron ageing. Solid curve: tangled field, pitch-angle scattering (JP model); dashed curve: tangled field, no scattering (KP model); dotted curve: uniform field at $\psi = 60°$, cutoff energy calculated assuming pitch-angle scattering. All models assume the same emitting geometry, magnetic field strength, and initial number density. For single burst injection.

area in which radiation loss occurs in a constant magnetic field; thus it should and does give a good fit to the integrated spectra of hotspots (Stephens 1987; Meisenheimer *et al.* 1989). On the other hand CI models steepen by only 0.5 in α between the low- and high-frequency regime, and the spectra of diffuse regions (bridges, tails, and plumes) often steepens by more than this, indicating that any reacceleration is not keeping up with radiation loss. Thus single-burst models are usually fitted to the spectra in these regions. The majority of spectral ageing work focusses on such regions because

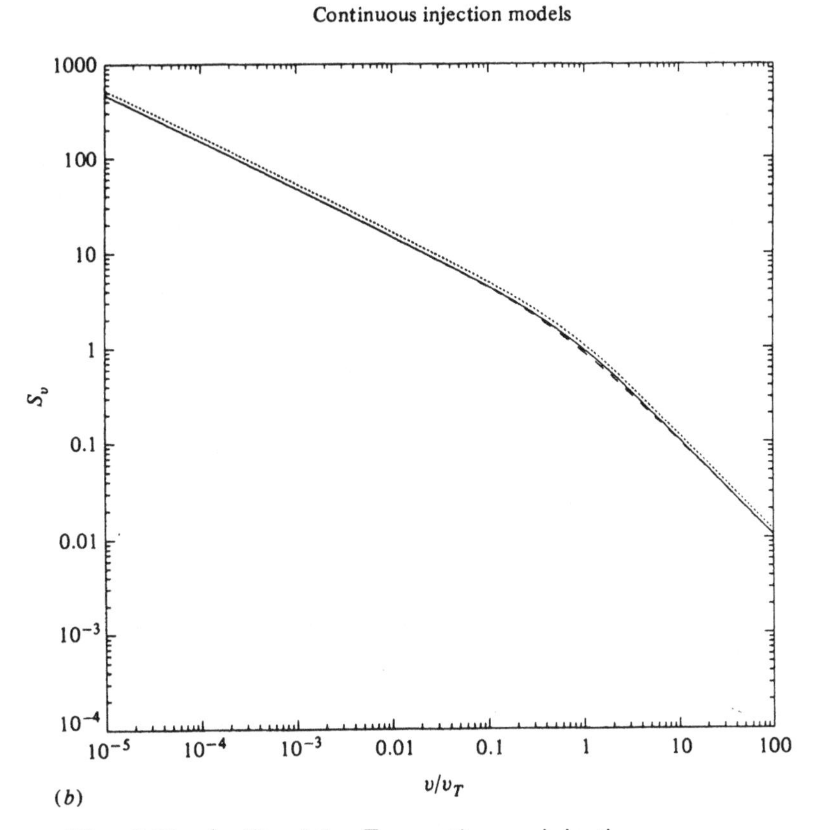

Fig. 3.3b. As Fig. 3.3a. For continuous injection.

they can be resolved at several frequencies, and also because they show the most dramatic spectral gradients.

In most cases the variation in high-frequency spectral index across a source is mainly due to variation in ν_T, while α_L is nearly constant (*e.g.*, Burch 1977; Winter *et al.* 1980; Alexander 1987) and therefore equal to the low-frequency spectral index of the integrated emission. If this result is assumed, then with the assumption of either the JP or KP model, images at two frequencies suffice to determine ν_T at each point.

Occasionally there is evidence that α_L is *not* constant across the source (*e.g.*, Stephens 1987). Also, in at least two cases (3C 20, Stephens 1987; Cyg A, Carilli *et al.* 1990) the hotspot spectra are flatter than the low-frequency spectral index of the lobes, which

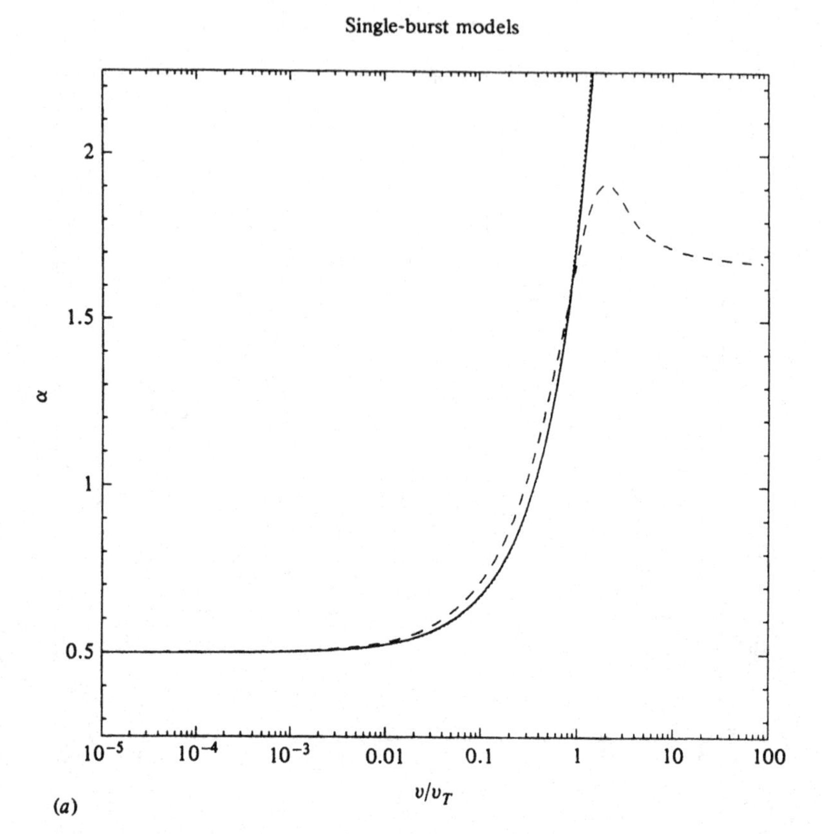

Fig. 3.4a. Spectral index as a function of frequency for the models of Fig. 3.3a.

clearly invalidates the simple model of ageing of a pure power law spectrum set up in the hotspots.

Such departures from the simple models should not surprise us: the variation in spectral index between sources suggests that the acceleration mechanism can produce a variety of α's, and there is no reason why it should not do so over a period of time within one source. Furthermore, expansion out of the hotspots will lower the CI break frequency (Table 3.3) so we might expect the low-frequency spectral index of the diffuse regions to lie between the low- and high-frequency spectral index of the hotspots. Such a curved injection spectrum should properly be taken into account when fitting spectra, but no spectral-age analysis has reached this level of sophistication. A more radical explanation for variation in low-frequency spectral index is particle acceleration in the diffuse

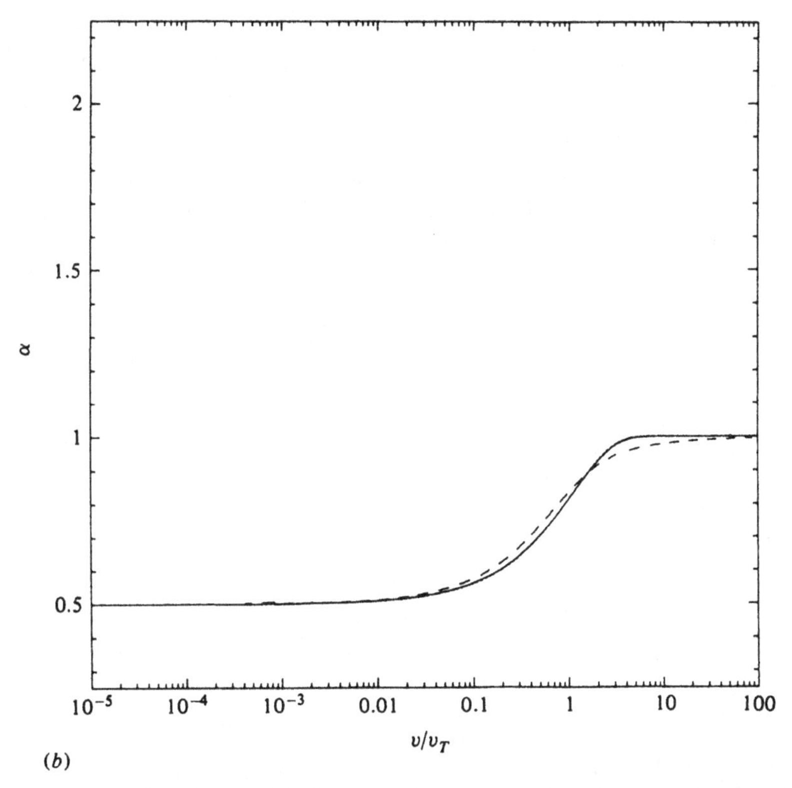

Fig. 3.4b. Spectral index as a function of frequency for the models of Fig. 3.3b.

lobes, which would largely invalidate the use of break frequencies to determine ages.

To derive ages we need a value for the magnetic field. Usually the equipartition field is used, but there is little reason to believe that this is close to the true field (§3.2.8). Even without a value for B, we can get a map of the relative ages across the source provided that the field is roughly constant. Fortunately in some FR II bridges $\langle j_\nu \rangle$ appears to be roughly constant at low frequencies, even when there is considerable variation in ν_T. The simplest interpretation is that the internal properties, in particular the magnetic field, are also roughly constant. The material in the bridges is presumed to have been through an initial expansion as it flowed out of the hotspots. Given that hotspot spectra show CI type ageing, the post-hotspot expansion naturally explains why the derived ages often do not tend to zero at the hotspots (*e.g.*, Alexander & Leahy 1987).

More detailed accounting for variation in the field is difficult and model-dependent. Generally, spatial variation in the source implies temporal variation in a fluid element as it flows through the source; therefore one cannot derive ages simply by combining ν_T with the assumed B at each point. Alexander (1987) shows that if the age is calculated using the *initial* magnetic field, the results are rather insensitive to subsequent steady expansion. This is because the errors incurred by ignoring the decrease in the radiation rate (caused by the decrease in B) nearly cancel those due to ignoring expansion losses.

For very low surface-brightness sources (*e.g.*, giant radio galaxies) the magnetic fields derived by most methods are so low that inverse-Compton losses should dominate, and then ages depend only on $\sqrt{B}$, *i.e.*, on the one-seventh power of most of the unknown parameters (§3.2.8). The transition from synchrotron to inverse-Compton loss implies a maximum spectral age at $u_B = u_{\rm MB}/3$. This is a true maximum provided that the plasma has not been compressed after the initial injection, that the particles have not diffused into a region of higher field strength (cf. Scheuer 1989), and, as for all spectral age determinations, that there has been no reacceleration. These assumptions are consistent with the usual dynamical models, in which the plasma tends to expand more-or-less monotonically after acceleration at regions of maximum compression (*i.e.*, hotspots, jet knots). Maximum ages provide an interesting limit at high z because of the rapid increase of $u_{\rm MB}$ with redshift. Leahy, Muxlow & Stephens (1989) find $\tau_{sa} \lesssim 10^7$ yr for a small sample of powerful FR II sources at $z \sim 1$.

One can also turn spectral ageing arguments around to derive a maximum magnetic field given the maximum expansion speeds for FR II sources ($\sim 0.2c$) derived in §3.2.5. Since the expansion speeds derived from spectral ageing and equipartition fields are of order $0.01 - 0.2c$ (*e.g.*, Alexander & Leahy 1987), the field strengths in the lobes must be less than about ten times the derived equipartition fields on average, unless there is significant re-acceleration.

3.2.7 The energy in relativistic electrons ($U_{\rm re}$)

As noted in §1.2.1.2, synchrotron emission at frequency ν is dominated by electrons with a critical frequency ν_c near ν. This approximately one-to-one relation means that the particle spectrum (as a function of B) can in principle be derived from the radio spectrum:

$$j_\nu d\nu \propto \gamma^2 B^2 n(\gamma) d\gamma. \tag{3.11}$$

This equation, like all those in this subsection, applies in the emitting frame. We can replace $n(\gamma)$ with the distribution as a function of ν_c:

$$n(\nu_c) = \frac{1}{c_3 c_8(\delta) c_9(\delta)} \frac{\tilde{\jmath}_{\nu_c}}{\nu_c B}. \tag{3.12}$$

The constant of proportionality is derived from Equation (6.27) of Pacholczyk (1970). Here $\tilde{\jmath}_\nu$ is the angle-averaged emission coefficient; it equals the observed emission coefficient if the field is fully-tangled. c_8 and c_9 are dimensionless functions which result from integrating respectively over the finite width of $F(x)$ and over angle. Strictly equation (3.12) applies only to a power law but it is fairly accurate if $n(\gamma)$ can be approximated by a power law over an octave or so. If $n(\gamma)$ contains strong features which are sharper than $F(x)$ then the radio spectrum will reflect the latter. In such cases equation (3.12) breaks down: one has to estimate $n(\nu_c)$ by fitting a model to the observed spectrum (as is done in spectral ageing). Usually spectra are very smooth and equation (3.12) is acceptable.

We can now integrate the spectrum to get the total energy density of the relativistic electrons, again as functions of B:

$$u_{\rm re} = m_e c^2 \int \gamma n(\nu_c) d\nu_c = c_3^{-1} c_{E1}^{-1/2} B^{-3/2} \int \frac{\tilde{\jmath}_\nu}{c_8(\delta) c_9(\delta) \nu^{1/2}} d\nu. \tag{3.13}$$

If the magnetic field is reasonably constant in the emitting regions, we can integrate both sides over the source, to give the total relativistic electron energy. Observed spectra are moderately well described by power laws, $L_\nu \propto \nu^{-\alpha}$; it is therefore convenient to normalise the integral at some reference frequency ν_0, giving

$$U_{\rm re} = c_3^{-1} c_{E1}^{-1/2} L_{\nu_0} \nu_0^\alpha B^{-3/2} \int (c_8 c_9)^{-1} \left(\frac{L_\nu}{L_{\nu_0} \nu_0^\alpha} \right) \nu^{-1/2} d\nu. \tag{3.14}$$

I will denote the integral on the right $\mathcal{I}_1$. An important feature of this equation is that it does not depend on the geometry of the emitting region, *i.e.*, it is independent of filling factor, line-of-sight depth, etc.

We can only observe over a limited frequency range, so formally we only have a lower limit to $\mathcal{I}_1$. In a few hotspots $\gamma n(\gamma)$ peaks in the observable window (Leahy *et al.* 1989), so reasonable estimates of $\mathcal{I}_1$ are possible. The best data is for the hotspots of Cyg A where the maximum is near $\nu_c \sim 200$ MHz. For a conventional magnetic field strength this implies $\gamma \sim 750$ (Carilli *et al.* 1990). If $\mathcal{I}_1$ for the hotspots can be determined, expansion relations (Table 3.3) can be used to estimate it in the post-hotspot flow. While far from foolproof, this approach seems more reasonable than the usual assumptions discussed below.

In most cases (including hotspots) $\gamma n(\gamma)$ declines faster than γ^{-1} (*i.e.*, $\alpha > 0.5$) throughout the observed spectrum, so extrapolation to zero energy leads to a divergence. Thus the value of $\mathcal{I}_1$ is critically dependent on its unobserved low-frequency limit. It is usual to parameterize the uncertainty in $\mathcal{I}_1$ in terms of an equivalent power law between cutoffs ν_L and ν_U:

$$\mathcal{I}_1 = (c_8 c_9)^{-1} \left(\frac{\nu_L^{\frac{1}{2}-\alpha} - \nu_U^{\frac{1}{2}-\alpha}}{\alpha - \frac{1}{2}} \right). \tag{3.15}$$

In principle we could set $\nu_U = \nu_T$ from §3.2.6, but the results are negligibly different if we use the conventional value of $\nu_U =$ 100 GHz. The brighter radio sources have been detected at 10 MHz (Bridle & Purton 1968), so it is customary to assume $\nu_L = 10$ MHz when a minimum value for $\mathcal{I}_1$ is required. These values are assumed to apply in the galaxy frame, to avoid introducing an obvious bias with redshift. Usually the dependence of $c_8 c_9$ on spectral index is neglected: instead the value for $\alpha = 1$ is used*, which is $16\pi/(27\sqrt{3}) = 1.075$. This approximation is accurate to about 30%.

Strictly, 10 MHz is a valid limit for ν_L only for the component brightest at low frequencies, usually the bridge in FR IIs and the plumes in FR Is. Hotspots have been detected at 81 MHz (Duffet-

* This is equivalent to equating the luminosity in a power law radio spectrum between ν_L and ν_U to the total radio power emitted by electrons with critical frequencies in this range.

Smith & Purvis 1982) and jets at 151 MHz (Junor 1987). Using these values for ν_L would reduce $\mathcal{I}_1$ by a factor of up to two.

This caveat apart, the conventional constant ν_L is reasonable for *minimum* energy, but implies that the minimum particle energy is *higher* in the faint regions, since the magnetic field there is presumably lower. In contrast, expansion arguments suggest that particle energies and critical frequencies should decrease with j_ν, and that ν_L in bridges is far below 10 MHz if hotspots are seen at 81 MHz.

To avoid this embarrassment, Myers & Spangler (1985, hereafter MS) suggest that instead of constant cutoff frequencies we assume constant cutoff energies. For a power law $n(\gamma)$ with n_0 given by equation (1.6) we have:

$$U_{\text{re}} = c_3^{-1} c_{E1}^{-\alpha} L_\nu \nu^\alpha B^{-(1+\alpha)} \int_{\gamma_L}^{\gamma_U} 2(c_8 c_9)^{-1} \gamma^{1-\delta} d\gamma. \tag{3.16}$$

The integral here will be labelled $\mathcal{I}_2$. This will also be changed by expansion, but very slowly (Table 3.3).

We do not understand particle acceleration well enough for theory to give a reliable γ_L (cf. Chapter 9), but two physically significant values of γ are 1 and m_p/m_e. Shock acceleration only works for $\gamma_{\text{electron}} \gtrsim \gamma_{\text{shock}} \beta_{\text{shock}} m_p/m_e$ if the fluid contains a significant energy density in protons (§9.3.5). For a sub-relativistic shock, this is suggestively close to the turnover γ derived for the Cyg A hotspots.

3.2.8 Minimum energy (U_{min}) and related quantities

Burbidge (1956) showed that there is a minimum value for the total energy in relativistic electrons and magnetic fields emitting a given synchrotron luminosity; this value corresponds roughly to equipartition of energy between particles and fields. Ever since, variants of his argument have been used to estimate many of the internal properties of synchrotron plasma. The widespread use of minimum energies means that the underlying assumptions need to be examined carefully: in fact they are very shaky.

As noted in §3.1.2, synchrotron plasma may contain several particle species other than the relativistic electrons/positrons responsible for the radiation. It is conventional to write the total energy in particles as $(1+k)u_{\text{re}}$ to allow for these components. We can find the value of B which minimises $u_{\text{tot}} = (1+k)u_{\text{re}} + u_B$ (note that this does *not* give a minimum magnetic field), and the corresponding minimum energy density u_{min}, pressure ($u_{\text{min}}/3$ for a tangled

Table 3.5. *Comparison of different energy estimates.*

Quantity	Minimum energy	Equipartition	Minimum pressure
$G(\alpha)$	$1+\alpha$	2	$\frac{(1+\alpha)}{3}$
$(1+k)u_{\rm re}/u_B$	$\frac{2}{(1+\alpha)}$	1	$\frac{6}{(1+\alpha)}$
$u_{\rm tot}/u_B$	$\frac{(3+\alpha)}{(1+\alpha)}$	2	$\frac{(7+\alpha)}{(1+\alpha)}$
$u_{\rm tot}/u_{\rm min}$ ($\alpha = 0.5$)	1	1.01	1.14

field), and total energy $U_{\rm min}$ from integrating $u_{\rm min}$ over the relevant volume.

Several other equivalent sets of values are in use. First one can simply assume equipartition of energy, $(1+k)u_{\rm re} = u_B$. Second, Owen and collaborators (*e.g.*, O'Dea & Owen 1987) have preferred to use "minimum pressure", where they define "pressure" as the sum of particle pressure ($(1+k)u_{\rm re}/3$) and the isotropic component of the magnetic stress (u_B). u_B is the effective pressure perpendicular to an organized field, but if the field is tangled the effective pressure is $u_B/3$, so minimum pressure is the same as minimum energy. For field geometries where the magnetic stress is zero or negative (see Table 3.2) the concept of "minimum pressure" is not useful.

All these variants can be summarized in the formulae:

$$B = \left(G(0.5)C_{me}(1+k)\tilde{\jmath}_{\nu_0}\nu^{\alpha}\mathcal{I}_1\right)^{2/7} \tag{3.17a}$$

$$B = \left(G(\alpha)C_{me}c_{E1}^{\frac{1}{2}-\alpha}(1+k)\tilde{\jmath}_{\nu_0}\nu^{\alpha}\mathcal{I}_2\right)^{\frac{1}{3+\alpha}} \tag{3.17b}$$

$$C_{me} = \mu_0/(c_3\sqrt{c_{E1}}) = c(4\pi m_e)^{5/2}/(3e^{7/2}) = 26.91 \text{ (SI units)}, \tag{3.17c}$$

for the conventional and Myers & Spangler versions of $u_{\rm re}$ respectively. The factor $G(\alpha)$ depends on whether B is derived using minimum energy, equipartition, or minimum pressure. The derived value depends weakly on H_0 through j_ν. Note that the MS formula reduces to the conventional one for $\alpha = 0.5$ (in which case $\mathcal{I}_1$ and $\mathcal{I}_2$ both reduce to $2\ln[\gamma_U/\gamma_L]$). The values for G and related quantities in our three cases are listed in Table 3.5. All variants imply that $\beta_P \sim 1$.

For minimum energy density the choice of the unknown parameters is fairly simple: $k = 0$, $\Phi = 1$, $\mathcal{I}_1$ evaluated only over the observed spectrum. As noted in §3.2.4 it is usually reasonable to set orientation-dependent factors to unity as well. The result is a minimum in the sense that most of the emission will come from a region with higher energy density. On the other hand the equivalent total energy depends on the unknown emitting volume $V_{\mathrm{em}} = \Phi V_{\mathrm{tot}}$, as $V_{\mathrm{em}}^{3/7}$. Since there are no good lower limits to Φ, the total energy could be much less than the conventional 'minimum' value, which assumes $\Phi = 1$. For many purposes, the dynamically interesting energy density is the average over the whole component, $\langle u \rangle = U_{\mathrm{tot}}/V_{\mathrm{tot}}$, which will also be reduced by a low Φ. Fortunately, we can get a hard lower limit to U_{re} and hence U_{tot} by using equation (3.14) together with an upper limit to B, *e.g.*, from inverse-Compton limits (§1.2.2.3). We have excluded from the minimum total energy the contribution of the non-emitting regions, since this could be small if, for instance, the components consist of magnetically self-confined filaments.

The use of u_{min} and U_{min} as limits requires a few caveats, but the situation is far worse if we want to use equation (3.17b) and its relatives to estimate the actual values of radio source parameters.

We have already discussed the large uncertainties in $\tilde{\jmath}_\nu$ due to unknown anisotropy, beaming, geometry and filling factor (§§3.2.3 and 3.2.4), and in $\mathcal{I}_2$ due to the fact that the integral does not converge in the observable frequency range (§3.2.7). In addition we have the unknown contribution of other pressure components, $(1 + k)$. Some particle acceleration mechanisms can produce 400 times more energy in protons than in electrons (see Chapter 9). Thermal material may also be present from the start and will almost certainly be entrained at some point in the flow (*e.g.*, Killeen, Bicknell & Ekers 1988). Entrainment will cause both Φ and k to vary substantially within the source. Finally the implicit assumption of equipartition in the form $\beta_P \sim 1$ has little empirical support: in well-studied Solar System plasmas a very wide range of β_P is found (cf. Chapter 9).

Despite these uncertainties many papers assume that minimum energy values are reasonable estimates of the true ones. Three arguments are commonly advanced to support this. Firstly, equipartition of energy is expected on general principles. As we have noted, however, in practice it is seldom found, at least to an accuracy of better than an order of magnitude. Furthermore, the uncertainty in $U_{\rm tot}$ is very large *even if equipartition holds exactly.* The main problems are Φ, k, and the low-energy cutoff in $\mathcal{I}_2$, each of which could change $U_{\rm tot}$ by an order of magnitude or more; but even the 'minor' uncertainties cumulatively amount to a factor of three or so. Secondly, there is a reluctance to assign more power than necessary to the active nucleus. This does not prevent $U_{\rm min}$ from being a systematic overestimate, but in any case the wide range in source luminosity means that there could be a fundamental problem only for the most powerful objects. Even there, current models for central engines can produce far more power than required (Chapter 8). Thirdly, minimum energy estimates are said to be consistent with other methods. This has little content since much effort has been devoted to explaining apparent discrepancies (*e.g.*, the 'confinement problem', §5.2; and its inverse, Killeen *et al.* 1988) even though 'other methods' generally provide rather loose constraints.

Perhaps the best attitude is that minimum energy estimates depend on a specific model for the emitting region, rather than being direct consequences of the observations. They should therefore be given no more weight than numbers derived from alternative sets of 'plausible assumptions', some of which are discussed next.

3.2.9 Alternatives to equipartition estimates

Magnetic fields: We have already encountered three methods giving values or limits: through synchrotron self-absorption, inverse-Compton emission (or rather, the lack of it) (§1.2.2.3), and from comparison of spectral ages with other estimates of the source age (§3.2.6). The first two limits are virtually model-independent.

Pressure: A straightforward approach is to assume that the source is in pressure balance with the external medium. This is strictly true if radio sources are bounded by fluid contact discontinuities, since pressure balance is a defining feature. If the source is inhomogeneous the mean pressure $\langle P_{\rm t}\rangle$ is the relevant quantity for comparison (cf. §3.2.8). Most models of FR II sources require a shock in the external medium, so that the pressure immediately surrounding the source is higher than in the undisturbed medium. The shocked region is expected to be small compared to the total size of the galaxy or cluster halo, so X-ray measurements of the latter give lower limits to the pressure within the source. So far a comparison has only been made for one source, Cygnus A (Arnaud *et al.* 1984). Their results show that in the lobes $\langle P \rangle > 3 \times 10^{-11} h^{1/2}\mathrm{Pa} \approx u_{\mathrm{eq}}$. In the future, observations in hard X-rays may detect the shocked region and hence allow direct measurements of pressure.

In FR I sources the models suggest that the central regions are near steady state and the overall expansion of the source is subsonic. In these cases we can be fairly certain that the jets are in contact with a more-or-less undisturbed external medium, in which case pressure-balance should be applicable.* Here u_{eq} seems to be less than or equal to the external pressure, as derived from X-rays (*e.g.*, Morganti *et al.* 1988) or emission-lines (*e.g.*, van Breugel *et al.* 1985). The only case where u_{eq} is much higher than the external pressure is in the jet in M 87; and here the filling factor is clearly rather low: Owen, Hardee & Cornwell (1989) argue that most of the emission is from the jet surface. The M 87 jet differs in many ways from those in typical FR Is and may be supersonic (Falle & Wilson 1985) rather than sub- or trans-sonic (cf. §3.5).

We can also derive an upper limit to $\langle P \rangle$ in FR II sources via ram-pressure balance (see §3.4.4), and the maximum expansion speeds (§3.2.5). The average pressure at the front of the lobe must be less than

$$\langle P \rangle_{\mathrm{max}} = \rho_a (\beta_{\mathrm{max}} c)^2, \tag{3.18}$$

where ρ_a is the density of the ambient medium. For Cyg A, $n_{e,a} = 8 \times 10^3 h^{\frac{1}{2}} \mathrm{m}^{-3}$ at the projected distance of the heads, giving $P \lesssim 5 \times 10^{-8}\mathrm{Pa}$. The method has the disadvantage that the limit derived is a time-average over the source's life-time; however simple models suggest that the pressure should decrease with time (since the source is expanding down a pressure gradient), so $\langle P \rangle_{\mathrm{max}}$ remains a model-dependent upper limit. Sure enough, values derived are larger than u_{eq} (the hotspot pressure should be regarded as a 'fluctuation' in this context, cf. §7.2.3).

A similar argument can be used for hotspots themselves, except that the hotspots are probably advancing faster than the overall

* Substantial pressure variations can occur at internal shocks (cf. §3.5.2) but on average the pressure within the jet is equal to the external pressure. Pressure-confinement arguments will also fail if magnetic self-confinement is important.

expansion speed (§7.2.3). Limits to the hotspot advance speed can be obtained from their M distribution (cf. §3.2.5). In fact hotspots are often rather asymmetric in brightness, but we can conclude from the fact that two hotspots are usually seen that the advance speed is not highly relativistic, so we can put $v_{\rm max} \sim c$ but ignore relativistic increase in ram pressure as $v \to c$, giving $P_{\rm HS} < \rho_a c^2$. For Cyg A we get $P_{\rm HS} \lesssim 10^{-6}$Pa. This limit could be reduced by an upper limit to the proper motion of the hotspots: in the next decade it should be possible to show that their motions are subluminal.

One common use of $u_{\rm eq}$ is to estimate pressure ratios within a source. However, the adiabatic relations (§3.3) are probably more reliable, except when relating the pre- and post-hotspot parameters. They can break down if there is significant entrainment into the flow or if further particle acceleration occurs. Unfortunately $u_{\rm eq}$ fares no better: entrainment changes Φ or k, and re-acceleration changes $\mathcal{I}_1$.

3.2.10 **Faraday rotation**

Both Faraday rotation and depolarization were introduced in §1.2.1.4. It was once common to use measurements of these effects, together with the minimum-energy magnetic field, to derive the density of thermal electrons ($n_{\rm th}$) within the radio source. It is now widely acknowledged that these densities are spurious, since much or all of the Faraday rotation takes place outside the source. This does not imply that the internal density is very low, since the densities derived from Faraday rotation are essentially lower limits (because the Faraday depth for a given $\langle n_{\rm th} B \rangle$ can be arbitrarily low if the field reverses many times on the line-of-sight through the source).

Modern analyses of Faraday rotation begin by attempting to distinguish rotation in front of the source (*foreground rotation*) from rotation within the radio-emitting regions (*internal rotation*). Internal rotation is characterised by depolarization and deviation from the λ^2 law before a rotation of $\pi/2$ is observed (*e.g.*, Laing 1984). Several cases are known where rotation angle goes accurately with λ^2 through much more than $\pi/2$ and in opposite directions in different parts of the source, so the bulk of the rotation is clearly external (*e.g.*, Perley *et al.* 1984; Dreher, Carilli & Perley 1987). While the interstellar medium in our own Galaxy can cause a large offset in the RM (and indeed dominates the average RM of most sources), small-scale fluctuations of detectable amplitude in the ISM only occur at low Galactic latitudes (Simonetti & Cordes 1986; Leahy 1987), so the foreground material in question is very likely the hot gas surrounding the radio source. Another possibility is a turbulent boundary layer where the relatively strong field in the radio plasma is amplified and mixed with the relatively high density of the external medium. Where an external medium is clearly identified, the Faraday rotation is either happening within 10 kpc of the centre of a large galaxy or in a dense cluster-type environment. Outside these regions, integrated polarization measurements imply that Faraday depths are much smaller (but not zero). In this rather different situation both internal and external rotation remain viable possibilities.

If depolarization and departures from the λ^2 law *are* observed, foreground rotation may still be the cause, since unresolved RM structure exactly mimics variation of Faraday depth through the source. However, we can set some limits on the plausible scale of structure in the X-ray medium. To achieve a given RM with fluctuations on scale size d when the column depth of the Faraday-active medium is D, requires

$$RM = C_F \langle n_{\rm th} B \rangle \sqrt{Dd/3}. \tag{3.19}$$

Most polarization mapping has been done at wavelengths of 22 cm or shorter, so large-angle rotation can only be observed if $RM \gtrsim 40$ rad m^{-2}. For typical parameters of an X-ray cluster (Table 3.1), such RMs require $d \gtrsim 1$ kpc, which is resolvable with the VLA in the nearest cases. Much smaller-scale fluctuations are possible if there is gas in the warm, high-density phase (10^4 K), but as a rule 'interesting' Faraday depths in warm gas are accompanied by detectable line emission (*e.g.*, Heckman *et al.* 1982), so this possibility can also be tested. As yet, there are no clear examples of depolarization at such high resolution except in the presence of line emission.

Leahy, Jägers & Pooley (1986) claimed detection of internal rotation on the basis of a correlation between depolarization and $|RM|$, but this could also arise in an external screen for some geometries. Much more compelling evidence is given by Jägers (1987), who finds a systematic tendency for depolarization to correlate with the line-of-sight depth in the plumes of some twin-jet sources: depolarization is concentrated along the axes of the plumes, with the edges being hardly depolarized at all. One reason why this effect has only recently been found is that the internal Faraday depths are so low in these cases that depolarization is only evident at a wavelength of 50 cm, much longer than used by most polarization studies. The Faraday depths involved are comparable to those through the ionosphere, which makes observations particularly difficult.

Given a measurement of the Faraday depth through either the source or the external medium, we can go on to constrain the fields in the external medium (knowing the density) or the density in the source (guessing the field). The latter is nowadays interpreted as giving information about entrainment processes (cf. the discussion of Bicknell's model in §3.5.4) rather than the original aim of determining the mass flux from the active nucleus. Perhaps more important, observations of coherent regions of Faraday rotation imply that the field is *not* very highly tangled, although the discovery of internal rotation in radio sources is too recent for this to have been used to try to constrain the scale of field reversals.

3.3 The environments of large-scale jets

The behaviour of jets is determined as much by the media through which they travel as by their internal properties, and so information on the environment is an essential ingredient in jet modelling. In this section I review the observational data on the 'hot' and 'warm' gas phases which interact directly with jets, and on the gravitational field. The latter affects jets largely through the quasi-hydrostatic equilibrium of the external medium, but perhaps also directly, as we shall see in §§3.5.4 and 3.5.5. The determination of the magnetic field in the external medium was discussed in §3.2.10.

3.3.1 The hot gas from X-rays

In principle X-ray observations allow one to determine the run of density, pressure and temperature in the hot medium filling and surrounding elliptical galaxies. The gas emits X-rays via bremsstrahlung, the spectrum giving the temperature and the emissivity being proportional to $n_{\rm th}^2$. The metallicity may be determined from the equivalent widths of recombination lines from highly-ionised species, notably various states of iron. Since the emission mechanism is isotropic and there is presumably no relativistic motion in the X-ray gas, the derivation of the X-ray luminosity is less ambiguous than in the radio: the accuracy is noise-limited except very close to the galactic plane when absorption by neutral hydrogen may be a slight problem.

A complication is that elliptical galaxies should contain a population of X-ray bright stars which may sometimes dominate the X-ray emissivity (Trinchieri & Fabbiano 1985). The large scatter of the X-ray to optical luminosity ratio suggests that, at least for galaxies overluminous in X-rays, hot gas is the main source, as is certainly the case in clusters of galaxies (as we know from the presence of line emission).

If the filling-factor of the hot gas were small the density in the emitting regions would be under-estimated and the total gaseous mass over-estimated. Fortunately there is some reason to believe that the gas is actually quite smooth. Observationally there is little evidence for small scale clumpiness ($<$ 10 kpc) in the well-resolved nearby clusters. Theoretically, following the discussion in §3.2.4, magnetic forces in the hot gas are probably too weak to confine it (§3.1.2), while only gas at $\gg 10^8$ K could provide pressure confinement without being detected. Such super-hot gas would be removed by buoyancy on the sound-crossing time, $\sim O(10^8$ yr), and since there is no obvious way to replenish it, it probably does not exist. (Of course jets produce super-hot gas, but not enough to fill the ICM.)

The detailed diagnostics promised in the first paragraph are not yet available. Current X-ray detectors with high angular resolution are broad-band and provide limited spectral resolution. Thus the observed count rate depends on temperature and density in a combination determined by the instrument's spectral response. Most available X-ray images were taken with the Imaging Proportional Counter (IPC) of the *Einstein* Observatory (Giacconi *et al.* 1979) and for this the count rate depended only very weakly on temperature* for temperatures between 3×10^6 and 10^8 K (*e.g.*,

* That this is true even for the lower end of the temperature range is due to the fact that the increasing contribution of iron recombination lines compensates somewhat for the fall-off of the continuum emission.

Trinchieri, Fabbiano & Canizares 1986). Spectral observations of hot gas in clusters (Jones & Forman 1984) and around elliptical galaxies (Trinchieri *et al.* 1986) show that almost always the temperature falls into this range. Thus IPC images to a zeroth approximation give a direct measure of the emission measure ($\int n_{\rm th}^2 dl$) of hot gas, with little direct information on the temperature.

Density distributions have been estimated from the surface-brightness profiles either by fitting models (Jones & Forman 1984), or by Abel deconvolution, constraining the solution to be in quasi-hydrostatic equilibrium (Fabian *et al.* 1981). The latter approach requires estimates of the gravitational potential and also produces a radial temperature profile. Since the density is determined virtually directly from the data, the derived temperature profile is quite sensitive to the assumed potential. Spherical symmetry is usually assumed, since the gas distributions are often within 10% of spherical symmetry, especially in the core region (Trinchieri *et al.* (1986), but see Fabricant, Rybicki & Gorenstein (1984) for a deprojection of a highly elliptical distribution).

Such analyses suggest that the temperature is roughly constant with radius in both clusters and galaxies, except that in some clusters and perhaps most galaxies there is a cool, dense region close to the centre. In a few nearby clusters these results are confirmed by direct measurement of the spectrum at different radii (*e.g.*, Virgo: Fabricant & Gorenstein 1983; Perseus: Fabian *et al.* 1981; Coma: Hughes *et al.* 1988). The central cooler region is probably due to radiative heat loss, which suggests the existence of a slow *cooling flow* (*e.g.*, Fabian, Nulsen & Canizares 1984).

With the exception of the cooling flow spike, the density distributions can usually be fitted with a *King approximation* profile:

$$n_{\rm th}(R) = n_0[1 + (R/R_{\rm core})^2]^{-3\beta_{\rm King}/2}. \tag{3.20}$$

Here $\beta_{\rm King}$ is an empirical constant, typically ~ 0.5, and $R_{\rm core}$ is the core radius. This equation is intended to be an analytic approximation to a *King model* truncated isothermal sphere, in which case $\beta_{\rm King}$ gives the ratio of the temperature of the gravitating matter to the gas temperature. However the approximation is very poor outside a few core radii, giving the wrong asymptotic power law at large radius (Trinchieri *et al.* 1986) and equation (3.20) is best thought of as a useful empirical fit. For clusters of galaxies the description is remarkably good. For galaxies the fit is adequate but the errors are large and only a small logarithmic range of radius is available for fitting, so this may have little physical significance.

Typical central densities and temperatures and core radii in Abell clusters are $2 \times 10^3 h^{1/2} \mathrm{m}^{-3}$, $5 \times 10^7 h^0 \mathrm{K}$, and $100h^{-1}$ kpc respectively (Jones & Forman 1984). For individual galaxies, Canizares, Fabbiano & Trinchieri (1987) find $n_0 \sim 3 \times 10^5 h^{1/2} \mathrm{m}^{-3}$, $T \sim 10^7 \mathrm{K}$ and $R_{\mathrm{core}} \sim 1h^{-1}$ kpc. These temperatures imply that the gas is gravitationally bound to the cluster or galaxy, unless there is a supersonic wind (cf. Mathews & Baker 1971).

We have focused on the nearest elliptical galaxies and clusters because they have the best data. How do these compare with regions actually containing jets? *Einstein* was used to observe around 100 radio galaxies and quasars. Only bright FR I sources were near enough for the diffuse emission to be detected and resolved (*e.g.*, Killeen, Bicknell & Carter 1986; Morganti *et al.* 1988). In X-rays, these radio galaxies appear to be typical large ellipticals. More distant and powerful radio sources were also detected (Fabbiano *et al.* 1984), but in these the X-ray emission is almost certainly due to the active nucleus and gives no information on the large-scale environment. Miller *et al.* (1985) analysed the upper limits to the extended emission around the nearest FR II sources and their results are consistent with the numbers for nearby ellipticals quoted above. There are only two cases of FR II sources in clusters of galaxies with dense, bright X-ray atmospheres that could be (and were) mapped by *Einstein*: Cygnus A (Arnaud *et al.* 1984) and 3C 295 (Henry & Henriksen 1986). It is hard to believe that it is a coincidence that these are the two most luminous radio sources closer than $z = 0.5$.

3.3.2 **Emission-line gas and other components**

The hot medium contains most of the gaseous mass and fills most of the volume in elliptical galaxies, but other components are also present. In particular filaments and clouds of emission-line gas on scales from 10 – 100 kpc (*e.g.*, van Breugel *et al.* 1985) are sometimes found, especially in radio galaxies and quasars. Large quantities of such 'warm' gas are almost always found in the most powerful sources at cosmological distances (Spinrad & Djorgovski 1987).

Typical densities and temperatures are $n_{\rm th} \sim 10^5 {\rm m}^{-3}$, $T \sim 10^4$ K. These parameters are derived from the intensity ratios of emission lines and so are independent of Hubble's constant. The temperatures are constrained by the cooling curve to lie within a factor of about 2 of the above value (Osterbrock 1988). Densities are usually based on the ratio of the S[II] lines ($\lambda\lambda 6716, 6731$) to eliminate uncertainties due to abundance variations and reddening, but systematic errors may still occur because the emitting region is likely to be inhomogeneous and there is a tendency for each line to be emitted predominantly in regions whose conditions make that line particularly strong, cf. Osterbrock (1988).

This material is expected to contain a negligible fraction of the total gaseous mass, and must have a very small filling factor ($\sim 10^{-6}$), so often its interaction with the jets will be minimal. However if the jet happens to collide with such a dense cloud the results should be spectacular: in fact van Breugel, Heckman and collaborators have shown that several of the most bizarrely distorted radio sources are indeed undergoing such interactions. The origin of the emission-line gas is currently a subject for debate: in most models its existence is independent of the radio source (if not the active nucleus) but Heckman *et al.* (1982) suggest that the gas is swept out of the narrow-line region by the radio jets, while the possibility that the gas is condensed by radiative shocks has been investigated in the context of Seyfert galaxies by Pedlar, Dyson & Unger (1985) (see also §3.5.6).

Colder ISM components such as atomic hydrogen and dust are also found in elliptical galaxies and the mass fraction of this material is correlated with enhanced radio activity (Knapp 1990). Direct interactions between HI clouds and jets have not been recorded: most likely the correlations reflect the conditions required to cause nuclear activity in the first place. This fascinating and rapidly-developing field is outside the scope of the current work.

3.3.3 The gravitational potential

Traditionally the mass distribution of elliptical galaxies was derived from the distribution of starlight, assuming a constant mass-to-light ratio and a rotationally-supported oblate spheroid geometry. This method is inaccurate for two reasons: first, large elliptical galaxies are actually triaxial systems supported by anisotropic velocity distributions, and second, as with spiral galaxies, there is probably a significant amount of dark matter, including both a large-scale halo and a massive object (or objects) associated with the active nucleus. Progress can be made in the inner regions by measuring the stellar velocity dispersion (σ_v) as a function of radius (or in principle by mapping it across the galaxy, but this is too time-consuming in practice). At large radii, where starlight is faint, the potential is probed in a few cases via radial velocities of globular clusters, HI rings, or polar rings, and from X-rays via the assumption of hydrostatic balance in the hot gas. Because hot atmospheres are believed to be present in most ellipticals, the latter method should eventually become the most useful, but at present it is plagued by poor sensitivity and resolution, and uncertainty in the temperature profile.

In some well-observed cases σ_v increases within the central core radius, $\sim O(1 \text{ kpc})$, which is consistent with either a central dark mass or highly radial orbits at the galaxy centre (Illingworth 1981). In general it seems that stars dominate the mass in the central 10-20 kpc (giving an M/L_B of about 6 solar units because of the old stellar population of ellipticals), while at larger radii a dark halo may become important (*e.g.*, Fabian *et al.* (1986) claim a *minimum* total M/L_B of about 40). In the best-studied case, M 87 (Stewart *et al.* 1984), the total mass density seems to fall off with radius as $\rho \propto R^{-2}$, which is the same relation deduced for the dark haloes of spirals (Sancisi & van Albeda 1987). However, as a cluster-centre galaxy, M 87 is probably not a typical elliptical.

3.4 Basic jet physics

A jet is a collimated fluid flow through a fluid medium. Everyday examples include jets of water from a garden hose, smoke rising from chimneys, and the exhaust from a jet engine. All of these cases have been used as models for astrophysical jets, along with a number of other ideas (such as magnetically-confined jets) for which there are no terrestrial analogues. The physics of these examples differ considerably from each other, but there are a few points which apply to jets in general, and these will be discussed in this section.

3.4.1 Jets vs. plasmons

One way to illustrate the basic features of jet flow is to compare jets with the previous plasmon model for radio sources. According to this, radio sources result from one or a few explosive events which eject a pair of plasma blobs (the plasmons) in opposite directions, which eventually expand to form the radio lobes. A plasmon will come to a halt in a finite distance by transferring all its momentum to the ambient medium. A little algebra shows that in a constant-density atmosphere the stopping length $D_{\rm max} \sim \eta s$, where η is the ratio of plasmon to external density and s is the scale-length of the plasmon. In most models the plasmon scale-length was of order its radius, so the appearance of compact objects far from the centre of the galaxy required very dense objects. In contrast a jet will continue to propagate as long as the central engine keeps running. Ram pressure still decelerates material at the end of the jet, but if the jet is supersonic, the deceleration zone is bounded by a shock, and propagation upstream is unaffected. If the jet is subsonic the situation is more complicated, but the basic point that the momentum is continuously re-supplied from the centre still applies. In the body of the jet, each fluid element is preceded by material moving with nearly the same velocity, and so ram pressure slowing is eliminated. In the language of plasmons, a jet has an infinite scale-length, and therefore an infinite stopping length even for a very low density ratio.

This fundamental difference seems paradoxical, since a sufficiently rapid stream of plasmons appears to be similar to a rather unsteady jet. The key point is the relaxation time of the ambient medium. The plasmon description applies if the ambient medium has time to fill in behind a plasmon before the next one comes along. If not, the plasmons will be separated by a low-density channel, and the system is equivalent to a pulsating jet. This time is at least $t_{\rm rel} = r/c_{sa}$, where c_{sa} is the external sound speed. If the channel is in pressure balance with its surroundings the relaxation time is set by turbulent break-up, $t_{\rm rel} = r/v_{\rm turb}$.

Another important difference between the plasmon and jet concepts is the 'continuous flow' behaviour at the end of the jet, discussed in §3.5.6.

3.4.2 Fundamental parameters

From one point of view jets are a method of getting mass, momentum and energy from the central engine to the outer reaches of the source. Important parameters are therefore:

$$\text{Mass per unit length} \quad m = \int \rho \, dS \tag{3.21}$$

$$\text{Mass flux } (\textit{Discharge}) \quad J = \int \rho \mathbf{v} \cdot \mathbf{dS} \tag{3.22}$$

$$\text{Momentum flux } (\textit{Thrust}) \quad \mathbf{T} = \int (\rho \mathbf{v}\mathbf{v} + \mathbf{\Psi}_{\text{tot}}) \cdot \mathbf{dS} \tag{3.23}$$

$$\text{Energy flux } (\textit{Power}) \quad K = \int [(u_{\text{tot}} + \rho\phi + \rho v^2/2)\mathbf{v} + \mathbf{q}_{\text{tot}} + \mathbf{\Psi}_{\text{tot}} \cdot \mathbf{v}] \cdot \mathbf{dS}. \tag{3.24}$$

Here u_{tot} is the total internal energy, $\mathbf{\Psi}_{\text{tot}}$ is the total stress tensor, $\mathbf{q}_{\text{tot}}$ is the total heat flux density, and ϕ the gravitational potential. $\mathbf{dS}$ is a vector element of area and the integral is over a surface S perpendicular to the jet. We define a unit vector normal to S as $\mathbf{n}$, the area as A, and the vector $\mathbf{A} = A\mathbf{n}$. Note that $\mathbf{T}$ is not necessarily parallel to $\mathbf{n}$, although it will be close to this in most jet-type flows. The main exception is the case of very dense jets, but for these it turns out that a fluid description is unnecessarily complicated, as the motion of fluid elements is essentially ballistic (see §3.5.5 below). In the rest of this section we therefore assume that the thrust is nearly parallel to the jet. We also need a rule to decide where the surface of the jet actually is, *i.e.*, which parts of the flow we choose to label 'jet' as opposed to 'external medium'. For a laminar jet we take the surrounding slip surface. For a turbulent jet we use the surface of the turbulent boundary layer; in this case streamlines will cross the jet surface (since turbulent jets entrain), but the 'external' flow is not itself turbulent. The problem of jet dynamics is then that of determining the jet shape (given to first approximation by $\mathbf{A}(l, t)$, where l is distance along the jet), and the flow through the jet as parameterised J, $\mathbf{T}$, and K. For a relativistic generalisation of these quantities, see Begelman, Blandford & Rees (1984).

The internal energy will include thermal, relativistic particle, magnetic, and turbulent components, and for each of these there is a corresponding stress tensor.* If we ignore the possible anisotropy of the particle distribution in a magnetised plasma (cf. §3.1.2), the thermal and relativistic stress tensors are isotropic and simply describe the pressure (*i.e.*, $\mathbf{\Psi} \cdot \mathbf{n} = P\mathbf{n}$). As noted in §3.1.2, the magnetic stress tensor is also isotropic if the field is fully tangled, giving $P_B = u_B/3$. These relations will break down at shock fronts and in turbulent flow, where viscosity and viscous dissipation become important.

Taking the simple (if perhaps contradictory) case of no turbulence and a tangled magnetic field, we can write the integrand in the energy flux equation as:

$$\rho(w + \phi + v^2/2)\mathbf{v} = \rho H \mathbf{v} \tag{3.25}$$

where $w = (u_{\rm tot} + P_{\rm tot})/\rho$ is the specific enthalpy, and H is the Bernoulli constant. If the jet fluid has an ideal gas equation of state with adiabatic index Γ, we can write the specific enthalpy as $c_s^2/(\Gamma - 1)$, where c_s is the sound speed. In the present case we have a mixture of thermal gas ($\Gamma_{\rm th} = 5/3$) and a relativistic component ($\Gamma_{\rm rel} = 4/3$) which includes both the particles and fields, but such a mixture can be fairly accurately represented as a single component with variable adiabatic index, given by $\Gamma_{\rm tot} P_{\rm tot} = (5/3)P_{\rm th} + (4/3)P_{\rm rel}$. If we also put gravity in terms of the escape speed $v_{\rm esc}$, the Bernoulli constant becomes:

$$H = \{v^2 + 2c_s^2/(\Gamma - 1) - v_{\rm esc}^2\}/2. \tag{3.26}$$

If the jet is to escape the galaxy, H must be positive. For the kinetic energy flux to dominate the power, we must have the Mach number $\mathcal{M} = v/c_s > \sqrt{2/(\Gamma - 1)}$. Similarly the thrust of the jet is dominated by thermal or magnetic pressure if $\mathcal{M}^2 < 1/\Gamma$ and by ram pressure (ρv^2) otherwise.

The advantage of the global parameters J, $\mathbf{T}$, and K is that they provide a simple characterisation of the jet even when the local parameters like ρ, P, and $\mathbf{v}$ vary substantially across the jet, for instance due to turbulence or internal shocks. On the other hand, knowledge of the profiles of such quantities across the jet is necessary to understand the dynamics in detail. Actually the physics depends not on the absolute values of these parameters

* The turbulent terms appear when the other terms represent mean quantities, as explained in more detail in §3.5.4.

but on the various dimensionless parameters which define the flow, notably the Mach number, Reynolds number, plasma β_P, density contrast $\eta = \rho_j/\rho_a$, pressure ratio (P_j/P_a), ratio of jet diameter to pressure scale-height, and Lorentz factor γ. (Here and in the following the subscript j refers to the jet and a to the external medium.)

The Reynolds number (Re) is discussed in §6.4.1. It parameterises the importance of viscosity, and in our case is always very large, meaning that mean-flow viscosity is negligible. High-Re flows tend to develop regions with small scale structure in which viscosity, and in particular viscous dissipation of energy, *is* important. In subsonic flow these are the cores of turbulent eddies; in supersonic flow shocks appear as well. Thus Re also parameterises the principle dissipative mechanism, viz: mean-flow viscosity, turbulence, or shocks.

3.4.3 **Equations for a thin jet**

This section gives the equations for the variation of J, $\mathbf{T}$, and K along a thin jet. While some simplifying assumptions about the external medium will be made, the only assumptions about the jet required are that its radius of curvature is larger than its width. Thus arbitrarily complicated and in particular turbulent jet profiles are covered, allowing us to assess the general validity of some standard results. For turbulent jets the equations apply formally to averages over ensembles of jets (cf. §3.5.4). The development here follows Henriksen, Vallée & Bridle (1981) and Bicknell (1984, 1986a), but is a little more general.

If we integrate the continuity equation over a test volume bounded by two surfaces S_1 and S_2 (as defined in §3.4.2), separated by an edge E of width Δl (see Fig. 3.5), we get:

$$\frac{\partial m}{\partial t} + \frac{\partial J}{\partial l} = -\frac{1}{\Delta l}\int_E \rho\mathbf{v}\cdot\mathbf{dS} \approx \rho_a v_{\rm in} r_p. \qquad (3.27)$$

The right-hand side represents entrainment into the jet. In the last form we have let $\Delta l \to 0$ and $v_{\rm in}$ is the average inflow speed around the jet of perimeter length r_p. This implicitly assumes that the radius of curvature is larger than the jet radius. In general ρ_a should be considered an average around the jet, although we shall usually assume that the variations of external quantities across the jet are small compared to their absolute values.

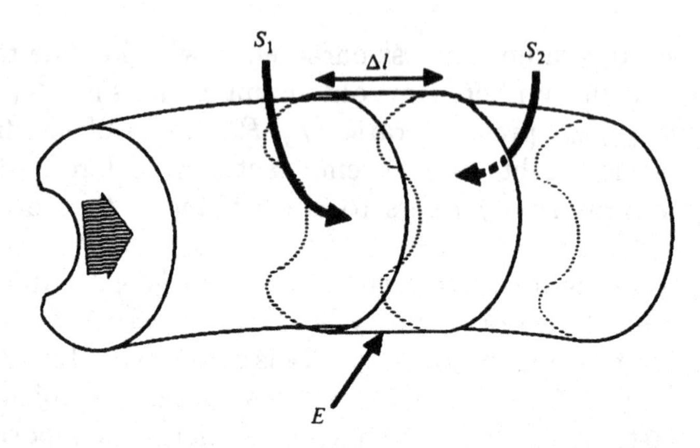

Fig. 3.5. Integration volume used to set up the integral equations for jet motion. The edge E follows the outer surface of the jet (which need not have any particular symmetry) but is considered to lie just outside the jet.

In this and the following equations, the effects of turbulent and magnetic forces within the jet appear solely in the form of a finite entrainment rate. An important corollary is that when entrainment can be neglected, the global dynamics of the jet are independent of the detailed internal structure. In particular the entrained momentum is usually negligible since velocities in the external medium are usually small (an exception is the case of a jet in a cross-wind), so internal magnetic and turbulent stresses usually have no effect on the momentum flux (equation (3.29) below).

Conservation of energy gives:

$$\frac{\partial}{\partial t}(\mathcal{E}_{\text{tot}}) + \frac{\partial K}{\partial l} = -\mathcal{L}_{\text{rad}} - \mathcal{L}_{\text{sound}} + H_a \rho_a v_{\text{in}} r_p. \tag{3.28}$$

Here $\mathcal{E}_{\text{tot}}$ is the total energy per unit length, and $\mathcal{L}_{\text{rad}}$ and $\mathcal{L}_{\text{sound}}$ are the luminosity per unit length in EM (principally radio) and sound waves respectively. Formally they result from the electro-magnetic and turbulent energy fluxes across the surface E, but it is usually more convenient to count the radio energy lost at source and never to include it in the energy flux. These terms are usually a small fraction of the total energy budget; the associated momentum loss is negligible (partly because of symmetry).

The momentum (thrust) equation is derived by integrating Euler's equation across the jet. It is convenient to write the pressure as the sum of an equilibrium pressure P_0 (*e.g.*, the pressure that would have been found in the absence of the jet) and a pressure P' caused by the jet. Outside the jet, P' contains ram pressure and magnetic pressure terms, the latter occurring where there is a current in the jet.

With $\mathbf{T}' = \mathbf{T} - \mathbf{A}P_0$ we have:

$$\frac{\partial J\mathbf{n}}{\partial t} + \frac{\partial \mathbf{T}'}{\partial l} = -m\nabla\phi - A\nabla P_0 + \left[\frac{\partial \mathbf{A}}{\partial l} + A\nabla_\perp\right] P_a' + \mathbf{v}_a \rho_a v_{\rm in} r_p. \quad (3.29)$$

Here we have assumed that $\nabla\phi$ and ∇P_0 are constant across the jet. The term $A\nabla_\perp P_a'$ can be considered a shorthand for $\oint_E P'\mathbf{n} \times \mathbf{dl}$; *i.e.*, it is the transverse pressure force per unit length on the jet, and does not depend on the variation of pressure within the jet. Since the edge E is in the external medium we ignore turbulent forces, relativistic particle pressure and externally generated magnetic fields which are assumed small outside the jet. The pressure terms seem rather asymmetric because the effect of stresses inside the jet, including P', is taken into account in the thrust term. As noted above, the entrainment term is usually negligible. If the equilibrium atmosphere is hydrostatic, the first two terms on the right give the usual Archimedes force $(m - m_0)\mathbf{g}$, where m_0 is the mass per unit length of the displaced fluid. If the jet is also steady and the current is small this is the only important term. This term will be negligible if either $\eta \rightarrow 1$ or $\mathcal{M} \rightarrow \infty$, and then the thrust of the jet is conserved.

In order to solve the above equations, we need to be able to express J, T', and K in terms of each other. A common approximation (often incorrect) is to assume that the jet is *transverse self-similar*, or self-similar for short. This means that the jet profile at any distance l along the jet has the same form but is scaled in amplitude and width. Thus defining $\tilde{r} = r/r_j(l)$, we can write $v = v_c V(\tilde{r})$, $w = w_c W(\tilde{r})$, etc., where subscript c indicates central values and the functions V, W, etc. are assumed independent of l. Often the precise form of these functions is not too important, nor do they have to be exactly constant as long as they change more slowly with l than the other parameters. Sometimes it is more reasonable to consider the *difference* between the actual value and the external value as self-similar.

3.4.4 Derivation of jet parameters

In this section we discuss the rather limited number of ways that jet parameters can be derived from observation without making very specific assumptions about the jet physics.

3.4.4.1 Collimation and Mach number

The collimation ($r_j(l)$) is nearly accessible from observation, since we can observe the width as a function of projected length along the jet. As in §3.2.2, we have to be careful in defining what we mean by 'width': an obvious point, occasionally overlooked, is that the opening angle of the half-intensity width is smaller than that of the full width. A little more subtle is that the half-widths for velocity, pressure, density, and radio emissivity may all be somewhat different, especially in a turbulent jet (cf. Bicknell 1986b). Fortunately in at least some jets (*e.g.*, M 87, Owen *et al.* 1989) there is a well-defined edge to the radio emission and it seems reasonable to assume that this also bounds the fluid parameters. In this and the following section I define the 'opening angle' $\xi = dr_j/dl$, where r_j indicates the full radius of the jet. If the jet is at an inclination angle i to the line of sight, we actually measure $\xi \csc i$.

The maximum opening angle for a jet of a given initial Mach number $\mathcal{M}_0$ is given by the free case (see §3.5.2), in which case ξ increases towards a limit determined by $\mathcal{M}_0$ (equation (3.36)). For all jets whose widths can be measured ξ varies along the jet, going through regions of both flaring and re-collimation. This is usually taken to show that jets are not free (cf. §5.2.1), and therefore have lower $\mathcal{M}_0$ than implied by equation (3.36). Since $\mathcal{M}$ may increase along a jet, this upper limit to an ill-defined 'initial' value is of little practical use.

Another route to the Mach number is through the bending of jets: in §5.3.3 it is shown that $\mathcal{M} \approx \sqrt{R_{\rm curv}/s}$, where $R_{\rm curv}$ is the radius of curvature and s is the scale-height of the pressure gradient responsible for bending the jet. This point is discussed further below: for jets in head-tail sources O'Dea (1985) finds $\mathcal{M} \sim$ 2-10 if s is the jet radius and $\mathcal{M} \sim 1$ if s is the scale-height of a galaxy atmosphere.

In general the collimation data can be fed into specific models and model-dependent parameters for the jet and/or atmosphere can be returned. Some such models are discussed in §3.5.

3.4.4.2 **Thrust**

Another 'nearly observable' is the radius of curvature of the jet ('nearly' because of projection effects). If $\mathbf{T} \parallel \mathbf{n}$, then the bending corresponds to a change of direction of the momentum of the jet fluid. The thrust is then given from the transverse component of equation (3.29) and

$$\frac{T}{R_{\text{curv}}} \approx \frac{\partial \mathbf{T}_\perp}{\partial l} \approx (m - m_0)\mathbf{g}_\perp + A\nabla_\perp P'_a. \qquad (3.30)$$

While the available bending force can be estimated with tolerable accuracy for specific models, the problem is deciding which case is actually occurring. For instance an S-type symmetry might suggest buoyant refraction (first term on right) of a jet ejected at an angle to the principle axes of an elliptical galaxy (Henriksen *et al.* 1982), ram pressure bending (second term), due to the precession of the central engine, (essentially a driven helical instability, cf. Hardee 1981), or precession in a ballistic jet (Gower *et al.* 1982), in which case the above equation does not apply. A more common case is C-symmetric bending in twin-tail radio galaxies. Here the pressure gradient ultimately comes from ram pressure due to the motion of the host galaxy through the intracluster medium. Since we have a good idea of the typical speeds of galaxies in clusters and of the density of the ICM, it might appear that we can immediately get a reasonable estimate for the thrust. Unfortunately there are two problems. First, we have $\nabla_\perp P \approx \rho_a v_{\text{gal}}^2/s$, where s is the length over which the pressure falls back to its ambient value. While an obvious value for s is the jet width (*e.g.*, Begelman, Rees & Blandford 1979), Jones & Owen (1979) have pointed out that the pressure may be mediated by a galactic atmosphere, in which case s is roughly twice the scale-height of the atmosphere, yielding a value for the thrust 5 – 10 times lower. The second problem is the possibility of bending at oblique shocks, in which case the flow is sharply deflected at the shock front (cf. §§5.3.7 and 7.2.2). Here the sharp turn does not imply that the bending force is applied at a point; instead it is spread out downstream, roughly over the length l_{bend} required for the pressure to return to its pre-shock value. In this case in equation (3.30) we should use $l_{\text{bend}}/\Delta\theta$ instead of the radius of curvature. Unfortunately l_{bend} may not be obvious from inspection of images. This point may help to explain very sharp bends which often occur in wide-angle-tail sources.

The curved jets in most twin-tails suggest steady bending and the results of O'Dea (1985) imply typical thrusts of 10^{28} N for bending in an atmosphere and closer to 10^{29} N if the jets are in direct contact with the streaming ICM.

Another approach to measuring the thrust is from ram-pressure balance at hotspots. The thrust is just the net force on an obstacle suddenly blocking the jet. In particular, if a supersonic jet runs into a wall, a shock forms and the flow is further decelerated to zero speed by a subsequent pressure gradient. The stagnation pressure is approximately T/A. This is essentially the Blandford & Rees (1974) model for a hotspot, provided we work in the frame in which the hotspot is stationary. Limits to the pressure can be estimated from the ram-pressure arguments of §3.2.9 or from equipartition, while the size of the hotspot can be measured from high-resolution maps. For Cyg A, the implied thrust is $3 \times 10^{31} > T' > 4 \times 10^{29}$ N respectively. Neither limit is cast-iron: the ram-pressure limit could be increased if the hotspots were advancing highly relativistically, while the equipartition limits can be reduced by reducing Φ in the hotspot. Note that a one-sided jet near the maximum thrust could propel a $10^9 M_\odot$ 'engine' out of its host galaxy in $\sim 10^7$ yr. §8.3.3 discusses how a moving engine could provide a 'flip-flop' power source.

3.4.4.3 Jet power and the energy budget

A common method of estimating the jet power, K is to assign the radio luminosity of part or the whole of the source to be some guessed fraction ϵ (the "conversion efficiency") of K. Popular values of ϵ are between 1% and 10%, owing to prejudice against very high energy output from the nucleus on the one hand and against very efficient emission mechanisms on the other. Implicit in this kind of argument is that the region considered is in fact in a steady state, with some fraction of the energy flow being tapped for radio emission. This seems plausible in jets and in hotspots, but, as noted in §3.1.1, is unlikely for the source as a whole. If applied to a region not in a steady-state, the value of ϵ depends on the history of the region and has little to do with any energy conversion process. It is not even required to remain less than unity: for instance if the jet cuts off, the lobe continues to radiate and formally $\epsilon = \infty$. Even when correctly applied, without a detailed model of both the flow and the particle acceleration mechanism there is no *a priori* way to obtain a meaningful value for the efficiency.

A slightly better method (*e.g.*, Saunders *et al.* 1981) attempts to estimate the source lifetime τ (cf. §§3.2.5 and 3.2.6) and the total energy $U_{\rm tot}$ supplied to the source over this period, then simply set $K = U_{\rm tot}/\tau$. $U_{\rm tot}$ contains the total energy within the source, the work done on the external medium, and the total energy radiated over the source lifetime. Initially we ignore entrainment, giving:

$$U_{\rm tot} = \int (u_{\rm tot} + \rho v^2 + \rho\phi)\, dV + \int_0^\tau \int (\mathbf{\Psi}_{\rm tot} \cdot \mathbf{v}) \cdot \mathbf{dS}\, dt + \int_0^\tau L_{\rm rad}\, dt. \quad (3.31)$$

The $P\,dV$ term (the second integral) will correspond mainly to excess heat in the gas surrounding the source, which should eventually be measurable through X-ray observations, at least for FR II sources, where the heated region is expected to be bounded by a shock front. Meanwhile, a rough estimate can be obtained by setting $\mathbf{\Psi}_{\rm tot} \cdot \mathbf{v} \approx P_{\rm tot}\mathbf{v}$ and noting that $\int\int \mathbf{v}.\mathbf{dS}\, dt = \int dV$; if we approximate $P_{\rm tot}$ as constant over each surface element with time the second integral then becomes $\int P_{\rm tot} dV$. While this may seem a rather crude approximation, note that if the radio source has been expanding roughly homologously, half of the volume has been created by the source expanding from 80 per cent of its current linear size. It may be reasonable to regard the pressure distribution on the surface as fixed over this period. Because the volume of the diffuse lobes usually greatly exceeds that of the jets and hotspots, the lobes dominate the first two integrals, even when they provide a small part of the luminosity. In these regions it is probably safe to assume very subsonic bulk motion (particularly in the bridges of FR IIs) and turbulence (on general principles), and the gravitational energy is also small compared to the thermal. The sum of the first and second integrals is therefore dominated by the total enthalpy in the diffuse regions.

For FR II sources the integrated radio output can usually be neglected compared to the enthalpy. This follows from the observation that the lowest break frequencies (about 1 GHz, Leahy *et al.* 1989) are much higher than the minimum frequencies over which the spectrum is well-fitted by a power law (below 10 MHz for diffuse emission). Since most of the relativistic electron energy injected into the source is at the bottom end of the energy distribution (cf. §3.2.8), the relatively high minimum break frequencies imply that radiation losses have not yet cut into the bulk of the particle energy. The argument is not affected by expansion losses from a region of higher emissivity. In principle the region of ultra-low break frequency could be hidden by further expansion from the faintest observed regions to an invisible reservoir with undetectably low luminosity. Because of its low energy density, such a reservoir would have to be large to be energetically important, and Leahy *et al.* argue that there is not enough room for this in FR II sources. In FR I sources with tails which fade into the noise the natural site for such reservoirs is as continuations of the tails, and there is no limit (other than the age of the Universe) to their length. Thus there may be radiative equilibrium in these sources for the relativistic electron energy. However a major portion of the energy in FR I jets may be in the very. low-grade form of thermal energy, which can never be converted into relativistic electron energy (*pace* Bicknell 1986a). Furthermore, while some magnetic and turbulent energy is available for particle acceleration (Chapter 9) most of it will end up as more heat. Therefore the value of ϵ for such sources is expected to be well below unity.

The total enthalpy is usually estimated from equipartition arguments, and the age from spectral ageing and equipartition fields. This is not nearly as bad as it seems however, because the largest uncertainty, the actual emitting volume, cancels out. This can be seen most clearly by deriving the jet power directly from the equation for the electron energy, equation (3.14), and the break frequency, equation (3.10). The energy flux that produces the relativistic electron energy in the lobes (and the associated work on the external medium) is $K_{\rm re} = (u_{\rm re} + P_{\rm re})V/\tau = (4/3)U_{\rm re}/\tau$. Entrained thermal energy cannot be converted to relativistic energy, and there are apparently few relativistic particles in the external medium, so $K_{\rm re}$ is independent of entrainment. For constant magnetic field, both B and the various constants cancel out, except for a numerical factor, giving:

$$K_{\rm re} = (2^8 3^{-9/2}\pi^2)\, L_{\nu_0}\nu_0^{\alpha}\mathcal{I}_1\, \nu_{T_{lobe}}^{1/2}\, [1 + (B_{\rm MB}/B)^2]. \qquad (3.32)$$

Here $\mathcal{I}_1$ is a weighted integral over the spectrum as described in §3.2.8. Provided $B_{\rm MB} < B$, $\mathcal{I}_1$ is the main uncertainty since it depends weakly on the unknown lower frequency cut-off ($\propto \nu_L^{\frac{1}{2}-\alpha}$). If the synchrotron plasma steadily expands as it ages, $K_{\rm re}$ is underestimated by a factor of roughly $(B_{\rm max}/\langle B\rangle)^{3/2}$. Because of these uncertainties the value of $K_{\rm re}$ derived is a lower limit (unless substantial particle acceleration occurs in the lobes). Evidently $K_{\rm tot} \geq K_{\rm re}$; for equipartition between magnetic and particle pressure, $K_{\rm tot} = 2(1+k)K_{\rm re}$. As noted in §3.2.8, entrainment may cause k to vary substantially along the jet. If we use a value for k thought to apply in the jet, entrainment into the lobes will not affect the estimated jet power, since it does not affect the estimate of $U_{\rm re}$. For Cyg A, the integrated spectrum (Baars *et al.* 1977) is dominated by the lobes, and with a minimum (JP) break frequency of 850 MHz (Carilli *et al.* 1990) I find $K_{\rm re} \gtrsim 5 \times 10^{38} h^{-2}$ W in each jet. This can be compared with a total radio luminosity of $9 \times 10^{37} h^{-2}$ W.

The value of the jet power returned by this global analysis is an average over the source lifetime or over the radiative lifetime of the overall flow, whichever is shorter. This is inevitable when the data used in the analysis refers mainly to the diffuse components with their very long response times to changes in the energy flow.

3.4.4.4 **Speed**

The literature is full of determinations of jet speeds; I have played this game myself (Leahy 1984). Apart from the upper limits derived from Doppler beaming arguments, none of these values can be accepted at face value. At best they are implications of specific models, but unfortunately most are simply meaningless. If this were not so, we would not still be arguing even about whether jets are relativistic or not. The vast majority of speeds quoted are based in one way or another on estimates of the jet density from Faraday depolarization; such 'densities' are currently worthless (§3.2.10), although long-wavelength observations may be beginning to detect authentic internal effects. Another method is to fit ballistic models (inappropriate, as argued in §3.5.5). I shall argue that most jets are lighter than the undisturbed external medium, and have Mach numbers not much less than unity. In this case they must be faster than the external sound speed, typically 300-1000 $\rm km\,s^{-1}$.

A mean speed for hypersonic jets can be derived by dividing

the jet power by the thrust. Doing this naively for Cyg A, the limits derived above imply $v_j > 0.06c$. But our ram pressure limits assumed a hotspot advance speed of near c; if we self-consistently require that $v_j > v_{\rm HS}$, we find $v_j > 0.4c > v_{\rm HS}$. If $\eta \ll 1$, as argued below, then $v_j \gg v_{\rm HS}$ from equation (3.41), so the jet is very likely relativistic.

3.5 A menagerie of model jets

In this section we review the main types of jet model which have been explored in the literature, and speculate on their applicability to various types of source. The approach is roughly in order of complexity, but in the end most of the corners of parameter space defined by the dimensionless parameters listed at the end of §3.4.2 will be covered. (On the other hand, the *middle* of parameter space, where all effects are important, has been avoided.) A major exception is the area of magnetically-dominated jets. Magnetic confinement is considered in §5.2.1, and force-free fields in §9.6.3. I consider only 'unperturbed' jet models: all jets suffer from various fluid instabilities to some degree, and their susceptibility to disruption by such instabilities is a major constraint on acceptable models. However a discussion of stability is deferred until Chapter 6.

3.5.1 Testing the models

Most of the models considered in this section have been applied to FR I jets, simply because the data for these is better and has been around longer than for jets in FR IIs. Not surprisingly, some cases are actually more appropriate to FR IIs. The present state of development is such that the reader should take detailed fits of models to individual sources with more than a pinch of salt: it turns out that a surprising variety of models can give a superficially good fit to some aspect or other of a typical FR I jet. Not uncommonly, the same models are flagrantly contradicted by other aspects of the evidence.

No model will *predict* every detail of the observations, but a minimum requirement is that it should be *consistent* with all the evidence. As well as the data on a given source, the evidence includes the general trends in the population as a whole, and the relationship of the source under consideration to that population. For instance, if the model implies that the source is in a particular phase of its life-cycle, it also implies that other phases will be represented in the population, and this can be tested.

3.5.2 Steady laminar jets

Equations (3.27) and (3.28) imply that the discharge and power are constant (apart from radiation losses) along a steady laminar jet. This case is simple enough to investigate the detailed structure of the jet, and in this section I will outline the approach taken by Sanders (1983) and developed by Wilson & Falle (1985). Laboratory experiments suggest that subsonic and mildly supersonic jets invariably become turbulent within a few nozzle diameters, because of short-wavelength Kelvin-Helmholtz instabilities on the jet surface. Stability analysis shows that the jets become more stable at higher Mach numbers and at higher η; since we usually assume that $\eta \ll 1$, the laminar jet theory discussed in this section is mainly relevant in the supersonic regime. We are usually interested in jets propagating through a hydrostatic atmosphere, so from equation (3.29) the jet will accelerate if $\eta < 1$ and decelerate if $\eta > 1$. The flow in a steady laminar jet is completely specified by the run of external pressure and gravitational potential along the jet, and by the distribution of $\mathcal{M}$, ρ_j, and P_j on a given cross-section. The flow is independent of the external density. If $v \gg v_{\rm esc}$, the internal density affects the flow only through the Mach number and pressure distributions; Bernoulli's equation then shows that the speed on each streamline tends to a constant with the average $\langle v \rangle \sim \sqrt{2K/J}$.

The evolution of the emissivity in laminar jets is given by the adiabatic relations of Table 3.3. The slowest brightness decline is given by the $B_\perp$ case, when (for a steady jet) we have

$$j_\nu \propto A^{-\frac{7\alpha+9}{6}} v^{-\frac{5\alpha+6}{3}}. \qquad (3.33)$$

It is a standard result, first pointed out by Burch (1979), that even this case predicts a much faster brightness decline along the jet than observed in FR Is, provided that the shape of the cross-section does not change systematically (so that $A \propto r_j^2$) and that the jets do not slow down. Hence light laminar models can only work if there is significant re-acceleration. Before discussing whether this is so, we introduce two idealised cases which illustrate important features of laminar jets.

3.5.2.1 **Equilibrium jets**

If a laminar jet is in pressure equilibrium with the external medium throughout its length it is said to be an equilibrium jet. As we shall see, this implies the jet is narrow and has a small opening angle, so we can reasonably assume that the fluid properties are constant across the jet ("top hat" profile). The small radial component of velocity appears only in the form of a variable cross-section. We assume the jet has a polytropic equation of state, $P_j \propto \rho_j^{\Pi_j}$. The assumptions of laminar flow and pressure balance suggest that there is little dissipation in the jet (unless the jet passes through a shock in the external medium), so the polytropic index Π_j is usually considered to be the adiabatic index Γ. By setting $\Pi_j < \Gamma$ we crudely allow for dissipation (Parker 1963; Ferrari *et al.* 1986), *e.g.*, low-level turbulence (§3.5.4), magnetic relaxation (Choudhuri & Königl 1986), or heat conduction in the guise of free streaming of electrons along the jet. The form of the jet is determined via the conservation of mass and Bernoulli's equation:

$$\rho v A = J \Rightarrow P^{\frac{\Pi_j+1}{2\Pi_j}} \mathcal{M} A = \text{const} \tag{3.34}$$

$$P^{\frac{\Pi_j-1}{\Pi_j}} \left[\frac{\mathcal{M}^2}{2} + \frac{\Pi_j/\Gamma}{\Pi_j - 1} + \frac{\phi}{c_s^2} \right] = \text{const.} \tag{3.35}$$

This form of Bernoulli's equation can be derived from Euler's equation and the polytropic equation of state; Π_j appears instead of Γ in the "enthalpy" term to allow for dissipation. If the jet is hypersonic, equation (3.35) reduces to $\mathcal{M} \propto P^{(1-\Pi_j)/(2\Pi_j)}$, while substitution of this into equation (3.34) gives $A \propto P^{-1/\Pi_j}$, implying a constant velocity. Combining these or assuming conservation of thrust also gives $\mathcal{M} \propto (PA)^{-1/2}$. Ferrari *et al.* (1986) give a more general solution to these equations valid for a power law mass distribution, and which assumes that $A(l)$ is given. The important point is that the collimation is determined by the pressure profile rather than by any intrinsic opening angle of the jet. If the jet is started off at a "nozzle" with a larger or smaller opening angle than the equilibrium jet, the jet will go out of pressure balance and the pressure gradients will tend to restore the jet to its equilibrium profile.

3.5.2.2 Free jets

The simplest non-equilibrium jet is the free jet, *i.e.*, a finite pressure jet expanding into a zero-pressure medium. A jet propagating down a very steep pressure gradient will approximate a free jet. The "nozzle" region is a classic expansion fan, and in such an expansion there is a maximum angle through which any given streamline can be bent, which is determined by the initial Mach number via the *Prandtl-Meyer function*, discussed more fully in §5.3.7 (see also Thompson 1972). The outer streamlines will be bent by this amount, hence the opening angle ξ_{max} of a free jet with an opening angle at the nozzle of ξ_0 is given by:

$$\xi_{max} - \xi_0 = \sqrt{\frac{\Gamma+1}{\Gamma-1}} \arctan\left[\sqrt{\frac{\Gamma+1}{\Gamma-1}}(\mathcal{M}_0^2-1)^{-\frac{1}{2}}\right] - \arcsin(1/\mathcal{M}_0)$$
$$\approx \frac{2}{\Gamma-1}\frac{1}{\mathcal{M}_0}. \tag{3.36}$$

The last form assumes $\mathcal{M}_0 \gg 1$. This equation shows that to look anything like a jet a free expansion must be supersonic. A simple explanation of this result is that the jet expands transversely at approximately its (initial) sound speed. Pressure continuity on the boundary ensures that the edge of the jet is at zero pressure; there is therefore a pressure gradient from the axis to the edge which initially drives the expansion. The expansion cools the jet and so the transverse expansion (and *a fortiori* the motion along the jet) becomes first super- and then hyper-sonic, in the sense that the remaining pressure gradient has negligible effect and the jet profile becomes self-similar, scaling linearly with distance from the nozzle (*i.e.*, far from the nozzle the streamlines are radial). The central density then scales as l^{-2} and the pressure as $l^{-2\Pi_j}$. The precise form of the profile is set by the effective nozzle (see below), but an important point is that the observed half-power opening angle of the jet will be substantially smaller than ξ_{max} because of the transverse gradients in u_{re} and B. Fig. 3.6 shows a numerical simulation of a jet becoming free.

3.5.2.3 Reconfinement

Although pressure forces push the jet towards the equilibrium profile, they do not necessarily succeed in making the jet track the equilibrium condition exactly, even in the steady state. If

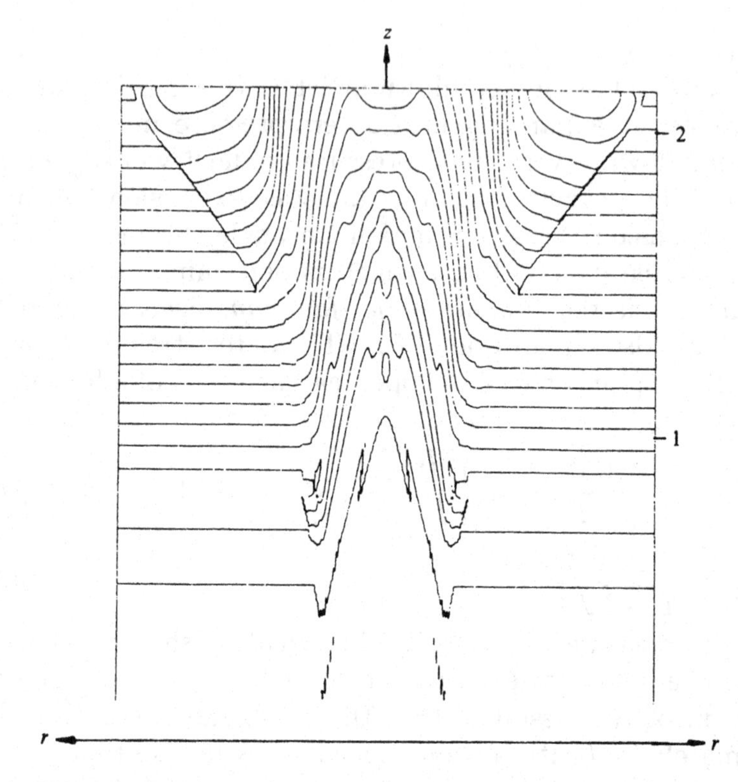

Fig. 3.6. Pressure contours of a steady-flow simulation of a jet in a rapidly-declining atmosphere (Wilson & Falle 1985). The jet is in equilibrium as it enters the grid at the bottom and is essentially free as it exits.

the external pressure changes abruptly the information about this change propagates into the jet at the sound speed. The jet will travel a distance $l_j = 2r_j\mathcal{M}$ before the information reaches the jet centre and the effects of the resulting pressure changes reach the surface again. l_j is thus the response length of the jet to variations in the external medium. We can define a characteristic length for variations in the external medium as $l_a = |P_a/(dP_a/dl)|$. Only if $l_j \ll l_a$ can the jet stay in equilibrium. Using the equilibrium relation $P \propto r_j^{-\Pi_j/2}$, this implies that $\xi \ll 1/(\mathcal{M}\Pi_j)$, *i.e.*, supersonic equilibrium jets have small opening angles. An initially equilibrium jet can depart from equilibrium even in a power law atmosphere ($P_a \propto R^{-a}$, so $l_a \propto R$). For the hypersonic case $l_j \propto P^{1/2}$, so if $a > 2$ the jet will pass out of equilibrium at some point determined by its initial Mach number. In a King model atmosphere, l_a has a minimum of $(2/3\beta_{\rm King})R_{\rm core}$ at the core radius, so a jet will go out of equilibrium near there unless $r_j\mathcal{M} \ll R_{\rm core}$.

What happens next depends on whether the opening angle of an equilibrium jet would be increasing ($a > 2\Pi_j$) or decreasing ($a < 2\Pi_j$). In the first case the jet cannot expand fast enough to reach the equilibrium configuration, so the jet will become free, and no shocks will form. In the second case, the centre of the jet will continue to expand at the upstream rate while the outer layers are turned inward. A shock will form at some intermediate radius, 'reconfining' the jet. The precise conditions for shocking are discussed in §5.3.7 in the context of jet bending, and can be applied to reconfinement shocks be identifying the curvature of the jet surface here with the curvature of the jet in that discussion.

Chan & Henriksen (1980; CH) captured some aspects of non-equilibrium jets in a simple axisymmetric self-similar model. They found a form which is an approximate solution to the MHD equations provided the velocity is mainly axial and the difference between the central and external pressure is much less than T/A. Bridle *et al.* (1981) showed that such a model will depart from equilibrium if the gradient of external pressure is steep enough. If the pressure gradient then flattens so the jet re-confines, the jet radius oscillates around its equilibrium value. However, the CH model does not include the shocks.

Wilson & Falle (1985) present numerical simulations of the full (axisymmetric) Euler equations to support the analysis sketched above. Fig. 3.7 shows one of their results. They find that jets will shock even if the maximum value of l_j/l_a is only 0.1. (If $l_j/l_a < 4\Pi_j$ they show that the shock will form only when l_j/l_a is decreasing after a peak.) The shock structures are conical, as originally found by Sanders (1983); the shock will reflect off the axis and the jet edge producing quasi-periodic structures. The pressure behind the shocks can be substantially higher than the ambient pressure: Falle (1987) claims the pressure can reach about 12 times ambient if the oblique shocks are strong. The point here is that shocks in steady-state jets can produce regions which are "unconfined" according to naive analysis: the fluid passing through these regions is indeed expanding but by the time it has expanded it is further down the (diverging) jet, and so there is no need to appeal to an additional confinement process.

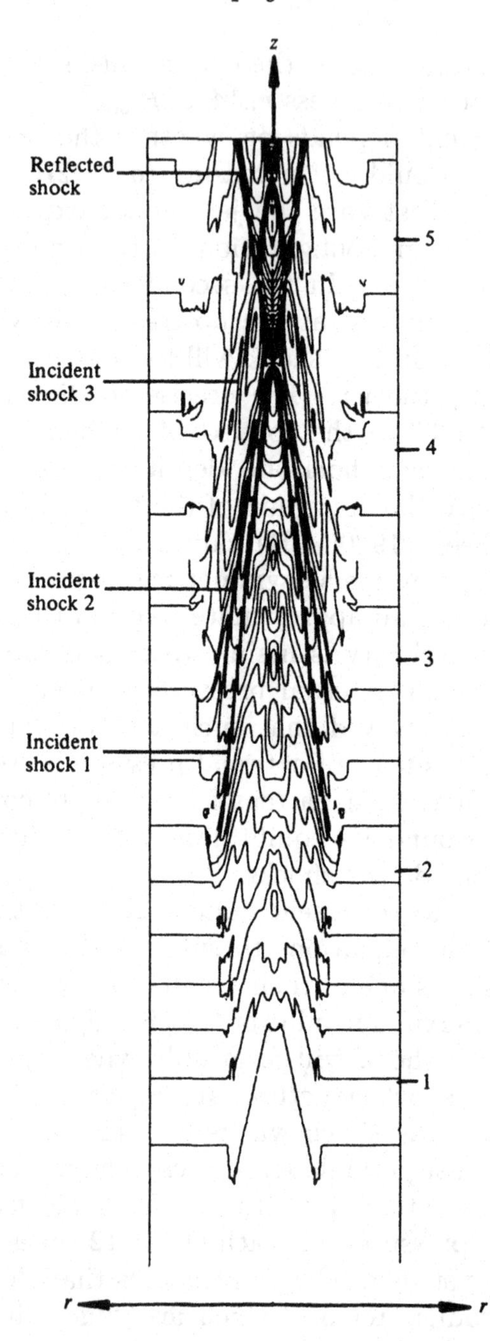

Fig. 3.7. Pressure contours of a jet in a reconfining atmosphere (Wilson & Falle 1985). The jet is in equilibrium as it enters the grid, goes out of equilibrium and is reconfined by three concentric conical shocks. Reflected shocks occur further down the jet.

It is tempting to identify the bright regions at the bases of FR I jets (the 'base knots') with quasi-stationary reconfinement shocks of this type, but at present a number of other possibilities cannot be excluded. Bicknell (1984) argues that these regions represent a transition to transonic rather than supersonic flow, and are accompanied by the development of strong turbulence. They may also represent blobs travelling outward along the jet, or non-stationary shocks (cf. §§4.3.3 and 7.2.4). This kind of 'random fluctuation' model for knots can be tested by searching for knot proper motion (apparently detected in 3C 120, Walker, Walker & Benson 1988) and also by examining the distribution of knot occurrence in a homogeneous sample of jets: systematic tendencies for knots to occur on particular scale sizes would support a quasi-stationary model.

3.5.2.4 Global models

Laminar models are sometimes used to deduce the run of pressure along the jet given the observed collimation, $r_j(l \sin i)$. The inclination angle i scales the derived lengths and Mach numbers ($\mathcal{M}_{\rm true} = \mathcal{M}_{\rm proj} \sec i$), but does not change the form of the pressure distribution (Wilson 1987a). One can adjust the initial overpressure and Mach number to obtain the most physically reasonable profile, although published models have assumed that the jet is initially in equilibrium. This has been done for both a CH model (for NGC 315, Krautter, Henriksen & Lake 1984), and a full gas-dynamic calculation (for NGC 6251, Wilson 1987a). An important constraint on the latter model is the spacing of major knots in the jet. To identify them with the pattern of reflected shocks, the knot spacing must be $O(r_j\mathcal{M})$, which typically gives $\mathcal{M} \sim 10$. These shocks turn out to be quite weak, and the flow is roughly adiabatic. It is therefore not surprising that the derived pressure profile is very similar to the one obtained by Perley, Bridle & Willis (1984), who fitted a CH jet with a simple analytic model for the pressure.

Eventually X-ray observations will determine the pressure profile, allowing direct testing of any laminar models. Existing data already shows that laminar jets cannot be in equilibrium prior to their initial flaring, as the above models assumed. This is because the models imply a very steep pressure gradient in the central regions ($a > 4$), and this is generally true for FR I jets because the opening angle increases in the flaring region. If the implied pressure peak is due to gas at $\sim 10^7$K, Perley *et al.* showed for NGC 6251 that its X-ray emission would exceed that detected by *Einstein* by at least an order of magnitude (the absolute pressure scale was set to equal the minimum pressure in the jet). Another problem with initial equilibrium is that the implied sonic point (where $\mathcal{M} = 1$) is several hundred parsecs from the nucleus, and hence that in the region between the VLBI and VLA scales the jet is subsonic. Perley *et al.* suggest instead (following Sanders 1983) that there is a large gap in scale-lengths between the initial (nuclear) collimation and the reconfinement regime at about 10 kpc, over which the jet is hypersonic and free. If the jet is nearly free its measured width will have little to do with the shape of the outer boundary and so cannot be used to deduce the external pressure.

Another advantage of assuming a free initial jet is that it may produce stronger reconfinement shocks than in Wilson's model, which will make it easier to re-accelerate particles and hence produce a 'sub-adiabatic' brightness decline. Strong reconfinement will substantially reduce the Mach number at the reconfinement shock, so the subsequent behaviour of the jet may be similar to that in Wilson's model.

In summary, shocks in steady laminar jets may provide enough dissipation to keep the surface-brightness up in some cases. To avoid a discrepancy with the X-ray data the jet must be initially collimated on a very small scale, break free, and be reconfined at $1-10$ kpc. The model may work best in 'knotty' jets where the spacing between major knots is several jet radii as in NGC 6251 and in the initial regime of M 87 (the criss-cross features in Fig. 2.18 are very suggestive of conical shocks), but more importantly as in most FR II jets. The overall dissipation in FR II jets seems to be much less than in FR Is which makes laminar models more plausible in the former. Unsteady laminar models may fit as well or better in the same cases. In jets whose surface-brightness declines smoothly and sub-adiabatically after an initial base knot, such as 3C 449 (Fig. 2.3) purely laminar models seem unattractive, although it is still plausible that the initial brightening at the base knot represents a reconfinement shock.

Choudhuri & Königl (1986) propose that magnetic reconnection, rather than shocks, provides continuous dissipation along quasi-laminar jets. Their model is discussed further in §9.6.4.

3.5.3 Relativistic jets

The evidence for and against bulk relativistic motion in large- and small-scale jets is presented in Chapters 2 and 4 respectively. Here I discuss the physical implications of relativistic motion, and also the possibility of subrelativistic large-scale jets when the parsec-scale jets are relativistic. For a discussion of the appearance of relativistic jets and possible observational diagnostics, see Chapter 4.

How fast is relativistic? Bridle & Perley (1984) define a jet as "one-sided" if the brightness ratio between jet and counterjet is greater than 4:1. Consider a population of randomly-oriented intrinsically symmetric twin-jet sources; half of them will be at less than 60° to the line of sight. From the discussion in §3.2.3, if the spectral index is 0.6 (typical for jets), then more than half will appear one-sided if $\beta > 0.52$, $\gamma > 1.2$. We might then take $\beta > 0.5$ as a definition of relativistic flow, although the jet would have to be somewhat faster before relativistic corrections have a large impact on the dynamics.

Bulk relativistic flows are qualitatively similar to ordinary ones. In fact the fully relativistic equations of steady fluid motion can be put into a form identical to the non-relativistic ones by a change of variables (Chiu 1973). (This is not true if MHD effects are important, or if the motion is unsteady.) Relativistic flows are essentially supersonic, since the kinetic energy density $(\gamma-1)(u_{\rm tot}+\rho c^2)$ is almost always greater than the internal energy density $u_{\rm tot}$. Wilson (1987b) presents a fluid-code simulation of a relativistic steady jet. Although his model contains typical X-type internal shocks, he shows that the simple Doppler beaming relations (§3.2.3) remain fairly accurate when averaged along the jet. This reflects the fact that for his boundary conditions the opening angle is small and the shocks are oblique and relatively weak.

The evidence for relativistic motion is much stronger for parsec than for kiloparsec jets, so there has been some interest in models in which only the former are relativistic. The simplest of these is just that the jet slows down between the two scales. Because of the conservation of thrust for highly supersonic jets, such slowing must be due to entrainment. This involves dissipating most of the initial energy (§§4.4.2 and 5.4.3); for FR IIs the power dissipated would be much larger than the bolometric luminosity of the galaxy. It is sometimes argued that it is hard to prevent a substantial fraction of the energy being radiated, so the lack of observed radiation implies that the jet is not slowing down. The argument is not watertight: if the energy is dissipated slowly and smoothly, buoyancy in the galactic core could reconvert much of it into kinetic energy, of course with a lower speed since the discharge has increased (cf. Begelman 1982).

In any case there is no problem for FR I sources since their jets may well be weak enough that the power dissipated could be equal to the observed nuclear line emission. Such a model could explain why FR I jets are more asymmetric on small scales (cf. §2.2.4). Since the small-scale brightness asymmetry rarely actually reverses further down the jet, one needs the large-scale jets to be sub-relativistic ($\beta \sim 0.2?$) also. VLBI data rarely shows superluminal motion in FR Is, but a prominent exception is 3C 120 (Walker, Benson & Unwin 1987) which, in its large scale structure and luminosity, resembles a wide-angle-tail.

Another possibility is the 'double jet' scenario (*e.g.*, Bridle 1984; Sol, Pelletier & Asséo 1989; see also Chapter 8) in which a weak but relativistic parsec scale jet is surrounded by a higher-power but sub-relativistic envelope which forms the large-scale jet. High-dynamic range VLBI and VLBI/MERLIN images are beginning to appear which show continuity between small and large-scale jets with no sign of such a core-sheath structure.

All "small-fast, large-slow" models predict a wide range of ratios between core and jet luminosity, since the large-scale jet will be seen at any angle while the core jet is weak outside its beaming cone. While there certainly is a large range in core-jet ratio in the literature (Lonsdale & Barthel 1987) the test needs to be done with a homogeneously-selected, randomly oriented sample, and such a study has not yet been made.

3.5.4 Steady turbulent jets

Turbulence is ubiquitous in jet-like flows in the everyday world, and is almost certainly important in astrophysical jets. This is unfortunate since it is also rather poorly understood, even in the laboratory. A classical approach is to average the Euler equations over an ensemble of statistically identical flows. This has been done for statistically-steady high-*Re* jets by Bicknell (1984), following Favre (1969). The averaging scheme preserves the continuity equation for the mean flow, so the concept of streamlines remains valid. The momentum equation contains an extra turbulent (Reynolds) stress $\boldsymbol{\Psi}_{Re} = \langle \rho \mathbf{v}'\mathbf{v}' \rangle$, (here primes indicate the fluctuating part of a quantity). Although the Reynolds stress cannot be calculated precisely it behaves a bit like viscous stress and therefore the simplest approximation to turbulent flow is viscous flow. This approach is developed in §5.4.1; see also Baan (1980) and Henriksen (1987). The Reynolds stress slows, widens, and entrains matter into the jet. The widening happens because by conservation of mass flux the averaged streamlines must diverge, while by conservation of momentum material at the edge of the jet is accelerated, further widening the velocity profile.

Turbulence also provides effective diffusion, and finally enhanced dissipation, since the Reynolds stress converts mean to turbulent kinetic energy and this can be dissipated effectively to heat or fast particle energy.

Turbulent stress has little dynamical effect on hypersonic jets, essentially because the turbulent velocities cannot substantially exceed the sound speed. Thus turbulence causes little entrainment, and its main effect is dissipation. Early work on turbulence in jets focussed on dissipation as the solution to the surface-brightness problem discussed in §3.5.2 (*e.g.*, Henriksen, Bridle & Chan 1982).

The rate of dissipation per unit mass is $O(v'^3/r_j)$ (*e.g.*, Landau & Lifshitz 1959), and dimensional analysis shows that even with $v' \sim c_s$ a jet can remain roughly in pressure balance with this heating rate, so the jet follows the equilibrium relations (cf. §3.5.2) with $\Pi_j < \Gamma$. For instance Begelman (1982) shows that in an α-model jet, *i.e.*, one where the shear stress is a constant fraction α of the pressure, the jet opening angle will be $\xi \sim \alpha$. With constant opening angle and $P \propto R^{-a}$, we have $\mathcal{M} \propto R^{(a-2)/2}$, which implies $\Pi_j = a/2$. For $a < 2$, which is consistent with observed X-ray distributions ($\beta_{\rm King} < 2/3$), the Mach number decreases along the jet. Eventually the jet will become transonic, and then the dynamical effects of the stress become important. Provided the jet remains light compared to the external medium, the pressure gradient along the jet will prevent the Mach number falling much below unity. Thus the combined effects of dissipation and pressure tend to force light jets towards the condition $\mathcal{M} \sim 1$. Note that this is the *internal* Mach number. Because the jet is light, its sound speed is above that in the confining medium, so the jet speed is substantially faster than the external sound speed.

Bicknell (1984, 1986b) has developed a simple model for transonic turbulent jets which is in some ways equivalent to the equilibrium approximation for laminar jets. The main difference is that turbulent jets have a finite entrainment rate which is an extra unknown parameter (cf. §3.4.3). An extra constraint is therefore needed, and Bicknell uses the observed spreading of the jet. The model is in the tradition of semi-empirical studies of laboratory turbulence: the Reynolds stress is not calculated from first principles (not possible anyway) but is derived from the observed structure. An important consequence is that the equations do not require that the shear stress is specifically due to turbulence: what is really derived is the overall entrainment rate.

The free parameters, the inclination angle i and the fiducial Mach number and density contrast ($\mathcal{M}_0$, η_0), can be determined by simultaneously fitting the brightness profile, which also provides a check on the model since arbitrary brightness profiles cannot be fitted. In the absence of independent data on the environment Bicknell fits simple models of the pressure distribution.

The basic assumptions are that the jet is statistically steady, nearly self-similar, and in pressure equilibrium. The last implies that $v' \ll c_s$. Bicknell claims that

$$\xi \rho v_c^2 = \xi \Gamma P_c \mathcal{M}_c^2 \approx \langle \rho v_z' v_r' \rangle \approx \langle \rho v_r'^2 \rangle. \tag{3.37}$$

Here v_c is the central velocity, and z and r are cylindrical coordinates. This implies that jets with $\mathcal{M}_c \lesssim 1$ are necessarily close to pressure balance. The first approximation in equation (3.37) is derived from dimensional analysis of the equations of motion, but the second is not justified in detail. In fact the left-hand side could be considerably smaller than the right because of decorrelation between the velocity components. If so, the turbulent kinetic energy would be comparable to the momentum density, and hence the pressure if $\mathcal{M} \gtrsim 1$. Thus pressure balance is not required in transonic turbulent jets, but must be explicitly assumed. The assumption is supported by laboratory studies which do not show substantial overpressures in such jets, in contrast to the model of Henriksen (1987).

The model is simple when the jet is hot and thermally-dominated (so that Γ is constant). Then the entrained mass is significant but the entrained energy is not (so the power K is roughly conserved). For a jet in pressure balance this is equivalent to the assumption $\eta \ll 1$ (the precise value of η is then irrelevant). In this special case of pressure balance and statistically steady flow (*i.e.*, $P' = 0$, $\partial/\partial t = 0$), the thrust equation (3.29) shows that T' is only affected by the pressure gradient. From the order-of-magnitude estimates above we can ignore the stress contribution to the thrust, so for a 'top-hat' profile (not a critical assumption) $T' \approx \rho v_c^2 A = \Gamma \mathcal{M}_c^2 P A$. Similarly $K \approx \rho w v_c A + T' v_c/2$, since the turbulent terms are small, as is the gravitational term since we are considering a light jet in a hydrostatic atmosphere. With equations (3.29) and (3.28) we get the following equation for the velocity:

$$\left(1 + F_1 \frac{\Gamma - 1}{2} \mathcal{M}_c^2\right) \frac{d \ln v_c}{dl} + \frac{1 + (1 - F_2/2)(\Gamma - 1)}{\Gamma} \frac{d \ln P}{dl} + \frac{d \ln A}{dl} = 0. \quad (3.38)$$

The factors F_1 and F_2 are ratios of various dimensionless integrals over the jet cross-section and are unity for top-hat profiles. The terms involving $(\Gamma - 1)$ are related to dissipation; for transonic jets they are both small. If they are ignored we recover Bicknell's equation $P^{-1/\Gamma} v_c A = \text{const}$, which he derives from flux conservation of the relativistic particles which produce the pressure, assuming no dissipation.

Bicknell (1986b) treats the general case when the entrained energy contributes significantly to the pressure (*i.e.*, when $\eta \to 1$). This is a little more complicated; for instance the relativistic and thermal pressures are comparable and have to be handled separately. In general one would need to know how dissipated energy (*i.e.*, the TdS term) is distributed between the two pressure components; Bicknell avoids the problem by neglecting dissipation, which is consistent with his other assumptions over short sections of transonic jets.

From the velocity profile, an adiabatic relation (Table 3.3) (depending on the observed polarization) gives the surface-brightness. If the jet is fully turbulent, the tangled-field relation seems the most appropriate (cf. §3.1.3), although Bicknell prefers the $B_\perp$ case; the point is discussed below.

A key feature of Bicknell's model is buoyancy, which allows the jets to propagate without disrupting for far more nozzle diameters than lab jets manage, by counteracting the slowing and flaring caused by entrainment. He suggests that some FR I jets terminate simply because the density contrast reaches unity, in which case the velocity quickly becomes negligible.

The most impressive result of this approach is that entraining models naturally reproduce the widespread effect that brightness decline is *anticorrelated* with opening angle (Parma *et al.* 1987), while laminar models predict the opposite. This works because equation (3.38) shows that for transonic jets the dependence of the velocity on the cross-section is $v \propto A^{-1}$. Inserting this into equation (3.33) we get $j_\nu \propto A^{(\alpha+1)/2} P^b$, where b is some constant. Thus if an increase in opening angle is due to increased turbulence rather than to faster pressure decline, we expect the rate of decline of j_ν to decrease, or even reverse.

In this analysis many terms of order ξ (~ 0.1) were neglected (*e.g.*, dissipation, transverse pressure gradients), and substantial errors may accrue if the model is applied over sections of the jet much longer than 10 (initial) radii.

Henriksen (1987) considers the case where turbulently-induced pressure gradients are significant. The pressure field set up is peaked on the jet axis (a standard result for 'viscous' jets, *e.g.*, Landau & Lifshitz 1959) and Henriksen suggests that the apparent overpressure in some jets may be due to this effect, which he calls "viscous collimation".

A serious problem for *any* turbulent model for FR I jets is the projected magnetic field configuration. Turbulent lab jets have a Gaussian-like velocity profile, indicating significant shear throughout the jet. Begelman *et al.* (1984) point out that a shear of $\partial v/\partial r > \xi v/r$ is sufficient to maintain a parallel field configuration despite the expansion and slowing of the jet. The shear in a Gaussian jet is $O(v/r)$ so we should expect a parallel magnetic field, although detailed simulations of the brightness and polarization across a turbulent jet have not been done. In fact the field in FR I jets is mainly perpendicular to the jet, with only a thin region of parallel field at the edge. For this reason Bicknell's later papers argue for 'top-hat' profiles in which shear is confined to a thin boundary layer; however it is not explained why there should be such a gross difference between astrophysical and laboratory jets.*

Actually there is a plausible agent: the magnetic field itself. In fact it is rather difficult to justify the neglect of magnetic forces in turbulent models, since the magnetic field will be amplified by the turbulence (§9.6) provided it is not too small to begin with (Parker 1979, p. 517). "Too small" here means several orders of magnitude below equipartition; in radio jets, such fields would usually imply embarassingly large pressures, so we are forced to conclude that u_B and $u_{\rm turb}$ are at least comparable. In this case the character of the turbulence completely changes: the wave spectrum goes from Kolmogorov to Kraichnan (cf. Eilek 1989a,b) and magnetic stress becomes at least as important as the Reynolds stress. For a frozen-in field not precisely aligned with the streamlines the magnetic stress in a shear layer is proportional to the amount of stretching, which increases linearly with distance along the jet, at least until reconnection takes place (after which the stress will remain constant). Thus magnetic forces tend to prevent shear except in a thin boundary layer where the velocity gradient is large (cf. Jackson 1975, p. 475). In astrophysical jets, such a boundary layer would be turbulent and reconnection would occur through turbulent diffusion.

Such a basically magnetic model seems rather far removed from a turbulent jet, but recall that Bicknell's analysis is independent of the origin of the stress. In fact it works somewhat better in the magnetic case since it is more plausible here that the off-diagonal terms in the stress tensor are of the same order as the diagonal terms (*i.e.*, equation (3.37)), justifying the neglect of the corresponding energy density compared to heat. In addition the central laminar flow justifies the use of the $B_\perp$ adiabatic relation to predict the surface brightness.

If turbulence is confined to a boundary layer, entrainment con-

* But note that the model includes a variable amount of turbulence and in some regions implies nearly laminar flow. These are often the more polarized regions.

sists of a convergence of external streamlines onto the jet. The new material contains no relativistic particles, and turbulent diffusion occurs only as material passes through the boundary layer, so some way down the jet there may be very few relativistic electrons in the outer layers. Essentially we would have an invisible sheath surrounding the observed jet with similar velocity but a substantially larger opening angle (cf. Henriksen 1987). Fitting the observed jet profile will then underestimate the entrainment (although conservation of relativistic particles will still give roughly the right velocity profile).

Another feature of boundary layer models is that the dissipation is confined to the surface. Unless this heat can be efficiently conducted into the jet or away (*e.g.*, by Alfvén waves), it will tend to create a hot, low density sheath surrounding the jet which may inhibit entrainment and further dissipation. If a substantial fraction of the dissipated energy goes into relativistic particles, the jet boundary could become the main emission region, as is apparently occurring in M 87 (Owen *et al.* 1989). Alternatively, if the emissivity were proportional to the density, the sheath might show up as diffuse emission surrounding a higher surface-brightness jet. Such emission has been noted in the twin-tail source NGC 1265 (O'Dea & Owen 1986), M 84 (Bridle & Perley 1984) and in unpublished data on several other jets.

In summary, the concept of buoyant, fully-turbulent jets may apply best to regions where the magnetic field is parallel to the source axis, such as the diffuse "plumes" in some twin-jet and wide-angle-tail sources. These regions often show large-scale filaments and other complex structure which are suggestive of massive entrainment of the (radio-quiet) external medium, implying that the entrainment process has nearly proceeded to $\eta = 1$. Weak turbulence in supersonic jets may be a more plausible source for sub-adiabatic brightness decline than large-scale shocks, since the re-acceleration can be more evenly distributed; this may be relevant in FR II jets. The well-studied FR I jets remain an anomaly because their perpendicular field configuration implies weak shear even though, as we shall see in §3.5.6, they are very likely subsonic as they end. Perhaps in these jets turbulence is suppressed by magnetic tension except in a boundary layer.

3.5.5 Wiggles

Steady jet models are a useful approximation, but real jets are certainly not completely steady. A sure sign is the wiggles in many of them, which range from low-amplitude oscillations (NGC 6251, Saunders *et al.* 1981) through sharp kinks (3C 449, Fig. 2.3) to complete loops (3C 40, O'Dea & Owen 1985). Such patterns cannot be fixed relative to the external medium, or even in some special frame of reference, unlike the idealised twin-tail source (§3.4.4).

The simplest class of model invoked to explain wiggles are ballistic jets (§5.3.1). Ballistic motion implies that the only significant force acting on a fluid element is gravity; in particular, ram pressure is neglected. A jet in a perfect vacuum is both ballistic and free, but the two approximations should not be confused: they correspond respectively to $P_a' \approx \rho_a v_\perp^2 \ll \rho_j v_c^2$ and to $P_{0a} \ll P_j$. Thus a dense jet in a hot medium can have a ballistically-determined path and yet be confined transversely by thermal pressure. Wiggles in ballistic jets result from variation in the direction of ejection, or from motion of the host galaxy. The latter case implies that the jet speeds must not be very much larger than the orbital speeds of galaxies (otherwise the jets would be nearly straight); consequently gravitational deceleration is important. Unfortunately there has been no investigation of whether this could account for the sub-adiabatic behaviour of FR I jets.

Ballistic models apply to jets much denser than their immediate surroundings (such as SS 433, and probably the jets in FR II sources relative to their cocoons, cf. §3.5.6) and for small-amplitude wiggles in high-thrust jets. Ram pressure drag will always become important if the jet propagates far enough. The onset of this regime is hastened because ballistic wiggles increase linearly in amplitude along the jet; this increases the angle between the jet axis and the velocity, and hence $v_\perp$.

Orbital ballistic models for FR I sources (*e.g.*, Blandford & Icke 1978) imply jet speeds of several hundred km s^{-1}, and $\eta \gtrsim 10$. The implied discharges are of order $1 M_\odot yr^{-1}$. These velocities are slower than most other estimates, and the discharge is on the high side, but not out of the question. The density contrast should increase along an equilibrium laminar jet. If the external medium is close to isothermal and the jet is adiabatic, then

$$\eta \propto \rho_a^{1/\Gamma_j - 1} \approx \rho_a^{-0.4}. \tag{3.39}$$

This might allow a transition to ballistic motion as the jet passes through the large pressure and density drop between broad-line and galactic scales.

Although in Chapter 5 Icke emphasizes the ability of ballistic models to reproduce source morphology, the main reason why such models are not popular (especially as applied to FR I jets) is that "the fit is not especially good" (to quote Blandford & Icke). Typical models have 5 – 10 free parameters and a polynomial of similar order would probably do as well. Real jets brighten on the outside edges of bends and the magnetic field is parallel to the edge, suggesting compression and/or shear, and hence that ram pressure is affecting the jets. The wiggles in real sources do not increase in amplitude linearly along the jet. The increased length per unit mass might be expected to produce a narrowing of the jet as the wiggles develop, which is the opposite of what is seen (although if the jet consisted of many independent plasmons, this would not happen). Finally the jet-lobe transition strongly suggests that $\eta < 1$ (§3.5.6).

There is often remarkable mirror or rotational symmetry between the two jets in a source, which strongly implies that, respectively, galaxy motion or precession is responsible for the wiggles. Rejection of ballistic models is often taken as rejection of these processes, but of course jets from orbiting or precessing engines do not *have* to propagate ballistically. Unfortunately the effect of ram pressure on such jets is practically impossible to calculate: one would need a high-resolution 3-D fluid code to do the job properly.* However one can make a few general comments. The qualitative structure and in particular the basic symmetries will remain the same as in the ballistic case, but the pattern speed of the wiggles will no longer be the same as that of the jet material. In most cases it will be slower, with the flow of material around the bend providing resistance against ram pressure and allowing a lower density contrast. Light jets will resonate when the driving frequency is close to one of their fundamental modes (Chapter 6). If light jets are allowed,

* Hence the true but empty claim that ballistic models give the best fits to the structure of complicated sources: there are no competitors.

buoyancy also becomes important. Buoyant acceleration away from the central galaxy will increase the wavelength of the wiggles in the outer jets.

Finally, it is quite possible that some wiggles represent growing (undriven) instabilities, as discussed in Chapter 6. In this case we do not expect detailed symmetry between the two jets, and indeed there are many cases where no symmetry is seen.

3.5.6 End effects: the standard model for FR IIs

So far in this section we have discussed only the problem of jet propagation.

Historically, in radio source physics, most attention has been focussed on termination of the jet; in fact 'beam models' were first introduced by Rees (1971) in order to produce the hotspots of FR II sources at the end of the beams. The current model for FR II sources is discussed in detail in Chapter 7, so here I will only review its main features, and comment on the way it can be extended to cope with FR Is and the double sources found in Seyfert galaxies.

We have already seen that the overall advance of the hotspots is determined by thrust balance (§§3.4.1 and 3.4.4). If we work in the frame where the hotspot is stationary, we can equate the sum of thermal and ram pressure in the jet and in the external medium:

$$\rho_j(v_j - v_{HS})^2 + P_j = \rho_a v_{HS}^2 + P_a. \tag{3.40}$$

If we express the thermal pressure as $P_{\rm th} = \rho v^2/\Gamma\mathcal{M}^2$, we obtain for the ratio of jet to hotspot speed:

$$(1 + (\Gamma_j\mathcal{M}_j^2)^{-1})v_j/v_{\rm HS} =$$
$$1 + \sqrt{\frac{1}{\eta}(1 + (\Gamma_j\mathcal{M}_j^2)^{-1})(1 + (\Gamma_a\mathcal{M}_{\rm HS}^2)^{-1}) - (\Gamma_j\mathcal{M}_j^2)^{-1}}. \tag{3.41}$$

Strictly this only applies when the jet is supersonic and non-relativistic, but the relation will not be far wrong if these conditions are relaxed somewhat. When both the jet and the bow shock are hypersonic, equation (3.41) reduces to the simple formula:

$$v_j = v_{\rm HS}(1 + \eta^{-1/2}). \tag{3.42}$$

If $\eta \ll 1$, this implies that the mass and energy currently in the jet are much less than the total mass and energy supplied over the source lifetime. Some of the energy is radiated away and some is used up doing work on the external medium, but most of it must be elsewhere. At first it was thought that the energy would pile up in the hotspot, but Longair, Ryle & Scheuer (1973) showed that this was not plausible, and instead the energy must flow through the hotspot to a "waste-energy basket". Scheuer (1974) showed that in a simple model this consisted of a very low density cavity surrounding the jet, and this basic picture has been confirmed by numerical simulations. Scheuer identified this cavity with the diffuse emission between hotspot and nucleus now referred to as the 'bridge'.

Thus light jets manufacture for themselves a very low density medium through which they can propagate, thereby stabilising themselves against surface instabilities and allowing nearly ballistic propagation until the jet runs into the cavity wall. The cavity therefore has similar beneficial effects to the possible hot sheath generated by friction on the jet surface, which we considered in §3.5.4.

Scheuer (1974) also pointed out that over time buoyancy would remove the cavity from the centre of the source, (see also §7.3.3). This would leave the beginning of the jet in direct contact (apart from the shear layer) with the dense external medium, so the discussion in §§3.5.2 and 3.5.4 will apply. Jets in FR II galaxies are occasionally bright near the core (*e.g.*, 3C 219, Bridle *et al.* 1986) which may indicate some such high-dissipation regime before the jets reach the safety of the cavity. Actually such jets appear to be surrounded by bridge emission, but this could be a projection effect.

A further problem for jet propagation is that the contact discontinuity between the shocked external medium and the backflowing jet material (see §7.1.2) is Kelvin-Helmholtz unstable. This will cause entrainment into the cavity, which will reduce the radio surface brightness by lowering either the filling factor of the relativistic material or its partial pressure, depending on the completeness of the mixing. Progressive entrainment is a possible explanation for the brightness gradient found in the bridges of powerful FR II galaxies (Leahy *et al.* 1989). Of course entrainment would partially nullify the protective effect of the cavity.

High-resolution simulations (*e.g.*, Smith *et al.* 1985; Lind *et al.* 1989) do show the development of large-scale eddies on the contact discontinuity, but this result cannot be taken at face value. Because the simulations are axisymmetric, the 'turbulence' is two dimensional and will behave very differently from true three-dimensional turbulence. In particular, there is an 'inverse cascade' of energy to small wave-numbers which results in most of the 'turbulent' energy winding up in large-scale eddies, in contrast to 3-D turbulence in which the Kolmogorov cascade transports energy to small scales where it is efficiently dissipated. Thus the 'turbulence' in high-resolution simulations is probably grossly exaggerated: paradoxically, low-resolution simulations in which numerical viscosity damps out such features may give a more accurate *qualitative* picture of radio source structure. The bottom line is, unfortunately, that fully three-dimensional simulations are required before a quantitative evaluation of the effects of entrainment into bridges is possible.

Where do FR I sources fit into the picture? The key observational peg on which the FR II model hangs is the existence of very bright, sharp-edged hotspots, which need to be confined by shocks and therefore imply light, supersonic jets. FR Is have no hotspots, and hence the outer regions of their jets are either heavy (*i.e.*, neo-ballistic models) or subsonic (*i.e.*, Bicknell's models and the variants discussed in §3.5.4). The predictions of the two for the head advance speed are quite different, as can be seen from equation (3.41). If $\eta \ll 1$ but $\mathcal{M}_j \sim 1$, we still have $v_{\rm HS} \ll v_j$ and we expect a bridge to be built up much as in FR II sources, but without the hotspots. On the other hand if $\mathcal{M}_j \gg 1$ and $\eta \gg 1$ then $v_{\rm HS} \sim v_j$ and no bridge is expected. In intermediate cases, a smallish lobe will form around the end of the jet.

The archetypical FR I source is 3C 31, in which no sign of bridge emission has ever been detected, and there is a general impression that this is typical, giving strong support to ballistic models. Unfortunately this is a case where the wrong archetype has been chosen. In a sample of 54 low-luminosity sources with jets (Parma *et al.* 1987) about half the jets are surrounded by diffuse emission, in most cases closely resembling the bridges in low-luminosity FR IIs. In most of the rest the jets are initially 'naked' and later enter diffuse 'plumes' with substantially poorer collimation, and (where the data is available) parallel magnetic field configurations. 3C 31 is a member of the latter class, as are wide-angle-tail sources and related objects. These objects are systematically larger than the sources with bridges, and also slightly more luminous. The remainder apparently consist entirely of naked jets, but these are generally the least luminous sources and the diffuse emission may simply be too faint to detect. High-sensitivity, low resolution observations of the FR I sources with bridges show that there is no diffuse emission extending much beyond the end of the jets, confirming the strong similarity with FR II sources. For the sources with plumes or tails, some fade out slowly as in 3C 31, while others show sharply-defined outer boundaries, as in NGC 315.

These results imply that at least in the bridged FR I sources, the jets are subsonic and light. Because of the similarity in brightness profiles, collimation, and magnetic field patterns between jets with bridges and those with plumes, we should avoid using radically different models for the two subclasses. This weighs against dense-jet models for the latter. Instead, plumed sources may be interpreted, following Bicknell, as objects in which entrainment has brought the density contrast up to unity. Alternatively, they may be sources in which buoyancy has removed the bridge from the central galaxy. In many cases the tailed sources are in regions of high galaxy density, and a cluster-scale atmosphere may explain the large distance over which buoyancy must be effective. Very likely both these processes are important in particular cases. Note that both processes should be more important in longer (hence older) jets.

Finally we consider the small radio sources in Seyfert galaxies. A feature of Seyfert galaxies which is likely to make the radio structure qualitatively different from that in radio ellipticals is that the external medium, *i.e.*, the interstellar medium near the centre of a spiral galaxy, is dense and cool enough that cooling shocks are possible. Such shocks are a standard feature of supernova theory (*e.g.*, Osterbrock 1988). Briefly, most of the heat generated by shock dissipation is quickly radiated away, so the pressure increase across the shock is achieved by a large increase in density. Thus the shocked external medium ahead of the jet forms a thin layer on the surface of the 'cavity', rather than a thick, hot region as found in the simulations of Chapter 7. Apart from this, the basic picture probably remains rather similar to the standard model. Note that Scheuer (1974) assumed such a thin layer of swept-up material without much affecting the applicability of his model to elliptical galaxy radio sources. Cooling shocks in Seyfert galaxies may produce the narrow-line gas, and hence explain its disturbed kinematics (Pedlar, Dyson & Unger 1985). The case for this is particularly strong in NGC 1068 (Wilson & Ulvestad 1983). Wilson & Ulvestad (1987) have also pointed out that in Seyferts the external medium contains significant energy density in relativistic electrons and magnetic fields (normal components of spiral interstellar media). Hence we should expect enhanced radio emission from the bow shock as well as the jet material. They make a convincing case that the edge-brightened NE lobe in NGC 1068 is in fact emission from such a bow shock.

There is one other class of radio sources interacting with dense, relatively cool environments: the very bright FR II sources found at high redshift. The very large amounts of forbidden line emission associated with these objects makes it attractive to assume that these interactions at the epoch of galaxy formation were similar to those now happening in Seyfert galaxies.

3.6 **Open questions**

This chapter may have given the impression that all questions are still open, and the vast increase in data quality produced recently, especially by the VLA, has not greatly improved our understanding of radio jets. In fact this is not true. To begin with, there is no longer serious doubt that the jet paradigm is fundamentally correct, even if current models are still highly over-simplified. Perhaps the most important advance in recent years is our understanding of cocoon production, which indicates that all large-scale jets initially have low density-contrast, and that the fundamental difference between FR I and II structures is that the latter are completely supersonic while the former are subsonic when they terminate. We should also be grateful that improved observational data has highlighted the weaknesses in the 'conventional arguments' which were previously applied indiscriminately (cf. §§3.2 and 3.4).

It remains true that large-scale jets remain an enigma in many ways. Most obvious is that we do not yet know the values of the basic physical parameters, notably the speeds of jets. Perhaps more interesting is the extent to which the simple fluid models discussed in §3.5 are incorrect. It should be clear that the detailed structure of real sources is far more complicated than the models. Such structure is often dismissed as "weather", but this is both premature and misleading. Premature because a number of features, such as the filaments observed in the lobes of Cygnus A, suggest large-scale order rather than random processes. Misleading because, to continue the analogy, a general understanding of the small-scale dynamics (weather) is essential if we are to understand the large-scale dynamics (climate), as any meteorologist will attest. Such structural details have been attributed to a number of complicating factors, most of which are almost certainly present. They include instabilities in the flow (Chapter 6), dynamically-important magnetic fields, and variability in the output of the central engine. Unfortunately we have no space to discuss the last two in detail. For magnetic fields, see the reviews by Asséo & Sol (1987) and Saikia & Salter (1988), and the conference proceedings edited by Asséo & Grésillon (1988). There is no overall review of the evidence for central-engine variability: multiple outbursts on various time scales are suggested by dramatically X-shaped sources (Leahy & Williams 1984), by 'pieces of jet' (Rudnick 1984; Bridle *et al.* 1986), and by the observed variability of central components in double sources. Evidence for precession of the ejection axis is provided by the common occurrence of some degree of rotational symmetry, occasionally dramatic (*e.g.*, Condon & Mitchell 1984). On the other hand the correlations between the luminosity of radio cores, nuclear and narrow-line emission regions, jets, hotspots, and lobes (Fig. 2.7), despite their very different response times to central-engine variability, set some limits to the typical amount of variability in the population (*e.g.*, Leahy *et al.* 1989). In particular, order-of-magnitude variability is restricted by such correlations to occur on the longest possible time scale, *i.e.*, the overall waxing and waning of nuclear activity in the host galaxy.

Finally, we are only beginning to think about the big picture, the overall relationships between the different sources we observe. The 'unified schemes' discussed in Chapters 2 and 4 provide at best a fragment of this. At first sight it seems impossible to hope that the whole picture can be uncovered. The structures of radio sources are influenced by the various jet parameters discussed in §3.4.2, by the details of the external medium, and by the age of the source. At least eight parameters are needed to specify even the simplest model jet and atmosphere. But the situation may actually be much simpler if a number of these parameters (or combinations of them) vary little from source to source. The most compelling evidence for such simplification is the relatively good correlation between source structure and integrated radio power, in particular the sharpness of the FR division: the population changes from almost all FR I to almost all FR II over a factor of less than 3 in power. This is quite mysterious if the division is essentially one in Mach number, since that is only one of several factors which should influence the radio power. The puzzle is even more intriguing because, at powers near the break, the two types of sources clearly exist in different environments (§2.2.4).

Baldwin (1982) has argued that the P-D diagram (*e.g.,* §2.4.1) represents the 'fundamental plane' of radio sources, equivalent in some ways to the Hertzsprung-Russell diagram for stars. The implication is that radio sources can be reduced to two important parameters, which for Baldwin are the jet thrust and the source age. The distribution of sources on the P-D plane then constrains both the P-D evolution of a source with a given jet power, and the distribution of powers. At present such a decomposition is highly speculative; in addition, the density of sources on the P-D plane is affected by Malmquist bias (trivially corrected) and by cosmological change in the population, which is known to be large but is extremely uncertain (*e.g.,* Peacock 1985).

Apart from such practical difficulties, the basic idea implies that sources with similar P and D should be similar in other respects. This is true to some degree, but mainly because of correlations with P rather than D; sources of very different size but similar power are often quite similar in structure and spectrum.* In particular there is little evidence that the FR division is a function of linear size (but see below). In fact, except for a few outliers, the scatter in linear size at a given power is hardly more than would be expected from the combined effects of projection and constant expansion speed for each source, with a small spread in maximum attainable size.

A possible way to make progress here is to focus on sources which clearly buck the major trends. By finding what is *different* about such sources, we may highlight the important factors which are similar for the rest. An obvious starting point is the FR division. A number of FR I sources have powers an order of magnitude or more

* Except that, for constant P and similar structure, small sources necessarily become optically-thick at higher frequency.

above the break. These objects are found in regions of high density such as clusters of galaxies, and can be classified as wide-angle-tails (§2.2.4) on the basis of their jet structure, although they often lack the C-symmetry which originally gave the class its name. It is not clear whether these sources are unusual because they contain relatively weak jets which create unusually bright radio emission (because of better confinement?), or because their jets would normally be powerful enough to remain supersonic but for some reason ('heavy weather'?) go subsonic unusually early.

FR II-like sources with powers far below the break are found in Seyfert galaxies. Since the densities here are also higher than in typical FR II sources, it is clear that density is not the only contributing factor. The fact that such sources are very small may be significant: presumably short jets are more likely to remain supersonic to their ends.

Prospects for answering some of these questions in the next decade are good. The ROSAT, XMM, and AXAF X-ray satellites should provide much-needed data on the direct environments of radio sources so that environmental effects can be investigated in detail. Continuing study of the cooler phases of the ISM of host galaxies and of radio-quiet control samples promises to give insight into the supply of fuel to the central engine and hence into the fundamental problem of what triggers radio activity. Meanwhile MERLIN and the VLA are at last producing high-quality images of large enough samples that systematic structural trends can be studied in detail, and distinguished from the peculiarities of individual sources. In addition, detailed morphological statistics can be built up, giving some teeth to the symmetry and variability arguments I have outlined.

Thanks are due to many colleagues who have read and commented on parts of this article, especially Philip Hughes, Jean Eilek, and Joan Wrobel. Thanks also to Peter Scheuer, Steve Gull, Mike Wilson, Robert Laing, Geoff Bicknell, Frazer Owen and my co-authors for many stimulating discussions of jet physics.

Symbols used in Chapter 3

Symbol	Meaning
$\mathbf{A}$, A	(vector) area of jet cross-section
a	index of power law pressure variation in an atmosphere
$\mathbf{B}$, B	magnetic field (strictly, magnetic flux density)
B_{MB}	magnetic field equivalent to the microwave background
b/a	axial ratio of elliptical cross-section
C_F	constant in Faraday rotation formula
C_{me}	constant in minimum energy and related formulae
c	speed of light
c_s	sound speed
c_{E1}	Pacholczyk constant in definition of ν_c
$c_{\gamma 1}$	equivalent to c_{E1} if energy is given by γ
c_3	Pacholczyk constant in expression for j_ν
c_7	Pacholczyk constant in definition of ν_T
$c_8(\delta)$, $c_9(\delta)$	Pacholczyk functions in expression for $\tilde{j}_\nu$
D	macroscopic scale size; projected linear size of radio source
$\mathcal{D}$	Doppler factor
d	small-scale correlation length for magnetic field
$\mathbf{E}$	electric field
E	'edge' surface of integration volume used in §3.4.3
e	electron charge
$G(\alpha)$	factor in 'minimum energy' and related expressions
$g(\theta,\alpha)$	anisotropy factor
H_0	Hubble's constant
H	Bernoulli constant
h	$H_0/100\mathrm{km\,s^{-1}\,Mpc^{-1}}$
$\mathcal{I}_1$, $\mathcal{I}_2$	integrals over the spectrum in expressions for U_{re}
i	angle between velocity of jet and the line of sight
J	mass flux (discharge) of jet
$\mathbf{j}$	electric current
j_ν	emission coefficient
$\tilde{j}_\nu$	emission coefficient averaged over all angles

K	energy flux (power) of jet
$(1+k)$	ratio of energy density due to all particles to that due to relativistic electrons
L_ν	spectral luminosity
l	length coordinate along streamline or jet
l_j	response length of jet
l_a	scale length for pressure change in the ambient medium
$\mathcal{M}$	Mach number
m	mass per unit length of jet
m_0	mass of ambient medium displaced by a unit length of jet
m_e, m_p	masses of electron, proton
$\mathbf{n}$	unit vector along jet (perpendicular to surface S)
n	number density
$n(\gamma)$	distribution function of relativistic electrons
P	pressure
P_0	undisturbed pressure
P'	pressure fluctuation due to a jet
P_ν	spectral power per unit solid angle
q_0	cosmological deceleration parameter
R	spherical radial coordinate around galaxy centre
$R_{\rm core}$	King model core radius
$R_{\rm curv}$	radius of curvature
$\mathcal{R}$	mean linear expansion factor
$\mathcal{R}_\parallel$, $\mathcal{R}_\perp$	expansion factors $\parallel$ and $\perp$ to streamline
Re	Reynolds number
r	cylindrical radial coordinate around jet axis
r_j	radius of jet
S	cross-sectional surface of jet
S_ν	radio flux density
s	scale length for pressure gradient in jet or plasmon
T	temperature
$\mathbf{T}$, T	(vector) momentum flux (thrust) of jet
$\mathbf{T}'$, T'	thrust excluding the undisturbed pressure
t	time coordinate
U	volume-integrated internal energy
u	internal energy density
V	volume

V_{em}	volume occupied by emitting material
$\mathbf{v}$, v	velocity
v_{esc}	escape velocity
v_{gal}	speed of galaxy
v_{in}	inflow speed of entrained material
v_{turb}, v'	velocity of turbulent motion
W	component width
w	specific enthalpy
z	redshift
α	spectral index ($S_\nu \propto \nu^{-\alpha}$)
β	v/c
β_P	ratio of particle pressure to magnetic energy density
β_{King}	parameter of King approximation density distribution
Γ	adiabatic index
γ	Lorentz factor
γ_T	characteristic Lorentz factor of aged spectrum
γ_L, γ_U	lower and upper cutoff γs of power law spectrum
δ	index of power law electron energy distribution
ϵ	efficiency of conversion of jet power to radio luminosity
η	density contrast (ρ_j/ρ_a)
θ	angle between principal axis of magnetic field and line of sight
λ	wavelength
μ_0	permeability of free space
ν	frequency
ν_c	critical frequency
ν_T	break frequency in aged spectrum
ν_L, ν_U	lower and upper ν_cs in power law spectrum
ξ	opening angle of jet (dr_j/dl)
Π_j	effective polytropic index of jet
ρ	density
τ, τ_{sa}	age, ... from spectral ageing
Φ	filling factor
ϕ	gravitational potential
ϕ_F	Faraday depth
$\mathbf{\Psi}$	stress tensor

Ψ_B, Ψ_{Re}	... magnetic, Reynolds (turbulent)
ψ	electron pitch angle

Subscripts	Meaning
a	ambient medium
B	magnetic
c	value on jet axis
e	electron
eq	equipartition
HS	hotspot
j	jet
max	maximum
MB	microwave background
min	minimum (nominal)
p	proton
re	relativistic electron
rel	all relativistic components (including a tangled field)
th	thermal (*i.e.*, non-relativistic)
tot	total

References

Alexander, P., 1987, *M. N. R. A. S.*, **255**, 27.

Alexander, P. & Leahy, J. P., 1987, *M. N. R. A. S.*, **225**, 1.

Alfvén, H. & Fälthammer, C.-G., 1963, *Cosmical Electrodynamics*, (Oxford University Press: Oxford).

Arnaud, K. A., Fabian, A. C., Eales, S. E., Jones, C. & Forman, W., 1984, *M. N. R. A. S.*, **211**, 981.

Asséo, E. & Grésillon, D., 1988, *Magnetic Fields in Extragalactic Objects*, (Editions de Physique: Les Ulis Cedex).

Asséo, E. & Sol, H., 1987, *Phys. Rep.*, **148**, 308.

Baan, W. A., 1980, *Ap. J.*, **239**, 433.

Baars, J. W. M., Genzel, R., Pauliny-Toth, I. I. K. & Witzel, A., 1977, *Astr. Ap.*, **61**, 99.

Baldwin, J. E., 1982, In *Extragalactic Radio Sources*, eds. Heeschen, D. S. & Wade, C. M. (Reidel: Dordrecht, Netherlands), p. 21.

Banhatti, D. G., 1980, *Astr. Ap.*, **84**, 112.

Begelman, M. C., 1982, In *Extragalactic Radio Sources*, eds. Heeschen, D. S. & Wade, C. M. (Reidel: Dordrecht, Netherlands), p. 223.

Begelman, M. C., Blandford, R. D. & Rees, M. J., 1984, *Rev. Mod. Phys.*, **56**, 255.

Begelman, M. C., Rees, M. J. & Blandford, R. D., 1979, *Nature*, **279**, 770.
Bicknell, G. V., 1984, *Ap. J.*, **286**, 68.
Bicknell, G. V., 1986a, *Ap. J.*, **300**, 591.
Bicknell, G. V., 1986b, *Ap. J.*, **305**, 109.
Blandford, R. D. & Icke, V., 1978, *M. N. R. A. S.*, **185**, 527.
Blandford, R. D. & Rees, M. J., 1974, *M. N. R. A. S.*, **169**, 395.
Bridle, A. H., 1984, In *Physics of Energy Transport in Extragalactic Radio Sources*, eds. Bridle, A. H. & Eilek, J. A. (N.R.A.O.: Green Bank, WV), p. 135.
Bridle, A. H., Chan, K. L. & Henriksen, R. N., 1981, *J. R. A. S. Canada*, **75**, 69.
Bridle, A. H. & Perley, R. A., 1984, *Ann. Rev. Astr. Ap.*, **22**, 319.
Bridle, A. H., Perley, R. A. & Henriksen, R. N., 1986, *Astron. J.*, **92**, 534.
Bridle, A. H. & Purton, C. R., 1968, *Astron. J.*, **73**, 717.
Burbidge, G., 1956, *Ap. J.*, **124**, 416.
Burch, S. F., 1977, *M. N. R. A. S.*, **180**, 623.
Burch, S. F., 1979, *M. N. R. A. S.*, **187**, 187.
Canizares, C. R., Fabbiano, G. & Trinchieri, G., 1987, *Ap. J.*, **312**, 503.
Carilli, C. L., 1989, Ph. D. Thesis, M. I. T.
Carilli, C. L., Perley, R. A., Dreher, J. W. & Leahy, J. P., 1990, in preparation.
Chan, K. L. & Henriksen, R. N., 1980, *Ap. J.*, **241**, 534.
Chiu, H. H., 1973, *Phys. Fluids*, **16**, 825.
Choudhuri, A. R. & Königl, A., 1986, *Ap. J.*, **310**, 96.
Condon, J. J. & Mitchell, K. J., 1984, *Ap. J.*, **276**, 472.
Dreher, J. W., Carilli, C. L. & Perley, R. A., 1987, *Ap. J.*, **316**, 611.
Duffet-Smith, P. J. & Purvis, A., 1982, In *Extragalactic Radio Sources*, eds. Heeschen, D. S. & Wade, C. M. (Reidel: Dordrecht, Netherlands), p. 59.
Eilek, J. A., 1989a, *Astron. J.*, **98**, 244.
Eilek, J. A., 1989b, *Astron. J.*, **98**, 256.
Fabbiano, G., Miller, L., Trinchieri, G., Longair, M. & Elvis, M., 1984, *Ap. J.*, **277**, 115.
Fabian, A. C., Hu, E. M., Cowie, L. L. & Grindlay J., 1981, *Ap. J.*, **248**, 47.
Fabian, A. C., Nulsen, P. E. J. & Canizares, C. R., 1984, *Nature*, **310**, 733.
Fabian, A. C., Thomas, P. A., Fall, S. M. & White III, R. E., 1986, *M. N. R. A. S.*, **221**, 1049.
Fabricant, D. & Gorenstein, P., 1983, *Ap. J*, **267**, 535.
Fabricant, D., Rybicki, G. & Gorenstein, P., 1984, *Ap. J.*, **286**, 186.
Falle, S. A. E. G., 1987, In *Astrophysical Jets and their Engines*, ed. Kundt, W. (Reidel: Dordrecht, Netherlands), p. 151.
Falle, S. A. E. G. & Wilson, M. J., 1985, *M. N. R. A. S.*, **216**, 79.
Favre, A., 1969, In *Problems of Hydrodynamics and Continuum Mechanics*, (Society for Industrial and Applied Mathematics: Philadelphia), p. 231.
Ferrari, A., Trussoni, E., Rosner, R. & Tsinganos, K., 1986, *Ap. J.*, **300**, 577.
Giacconi, R., Branduardi, G., Briel, U., Epstein, A., Fabricant, D., Feigelson, E., Forman, W., Gorenstein, P., Grindlay, J., Gursky, H., Harnden Jr., F. R., Henry, J. P., Jones, C., Kellogg, E., Koch, D., Murray, S., Schreier, E., Seward, F., Tananbaum, H., Topka, K., van Speybroeck, L., Holt, S. S., Becker, R. H., Boldt, E.

A., Serlemitsos, P. J., Clark, G., Canizares, C., Markert, T., Novick, R., Helfand, D. & Long, K., 1979, *Ap. J.*, **230**, 540.
Gower, A. C., Gregory, P. C., Hutchings, J. B. & Unruh, W. G., 1982, *Ap. J.*, **262**, 478.
Gunn, J. E., 1978, In *Observational Cosmology*, eds. Maeder, A., Martinet, L. & Tammann, G. (Geneva Observatory: Sauverny), p. 1.
Hardee, P. E., 1981, *Ap. J. Lett.*, **250**, L9.
Heckman, T. M., Miley, G. K., Balick, B., van Breugel, W. J. M. & Butcher, H. R., 1982, *Ap. J.*, **262**, 529.
Henriksen, R. N., 1987, *Ap. J.*, **314**, 33.
Henriksen, R. N., Bridle, A. H. & Chan, K. L., 1982, *Ap. J.*, **257**, 63.
Henriksen, R. N., Vallée, J. P. & Bridle, A. H., 1981, *Ap. J.*, **249**, 40.
Henry, J. P. & Henriksen, M. J., 1986, *Ap. J.*, **301**, 689.
Hines, D. C., Eilek, J. A. & Owen, F. N., 1989, *Ap. J.*, in press.
Hughes, J. P., Gorenstein, P. & Fabricant, D., 1988, *Ap. J.*, **329**, 82.
Illingworth, G., 1981, In *The Structure and Evolution of Normal Galaxies*, eds. Fall. S. M. & Lynden-Bell, D. (Cambridge University Press: Cambridge), p. 27.
Jackson, J. D., 1975, *Classical Electrodynamics*, (Wiley: New York).
Jaffe, W. J. & Perola, G. C., 1973, *Astr. Ap.*, **26**, 423.
Jägers, W. J., 1987, *Astr. Ap. Suppl.*, **71**, 75.
Jones, C. & Forman, W., 1984, *Ap. J.*, **276**, 38.
Jones, T. W. & Owen, F. N., 1979, *Ap. J.*, **234**, 818.
Junor, W., 1987, Ph. D. Thesis, University of Manchester.
Kardashev, N. S., 1962, *Soviet Astr. - AJ*, **6**, 317.
Killeen, N. E. B., Bicknell, G. V. & Carter, D., 1986, *Ap. J.*, **309**, 45.
Killeen, N. E. B., Bicknell, G. V. & Ekers, R. D., 1986, *Ap. J.*, **302**, 306.
Killeen, N. E. B., Bicknell, G. V. & Ekers, R. D., 1988, *Ap. J.*, **325**, 180.
Kim, K.-T., Kronberg, P. P., Dewdney, P. E. & Landecker, T. L., 1986, In *Radio Continuum Processes in Clusters of Galaxies*, eds. O'Dea, C. P. & Uson, J. M. (N.R.A.O.: Green Bank, WV), p. 199.
Knapp, G. R., 1990, In *The Interstellar Medium in External Galaxies*, eds. Thronsen, H. A. & Shull, J. M. (Kluwer: Dordrecht, Netherlands).
Krautter, A., Henriksen, R. N. & Lake, K., 1984, *Ap. J.*, **269**, 81.
Laing, R. A., 1984, In *Physics of Energy Transport in Extragalactic Radio Sources*, eds. Bridle, A. H. & Eilek, J. A. (N.R.A.O.: Green Bank, WV), p. 90.
Laing, R., 1989, In *Hotspots in Extragalactic Radio Sources*, eds. Meisenheimer, K. & Röser, H.-J. (Springer-Verlag: Berlin), p. 27.
Laing, R. A., Riley, J. M. & Longair, M. S., 1983, *M. N. R. A. S.*, **204**, 151.
Landau, L. D. & Lifshitz, E. M., 1959, *Fluid Mechanics*, (Pergamon: London).
Leahy, J. P., 1984, *M. N. R. A. S.*, **208**, 323.
Leahy, J. P., 1987, *M. N. R. A. S.*, **226**, 433.
Leahy, J. P., Jägers, W. J. & Pooley, G. G., 1986, *Astr. Ap.*, **156**, 234.
Leahy, J. P., Muxlow, T. W. B. & Stephens P. W., 1989, *M. N. R. A. S.*, **239**, 401.
Leahy, J. P. & Williams, A. G., 1984, *M. N. R. A. S.*, **210**, 929.
Lind, K. R., Payne, D. G., Meier, D. L. & Blandford, R. D., 1989, *Ap. J.*, **344**, 89.

Longair, M. S., 1987, In *Observational Cosmology*, eds. Hewitt, A., Burbidge, G. & Fang, L.-Z. (Reidel: Dordrecht, Netherlands), p. 823.
Longair, M. S. & Riley, J. M., 1979, *M. N. R. A. S.*, **188**, 625.
Longair, M. S., Ryle, M. & Scheuer, P. A. G., 1973, *M. N. R. A. S*, **164**, 243.
Lonsdale, C. J. & Barthel, P. D., 1987, *Astron. J.*, **94**, 1487.
Mackay, C. D., 1973, *M. N. R. A. S.*, **162**, 1.
Macklin, J. T., 1981, *M. N. R. A. S.*, **196**, 967.
Mathews, W. G. & Baker, J. C., 1971, *Ap. J.*, **170**, 241.
McCarthy, P. J., 1989, Ph. D. Thesis, University of California (Berkeley).
Meisenheimer, K., Röser, H.-J., Hiltner, P. R., Yates, M. G., Longair, M. S., Chini, R. & Perley, R. A., 1989, *Astr. Ap.*, **219**, 63.
Melrose, D. B., 1980, *Plasma Astrophysics*, (Gordon and Breach: New York).
Miller, L., Longair, M. S., Fabbiano, G., Trinchieri, G. & Elvis, M., 1985, *M. N. R. A. S.*, **215**, 799.
Morganti, R., Fanti, R., Gioia, I. M., Harris, D. E., Parma, P. & de Ruiter, H., 1988, *Astr. Ap.*, **189**, 11.
Myers, S. T. & Spangler, S. R., 1985, *Ap. J.*, **291**, 52.
O'Dea, C. P., 1985, *Ap. J.*, **295**, 80.
O'Dea, C. P. & Owen, F. N., 1985, *Astron. J.*, **90**, 927.
O'Dea, C. P. & Owen, F. N., 1986, *Ap. J.*, **301**, 841.
O'Dea, C. P. & Owen, F. N., 1987, *Ap. J.*, **316**, 95.
Osterbrock, D. E., 1988, *Astrophysics of Gaseous Nebulae and Active Galactic Nuclei*, (University Science Books: Mill Valley, California).
Owen, F. N. & Laing, R. A., 1989, *M. N. R. A. S.*, **238**, 357.
Owen, F. N., Hardee, P. E. & Cornwell, T. J., 1989, *Ap. J.*, **340**, 698.
Pacholczyk, A. G., 1970, *Radio Astrophysics*, (Freeman: San Francisco).
Parker, E. N., 1963, *Interplanetary Dynamical Processes*, (Interscience: New York).
Parker, E. N., 1979, *Cosmical Magnetic fields*, (Oxford University Press: Oxford).
Parma, P., Fanti, C., Fanti, R., Morganti, R. & de Ruiter, H. R., 1987, *Astr. Ap.*, **181**, 244.
Peacock, J. A., 1985, *M. N. R. A. S.*, **217**, 601.
Pedlar, A., Dyson, J. E., & Unger, S. W., 1985, *M. N. R. A. S.*, **214**, 463.
Perley, R. A., Bridle, A. H. & Willis, A. G., 1984, *Ap. J. Suppl.*, **54**, 291.
Perley, R. A., Dreher, J. W. & Cowan, J. J., 1984, *Ap. J.*, **285**, L35.
Rees, M. J., 1971, *Nature*, **229**, 312.
Rudnick, L., 1984, In *Physics of Energy Transport in Extragalactic Radio Sources*, eds. Bridle, A. H. & Eilek, J. A. (N.R.A.O.: Green Bank, WV), p. 35.
Saikia, D. J., 1984, *M. N. R. A. S.*, **209**, 525.
Saikia, D. J. & Salter, C. J., 1988, *Ann. Rev. Astr. Ap.*, **26**, 93.
Sancisi, R. & van Albeda, T. S., 1987, In *Dark Matter in the Universe*, eds. Kormandy, J. & Knapp, G. R. (Reidel: Dordrecht, Netherlands), p. 67.
Sandage, A., 1987, In *Observational Cosmology*, eds. Hewitt, A., Burbidge, G. & Fang, L.-Z. (Reidel: Dordrecht, Netherlands), p. 1.
Sanders, R. H., 1983, *Ap. J.*, **266**, 73.

Saunders, R., Baldwin, J. E., Pooley, G. G. & Warner, P. J., 1981, *M. N. R. A. S.*, **197**, 287.

Scheuer, P. A. G., 1974, *M. N. R. A. S.*, **166**, 513.

Scheuer, P., 1989, In *Hotspots in Extragalactic Radio Sources*, eds. Meisenheimer, K. & Röser, H.-J. (Springer-Verlag: Berlin), p. 159.

Simonetti, J. H. & Cordes, J. M., 1986, *Ap. J.*, **310**, 160.

Smith, M. D., 1984, *M. N. R. A. S.*, **207**, 41p.

Smith, M. D., Norman, M. L., Winkler, K.-H. A. & Smarr, L., 1985, *M. N. R. A. S.*, **214**, 67.

Sol, H., Pelletier, G. & Asséo, E., 1989, *M. N. R. A. S.*, **237**, 411.

Spinrad, H. & Djorgovski, S., 1987, In *Observational Cosmology*, eds. Hewitt, A., Burbidge, G. & Fang, L.-Z. (Reidel: Dordrecht, Netherlands), p. 129.

Spitzer Jr., L., 1962, *Physics of Fully Ionised Gases*, (Interscience: New York).

Stephens, P. W., 1987, Ph. D. Thesis, University of Manchester.

Stewart, G. C., Canizares, C. R., Fabian, A. C. & Nulsen, P. E. J., 1984, *Ap. J.*, **278**, 536.

Thompson, P. A., 1972, *Compressible-Fluid Dynamics*, (McGraw-Hill: New York).

Trinchieri, G. & Fabbiano, G., 1985, *Ap. J.*, **296**, 447.

Trinchieri, G., Fabbiano, G. & Canizares, C. R., 1986, *Ap. J.*, **310**, 637.

van Breugel W., Miley, G., Heckman, T., Butcher, H. & Bridle, A., 1985, *Ap. J.*, **290**, 496.

Walker, R. C., Benson, J. M. & Unwin, S. C., 1987, *Ap. J.*, **316**, 546.

Walker, R. C., Walker, M. A. & Benson, J. M., 1988, *Ap. J.*, **335**, 668.

Wilson, A. S. & Ulvestad, J. S., 1983, *Ap. J.*, **275**, 8.

Wilson, A. S. & Ulvestad, J. S., 1987, *Ap. J.*, **319**, 105.

Wilson, M. J., 1987a, *M. N. R. A. S.*, **224**, 155.

Wilson, M. J., 1987b, *M. N. R. A. S.*, **226**, 447.

Wilson, M. J. & Falle, S. A. E. G., 1985, *M. N. R. A. S.*, **216**, 971.

Winter, A. J. B., Wilson, D. M. A., Warner, P. J., Waldram, E. M., Routledge, D., Nicol, A. T., Boysen, R. C., Bly, D. W. J. & Baldwin, J. E., 1980, *M. N. R. A. S.*, **192**, 931.

4

Interpretation of Parsec Scale Jets

T. V. CAWTHORNE
Department of Physics and Astronomy, University of Glasgow, UK, and Department of Physics, Brandeis University, Waltham, MA 02154, USA.

4.1 Introduction

Flat spectrum nuclei are found at the centre of many types of extragalactic radio source, including the powerful classical doubles, and the 'isolated' compact sources (that have comparatively weak extended structure). They turn out (when examined with sufficient resolution) to be the self-absorbed bases of jets that feed the extended structure, whose continuations are often seen on much larger scales. The systematic properties of VLBI jets have been reviewed frequently and can be summarised thus:

1. The jets are nearly always seen on only one side of the nucleus. 'Counterjets' have only been seen in a few sources (for example 3C 236, Schilizzi *et al.* 1988).
2. The jets contain bright 'knots' of emission that are often seen to move outwards, away from the base at speeds commonly in the region of 5 to 10 times the speed of light ($H_0 = 100$ km s^{-1} Mpc^{-1}). Although some apparently stationary knots have been observed, for example the outer component in 4C 39.25 (Shaffer & Marscher 1988), none has ever been observed to move inwards (Marcaide *et al.* 1985).
3. The speeds and trajectories of the individual knots do not in general vary greatly as the knots move out; it is difficult to place very severe limits on acceleration because of the limited accuracy of positional measurements and the relatively short distances over which the knots are observed to move. Significant changes in speed and direction have been detected in 3C 345 (Biretta *et al.* 1986).
4. Substantial misalignments between VLBI jets and those on larger scales are common. 3C 120 is a fine example in which

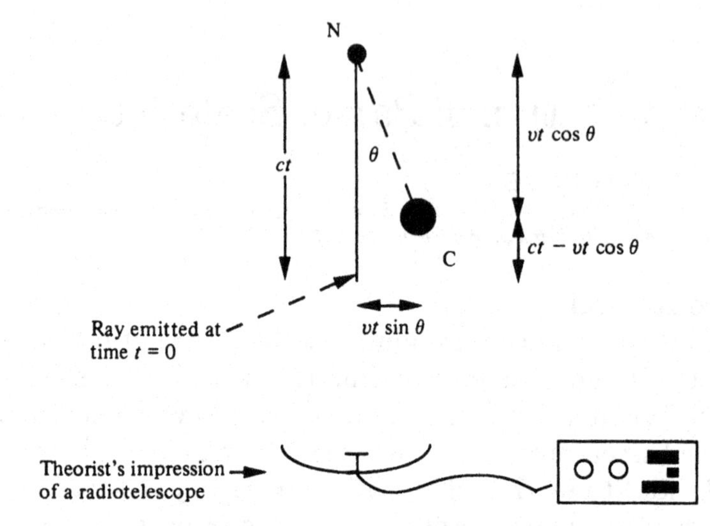

Fig. 4.1. Illustrating how superluminal motion might occur as a result of relativistic motion of the component C directed at an angle θ to the observer (O)'s line of sight with speed $v = c\beta$. On being ejected from the nucleus N, component C emits radiation towards the observer. At time t later, C has moved across the sky a distance $ct\beta \sin\theta$, and is only a distance $ct\beta(1 - \beta\cos\theta)$ behind the radiation it emitted on leaving N. Thus O sees the travel time of C compressed to $t(1 - \beta\cos\theta)$ and deduces an apparent speed $c\beta\sin\theta/(1 - \beta\cos\theta)$ as in equation (4.1).

the curvature of the jet has been followed continuously from parsec to kiloparsec scales (Walker, Benson & Unwin 1987).

5. The X-ray luminosities are generally rather lower than those expected as inverse-Compton emission from a stationary, incoherent synchrotron source (cf. §1.2.2).

Previous chapters have described how these phenomena may be explained if the observed knots move at speeds $c\beta$ very close to that of light, c, along trajectories making small angles θ with our lines of sight towards them. (Fig. 4.1 illustrates how apparent superluminal motion arises, and §1.2.2 discusses how bulk relativistic motion eases problems associated with inverse-Compton radiation and variability.) The apparent speed is then given by the well known formula

$$\beta_{\text{app}} = \beta\sin\theta/(1 - \beta\cos\theta). \tag{4.1}$$

Nowadays, this model is implicit in many (if not most) discussions of apparent superluminal motion. Other mechanisms such as gravitational lensing have been advanced to account for these phenomena, and have been reviewed by Scheuer (1984a). However although the relativistic jet picture is far from perfect, it raises fewer serious objections than the alternatives, and in keeping with the spirit of this volume they are not discussed here. Interestingly, one explanation that can be ruled out fairly easily is genuinely superluminal motion (as might result from the passage of a spot of light over a screen when the beam makes a small angle to the screen). In that case if $\beta \cos\theta > 1$ as it is likely to be for some orientations, then $\beta_{\rm app}$ is negative and the apparent motion is inwards, rather than outwards as observed.

Having explained the empirical facts in terms of a model like that displayed in Fig. 4.1, it is necessary to try to devise experiments that will probe the validity of the proposed interpretation, and unfortunately this proves rather difficult. Such models are usually devised to explain a particular aspect of the behaviour of sources, and other observational features that one might conceivably hope to use to test them depend as much on other, unspecified properties as on the basic model under scrutiny. The results of such comparisons may depend upon the range of sources that one attempts to describe with a given model. For example, should radio sources identified with quasars be included alongside those identified with galaxies? The answer depends upon how the species are related; there are often several possibilities. Thus, attempts to constrain physical models can easily lead to a complicated labyrinth of arguments at whose many bifurcations there is no clear way forward. The aim of this chapter is to describe and contrast the predictions of elementary models for VLBI jets that might be used (in conjunction with observational data) to shed light on how well (or badly) they mimic real radio sources. Obviously, it is sensible to look for predictions that are, as far as possible, independent of the fine details of the model. §4.2 deals with a simple prediction of the distribution of the apparent velocities suggested by equation (4.1) which probably offers the best chance of constraining the simple framework on which it is based. §4.3 deals with models for the emitting regions and particularly how they influence the distribution of nuclear flux in samples of sources. Account is taken of opacity, and whether the moving features are simple components, or regions bounded by shock waves that propagate through the fabric of the jet. §4.4 deals with the relationship between structures on small (parsec) and large (kiloparsec) scales that are observed by VLBI and conventional aperture synthesis, respectively. The first part discusses the relationship between the extended lobes and the nuclei, and how its interpretation in different ways over the radio source population can influence the outcome of comparisons between models and observational data. The second part is concerned with how systematic observational properties illuminate the relationship between parsec- and kiloparsec-scale jets.

4.2 The distribution of apparent velocities

4.2.1 Predictions from the velocity formula

If a model for extragalactic radio sources has been designed to explain a particular feature of their behaviour, then the distribution of that feature over an ensemble of sources is very often the prediction most easily accessible to observational scrutiny. For the simple model leading to equation (4.1) that seeks to account for the apparently superluminal sources, there are two variables, $v = c\beta$, the true speed of the knots, and θ, the inclination of their direction of motion to the line of sight. Thus, if the effect on β_{app} of the range in v is much less than that of the range in θ, the distribution of β_{app} is known for any sample in which the jet axes are randomly oriented. This section considers the possibility of using the distribution of β_{app} to test the basic framework of the model (Fig. 4.1). §4.2.2 is devoted to practical considerations, while this section reviews what one might hope to achieve in principle.

Assuming that the jets are intrinsically two-sided, and that their apparent one-sidedness is due to the mixture of Doppler-shifting and relativistic aberration that makes the approaching jet by far the brightest, then the fraction of sources in a sample with no bias in θ having apparent speed greater than β'_{app} is

$$F(\beta_{\mathrm{app}} > \beta'_{\mathrm{app}}) = \cos\theta_1 - \cos\theta_2$$
$$= 2\sqrt{[1 - \beta'^2_{\mathrm{app}}/(\gamma^2 - 1)]}(1 + \beta'^2_{\mathrm{app}})^{-1}, \quad (\beta < \beta'_{\mathrm{app}} < \gamma\beta) \quad (4.2)$$

where θ_1 and θ_2 are the two solutions to $\beta'_{\mathrm{app}} = \beta \sin\theta/(1 - \beta\cos\theta)$, and γ is the Lorentz factor of the moving features, $(1 - \beta^2)^{-1/2}$. The range of β'_{app} specified in equation (4.2) corresponds to the interval between the largest possible speed at $\cos\theta = \beta$ and that when the jet lies in the plane of the sky. Apparent speeds less than

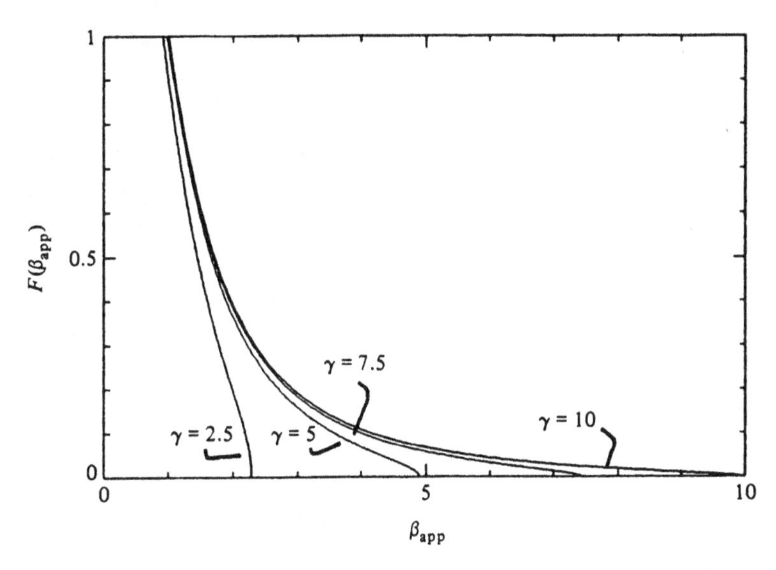

Fig. 4.2. The fraction $F(\beta_{\text{app}})$ of sources having two-sided jets (of no preferred orientation) in which the apparent speed exceeds $c\beta_{\text{app}}$ is plotted as a function of β_{app} for various γ, the Lorentz factor of the jet.

$c\beta$ are obtained for small angles much less than $\cos^{-1}\beta$, but the fraction of these is $(1-\beta)^2/(1+\beta^2) \approx 1/8\gamma^4$ when $\gamma \gg 1$. Even for $\gamma = 2$, this is only about 0.01, and so in sources having two-sided relativistic jets, such speeds are likely to be rare. However, if jets are intrinsically one-sided, only one in two sources has a jet pointing into the observer's hemisphere, and so equation (4.2) is multiplied by one half, while its validity is extended to cover the interval $0 < \beta'_{\text{app}} < \gamma\beta$. Then, a little more than half the sources have $\beta_{\text{app}} < \beta$.

Fig. 4.2 shows the fraction of (two-sided) jets with apparent component speeds greater than β_{app} for various true Lorentz factors γ. Perhaps the most important feature of this diagram is that for γ larger than about 3, the curves tend to bunch together very tightly below $\beta_{\text{app}} = \gamma\beta$. Of course the reason is that when $\gamma \gg 1$ and $\theta \gg 1/\gamma$, as it will be for most sources in a sample whose jets have no preferred orientation, equation (4.1) shows that to all intents and purposes β_{app} depends only on θ, not γ, and consequently F given by equation (4.2) depends little on the exact value of γ. This result leads to a prediction that is independent of the distribution of true speeds likely to be present in any real sample of sources: the largest fraction of sources expected to have apparent speed greater than β_{app} is $2(1+\beta^2_{app})^{-1}$, which occurs in the formal limit where $\beta = 1$ in all sources. Thus not more than 8% of sources in a randomly oriented sample should have apparent speeds greater than 5c, while in fewer than 2% do they exceed 10c. If much greater fractions of sources were found to show large apparent speeds, the simple ideas leading to equation (4.1) would have to be revised.

One corollary of this result is that the distribution of apparent speeds tells us little about the distribution of Lorentz factors in such a sample. To emphasize the point, consider the differential distribution of apparent speeds $p(\beta_{app})$ so the fraction of speeds in the range $(\beta_{app}, \beta_{app} + d\beta_{app})$ is $p(\beta_{app})d\beta_{app}$.

$$p(\beta_{app}) = |\frac{d}{d\beta_{app}}F(\beta_{app})| = d/d\beta_{app}(\cos\theta_2 - \cos\theta_1), \qquad (4.3)$$

which is shown in Fig. 4.3. (The infinities at $\gamma\beta$ occur because β_{app} is a maximum there – Fig. 4.2 shows that very few apparent speeds are found in that region.) Except at very low values of β_{app}, any two curves corresponding to various γ lie very close together until one of them reaches its value of $\gamma\beta$. It would be very difficult to distinguish between the distribution of apparent speeds for large values of γ unless the sample of sources monitored was very large. The most that can be hoped for is to obtain some idea of the fraction of sources with small Lorentz factors ($\gamma \leq 3$), and of the smallest possible upper limit to γ from the largest values of β_{app} observed. However, in order to attribute an absence of sources showing apparent speeds in excess of β'_{app} to an absence of sources having $\gamma \gg \beta'_{app}$, the probability of finding no sources with such large apparent velocities would have to be quite small, say, less than ϵ. If the sample contains N sources then this probability is

$$[1 - F(\beta_{app} > \beta'_{app})]^N < \epsilon \qquad (4.4a)$$

or

$$N > \log\epsilon / \log(1 - 2(1+\beta'^2_{app})^{-1}) \qquad (4.4b)$$

since $F(\beta_{app} > \beta'_{app}) < 2(1+\beta'^2_{app})^{-1}$. Hence if in a sample no sources were found showing apparent speeds in excess of $7c$, then to infer that that could occur in a sample of sources in which $\gamma \gg 7$ with likelihood less than 0.1, the sample would have to contain more than 56 sources. VLBI monitoring of all members of such a large sample of sources over an extended period of time would be a con-

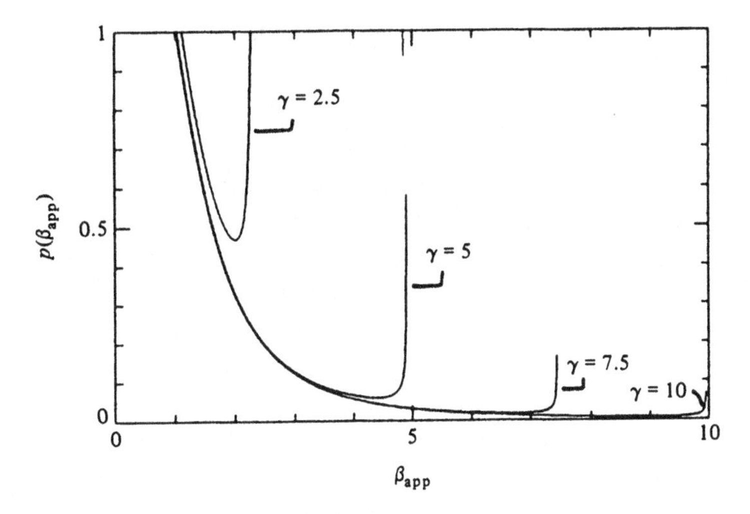

Fig. 4.3. The differential distribution of apparent speeds $p(\beta_{\rm app}) = -dF/d\beta_{\rm app}$ (see equation (4.3) and Fig. 4.2). At $\beta_{\rm app} = \gamma\beta$, $p(\beta_{\rm app})$ tends to infinity. (The curves on this diagram are drawn as far as the sharp turn up in $p(\beta)$.)

siderable undertaking, and furthermore, since one can always argue that some jets may happen to be particularly stable and therefore show no moving components, caution is required in deducing anything from a range of apparent speeds that are *not* detected.

Whether the distribution of apparent speeds is compatible with the simple model that leads to equation (4.1) could depend upon whether one regards VLBI jets as intrinsically one or two-sided, since for the latter, twice as many jets with apparent speeds exceeding some appropriate limit are allowed. A population of two-sided jets among the (so far unobserved) sources with relatively dim nuclei would argue in favour of two-sided jets in apparently one-sided sources. However the faintness of the dimmest nuclei makes them difficult to observe; furthermore, they may require higher resolution observations than those that are readily available, since the dimmer jets may well appear shorter (cf. §4.3.2). Roberts & Wardle (1986) have observed that VLBI polarization maps could well help to resolve this issue: the bright point-like 'cores' that dominate total intensity maps are usually comparatively faint on maps of polarized intensity. As a result, on maps of limited dynamic range, weak but highly polarized components (*e.g.*, the outer parts of a counterjet) should be more easily visible in polarized than in total intensity.

4.2.2 Constraints on $p(\beta_{\rm app})$ in practice

There are two main requirements of any sample of sources to be monitored as a means of constraining $p(\beta_{\rm app})$: first there must be no preferred orientation of the jet axes, and second, it must contain a sufficiently large fraction of sources that have bright central components (accessible to VLBI observations) to place useful limits on $p(\beta_{\rm app})$.

The fact that radiation emitted by a highly relativistic jet is likely to be quite anisotropic (enhanced along the direction of flow) means that a complete sample of radio sources (*i.e.*, one containing all the sources in a certain patch of sky having total flux above some limit at one specified frequency) could well be over-representative of sources whose jets are viewed at small inclinations, θ; there will be some sources lifted above the sample limit by beamed flux from an appropriately aligned jet, while their unaligned counterparts will not be included. In order to remove this kind of bias, only the flux in extended lobe-like structure, which as shown in Chapter 3 is unlikely to be moving much faster than 0.1c, should be included to determine membership of the sample to be monitored. However, before tackling this problem, some preliminary pruning of the complete sample is desirable in order to ensure that it is reasonably homogeneous, rather than a mix of very different types of source.

It is probably best to choose powerful sources from the parent sample so as to obtain a subsample containing a reasonable proportion of sources with bright central components, but in which a large part of the total flux originates in extended structure. This might be achieved by choosing only quasars, or by computing the power directly if the redshifts are known. The next step is to look at maps of the remaining sources to assess which meet the limits on flux (and power – if appropriate) by virtue of flux from extended emission alone. For most of the sources likely to fall into the classical double or compact, flat-spectrum ('core-jet') categories, this poses no great problem.

Most if not all the latter and a few of the former will be axed from the sample. However there are often a number of sources whose structure is harder to interpret: the steep-spectrum compact sources (*e.g.*, Fanti *et al.* 1985) provide good examples. These have spectral indices greater than about 0.7, sizes of order 10kpc (or a few arcseconds at typical redshifts – rather smaller than most classical doubles) and irregular structure. It has not as yet been possible to disentangle core, jet and lobes, and so it is not clear whether they should be kept in a sample of this kind. Excluding them by simply imposing a lower limit on the angular size is probably not a good idea: the average angular size of powerful classical double sources is about 10 arcseconds, and so any useful limit could prejudice the orientation of the sample members. For these sources, VLBI observations, or pursuit of their spectra to high frequencies where flat spectrum components dominate might help to clarify their structure.

The size of the sample need not be so large that the task of regular VLBI monitoring would place a heavy burden on existing resources. The number of sources in a sample of N having apparent speeds greater than $c\beta_{\rm app}$ should be Poisson distributed with mean no greater than $\mu = 2N(1 + \beta_{app}^2)^{-1}$ (from equation (4.2)). The chance of finding r or more sources with apparent speed greater than $\beta_{\rm app}$ is thus approximately $\Sigma_{i=r}^{i=N} e^{-\mu}\mu^i/i!$. In a sample of (say) 30 sources, the mean number having apparent speed greater than $5c$ is thus less than 2.3, and so the chance of finding 6 or more sources with apparent speed in excess of 5c is only 0.03. The mean number expected to exceed 7c would be 1.2, but the chance of finding as many as five or more is only about 0.01. Thus as few as ten sources with observable nuclei out of a sample of thirty could be a sufficient number to place useful limits on $p(\beta_{\rm app})$, and if, as is the case with the well-observed flat-spectrum sources whose flux is dominated by that of the nucleus, those with powerful, extended structure were commonly found to contain knots that move at about $\beta_{\rm app} \sim 5$, the simple time-delay model leading to equation (4.1) would have to be revised.

The problem of whether to include radio galaxies as well as quasars in such a sample is important because the radio emission from the nuclei of galaxies is much weaker than that from quasars with similar luminosities from extended lobes. For example, in the subsample of the thirty most powerful extended radio sources from the 3C complete sample (Laing, Riley & Longair 1983) no galaxy has a nucleus brighter that 2mJy at a frequency of 5 GHz, and so the powerful sources whose nuclei can be mapped are essentially all quasars. Nevertheless, galaxies constitute some 50% of that subsample, and so their inclusion in the final, edited sample will dilute the fraction of sources found in a given range of apparent speed by a factor of about two. Whether or not galaxies should be included depends on their relationship to the quasars, and there are at least two points of view one might take. One is that galaxies, while being similar objects to quasars, have intrinsically much dimmer nuclei. In this case, the optical appearance of a source is independent of its orientation, and so excluding galaxies would be quite acceptable. However it is often envisaged (*e.g.*, Saunders 1984) that the quasars are surrounded by clouds of matter that obscure the optical continuum source and broad line regions, but which are punctured by the escaping jets. In that case it is possible that some quasars masquerade as narrow-line radio galaxies when viewed through the obscuring matter. If such radio galaxies are misaligned quasars then they must be retained in order to preserve random orientation of the nuclear jets.

There is some indirect evidence for such a model. For example, the largest angular size of the extended structure of quasars is systematically smaller than that of radio galaxies (by a factor of roughly two) as one expects if quasars are only identified as such when seen in projection (Wardle, private communication; Miller 1983; Hough & Readhead 1989). Furthermore, IRAS observations of several bright, nearby narrow-line radio galaxies (*e.g.*, Cygnus A) reveal far infra-red luminosities comparable to those of quasars ($\sim 10^{40} W$), (Cawthorne 1985; Yates & Longair 1989). This emission could be energy derived from an obscured quasar, subsequently re-emitted at the temperature of the intervening matter. However it could also be a result of star formation on a grand scale, so that this evidence is, again, only circumstantial.

Several authors have discussed samples that could be monitored profitably for component proper motion (*e.g.*, Zensus & Porcas 1987; Hough 1986; Hough & Readhead 1987a,b, 1989; Cawthorne *et al.* 1986). A number of classical double sources with relatively bright nuclei have already been monitored extensively, (*e.g.*, 3C 179 and 3C 263, Zensus & Porcas 1987) and among these, superluminal motion is found to be quite common. $\beta_{\rm app}$ ranges from 1.3 in the case of 3C 263 (Hough & Readhead 1987a) to 4 in 3C 179

(Zensus & Porcas 1987) (for $H_0 = 100\,\mathrm{km\,s^{-1}\,Mpc^{-1}}$). In order to obtain strong constraints, it is necessary to extend such observations to cover sources with much weaker nuclei, perhaps as weak as 50mJy. Recent work by Hough & Readhead (1987a), who have mapped sources with nuclei as faint as 80mJy, shows that this is well within the bounds of feasibility. To obtain sufficient sensitivity, the Mark III recording system must be used. Unfortunately Mark III resources are strictly limited, and the pace of this work is necessarily restricted. The completion and commissioning of the NRAO Very Long Baseline array (VLBA), an array of antennae dedicated to VLBI, should greatly accelerate matters at some point in the not-too-distant future.

4.2.3 Models giving a better chance of superluminal motion

What other models can be employed should more sources turn out to have large apparent component speeds than can be accommodated by equation (4.2)? One model employs components that are ejected in several directions, large apparent speeds being seen from close to each. An objection usually raised against this model is that one might expect to see some sources with several jets in apparently random directions. However if there were only a few highly beamed and well separated jets, that could have a significant impact on the likelihood of superluminal motion without the risk of seeing a second jet. Perhaps a more serious worry is that the evidence from the large scale structure in the powerful classical doubles is that, here at least, the ejection is in one dominant direction over very long periods of time ($> 10^7$ years). Some means of collimating the different jets into one direction on kpc scales would presumably be required.

A more subtle approach to the problem involves radiating matter moving along a cluster of bent trajectories such as that shown in Fig. 4.4. Matter ejected at the base (where they connect) moving outwards along a trajectory at speed approaching c will be seen by a distant observer when the velocity on that trajectory points towards him. Thus there is no causal connection between the series of events that constitutes apparent outwards motion of a single component. A given component can be observed at some point on its path from a wide range of angles, and so it comes as no surprise that

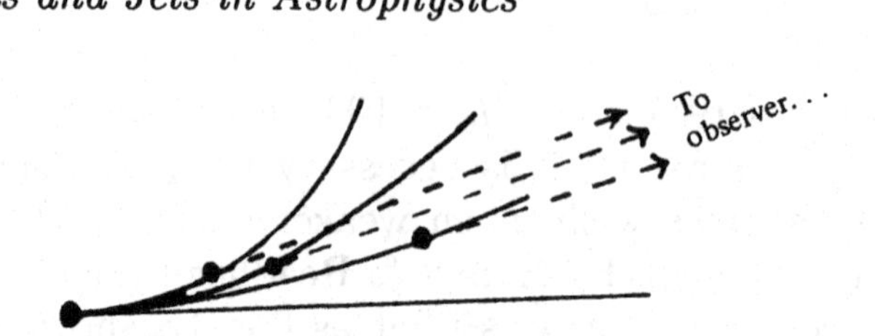

Fig. 4.4. Sketch illustrating the model for superluminals discussed by Sanders (1974) and developed by Scheuer (1984a). At some instant emitting components are ejected from a point and move along a group of trajectories (the continuous lines). If the components have Lorentz factor $\gamma >> 1$ then they will be seen only when moving within an angle $1/\gamma$ to the observer's line of sight (the dashed lines). Therefore they are seen at different times and at points on the trajectories marked by the dots.

apparent superluminal motion can be seen over a much larger solid angle than was possible for the simple model involving rectilinear motion. This kind of explanation was first discussed by Sanders (1974), who envisaged that the trajectories could be dictated by the field lines of a magnetic dipole along which radiating electrons are constrained to flow. However Scheuer (1984a) has shown that the precise form of the trajectories is relatively unimportant, and that synchrotron emitting components moving along any similar cluster of curved paths are likely to produce this effect.

4.3 Jet models and their observational consequences

4.3.1 Simple jets

In the introduction to this volume it has been observed that a single source (such as an opaque sphere or an optically thin synchrotron source with a tangled magnetic field) that emits isotropically in its rest frame (Σ') has a radiation pattern in another frame (Σ), with respect to which it moves at speed $c\beta$, given by $S(\nu,\theta) = S'(\nu)\mathcal{D}^{3+\alpha}$, where $\mathcal{D} = [\gamma(1-\beta\cos\theta)]^{-1}$ (the Doppler shift) and $\alpha = -d\ln S'/d\ln\nu'$ is the spectral index, assumed constant over the entire range of $\nu' = \nu/\mathcal{D}$. θ is the angle between the observer's line of sight and the direction of relative motion. (A clear derivation of this result is given by Rybicki & Lightman (1979), §4.9.) For simple jets (such as a steady stream of isotropically emitting optically thin material or a series of optically thick spheres – so long as they don't appear to overlap) the radiation pattern emitted in Σ is

$$S(\nu,\theta) = S'(\nu)\mathcal{D}^{2+\alpha}. \tag{4.5}$$

One Doppler shift is lost because the lifetime of any one emitting particle is smaller as seen from θ in Σ than in Σ' by a factor of $\mathcal{D}$, and *that* many fewer emitting components are seen in an approaching jet at any one time. Equation (4.5) implies that the radiation is brightest along the direction of motion and for highly relativistic jets, has fallen by a factor of $2^{2+\alpha}$ at $\theta \approx 1/\gamma$, the familiar $1/\gamma$ cone angle. From an angle θ, a given source will appear brightest when its speed $c\beta = c\cos\theta$ ($\gamma \approx 1/\theta$). At larger speeds the radiation cone collapses further, excluding the line of sight to the observer.

It is worth emphasizing that equation (4.5) will describe $S(\theta)$ *provided* the jet emits isotropically in its rest frame (Σ') and appears as a stationary pattern (through which emitting material flows) in Σ. If the radiation were anisotropic in Σ' then $S(\theta) = S'(\theta')\mathcal{D}^{2+\alpha}$, so that if, for instance, the source is an optically thin synchrotron emitter with magnetic field parallel to the axis of the jet, then in Σ' the radiation pattern is $S'(\theta') \propto (\sin\theta')^{1+\alpha}$ (cf. equation (1.6a)). Therefore from Σ, $S \propto \mathcal{D}^{2+\alpha}(\sin\theta')^{1+\alpha} = \mathcal{D}^{3+2\alpha}(\sin\theta)^{1+\alpha}$ since $\sin\theta' = \mathcal{D}\sin\theta$. This gives a much more pronounced anisotropy than equation (4.5); S no longer increases monotonically as θ falls. (As with most of the simple formulae derived here, this form for $S(\theta)$ will break down at very small θ where the line of sight passes inside the jet. Then it is necessary to take account of the different Doppler shifts of its constituent parts.) Nevertheless, in the absence of knowledge of the detailed properties of jets that affect $S'(\theta')$, equation (4.5) has proved a useful first guess at $S(\theta)$ and the remainder of this subsection will examine some of the observational consequences that follow from assuming equation (4.5) for a simple jet. This form will also serve as a 'benchmark' against which to compare slightly more complicated models encountered later on.

As with the velocity formula (4.1) the form for equation (4.5) may be inverted to yield a distribution for S, *provided* there is some way to normalise the flux, allowing for the likely range in S'. This may be done conveniently by taking a sample of sources belonging to one family (the powerful classical doubles, for instance) and assuming that the spread in the ratio of the *intrinsic* flux of the nuclear jet to that of the extended lobes is much smaller than that found from the *observed* nuclear flux due to changes in θ. Then

$$R(\theta) = S_{\rm nuclear}/S_{\rm ext} = R_0(1-\beta\cos\theta)^{-(2+\alpha)} \tag{4.6}$$

for one-sided jets where $0 < \theta < \pi$ and

$$R(\theta) = (R_0/2)(1-\beta\cos\theta)^{-(2+\alpha)} \tag{4.7}$$

for two-sided jets at sufficiently small angles that the flux from the receding jet may be ignored. $R_0 = R(\pi/2)$. Most nuclei have fairly flat spectra, so setting $\alpha = 0$, the fraction of sources having $R > R'$ in a sample whose (two-sided) jets are randomly oriented becomes

$$F(R > R') = (1-\cos\theta(R')) \approx \beta^{-1}[(R_0/2R)^{1/2} - 1/(2\gamma^2)] \tag{4.8}$$

and the ratio of the largest to the smallest R, $R_{\rm max}/R_0$, is approximately $2\gamma^4$.

Comparison with observations, a task first tackled by Scheuer & Readhead (1979), is similar to the procedure with the apparent speeds encountered in §4.1.3, and suffers from many of the same problems. For instance the value of γ can only be determined with any confidence from the largest and smallest values of R present in a sample *if* the sample is sufficiently large that sources with $R > R_{\rm max}$ would almost certainly be included were $\gamma > (R_{\rm max}/2R_0)^{1/4}$ (again assuming two-sided jets).

This problem was circumvented by Orr & Browne (1982) but at the expense of further model assumptions. They hypothesised that R_0 and γ take the same values in the compact, flat-spectrum radio sources whose extended structure is relatively faint as they do in the powerful, classical doubles. That enabled them to relate the proportion of flat-spectrum sources to R_0 and γ in a purely flux-limited sample that does not have randomly oriented nuclear jets (*i.e.*, one which has not undergone the kind of pruning described in §4.1.3). Their procedure may be illustrated by a crude estimate (which does little justice to their painstaking calculations).

Orr & Browne assumed that the spectral indices of the nuclei and extended structure were 0 and 1, respectively. It follows that the ratio of Rs at different frequencies (but the same angle) is $R(\nu_h)/R(\nu_l) = (\nu_h/\nu_l)$. They further defined a 'flat-spectrum' source to have a two point spectral index $\alpha(\nu_l,\ \nu_h) = -\ln[S(\nu_h)/S(\nu_l)]/\ln(\nu_h/\nu_l)$ less than 0.5. This can be rewritten as a constraint on $R(\nu_h)$ since $S(\nu) = S_{\rm ext}(\nu)[1 + R(\nu)]$ and

$$\alpha(\nu_l, \nu_h) = -\ln \frac{[1 + R(\nu_h)]}{\nu_h/\nu_l + R(\nu_h)} / \ln(\nu_h/\nu_l) < 1/2 \tag{4.9}$$

requires

$$R(\nu_h) > (\nu_h/\nu_l)^{1/2}. \tag{4.10}$$

Finally, they assume that the number of sources having flux in extended structure in the range $S_{\rm ext}$ to $S_{\rm ext} + dS_{\rm ext}$ is proportional to $S_{\rm ext}^{-m}$. In a sample chosen at frequency ν_h and with lower limit on flux density limit S_0, sources with a particular value of $R(\nu_h)$ must have $S_{\rm ext}$ greater than $S_0/[1 + R(\nu_h)]$. Therefore, the fraction of sources in such a sample having $\alpha(\nu_l, \nu_h) < 1/2$, or $R(\nu_h) > (\nu_h/\nu_l)^{1/2}$ is

$$f = \frac{\int_{\sqrt{(\nu_h/\nu_l)}}^{R_{\rm max}} \int_{S_0/(1+R)}^{\infty} S_{\rm ext}^{-m} p(R) dR dS_{\rm ext}}{\int_{R_0}^{R_{\rm max}} \int_{S_0/(1+R)}^{\infty} S_{\rm ext}^{-m} p(R) dR dS_{\rm ext}}. \tag{4.11}$$

p(R)dR is the fraction of sources in the parent population between R and R+dR, and since *their* nuclear jets should be randomly oriented, p(R) is related to F (equation (4.8)) by $p(R) = -dF/dR$. Therefore $p(R) \propto R^{-3/2}$. For many samples with fairly high S_0, m is fairly close to its Euclidean value, 5/2, and for convenience this value will be assumed here. Then performing the integration over $S_{\rm ext}$ yields

$$f = \frac{\int_{\sqrt{(\nu_h/\nu_l)}}^{R_{\rm max}} (1 + R)^{3/2} R^{-3/2} dR}{\int_{R_0}^{R_{\rm max}} (1 + R)^{3/2} R^{-3/2} dR}. \tag{4.12}$$

R_0 is deduced from the smallest R observed, and recalling that $R_{\rm max}/R_0 \approx 2\gamma^4$, the observed value of f yields an estimate of γ. To obtain a simple result analytically, it is convenient to approximate $(1 + R)$ in f by R when $R > 1$, and by 1 when $R < 1$. This yields

$$f = [R_{\rm max} - (\nu_h/\nu_l)^{1/2}]/(R_{\rm max} + 2R_0^{-1/2} - 3) \tag{4.13}$$

from which

$$R_{\rm max}/R_0 = 2\gamma^4 = [(\nu_h/\nu_l)^{1/2} + f(2R_0^{-1/2} - 3)]/[R_0(1 - f)]. \tag{4.14}$$

For the complete sample of sources selected at 5 GHz by Kühr *et al.* (1981), Orr & Browne found that 82% had spectral indices between 2.7 and 5 GHz smaller than 0.5. Thus taking $f = 0.82$, $(\nu_h/\nu_l)^{1/2} = 1.36$ and an estimate of R_0 (~ 0.036 for a mean redshift of 0.5), equation (4.14) yields $\gamma \sim 5.0$. In their more careful calculation

Orr & Browne found $\gamma \sim 5.2$ for this sample, and γ in the range 3 to 5 for a number of other samples.

If the extended structure around the compact, flat-spectrum quasars really does bear the same quantitative relation to their nuclear jets as do their counterparts in the *intrinsically* more powerful classical double sources, and if their nuclear jets really are simple jets having the same γ of bulk flow and radiation pattern given by equation (4.5), then it would seem that the data are broadly compatible with $\gamma \sim 5$. However, there is not so much beaming as one would then expect for simple jets characterised by the Lorentz factors of bulk flow as high as 10 that are required to explain the apparent superluminal motion observed in a number of sources (Zensus & Pearson 1988); this need not be surprising. Even setting aside the possibility of making the parameters of Orr & Browne's unified model more flexible, there are other models for the emitting regions in which superluminal motion occurs exactly as outlined in §4.1 (for instance) and to which all the statistical considerations of that section still apply, but in which the emitting material may be stationary (as in the case of the light-echo model, §4.3.5) or moving more slowly than the observed features (as in the case of emission from material excited by a relativistic shock that is propagating forwards through the fabric of the jet, §4.3.4). Furthermore, much of the total flux from observed jets comes from the inner, optically thick part which may well not emit isotropically in Σ'; thus opacity is also likely to cause $S(\theta)$ to deviate from the form (4.5).

The remainder of §4.3 will be concerned with simple models incorporating these effects through which their general influence on observational properties may be examined. There will be more to say about the relationship between nuclear and extended structure, and 'unified models' in §4.4.1.

4.3.2 Partially opaque jets

Model jets discussed hitherto in this section have been either entirely optically thick, or entirely optically thin. It is far more likely that real jets are partially opaque, since near the base, their spectra are seen to flatten as is expected in the transition from optically thin to optically thick emission. The flux of a completely opaque jet would be proportional to the product of its surface brightness and projected area. The latter is proportional to

$\sin\theta$, and the former would be proportional to $\mathcal{D}^{-1/2}$ since $\alpha = 5/2$ for an opaque synchrotron source. Hence the flux of an optically thick source is likely to decrease as θ falls. On the other hand, in a jet which is opaque at the base becoming optically thin further out, the forward beaming effect will be somewhat restored, since as θ decreases and the emission is at lower radiated frequencies, so the optically thick region extends further out into the jet. It is therefore of interest to explore the consequences of a simple model allowing for the effects of opacity in a jet emitting synchrotron radiation. The example given in this section is similar to (though less detailed than) those worked by a number of authors (*e.g.*, Marscher 1980; Königl 1981; Reynolds 1982).

Consider a conical jet symmetrically disposed about the z axis through which material flows as speed $c\beta$ (corresponding to a Lorentz factor γ), expanding sideways making a small opening angle ψ. The jet is composed of synchrotron emitting plasma with a tangled magnetic field so that in its rest frame Σ' the emission and absorption coefficients are independent of θ'. The magnitude of the magnetic field in Σ', B', may be characterised by the corresponding Larmor frequency $\nu_g' = eB'/(2\pi m_e)$, and it is assumed to vary with z as $\nu_g' = \nu_{g0}(z/z_0)^{-\zeta}$. Likewise, the number density of relativistic electrons in Σ' in the energy interval $m_e c^2 d\gamma_e$ at z and γ_e is $n'(\gamma_e, z)d\gamma_e = n_0\gamma_e^{-(2\alpha+1)}(z/z_0)^{-\epsilon}d\gamma_e$. Note that this describes quantities measured in Σ' at points corresponding to z in Σ. α is the spectral index of optically thin material. The power-law forms adopted for ν_g' and n' do not in general conserve energy in magnetic field and relativistic electrons, and so some source of such energy (see Chapter 9) is required. Furthermore, the assumption of constant velocity means that any piece of plasma is stretched transversely as it passes along the jet, so that some process, such as random motion of the plasma into which the magnetic field is frozen, is required to maintain the isotropy of the field. Lastly, the angle θ between the z axis and the direction from which the jet is observed is assumed to be sufficiently large that the values of n' and ν_g' are reasonably constant along any one line of sight. The model is summarised in Fig. 4.5.

A line of sight that intersects the jet axis at a distance z from the base encounters an opacity of approximately (Rybicki & Lightman

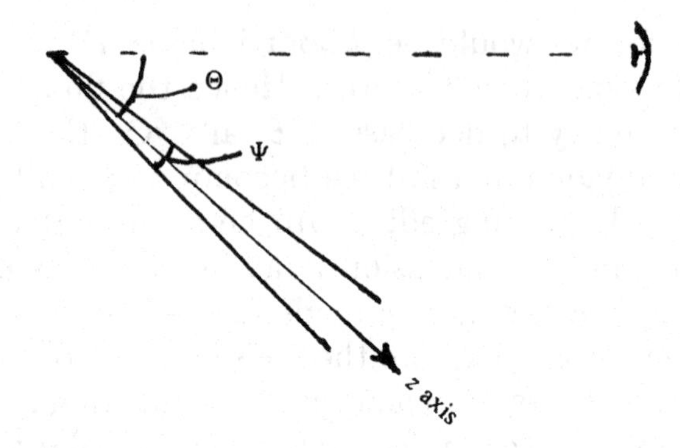

Fig. 4.5. Sketch showing the framework for the simple model for partially opaque jets discussed in §4.3.2.

1979)

$$\tau(z) = Cn_0\nu_{g0}^{3/2+\alpha}\left(\frac{z}{z_0}\right)^{-(\epsilon+\zeta(3/2+\alpha))}\left(\frac{\nu}{\mathcal{D}}\right)^{-(5/2+\alpha)}\left(\frac{z\psi}{\mathcal{D}\sin\theta}\right) \qquad (4.15)$$

where the path length of a ray measured in Σ' $(z\psi/\mathcal{D}\sin\theta)$ is used because the field and electron density are also measured in that frame. τ itself is, of course, an invariant quantity. The constant C in the equation for τ above has a weak dependence on α which will be ignored.

Solving for $z(\tau = 1)$, the jet is divided into two parts, one at $z < z(\tau = 1)$ will be assumed optically thick with spectral index $-5/2$ and the other, at $z > z(\tau = 1)$ optically thin, where the spectral index is α. The smooth transition from opaque to transparent emission has been approximated by a sharp transition at $z(\tau = 1)$. It follows that

$$\frac{z(\tau=1)}{z_0} = \left[Cn_0\nu_{g0}^{\alpha+3/2}\left(\frac{\nu}{\mathcal{D}}\right)^{-(5/2+\alpha)}\frac{z_0\psi}{\mathcal{D}\sin\theta}\right]^{1/(\epsilon+\zeta(3/2+\alpha)-1)} . \qquad (4.16)$$

Defining z_0 to be the distance along the jet at which $\tau = 1$ when viewed at $\theta = \pi/2$, $\nu = \nu_0$ and $\gamma = 1$ gives

$$z(\tau = 1) = z_0[(\nu/\nu_0)^{-(5/2+\alpha)}\mathcal{D}^{3/2+\alpha}/\sin\theta]^{1/(\epsilon+\zeta(3/2+\alpha)-1)}. \qquad (4.17)$$

The intensity at the surface of an optically thick synchrotron source is approximately $(2/3)m_e\nu^{5/2}\nu_g^{-1/2}$ (Longair 1981), with local spectral index $-5/2$. Therefore the intensity transforms according to $I(\nu) = I'(\nu)\mathcal{D}^{3+\alpha} = I'(\nu)\mathcal{D}^{1/2}$ (Rybicki & Lightman 1979), and so integrating the *observed* intensity over the *projected* area of the jet yields the total optically thick flux

$$S(z < z(\tau = 1)) = [(2/3) m_e \nu^{5/2} \nu_{g0}^{-1/2} \mathcal{D}^{1/2} (z(\tau = 1)/z_0)^{\zeta/2}] \times \int_0^{z(\tau=1)} (z/z(\tau = 1))^{\zeta/2} z\psi \sin\theta dz$$
$$= \frac{2}{3(2+\zeta/2)} m_e \nu_0^{5/2} \nu_{g0}^{-1/2} \mathcal{D}^{1/2} (z_0^2 \psi \sin\theta)(\nu/\nu_0)^{5/2} \times \{(\nu/\nu_0)^{-(5/2+\alpha)} \mathcal{D}^{3/2+\alpha} / \sin\theta\}^{(2+\zeta/2)/(\epsilon+\zeta(\alpha+3/2)-1)}. \quad (4.18)$$

The intensity of an optically thin synchrotron source is equal to $\tau I_{\substack{optically \\ thick}} \propto n\nu_g^{1+\alpha} z$, so that it varies with z as $z^{1-\epsilon-\zeta(1+\alpha)}$. Thus the optically thin intensity may be written in terms of the optically thick intensity at $z(\tau = 1)$

$$I_{\substack{optically \\ thin}}(z) = I_{\substack{optically \\ thick}}(z(\tau = 1)) \left(\frac{z}{z(\tau = 1)} \right)^{(1-\epsilon-\zeta(1+\alpha))} \quad (4.19)$$

and so the optically thin flux is given by

$$S(z > z(\tau = 1)) = I_{\substack{optically \\ thick}}(z(\tau = 1)) \int_{z(\tau=1)}^{\infty} (z/z(\tau = 1))^{1-\epsilon-\zeta(1+\alpha)} z\psi \sin\theta dz. \quad (4.20)$$

Since $I_{\substack{optically \\ thick}}(z(\tau = 1))$ is given by the term in square brackets in the first form of equation (4.18) it follows that

$$S(z > z(\tau = 1)) = S(z < z(\tau = 1)) \left(\frac{2+\zeta/2}{\epsilon + \zeta(1+\alpha) - 3} \right). \quad (4.21)$$

The two components have the same dependence upon ν, $\mathcal{D}$ and $\sin\theta$. The spectral index of the total flux is

$$s = -5/2 + (\alpha + 5/2)(2 + \zeta/2)/(\epsilon + \zeta(\alpha + 3/2) - 1). \quad (4.22)$$

The index of $\mathcal{D}$ in the expression for the total flux is therefore

$$1/2 + (\alpha + 3/2)(2 + \zeta/2)/(\epsilon + \zeta(\alpha + 3/2) - 1) = 2 + s + (\alpha - s)/(\alpha + 5/2), \quad (4.23)$$

the index of $\sin\theta$ is

$$1 - (2 + \zeta/2)/(\epsilon + \zeta(\alpha + 3/2) - 1) = (\alpha - s)/(\alpha + 5/2) \quad (4.24)$$

and the ratio of the optically thick to optically thin fluxes is

$$(\epsilon + \zeta(\alpha + 1) - 3)/(2 + \zeta/2) = (\alpha - s)/(s + 5/2). \quad (4.25)$$

(This latter ratio may well be increased by synchrotron losses in the optically thin region. Such losses cause a steepening of the emission spectrum far out in the jet (Königl 1981).) The total flux of the jet therefore comes to

$$S \approx \frac{2}{3(2+\zeta/2)} m_e \nu_0^{5/2} \nu_{g0}^{-1/2} (z_0^2 \psi)(\nu/\nu_0)^{-s}[1+(s+5/2)/(\alpha-s)]$$
$$\times \mathcal{D}^{2+s+(\alpha-s)/(\alpha+5/2)}(\sin\theta)^{(\alpha-s)/(\alpha+5/2)}. \qquad (4.26)$$

The important dependences on ν, $\mathcal{D}$ and θ can therefore be written in terms of only quantities that are (in principle) observable: the spectral index of the total flux and the spectral index in the region where the jet is optically thin. These dependences may be shown to agree with those derived by Königl (1981) and Marscher (1980).

Note that $\epsilon + \zeta(1+\alpha) - 3$ has been assumed positive in order that the optically thin flux remains finite. From equation (4.25), this requires only that $s < \alpha$, which is an obvious requirement anyway since the jet is a blend of material with spectral indices $-5/2$ and α, where the latter is most certainly greater than the former.

Equation (4.25) shows that the dependence of S upon $\mathcal{D}$ and θ comes remarkably close to that which would have obtained from setting α, in the form (4.5) for the 'simple jets' described in the preceding subsection, equal to s. For typical values of the spectral indices ($\alpha \sim 0.5$ and $s \sim 0$), the additional terms in $\mathcal{D}$ and $\sin\theta$ are raised to only quite small powers ($\sim 1/6$). The largest value of S (found by maximising expression (4.26) with respect to θ) occurs at

$$\cos\theta_{\max} = \sqrt{\{1+q+q^2/4\beta^2\}} - q/(2\beta), \qquad (4.27a)$$

where $q = (\alpha - s)/[(\alpha+5/2)(2+s)]$ and for $q \ll 1$ and $1-\beta \ll 1$

$$\cos\theta_{\max} \approx 1 - (1/4\gamma^2)\{q + O[q^2] + \ldots\}. \qquad (4.27b)$$

In that case, $\theta_{\max} \approx (1/\sqrt{2}\gamma)[(\alpha-s)/(\alpha+5/2)(2+s)]^{1/2}$. The ratio of the corresponding maximum flux to that at $\theta = \pi/2$ is

$$S_{\max}/S(\pi/2) \approx (2\gamma^2)^{2+s}$$
$$\times \left[\left(2\gamma^2 \frac{(\alpha-s)}{(\alpha+5/2)(2+s)}\right)^{(\alpha-s)/2(\alpha+5/2)}\right.$$
$$\left.\times \left(1 + \frac{\alpha-s}{2(\alpha+5/2)(2+s)}\right)^{-(2+s)-(\alpha-s)/(\alpha+5/2)}\right]. \qquad (4.28)$$

The term in square brackets multiplies the value found by putting

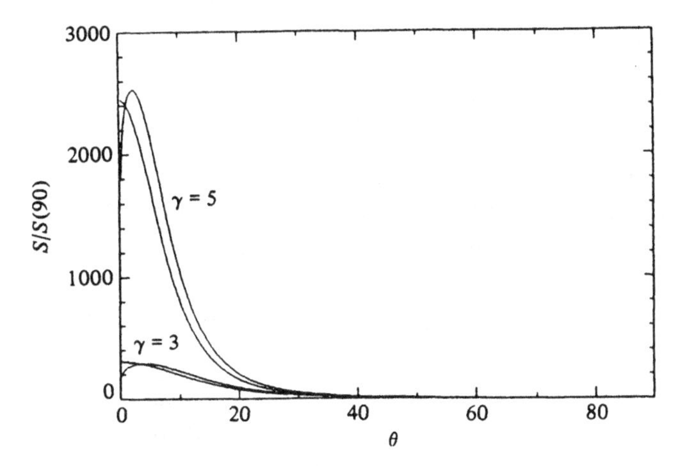

Fig. 4.6. Comparison of $S(\theta)$ for the simple jet model (equation (4.5) – the monotonically falling curves) and for the partially opaque jet (equation (4.26) – the curves having maxima). The observed flux S when viewed at inclination θ is plotted against θ (in degrees) for various values of γ, the jet Lorentz factor.

$\alpha = s$ in the form (4.5) for a simple jet, and since $(\alpha - s)/(\alpha + 5/2)$ is likely to be quite small, the corresponding ratio $S_{\max}/S(\pi/2)$ will be quite similar. However the partially opaque model gives significantly less anisotropy than would be obtained from an optically thin jet made from the same material (*i.e.*, having the same α).

The similarity of the forms (4.5) and (4.26) for S (Fig. 4.6) means that inclusion of opacity in a simple uniform jet model does not alter greatly the statistical conclusions of the previous section. One other important feature of the model is that it permits an estimate of the scaling of the length of a jet with $\mathcal{D}$ and θ. Let $z(f)$ be the length along a jet that contains a fraction f of the total flux, assumed to lie in the optically thin region. Since the ratio of optically thick to optically thin flux is independent of $\mathcal{D}$ and θ (equation (4.25)), f will correspond to some fraction of the optically thin flux which can be determined by integrating the observed intensity, equation (4.19), out to $z(f)$, and then dividing by the total flux. Therefore $z(f)$ scales with $\mathcal{D}$ and θ in the same way as $z(\tau = 1)$ – equation (4.17). From the first form of equation (4.18),

$$z(f) \propto (S\mathcal{D}^{-1/2}/\sin\theta)^{1/(2+\zeta/2)} \tag{4.29}$$

and the projected distance that contains a fraction f of the total flux is

$$r(f) = z(f)\sin\theta$$
$$\propto \left(\mathcal{D}^{3/2+s+(\alpha-s)/(\alpha+5/2)}(\sin\theta)^{1+\zeta/2+(\alpha-s)/(\alpha+5/2)}\right)^{1/(2+\zeta/2)}. \quad (4.30)$$

As S falls, so does $r(f)$ (except for very small θ), and so fainter jets appear shorter. (For example if $s = 0$ and $\zeta = 1$, $r(f)$ depends on $\mathcal{D}$ and $\sin\theta$ through $\mathcal{D}\sin\theta$ as does the velocity formula (4.1). Then, r is a maximum when $\beta = \cos\theta$.) This means that in order to monitor the faint nuclei of classical double sources for proper motion (as would be required to constrain the distribution of apparent speeds, §4.2), VLBI arrays of higher resolution than those used currently for sources with bright nuclei may well be required. Since there is little scope for increasing further the baselines between telescopes attached to the Earth's surface, this would require either observing at higher frequencies, or using antennae in orbit around the Earth. Although both these avenues are being explored at present, for this particular application satellite interferometry is likely to be the most profitable technique, since jets are also likely to shorten with increasing frequency as the $\tau = 1$ surface moves inward. Already, Levy *et al.* (1989) and Linfield *et al.* (1989) have reported the results of some early satellite interferometry, and there are proposals by a number of consortia to launch dedicated orbiting VLBI antennae, for example RADIOASTRON (Kardashev & Slysh 1988).

(Note that the shortening of jets with falling flux density is not a *unique* feature of models such as this. Scheuer (1984b) has pointed out that the same *qualitative* result holds for a jet made of material moving at speed well below c, whose brightness temperature is limited by inverse-Compton scattering throughout its length to $10^{12}K$ (cf. Chapter 1)).

Among the criticisms that one might level at the model outlined above is that it fails to reproduce two observed features of VLBI jets. The first is that, since the magnetic field is tangled, the emergent radiation is unpolarized. A tangled magnetic field was adopted in ignorance of the real structure, and is consistent with the fact that low resolution observations of the compact radio sources often show rather low percentage polarization.

The second criticism, that there are no knots in these model jets, might well be unimportant if the knots are simply a series of bright components whose boundaries move along with the bulk of the material. On the other hand, if they represent material excited to emit by shocks in the flow, interesting consequences ensue that are explored in the next subsection.

4.3.3 Knots as relativistic shocks

If a piston advances into a gas at rest at a speed less than the sound speed in the gas, then a smooth pressure gradient builds up in front of the piston, in which each element of gas is in dynamic equilibrium with those on either side. This happens because changes in pressure can be communicated throughout the compressed gas before a significant amount of new gas has been 'swept up'. However, if the piston advances supersonically, pressure waves in the rapidly moving, hot gas near the piston slow down as they reach the colder material, tending to pile up in that region, until the pressure gradient steepens into a shock which is a region of thickness one mean free path for interaction between the gases. Hence the shock separates the cold gas which is not yet affected by the piston and the hot high pressure material. In astrophysical jets, shocks may occur due to sudden changes in speed causing one body of gas to advance into another. That might be brought about by variations in the speed at which material is injected into the jet, or by variations in the pressure of the surrounding medium (Falle & Wilson 1985). In such a case there will of course be two shocks, one in the 'piston' gas and the other in the gas 'at rest'. The two components of shocked gas (assumed immiscible) are separated by a contact discontinuity. For simplicity it is often assumed that one shock is usually much stronger than the other (in the sense that it undergoes a larger pressure jump) and the weaker shock may be ignored (Hughes, Aller & Aller 1985). Then the emitting region can be described by a single compression ratio. Under the conditions relevant here, the mean free path for particle collisions is, of course, far longer than any scale size of interest, and the energy transport through the shock is provided by plasma waves; the structure of these collisionless shocks is discussed extensively in §9.3. In this chapter shocks are viewed in terms of jump conditions, *i.e.*, conservation laws relating the up- and down-stream fluids.

It is quite tempting to identify the boundaries of knots that appear on VLBI maps of nuclear jets with shock waves. Such shocks could provide a source of highly relativistic electrons (through sharing of energy between the post-shock electrons and protons, or through particle acceleration (Chapter 9)). Compression of the plasma will also lead to an enhancement of the component of magnetic field in the plane of the shock, and so shocks provide conditions propitious to synchrotron radiation. Furthermore, as discussed in §1.2.1.3, if the magnetic field is tangled, then even quite modest compressions lead to significant polarization when viewed within about 30° of the plane of the shock. For emission from a relativistic shock that would correspond to a viewing angle close to (Lorentz factor of the shocked gas)$^{-1}$ in the observer's frame, VLBI polarization measurements show that the knots are indeed polarized (Wardle & Roberts 1988) though to be fair, it is not yet clear whether the percentage polarization is greater at the knots than in the dimmer regions between them. The polarized outbursts at radio frequencies have also been modelled successfully in terms of shocks (Hughes, Aller & Aller 1985, 1989a,b).

Lind & Blandford (1985) have drawn attention to an important consequence of emission by shocked gas, namely that if the emitting material is moving relativistically in the rest frame of the shock, then the Lorentz factors that characterise the apparent component speed $c\beta_{\rm app}$ and the radiation pattern $S(\theta)$ may be quite different. Following Lind & Blandford (1985), consider a jet model in which the jet material (through which shocks propagate at speed β_u) is stationary (or rather, moves non-relativistically) in the observer's rest frame, Σ, and in which the shocked gas is optically thin and emits isotropically in *its* rest frame, Σ''. Each element of shocked gas is then assumed to emit uniformly as the shock recedes until it begins to cool and expand, at which point it ceases to emit. Let Σ' be the rest frame of the shock itself, then $\mathcal{D} = \nu/\nu'$ is the Doppler shift relating frequencies in Σ and Σ' while $\mathcal{D}' = \nu'/\nu''$ is the corresponding Doppler shift between Σ' and Σ''. The Doppler shift between Σ and Σ'' is obviously $\mathcal{D}'' = \nu/\nu'' = \mathcal{D}\mathcal{D}'$. In the rest frame of the shock Σ', the emission arises from a region of fixed volume and emission coefficient $j' \propto \mathcal{D}'^{2+\alpha}$ (Rybicki & Lightman 1979) so that $S' \propto \mathcal{D}'^{2+\alpha}$. In the observer's frame, Σ', the flux of a region excited by a single plane shock would therefore vary with θ as $\mathcal{D}'^{2+\alpha}\mathcal{D}^{3+\alpha}$. However, for a series of shocks that fade as they move outwards leading to a steady flux in Σ, $S(\theta) = \mathcal{D}^{2+\alpha}S'(\theta')$ as discussed in §4.3.1 so that

$$S(\theta) \propto (\mathcal{D}\mathcal{D}')^{2+\alpha} = \mathcal{D}''^{2+\alpha} = [\gamma_d(1-\beta_d\cos\theta)]^{-(2+\alpha)}, \tag{4.31}$$

where $c\beta_d$ and γ_d are the speed and Lorentz factor of the shocked (downstream) gas measured in Σ. This means that the flux of a jet consisting of a series of plane shocks has the same dependence on viewing angle θ as the simple continuous jet described in §4.3.1. This is not surprising since, in the limit in which the shocks become sufficiently close that the region of shocked gas behind one shock just touches the following shock, the two models are identical.

The results relating the conditions on either side of a non-relativistic shock are well known (*e.g.*, Dyson & Williams 1980). However the corresponding results for relativistic shocks (Blandford & McKee 1976; Königl 1980; Peacock 1981) are perhaps less so, and are therefore summarised here. In their relativistic form the Rankine-Hugoniot conditions can be expressed in the rest frame of the shock in terms of total energy density u, pressure P and particle number density n (measured in the respective rest frames of the fluids) as conservation of particles, $\gamma'\beta' n$, energy $\gamma'^2\beta'(u+P)$ and momentum $\gamma'^2\beta'^2(u+P)+P$ (Rindler 1982). For particles of mean rest mass m, u can be eliminated in favour of a variable Γ using

$$P = (\Gamma - 1)(u - mnc^2) \tag{4.32}$$

where in the non-relativistic and extreme relativistic limits, Γ (= 5/3 and 4/3, respectively) is the familiar ratio of specific heat capacities. In the intermediate region Γ is a (slowly varying) function of $u/(mnc^2)$. Using the particle and energy equations to eliminate n_d and P_d from the momentum equation yields (Königl 1980)

$$\frac{\Gamma_u}{\Gamma_u - 1}\gamma_u'^2\beta_u'^2(1-\beta_d'/\beta_u') - \frac{\gamma_u'^2\beta_u'\Gamma_u(\Gamma_d-1)}{\gamma_d'^2\beta_d'\Gamma_d(\Gamma_u-1)} + 1 =$$
$$\frac{mn_uc^2}{P_u}\left[\gamma_u'^2\beta_u'^2(\beta_d'/\beta_u' - 1) + \frac{\Gamma_d - 1}{\Gamma_d}\frac{\gamma_u'\beta_u'}{\gamma_d'\beta_d'}(\frac{\gamma_u'}{\gamma_d'} - 1)\right] \tag{4.33}$$

where the suffices u and d refer to the up- and down-stream gases, respectively.

There are three simple (relativistic) cases for which equation (4.33) can be solved analytically:

(a) Both pre- and post-shock gases consist only of relativistic particles so that $\Gamma_u = \Gamma_d = 4/3$. In that case $P_u \gg mn_uc^2$ and one is left with the equation [LHS of (4.33)] = 0. That equation yields the (non-trivial) result

$$\beta_d' = 1/(3\beta_u') \tag{4.34}$$

which may be cast into the rest frame of the upstream gas Σ to yield

$$\beta_d = 1/(2\beta_u') - 3\beta_u'/2 \tag{4.35}$$

or

$$-\beta_u' = (\beta_d + [\beta_d^2 + 3]^{1/2})/3 \tag{4.36}$$

which gives the shock speed in terms of the speed of the shocked gas in the rest frame of the upstream gas. (β_u' and β_d' are here regarded as negative, being directed towards the nucleus in this example.)

(b) The strong shock limit, $P_d/n_d \gg P_u/n_u$. In this case equation (4.33) can be simplified resulting in the equation [term in square brackets on RHS of (4.33)] = 0. That equation can be cast conveniently into Σ yielding (*after some effort!*)

$$\gamma_u'^2 = \frac{(\gamma_d + 1)(\Gamma_d(\gamma_d - 1) + 1)^2}{\Gamma_d(2 - \Gamma_d)(\gamma_d - 1) + 2} \tag{4.37a}$$

which gives the shock Lorentz factor in terms of the Lorentz factor of the shocked gas, both in the rest frame of the unshocked gas, and is valid for mildly relativistic as well as highly relativistic shocks (Blandford & McKee 1976). The speed of the shocked gas in the rest frame of the shock may be found conveniently from

$$\gamma_d' = \gamma_u'/[\Gamma_d(\gamma_d - 1) + 1] \tag{4.37b}$$

(Peacock 1981). The strong shock condition $P_d/n_d \gg P_u/n_u$ can be written (Blandford & McKee 1976)

$$\gamma_d - 1 \gg \frac{P_u}{(\Gamma_d - 1)mn_uc^2}\left[1 + \frac{\Gamma_uP_u}{(\Gamma_u - 1)mn_uc^2}\right]^{-1} \tag{4.38}$$

so that if the upstream gas is highly relativistic, *i.e.*, $\Gamma_u = \Gamma_d = 4/3$, equation (4.37) holds when $\gamma_d - 1$ is large. In that limit γ_u' deduced from equation (4.36) tends to $\sqrt{2}\gamma_d$ while that from equation (4.37) approaches $\sqrt{2}\gamma_d - 3/(4\sqrt{2})$ and their ratio tends to unity as required. Except in the extreme- and non-relativistic limits, equation (4.32) is not strictly an equation of state since Γ depends upon P; the relationship between Γ and P is then rather complicated, so that in general when Γ_u lies between 5/3 and 4/3 and Γ_d is an unknown, the relativistic Rankine-Hugoniot equations must be solved numerically (Peacock 1981). Heavens & Drury (1989) have provided useful polynomial approximations to such solutions in the case for which the upstream gas is cold ($P_u/mn_uc^2 \ll 1$) and more complicated, but nonetheless tractable expressions directly relating up- and down-stream speeds for upstream gas of arbitrary temperature.

(c) Finally, in the limit of very weak shocks where $\gamma_u'^2\beta_u'^2[1-(\beta_d'/\beta_u')^2] \ll 1$ and $\Gamma_u = \Gamma_d$, equation (4.33) can be solved (Königl 1980) whence

$$\frac{\beta_d'}{\beta_u'} = \frac{2+(\Gamma\gamma_s^2-1)\mathcal{M}^2}{\mathcal{M}^2(\Gamma\gamma_s^2+1)} \tag{4.39}$$

where $\gamma_s = (1-\beta_s^2)^{-1/2}$, β_s is the sound speed in the unshocked gas, and $\mathcal{M} = (\gamma_u'\beta_u'/\gamma_s\beta_s)$. (This reduces to the general non-relativistic result as γ_u' and γ_s tend to unity.)

Since a slowly moving jet that consists of a series of components of (optically thin) shocked gas has $S(\theta) \propto [\gamma_d(1-\beta_d\cos\theta)]^{-(2+\alpha)}$, these results contrast the Lorentz factors that describe the apparent velocity of the moving components and the radiation pattern $S(\theta)$; the latter can be smaller than the former. For example, using case (a) above, a slow jet illuminated by a series of plane relativistic shocks whose apparent speed is characterised by $\gamma = 5$ in equation (4.1), would have $S(\theta)$ characterised by $\gamma \sim 3.5$ in equation (4.5).

Not only the speeds of the radiating material and its boundary but also their directions of motion may differ. Lind & Blandford (1985) suggest a possible realization of this effect: suppose again that the jet is moving slowly in the observer's frame, and that some dense material travels along the jet at a speed closely approaching c generating a shock wave that spreads out behind it. They idealize such a shock conveniently as a conical shock wave that makes an angle η with the jet axis. In the observer's frame (that of the unshocked gas) the shock wave moves along the axis of symmetry at speed β_p compressing material and imparting momentum perpendicular to the shock so that its velocity makes an angle $(90-\eta)^\circ$ with the jet axis (Fig. 4.7).

The conical shock can be thought of as the surface formed by a sheaf of plane shocks each inclined at η to the jet axis. The speed of post shock flow, which is just that of any one such plane shock, can be deduced by transforming into a frame (Σ^*) that travels parallel to the shocked front at speed $\beta_p \sin\eta$. In that

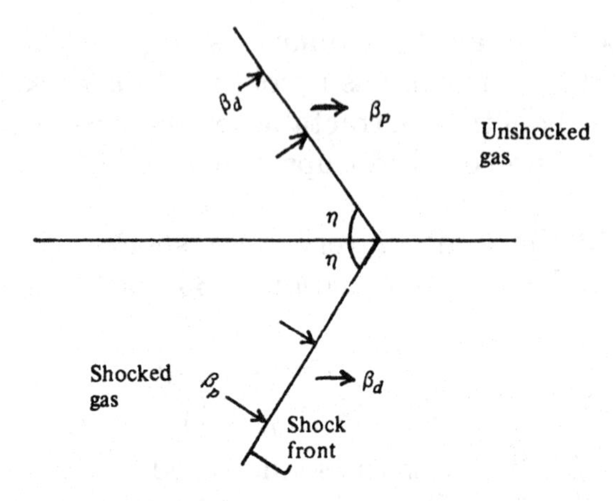

Fig. 4.7. Sketch illustrating the conical shock model of Lind & Blandford (1985). The shock front propagates along the axis of a jet composed of slowly moving material at speed $c\beta_p$ and everywhere makes an angle η with the jet axis. The shocked gas moves at speed β_d at right angles to the shock front.

frame the shock is stationary, and perpendicular to the velocities of the upstream and downstream gases. Again assuming that both have highly relativistic equations of state, equation (4.34) yields $\beta_d^* = 1/(3\beta_u^*) = -1/(3\beta_p \sin\eta)$. Then transforming back into Σ:

$$\beta_d = \left(\frac{\beta_d^* - \beta_u^*}{1 - \beta_d^*\beta_u^*}\right) = 3\beta_p \sin\eta/2 - 1/(2\beta_p \sin\eta). \qquad (4.40)$$

The component of velocity of the pre-shock gas normal to the shock itself must exceed the sound speed ($c/\sqrt{3}$ in this case) so that $\sin\eta > \sin\eta_{min} = 1/(\sqrt{3}\beta_p) > 1/\sqrt{3}$. The smallest allowed value for η (as β_p approaches unity) is approximately 35.5°. When $\eta = 90°$, equation (4.40) is equivalent to equation (4.35). In Fig. 4.8 the resulting Lorentz factor γ_d is plotted as a function of η for $\gamma_p = 3$, 5 and 10. The Lorentz factor of the radiating gas falls markedly (and monotonically) as its direction of motion deviates from the jet axis. Each element of the shocked gas emits most brightly at an angle η to the jet axis, but with a radiation pattern characterised by a Lorentz factor considerably less than that of the shock along the jet, $(1 - \beta_p^2)^{-1/2}$, which characterises the apparent speed. Furthermore, in a conical shock all elements radiate in different directions, so that the beaming effect, already reduced, is still further diluted over the surface of a cone making an angle $(90 - \eta)°$ with the jet

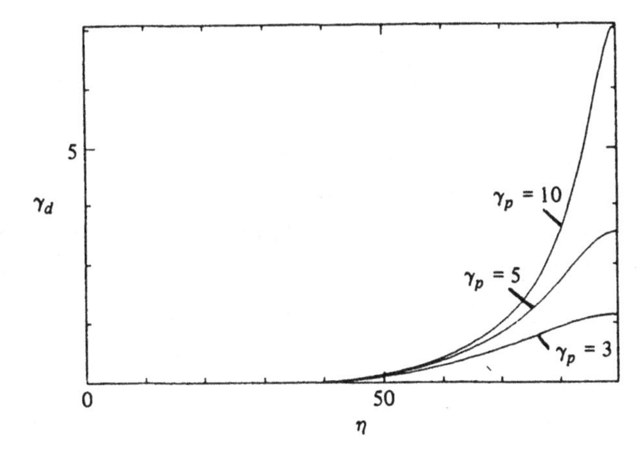

Fig. 4.8. The Lorentz factor (measured in the rest frame of the downstream gas) of gas compressed by an oblique shock front making an angle η with its direction of motion is plotted against η (in degrees), for several values of the shock Lorentz factor γ_p.

axis. For a given speed of the component boundary (or pattern) a jet consisting of essentially static material illuminated by a series of conical shocks with speed approaching c will therefore emit less anisotropically than a jet with plane shocks, which in turn emits less anisotropically than the jets consisting of just a steady flow or simple components discussed in §4.3.1. (The same qualitative result holds if the jet moves relativistically in the observer's frame but the shock speed is greater than the jet speed. However if the shock speed is less than the jet speed, the converse applies and quite slowly moving components could be seen in very highly beamed emission. What little evidence there is from beaming statistics, such as that from Orr & Browne (1982) discussed in §4.3.1, suggests that for the nuclei of powerful classical-double quasars, the speed of the radiating material in the observer's frame should not greatly exceed that of the components required to account for the apparent superluminal velocities.)

Models for the emitting region should yield predictions for the evolution of the spectrum of newly created components, and comparison with observations should provide important clues as to their structure which can so strongly affect the relationships between observable quantities. At very high frequencies the lifetime of electrons emitting synchrotron radiation is likely to be much shorter than the time between the outbursts, so that it is possible through simultaneous measurements of only the total flux at several frequencies to follow the component evolution. An example of such work was provided by Robson *et al.* (1983) who followed the evolution of a new, sharply peaked component in the infra-red and millimetre-wave spectrum of 3C 273. One surprising result of their work was that, in the later stages of the component evolution, as the peak in the spectrum of the new component moved to lower frequencies, its maximum flux density remained approximately constant. In a model that utilises a simple expanding component, losses of energy in magnetic field and relativistic electrons would normally lead to a peak flux density that falls. A peak of constant flux would require a substantial energy input. However, Marscher & Gear (1985) pointed out that this behaviour is far more likely if matter is excited to radiate by a plane shock. If, at high frequencies, the synchrotron lifetime of the excited electrons is less than the time they take to reach the back of the emitting region, then firstly, as the source expands, the lifetime of the electrons increases as the field goes down, resulting in an increasing volume of emitting material; and secondly, the volume decreases with increasing frequency leading to a smaller inferred index for the electron energy distribution, which in turn reduces the rate at which the synchrotron emissivity is predicted to fall with time. Quantitatively, the flux of the component may be written

$$S \propto n_0 B^{(\delta+1)/2} \nu^{-(\delta-1)/2} R^2 x, \tag{4.41}$$

where $n_0 \gamma_e^{-\delta} d\gamma_e$ is the number of relativistic electrons in the energy interval $m_e c^2 d\gamma_e$, R is the width of the component normal to the jet axis, and x (assumed $\ll R$) is the length of the component along the jet axis. Everything is measured in the rest frame of the shocked gas, but the Lorentz factor and Doppler shift are assumed constant. The frequency at the peak in the spectrum ν_m, which is always close to that at which the optical depth is unity, can be found from

$$S(\nu_m) \propto B^{-1/2} \nu_m^{5/2} R^2. \tag{4.42}$$

Let B vary with R as $R^{-\varsigma}$. Then for a simple component expanding sideways as it moves outwards, adiabatic expansion yields $n_0 \propto R^{-2(\delta+2)/3}$ (*i.e.*, x assumed constant). Then equations (4.41) and

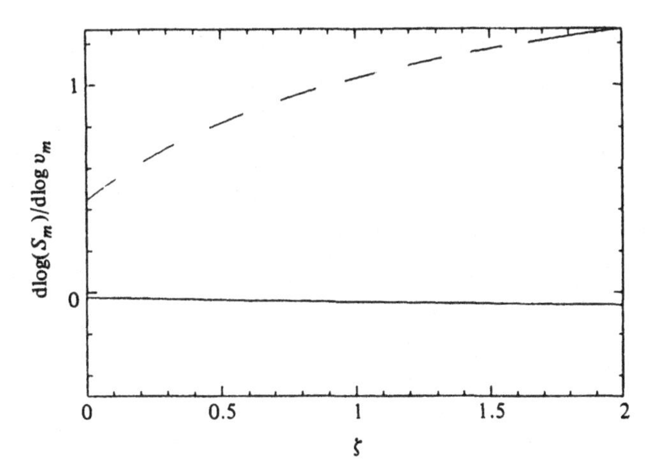

Fig. 4.9. The index for the variation of S_m, the peak flux density, with ν_m, the peak frequency, for the expanding component model (dashed line – equation (4.45)) and for the shock model of Marscher & Gear (1985) (continuous line – equation (4.49)), plotted against ζ, the power law index governing the decay of the magnetic field. (Adiabatic expansion is assumed.)

(4.42) give

$$S \propto R^{[4(1-\delta)-3\zeta(\delta+1)]/6}\nu^{-(\delta-1)/2}, \tag{4.43}$$

$$\nu_m \propto R^{-(2+\delta)(3\zeta+4)/[3(4+\delta)]}. \tag{4.44}$$

Setting α, the observed spectral index in the optically thin region above ν_m, equal to $(s-1)/2$

$$S(\nu_m) \propto \nu_m^{(12\alpha\zeta+15\zeta+8\alpha)/(2\alpha+3)(3\zeta+4)}. \tag{4.45}$$

The index of ν_m in equation (4.45) is plotted (dashed line) on Fig. 4.9 as a function of ζ, using the value $\alpha = 1.2$ found by Robson *et al.* (1983). This shows that for plausible values of ζ, the peak flux should vary noticeably with ν_m. If the component were to expand in three dimensions (along the x direction as well) matters would be still worse.

If however the emitting material is excited by a shock, then electrons that radiate predominantly at a frequency ν have lifetime $\propto \gamma/(\gamma^2 B^2) \propto B^{-3/2}\nu^{-1/2}$ as $\gamma \approx (\nu/\nu_g)^{1/2}$. Therefore, as the shock recedes, the distance along the line of sight occupied by emitting plasma is

$$x \propto \beta_d' B^{-3/2}\nu^{-1/2} \tag{4.46}$$

where β_d' is assumed constant. Provided the speed of the shock is constant, the compression ratio n_d/n_u ($= \gamma'_u\beta'_u/\gamma'_d\beta'_d$) is constant. If the energy of each emitting electron is increased in proportion to the mean particle energy (proportional to $(n_d/n_u)^{1/3}$ for a relativistic gas) then the value of n_0 differs from its value in the unshocked jet by only a constant factor. Then if each element of the jet is travelling at constant speed (until it is shocked) and expanding sideways adiabatically, $n_0 \propto R^{-2(\delta+2)/3}$ as before. Then equations (4.41), (4.42) and (4.46) lead to

$$S \propto R^{[4(1-\delta)-3\zeta(\delta-2)]/6}\nu^{-\delta/2} \tag{4.47}$$

$$\nu_m \propto R^{-[4(2+\delta)+3\zeta(\delta-1)]/3(\delta+5)}. \tag{4.48}$$

Then putting $\alpha = \delta/2$ yields

$$S(\nu_m) \propto \nu_m^{(4\alpha-5)(2+3\zeta)/[8(1+\alpha)+3\zeta(2\alpha-1)]}. \tag{4.49}$$

Again using the observed spectral index of the optically thin emission, $\alpha = \delta/2 = 1.2$, the index of ν_m in equation (4.49) is plotted as a function of ζ in Fig. 4.9 (continuous line). Over a wide range of plausible values for ζ ($1 < \zeta < 2$), the peak flux varies only weakly with ν_m, as was required. Marscher & Gear (1985) also demonstrate that the shock model mimics the observed $S(\nu_m)$ relationship in the earlier phase, when electron energy losses are dominated by inverse-Compton scattering, and $S(\nu_m)$ falls as ν_m falls.

Thus it seems that without the need to specify the model parameters very precisely, the shock model can reproduce the evolution of the flares observed in 3C 273 by Robson *et al.* (1983). Studies of component evolution in this way should, in future, reveal how widespread this type of behaviour is. As Marscher & Gear observe, such flares could be detected in their very early stages when they are visible in the X-ray waveband, giving observers about a month's warning of their appearance in the infra-red.

Recent work by Hughes, Aller & Aller (1985, 1989a,b) has demonstrated that simple shock models reproduce the high radio frequency (*i.e.*, 5 – 15 GHz) flux, polarization and spectral variations for a number of sources remarkably well. Their model employs shocks that compress the gas producing a tangled web of magnetic field in the plane of compression. When viewed at small angles to that plane in the rest frame of the emitting gas, the projected field will appear predominantly normal to the jet, and quite substantial degrees of polarization can be observed (Laing 1980). Hughes *et al.* use models derived from the variations in total intensity to predict variations in the polarized flux, which can then be compared to the results of their monitoring programme.

4.3.4 Polarization of the knots

One of the most interesting and remarkable of recent developments in the field of very long baseline interferometry has been the recovery of polarization data from phase unstable arrays. This work has been pioneered by the Astrophysics group at Brandeis University (Wardle & Roberts 1988), and so far many maps of the polarization structures of quasars and BL Lac objects have been analysed and published. One of the most striking results of this work is the difference that has emerged between the structures of strong and weak emission line objects.

Weak emission line objects (the BL Lac objects) have nuclei and knots that can be quite highly polarized, and while the polarization of the nuclei can take on quite a wide range of orientations, that of the outer knots is consistently such that the E-vectors are normal to the direction of separation from the nucleus, indicating a preferred direction of magnetic field perpendicular to the jet. (See Gabuzda *et al.* (1989) for a particularly lucid account of this phenomenon.) The most likely explanation is that an initially tangled field in the jet is compressed by a shock wave that propagates along its axis. If the shock wave is perpendicular to the axis the field forms a 'tangled web' with the same orientation and the emergent synchrotron radiation has (E-vector) polarization parallel to the jet. Polarization by partial reordering of the field is discussed more extensively in §1.2.1.3. Other structures, such as conical shocks (§4.3.3), can also yield polarization having predominantly the same polarization (Cawthorne & Cobb 1990).

Although this kind of polarization structure is circumstantial evidence that shocks are responsible for the knots in the jets of BL Lac objects, a demonstration that the polarization of emission *increases* at the knots would be far more convincing. Unfortunately, this presents serious difficulties in image processing which have not yet been solved. However in the BL Lac object OJ 287 (0851+202), one knot was observed to be over 60% polarized, close to the theoretical maximum, and attempts to account for both the high degree of polarization and the apparent superluminal motion in terms of a simple ballistic model (using partially ordered fields) were unsuccessful. Allowing the knots to be excited by a simple plane shock wave greatly alleviated these difficulties (Cawthorne & Wardle 1988).

In contrast to the BL Lacs, the strong emission line objects (principally quasars) have very weakly polarized nuclei on milliarcsecond scales (usually less than 1%). The emerging jets, however, can be quite highly polarized, often 20 or 30%. It is well known that on arcsecond scales, the jets are polarized (with E-vectors) perpendicular to the jet, indicative of a longitudinal magnetic field. On VLBI scales, misalignments between the inferred direction of the field and the local jet direction seem to be fairly common, although the orientation of the polarization can be fairly uniform in the outer parts of the jet. So far all the VLBI polarization maps made have been at wavelength 6cm so that although these misalignments may be due to Faraday rotation of intervening material, future observations at other wavelengths are necessary to see if this really is so. The Faraday rotation measurements made on arcsecond scales may well not apply to the milliarcsecond scales of interest here. In one of the most detailed observations made so far, that of 3C 273, the inferred magnetic field is aligned with the jet at large distances from the nucleus, but swings dramatically closer in (Kollgaard *et al.* 1990). If such behaviour is due to Faraday rotation (presumably by the gas surrounding the line emitting clouds) then VLBI polarimetry may well become a sensitive probe of its distribution.

The low polarization of the nuclei of quasars (less than 1%) contrasts with the figure of 5% typical for the nuclei of BL Lac objects, and it is very interesting to ask why the degree of polarization should be so low. Internal Faraday depolarization and opacity are easily ruled out as the cause, for in each case, no matter how much optical thickness or Faraday depth, simple jets yield most of the polarized flux from the outermost parts where opacity and Faraday rotation must become unimportant (Cobb, Wardle & Roberts 1988; Cobb & Wardle 1990). Tangled fields and external Faraday rotation remain possible explanations for the very low polarization. External Faraday depolarization is an attractive possibility since it accounts for the distinction between the quasars and the BL Lacs (which have little line emitting gas to depolarize their nuclei). However there are worries that the rapid increase in polarization with increasing frequency that accompanies this effect would be incon-

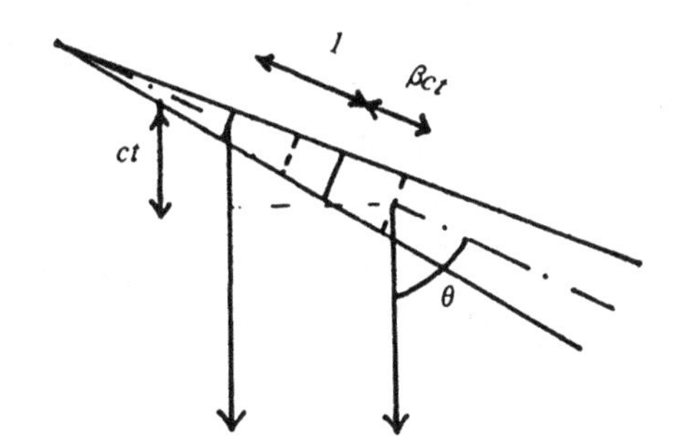

Fig. 4.10. Sketch illustrating the light-echo model of Lynden-Bell (1977). A pulse that 'excites' a jet to emit isotropically has length l, travels at speed $c\beta$ and is viewed from direction θ to its direction of motion. That observer sees scattered light from the back of the pulse when it is at the position shown by continuous lines, and from the front of the pulse when at that shown by the dashed lines. Therefore the apparent length exceeds the true length by a factor $(l + \beta ct)/l = (1 - \beta \cos\theta)^{-1}$, and so the apparent brightness is proportional to that factor. If the pulse is a beam of light observed by virtue of scattering on an essentially static medium, then $\beta = 1$.

sistent with the total polarization observed at higher frequencies. It may be simply that the field is very tangled on small scales and is only ordered by shearing much farther out in the jet. Again, observations at several wavelengths will distinguish between these possibilities.

4.3.5 The light-echo model

The aim of this section has been to present a few of the likely ingredients for jet models, and to describe their observational consequences. The most important conclusions are qualitative; the specific examples given are simple idealizations of what might really occur and are intended to show that even a small collection of simple models can give rise to quite diverse properties (the form of $S(\theta)$, for example). The models described have been those that are most easily accommodated by the notion that the VLBI knots are related to the jets required to feed the extended (kpc scale) structure. However the knots themselves are usually unresolved or only poorly resolved, so that these observations do not uniquely specify the relationship between these features and the underlying flow. In view of this uncertainty, it seems a good idea to remain open minded to the range of possible jet models, and to be as wary of treating simple models too seriously as of equating complexity with realism. One should not forget models in which the knots are neither plasma bullets nor shocks; light-echo models (Lynden-Bell 1977) are a good example. These involve a light beam (or other signal) that propagates along a slowly moving jet from which radiation is either scattered out of the jet or emitted by particles excited by the signal. The result is a visible feature moving along the jet whose apparent velocity is given by the familiar form (4.1). If the signal is a light pulse then $\beta = 1$ and $v_{\rm app}$ tends to infinity as θ tends to zero. If the emission is more or less isotropic in the rest frame of the jet (and observer) (as for a light signal scattered by particles in the jet), then the flux S is simply proportional to the apparent volume. For a signal speed $c\beta$, the apparent length of the pulse of length l extends to $l + tc\beta$ during the time delay t between seeing emission from the front and back (Fig. 4.10). Clearly $(l + tc\beta)\cos\theta = ct$, so that $S(\theta) \propto l_{\rm app} = l/(1 - \beta\cos\theta)$. For a *light* echo, $S \propto (1 - \cos\theta)^{-1}$.

4.4 The relationship between nuclear jets and kiloparsec-scale radio structure

4.4.1 Nuclear jets and the extended lobes

Long before radio jets were ever observed in powerful classical double sources, Longair, Ryle & Scheuer (1973) demonstrated that the only reasonable form in which to transport energy from the nucleus to the lobes is a flow of bulk kinetic energy. Thus these early observations of the lobes gave the first, and in many ways still the most convincing evidence that jets in some form must exist in these sources. In Chapters 2 and 3 the reader has seen how further studies of the lobes, in terms of their energy content and the structure of the hotspots, continue to yield information on the properties and content of the jets that feed them, and therefore on processes that occur on scales deep down in the nucleus itself. The first part of this section raises the question "What light do the extended lobes shed in particular on the parsec-scale nuclear jets?"

§4.3.1 described how the lobe flux has been used as a measure of the *intrinsic* brightness of the nuclear jets: the assumption is that the lobe flux is proportional to the rest-frame flux of the nuclear jets that feed them. This could be a poor assumption: the synchrotron emissivity in both these locations depends very strongly on local particle acceleration and evolution of the magnetic field (both only poorly understood – cf. Chapter 9), and there seems to be no *compelling* reason for such a small spread in the ratio of two flux densities. Scheuer & Readhead were the first to suggest using this assumption to constrain the Lorentz factors characterising $S(\theta)$, in their example for the nuclei of powerful classical double quasars. Numerous other attempts have been made since, for example that by Orr & Browne (1982) discussed in §4.3.1.

The apparent differences between the extended structure seen around the classical doubles and the compact, flat-spectrum sources argue against them being intrinsically similar: the extended structure around the latter type shows no clearly defined lobes, and while hotspot-like features are seen at the end of the bright arc-second jets (having polarization E-vectors parallel to the jet axis as in the classical doubles) there are no bright counter-hotspots. However these arguments are weakened if the extended structures around the compact, flat-spectrum sources are, like their nuclear jets, seen in projection at a small angle to the line of sight. That could mask the double lobe structure, and the absence of a bright counter-hotspot might be explained if the emitting material in the hotspot itself is mildly relativistic in velocity and its emission suffers some degree of relativistic beaming. Until recently there were few maps with sufficient dynamic range to place stringent limits on the flux of counter-hotspots. However recent work by Kollgaard, Wardle & Roberts (1989) has shown that in at least one source there is a faint component that lies just where one would expect the counter-hotspot. This component has polarization E-vector parallel to the jet axis just like a typical hotspot consisting of a sheet of compressed plasma.

Despite questions over sample membership, most authors seem to agree that the Lorentz factors that characterise the beaming pattern should not be so large as those required to explain the largest apparent speeds observed in the superluminal quasars, particularly if one assumes H_0 closer to 50 than to 100 km s^{-1} Mpc^{-1}. Thus some model more sophisticated than the simplest ballistic treatment seems desirable. That mentioned in §4.4.3 in which shocks propagate faster (in the observer's frame) than the emitting fluid is one possibility.

Apparent superluminal motion has been found among some sources having extended structure of very large linear size. Some of these observations refer to the compact, flat-spectrum sources (*e.g.*, Schilizzi & deBruyn 1983) and it is not wholly clear with what sample one should compare those as their 'unaligned counterparts' may not appear in the high flux samples. However Barthel (1987) has recently observed apparent superluminal motion with $v_{\mathrm{app}} \sim 3c$ ($H_0 = 100$ km s^{-1} Mpc^{-1}) in the second largest radio source associated with a quasar, 4C 34.47 (1721+343), a classical double whose flux density is dominated by its lobes. Using the largest angle allowed by equation (4.1) of $\sim 36°$ to deproject the apparent size of 560 kpc, leads to an overall angular extent of more than 1 Mpc. It is mildly worrying that a quasar with such a large projected size requires an orientation that takes its linear size even further away from the mean. This may be a further indication that models (such as those discussed by Scheuer (1984a) in which apparent superluminal motion is less critically dependent upon orientation, are relevant. The problem could also be eased by the model alluded to in §4.2.2, in which some quasars may appear as galaxies when viewed through obscuring material that partly surrounds the nucleus: the counterparts to 4C 34.47 whose jets point away from the line of sight could then be found among the population of radiogalaxies.

The largest radio source associated with a *bona fide* quasar is 4C 74.26, which has an apparent linear size of 1.6 Mpc and was discovered by Riley *et al.* (1989). So provided that 4C 74.26 does not also display apparent superluminal motion with a high β_{app}, a true size of more than 1 Mpc for Barthel's source is certainly not inconceivable.

4.4.2 Parsec and kiloparsec scale jets

Kiloparsec-scale jets have been observed in both flat-spectrum compact and classical-double quasars. The discussion of Chapters 2 and 3 showed that arguments based on the energy requirements of the lobes lead to some ambiguity in the determination of jets speeds, and it is natural to ask whether the nuclear jets shed any light on the properties of their large scale continuations.

An important observation in this regard is that all sources in which kiloparsec-scale jets have been found also have bright nuclear jets, and Saikia (1984) has shown that quasars in which large scale jets are visible have relatively brighter nuclei than those in which none have been found. Thus if we believe that nuclear jets are relativistic flows and that their emission suffers some degree of 'beaming', then an argument due to P. K. Moore (quoted by Scheuer 1984a) shows that large scale jets must be at least mildly relativistic, otherwise in cases where the nuclear jets are rendered invisible by beaming away from the observer one would still expect to see the large scale jet; yet there are no examples of 'disembodied' large scale jets, *i.e.*, jets without a bright 'core' at their base. All of this is confirmed quite independently in recent work of Garrington & Laing (1988). Their multi-frequency maps of classical double sources show that in those sources in which jets are found, the emission from the lobe containing the visible jet undergoes a significantly smaller Faraday depolarization than that from the lobe opposite. Since this depolarization is thought to be due to the halo of gas surrounding the source rather than gas inside the lobes themselves, this result implies a shorter path length through the halo for emission from the lobe containing the jet than for that from its (apparently) jetless counterpart: *i.e.*, the lobe containing the jet is pointing towards us. This establishes a link between the brightness of the large scale jets and their orientation in a manner that is straightforwardly accounted for if the large scale jets undergo some degree of relativistic beaming. However they do not *necessarily* suffer a very strong beaming effect: in fact the high rate of occurrence of large scale jets in classical double sources (apparently approaching 50% (Bridle & Perley 1984)) suggests that this is the case (*i.e.*, the number of sources with large scale jets in the top decade of relative surface brightness is a large fraction of the total; see Scheuer (1987) for a discussion of some caveats).

All of this raises a question addressed by Scheuer (1982): can a jet slow down on going from parsec to kiloparsec scales without losing kinetic energy so that it can still form a hotspot at its end? Slowing by entrainment of external matter turns kinetic energy into heat, and preserving the kinetic energy requires some external force, provided perhaps by the forward pressure of diverging walls. If we ignore relativistic effects, Newton's second law applied to a volume V of gas moving at speed v in pressure $P(x)$, $v d(mv)/dx = -V dP/dx$, and the condition that the kinetic energy should not decrease, $d(mv^2/2)/dx \geq 0$, yields

$$-\frac{\Delta v}{v} \leq -\frac{1}{\Gamma \mathcal{M}^2}\frac{\Delta P}{P} \tag{4.50}$$

where Γ here is the ratio of specific heats and $\mathcal{M}$ is the Mach number. Therefore the velocity varies everywhere more weakly than the $1/\Gamma\mathcal{M}^2$ power of the jet pressure, and so despite a dramatic fall in pressure between parsec and kiloparsec scales, the speed of a highly supersonic jet can vary little if it is not to dissipate large amounts of kinetic energy in the form of heat.

4.5 Conclusions

Most of the discussion in this chapter has centred around the simple model in which the radio sources associated with active nuclei are described as a continuous or nearly continuous flow of synchrotron emitting plasma. Levels of sophistication built around this elementary framework are proposed either as interpretations strongly implied by features in the maps, or as solutions to problems hinted at by data, such as statistical data on relative core brightness. Applying data to constrain these models is a difficult and arduous procedure. The best hope of constraining the elementary framework would appear to come from the distribution of apparent speeds (§4.2) in unbiased samples. Apparent speeds in excess of (say) 5c in a large fraction of sources would be a clear indication that that framework requires revision (Fig. 4.2). Using relative central component brightness as a test statistic is far less straightforward because of uncertainties over what to use for the radiation pattern $S(\theta)$ (§4.3). However, when results of VLBI monitoring become available for a large number of sources with faint nuclei, comparison of apparent speeds and relative nuclear brightness could well shed light on the nature of $S(\theta)$. Consideration of opacity effects (§4.3.2) leads to the conclusion that relatively fainter nuclei are likely to be associated with shorter jets, and higher resolution (satellite VLBI) may be required to follow the evolution of components in such sources.

The relationship between the relatively weak extended structure around the compact, flat-spectrum sources and the lobes of classical doubles remains unclear. Some of the former undoubtedly show considerable luminosity in extended emission, and there is some evidence for faint counter-hotspots. Despite these results, it is not certain that the two species should be described *en masse* by the same quantitative model. The ubiquitous bright nuclei that are always found at the base of kiloparsec-scale jets imply that they are at least mildly relativistic. In fact, theoretically it seems difficult for a jet to slow drastically between parsec and kiloparsec scales without losing a large part of its kinetic energy, impairing its ability to form a hotspot.

L. Miller, J. A. Peacock and P. A. Hughes have provided much helpful advice and criticism during the completion of this work. The number of errors has been greatly reduced thanks to a critical reading of the manuscript by D. C. Gabuzda.

Commonly used symbols in Chapter 4

Symbol	Meaning
B	the magnetic flux density
$\mathcal{D}$	doppler shift $= \nu/\nu'$
$\mathcal{D}'$	ν'/ν''
$\mathcal{D}''$	ν/ν''
e	electron charge
$F(x > x')$	the fraction of a population having $x > x'$
H_0	Hubble's constant
I	radiative intensity (emitted power per unit area per unit solid angle per unit bandwidth)
m_e	electron mass
$\mathcal{M}$	$\gamma\beta/(\gamma_s\beta_s)$, generalized Mach number
n	particle density
$n(\gamma_e, z)$	$n_0\gamma_e^{-(2\alpha+1)}(z/z_0)^{-\epsilon}$ – the distribution of relativistic electrons
$p(x)$	$-dF/dx$ – differential distribution of population with x
P	pressure
R	$S_{\text{nuclear}}/S_{\text{extended}}$
R_0	$R(\theta = \pi/2)$
s	spectral index of total flux
S	flux density (received power per unit area per

	unit bandwidth)
S_0	the lower limit on flux density for a complete sample
u	relativistic particle energy density
v	component speed
$v_d = c\beta_d$	speed of post-shock (downstream) fluid
$v_u = c\beta_u$	speed of pre-shock (upstream) fluid
z	distance along the jet
z_0	reference distance along jet
α	spectral index of optically thin emission
β	v/c
$\beta_{\rm app}$	(apparent component speed)$/c$
β_s	(sound speed)$/c$
γ	$(1 - v^2/c^2)^{-1/2}$, the Lorentz factor of bulk flow
γ_e	the electron Lorentz factor
γ_s	$(1 - \beta_s^2)^{-1/2}$
Γ	$1 + p/(u - mnc^2)$, "ratio of specific heats"
η	cone semi-angle for a conical shock
θ	angle of inclination of jet to observer's line of sight
ν	radiation frequency
ν_g	the electron Larmor frequency
$\nu_g(z)$	$\nu_{g0}(z/z_0)^{-\varsigma}$
ν_m	the peak frequency of a spectrum
Σ	label for the observer's frame
Σ'	label for rest frame of component boundary or 'pattern'
Σ''	label for rest frame of emitting gas (when different from Σ')
τ	optical depth
ψ	jet opening angle
$[]_u$	appends to property of gas upstream in relation to a shock wave
$[]_d$	appends to property of gas downstream in relation to a shock wave

References

Barthel, P. D., 1987, In *Superluminal Radio Sources*, eds. Zensus, J. A. & Pearson, T. J. (Cambridge University Press: Cambridge), p. 148.

Biretta, J., Moore, R. L. & Cohen, M. H., 1986, *Ap. J.*, **308**, 93.

Blandford, R. D. & McKee, C. F., 1976, *Phys. Fluids*, **19**, 1130.
Bridle, A. H. & Perley, R. A., 1984, *Ann. Rev. Astr. Ap.*, **22**, 319.
Cawthorne, T. V., 1985, Ph. D. Thesis, University of Cambridge.
Cawthorne, T. V. & Cobb, W. K., 1990, *Ap. J.*, in press.
Cawthorne, T. V., Scheuer, P. A. G., Morison, I. & Muxlow, T. W. B., 1986, *M. N. R. A. S.*, **219**, 883. (See also erratum: *M. N. R. A. S.*, **222**, 895.)
Cawthorne, T. V. & Wardle, J. F. C., 1988, *Ap. J.*, **332**, 696.
Cobb, W. K. & Wardle, J. F. C., 1990, *Ap. J.*, in press.
Cobb, W. K., Wardle, J. F. C. & Roberts, D. H., 1988, In *The Impact of VLBI on Astrophysics and Geophysics*, eds. Reid, M. J. & Moran, J. M. (Reidel: Dordrecht, Netherlands), p. 153.
Dyson, J. E. & Williams, D. A., 1980, *Physics of the Interstellar Medium*, (J. Wiley: New York).
Falle, S. A. E. G. & Wilson, M. J., 1985, *M. N. R. A. S.*, **216**, 79.
Fanti, C., Fanti, R., Parma, P. & Schilizzi, R. T., 1985, *Astr. Ap.*, **143**, 292.
Gabuzda, D. C., Cawthorne T. V., Roberts, D. H. & Wardle, J. F. C., 1989, *Ap. J.*, **347**, 701.
Garrington, S. T. & Laing, R. A., 1988, *Nature*, **331**, 147.
Heavens, A. F. & Drury, L. O'C., 1989, *M. N. R. A. S.*, **235**, 997.
Hough, D. H., 1986, Ph. D. Thesis, California Institute of Technology.
Hough, D. H. & Readhead, A. C. S., 1987a, In *Superluminal Radio Sources*, eds. Zensus, J. A. & Pearson, T. J. (Cambridge University Press: Cambridge), p. 114.
Hough, D. H. & Readhead, A. C. S., 1987b, *Ap. J. Lett.*, **321**, L11.
Hough, D. H. & Readhead, A. C. S., 1989, *Astron. J.*, **98**, 1208.
Hughes, P. A., Aller, H. D. & Aller, M. F., 1985, *Ap. J.*, **298**, 301.
Hughes, P. A., Aller, H. D. & Aller, M. F., 1989a, *Ap. J.*, **341**, 54.
Hughes, P. A., Aller, H. D. & Aller, M. F., 1989b, *Ap. J.*, **341**, 68.
Kardashev, N. S. & Slysh, V. I., 1988, In *The Impact of VLBI on Astrophysics and Geophysics*, eds. Reid, M. J. & Moran, J. M. (Reidel: Dordrecht, Netherlands), p. 433.
Kollgaard, R., Gabuzda, D. C., Brown, L. F., Wardle, J. F. C. & Roberts, D. H., 1990, *Ap. J.*, in press.
Kollgaard, R., Wardle, J. F. C. & Roberts, D. H., 1989, *Astron. J.* , **97**, 1550.
Königl, A., 1980, *Phys. Fluids*, **23**, 1083.
Königl, A., 1981, *Ap. J.*, **243**, 700.
Kühr, H., Witzel, A., Pauliny-Toth, I. K. K. & Nauber, U., 1981, *Astr. Ap. Suppl.* , **45**, 367.
Laing, R. A., 1980, *M. N. R. A. S.*, **193**, 439.
Laing, R. A., Riley, J. M. & Longair, M. S., 1983, *M. N. R. A. S.*, **204**, 151.
Levy, G. S., Linfield, R. P., Edwards, C. D., Ulvestad, J. S., Jordan Jr., J. F., DiNardo, S. J., Christensen, C. S., Preston, R. A., Skjerve, L. J., Stavert, L. R., Burke, B. F., Whitney, A. R., Cappallo, R. J., Rogers, A. E. E., Blaney, K. B., Maher, M. J., Ottenhoff, C. H., Jauncey, D. L., Peters, W. L., Reynolds, J., Nishimura, T., Hayashi, T., Takano, T., Yamada, T., Hirabayashi, H., Morimoto, M., Inoue, M.,

Shiomi, T., Kawaguchi, N., Kunimori, H., Tokumaru, M. & Takahashi, F., 1989, *Ap. J.*, **336**, 1098.

Lind, K. & Blandford, R. D., 1985, *Ap. J.*, **295**, 358.

Linfield, R. P., Levy, G. S., Ulvestad, J. S., Edwards, C. D., DiNardo, S. J., Stavert, L. R., Ottenhoff, C. H., Whitney, A. R., Cappallo, R. J., Rogers, A. E. E., Hirabayashi, H., Morimoto, M., Inoue, M., Jauncey, D. L. & Nishimura, T., 1989, *Ap. J.*, **336**, 1105.

Longair, M. S., 1981, *High Energy Astrophysics*, (Cambridge University Press: Cambridge).

Longair, M. S., Ryle, M. & Scheuer, P. A. G., 1973, *M. N. R. A. S.*, **164**, 243.

Lynden-Bell, D., 1977, *Nature*, **270**, 396.

Marcaide, J. M., Bartel, N., Gorenstein, M. V., Shapiro, I. I., Corey, B. E., Rogers, A. E. E., Webber, J. C., Clark, T. A., Romney, J. D. & Preston, R. A., 1985, *Nature*, **314**, 424.

Marscher, A. P., 1980, *Ap. J.*, **235**, 386.

Marscher, A. P. & Gear, W. K., 1985, *Ap. J.*, **298**, 114.

Miller, L., 1983, Ph. D. Thesis, University of Cambridge.

Orr, M. J. L. & Browne, I. W. A., 1982, *M. N. R. A. S.*, **200**, 1067.

Peacock, J. A., 1981, *M. N. R. A. S.*, **196**, 135.

Reynolds, S. P., 1982, *Ap. J.*, **256**, 13.

Riley, J. M., Warner, P. J., Rawlings, S., Saunders, R. D. E. & Pooley, G. G., 1989, *M. N. R. A. S.*, **236**, 13p.

Rindler, W., 1982, *Introduction to Special Relativity* (Chapter 7), (Oxford University Press: Oxford).

Robson, E. I., Gear, W. K., Clegg, P. E., Ade, P. A. R., Smith, M. G., Griffin, M. J., Nolt, I. G., Radostitz, J. V. & Howard, R. J., 1983, *Nature*, **305**, 194.

Roberts, D. H. & Wardle, J. F. C., 1986, *Quasars*, eds. Swarup, G. & Kapahi, V. K. (Reidel: Dordrecht, Netherlands), p. 141.

Rybicki, G. B. & Lightman, A. P., 1979, *Radiative Processes in Astrophysics*, (J. Wiley: New York).

Saikia, D. J., 1984, *M. N. R. A. S.*, **208**, 231.

Sanders, R. H., 1974, *Nature*, **248**, 390.

Saunders, R. D. E., 1984, In *VLBI and Compact Radio Sources*, eds. Fanti, R., Kellermann, K. & Setti, G. (Reidel: Dordrecht, Netherlands), p. 193.

Scheuer, P. A. G., 1982, *Highlights in Astronomy*, **6**, 735.

Scheuer, P. A. G., 1984a, In *VLBI and Compact Radio Sources*, eds. Fanti, R., Kellermann, K. & Setti, G. (Reidel: Dordrecht, Netherlands), p. 197.

Scheuer, P. A. G., 1984b, In *Proc. of Workshop on QUASAT, Gr Enzerdorf, Austria*, ESA SP-213.

Scheuer, P. A. G., 1987, In *Superluminal Radio Sources*, eds. Zensus, J. A. & Pearson, T. J. (Cambridge University Press: Cambridge), p. 104.

Scheuer, P. A. G. & Readhead, A. C. S., 1979, *Nature*, **277**, 182.

Schilizzi, R. T. & deBruyn, A. G., 1983, *Nature*, **303**, 26.

Schilizzi, R. T., Skillman, E. D., Miley, G. K., Barthel, P. D., Benson, J. M. & Muxlow,

T. W. B., 1988, In *The Impact of VLBI on Astrophysics and Geophysics*, eds. Reid, M. J. & Moran, J. M. (Reidel: Dordrecht, Netherlands), p. 127.

Shaffer, D. B. & Marscher, A. P., 1988, In *The Impact of VLBI on Astrophysics and Geophysics*, eds. Reid, M. J. & Moran, J. M. (Reidel: Dordrecht, Netherlands), p. 43.

Walker, R. C., Benson, J. M. & Unwin, S. C., 1987, *Ap. J.*, **316**, 546.

Wardle, J. F. C. & Roberts, D. H., 1988, In *The Impact of VLBI on Astrophysics and Geophysics*, eds. Reid, M. J. & Moran, J. M. (Reidel: Dordrecht, Netherlands), p. 143.

Yates, M. G. & Longair, M. S., 1989, *M. N. R. A. S.*, **241**, 29.

Zensus, J. A. & Pearson, T. J., 1988, In *The Impact of VLBI on Astrophysics and Geophysics*, eds. Reid, M. J. & Moran, J. M. (Reidel: Dordrecht, Netherlands), p. 7.

Zensus, J. A. & Porcas, R. W., 1987, In *Superluminal Radio Sources*, eds. Zensus, J. A. & Pearson, T. J. (Cambridge University Press: Cambridge), p. 126.

5

From Nucleus to Hotspot: Nine Powers of Ten

VINCENT ICKE
Sterrewacht Leiden, Postbus 9513, 2300 RA Leiden, Netherlands.

5.1 Introduction

In Chapters 1 through 4, we saw that the outer radio lobes associated with active galaxies receive their energy from a bulk hydrodynamic flow which emanates from the galactic nucleus. Bisymmetric outflow occurs on a wide range of scales in less energetic objects as well, as will be shown in Chapter 10.

Perhaps the most astonishing feature of cosmic jets is their ability to stay together over a very large range of distance scales. On their way from the black hole in the galactic nucleus to a radio lobe, cosmic jets cover a stupendous factor 10^9 in length scale. That is as if, exhaling forcefully at my desk in Leiden, I could blow about the papers on the desk of a colleague in Minneapolis. This should be a caution against off-the-cuff comparison between jets in radio galaxies and such comparatively easily understood items as laboratory jets, rocket exhausts, and numerical simulations.

The most natural explanation of the coherence of jets is that they are not jets at all, but gaseous cannonballs with a density that is much *higher* than that of their surroundings. What we perceive as jets would be a mixture of gas ablated from these "tracer bullets" and surrounding gas set aglow by their passage. Even though there is observational evidence to support such a picture (directly in the famous case of SS 433, indirectly in the fact that the most successful models for large-scale radio source structure are all *ballistic* models), the gasball hypothesis has not found favour in the community.

But the fact of a journey through a scale of 10^9 remains, and it poses a wide range of problems. Naturally, such a trip is very hazardous to the integrity of the jet. Thus, it is surprising that the observed radio jets (Cygnus A provides the most remarkable example) remain so narrow over such long distances. It appears that the physical cross-section of the Cygnus A jet is almost constant; that, too, is remarkable because it indicates either an effective confinement mechanism or a persistent high internal Mach number.

In either case, it is to be expected that the jet interacts with gas in its environment, and numerous observations (such as the bending of head-tail sources) demand such interactions. Thus, it becomes necessary to understand how supersonic, and maybe even relativistic, streams of gas can survive their journey through nine powers of ten.

To a jet, the surrounding medium may be an impediment, but to us the interaction serves as a tracer. The phenomena to be discussed presently have been used to estimate such parameters as the speed and the density of the jets. Curiously, these estimates all indicate comparatively low, and in particular nonrelativistic, velocities. Likewise, velocity estimates based on the ballistic modelling of radio trails fall far short of the $\beta \sim 0.99$ which is implied by the more extreme superluminal sources. The meaning of this discrepancy is at present unclear. Given the likelihood that jets are subrelativistic, (or at worst marginally relativistic) on the kiloparsec scale, and to make the analysis both tractable and transparent, I formulate most of the following discussion in terms of nonrelativistic flow speeds.

Numerical hydrodynamic calculations have provided much insight into the phenomena associated with the head of the jet, but they cannot cover the required range of length scales; therefore, analytical estimates must serve to indicate what might happen during the traverse from the central black hole to the outermost lobe. I have attempted to apply some of the well established techniques of classical mechanics and hydrodynamics to the propagation of gaseous jets. My aim is to emphasize the global consequences of the interaction between a jet and its environment.

Thus, local phenomena, such as linear, small scale instabilities, are not considered here; the enormity of the journey from nucleus to hotspot makes it quite improbable that perturbations remain small enough to be treated linearly. It is possible that stability analysis says much more about the choice of initial conditions than about the real jet.

The majority of the analytical work that has been done on the problem of jet propagation is restricted to regimes that are almost certainly inapplicable in reality (*e.g.*, linear stability analysis of initially perfectly smooth and symmetrically separated fluids), and all the numerical work is similarly restricted (*e.g.*, cylindrical symmetry, perfect and stationary inflow boundary conditions, and, above all, a ratio of at most 30 between propagation length and initial diameter, as opposed to the one billion that is required). Because I intend to obtain generic constraints where possible, I will not follow these inquiries, and trade precision for generality. Accordingly, I merely consider the most general sources of constraint, namely those based on overall conservation laws (momentum, energy, Lorentz symmetry) and on the most basic hydrodynamic entities (simple waves).

First, I make a few brief remarks on the case for jet confinement (§5.2). The most important evidence is, that jets can maintain a virtually unchanged cross-section over as much as a hundred kiloparsec, and that they retain a clear memory of their previous direction even after sudden broadening or other forms of interference.

Second, I consider the bending of jets. The most dramatic bends, such as in precessing sources, can be explained without any hydrodynamics at all (§5.3.1). Pressure gradients are effective jet bending agents (§§5.3.2-3); several examples of this are considered. When the gradients are steep, we can almost speak about obstacles in the way of the jets (§§5.3.4-5). Bending, however, disturbs the inner structure of jets in dangerous ways (§§5.3.6-7).

Third, I consider entrainment of external gas. When the Mach number of mixing jets is considered (§§5.4.2-3), it turns out that classical jets change only mildly compared with relativistic jets, which suffer an enormous decrease in Mach number upon admixing even trace amounts of surrounding gas.

Fourth, I summarize the difficulties which the jets encounter when propagating through an external medium, and consider what observational diagnostic is best (§5.5). It is argued that thermal and non-thermal emission must come from disjoint volumes, which makes it difficult to observe the desperately needed velocity fields.

5.2 Confinement

5.2.1 Evidence and conjectures

The opening angle of a free jet was discussed in §3.5.2.2. The broadening expected on the basis of equation (3.36) is in blatant contradiction with the observations: Sometimes the jets become gradually wider, but often the cross-section remains constant, or there is a sudden broadening, as in 3C 31, 3C 449, and M 87. Consequently, the jets are either confined by external agents, or they cool faster than adiabatic (*e.g.*, through radiative losses), or a host of other effects may play a role. Bridle & Perley (1984) conclude that "The [opening angle] data for well-resolved jets show that few are free at all [distances from the nucleus]." See also Bridle & Eilek (1984) and Henriksen (1986), and contributions therein. The energy density (which is a measure of the effective pressure) inferred for many jets is comparable with that which is found in the ordinary interstellar medium (see *e.g.*, Begelman, Blandford & Rees 1984, §II B4), so that interaction with the environment should be noticeable.

Precisely what happens is at present quite unclear, even though the environment of at least some galaxies contains high pressure gas (Forman, Jones & Tucker 1985). A false feeling of understanding is conveyed by statements in the observational literature to the effect that "...the jet expands ... is recollimated ... passes through a shock ..." and so forth. These are *morphological* statements, which do not solve any equations of motion. The fact remains that all we know for certain is that free-expansion is observed rarely, and hence it seems equally certain that the jets interact (occasionally vigorously) with their environment (see also Chapter 2 on the variation of the jet width).

At present, it would appear that the most likely agent in governing jet propagation is the "cocoon" or "backflow" of the jet material itself (Norman *et al.* 1983). Even though the interface between the jet and the cocoon gas is ragged (see also Chapters 6 and 7), at least the cocoon material is the jet gas itself: low density and hot. Like an oil drill string, a cosmic jet might well be lubricated by its own waste material. Even this plausible picture, however, has not yet produced quantitative understanding of the observed dependence of the cross-section of a jet on the distance from its source. Dynamic pressure due to the backflow might help to keep at least the cores of the jets together, but the numerical calculations of this effect cover only a range of 1:30 in jet diameter/length ratio.

Magnetic "cages" might be present, and indeed filamentary structure is seen, but it is unclear whether or not this is dynamically important (*e.g.*, Begelman, Blandford & Rees 1984, §II C4). Lind *et al.* (1989) have shown that a toroidal field may stiffen the working surface of the jet into a quasi-solid object which they call a "nose cone". Dynamically, this sounds plausible, but again the quantitative explanation of the behaviour of the jet cross-section is lacking. Moreover, as these authors themselves are fully aware, the necessary restriction to two dimensions suppresses a host of possible instabilities, which is unfortunate because in the world of magnetohydrodynamics virtually anything seems to be unstable (see *e.g.*, Chen 1984).

5.2.2 Confinement by an evaporating disc

It would be nice if the confinement of the jet could be provided by something which is likely to be present in the active galaxy anyway, such as the interstellar medium. There is evidence that galactic gas is, in fact, important in shaping jets: the stubby jets in some Seyfert galaxies, the anomalous arms of NGC 4258, and the outlying emission knots in Centaurus A are cases in point. Also, there is evidence of interference by gas in quasars: those at higher redshift have jets that are more bent than nearby ones, presumably because distant quasars have a higher gas content. Likewise, galaxies at redshift larger than 2 which have a steep radio spectrum show a remarkable alignment of their optical and radio axes, which could well be due to a jet propagating through the protogalactic cloud. Unfortunately, the irregular structures that are observed indicate that interstellar matter is a hindrance rather than a help in keeping jets together.

Another possibility is relevant only in the innermost parts of the active galactic nucleus. If indeed such sources are powered by accretion, it is natural to suspect that the accretion disc can help to confine a jet: any wind emanating from the surface of the disc, or outflow of a disc corona (Icke 1976), would squeeze the jet towards the symmetry axis of the disc (Icke & Choe 1983; Mobasher & Raine 1989). If the disc atmosphere is sufficiently extended to reach the jet, its rotation is negligible; if the flow is stationary, the momentum density $\rho\mathbf{v}$ of the disc gas can be derived from a potential ϕ:

$$\frac{\partial \rho}{\partial t} + \nabla \cdot \rho \mathbf{v} = 0 \ , \tag{5.1}$$

$$\rho \mathbf{v} \equiv \nabla \phi \quad ; \qquad \Delta \phi = 0 \ . \tag{5.2}$$

Because of its cylindrical symmetry, ϕ can be found (by separation of variables) to obey

$$\phi = \begin{cases} \sum_{n=1}^{\infty} a_n r^n P_n(x) \ , & r < 1 \ ; \\ \sum_{n=1}^{\infty} b_n r^{-n-1} P_n(x) \ , & r \geq 1 \ . \end{cases} \tag{5.3}$$

Here P_n is the n-th Legendre polynomial, (r, θ) are the radius and the polar angle in spherical coordinates (because of cylindrical symmetry, the azimuthal angle ϕ does not appear), and $x \equiv \cos\theta$. The flow lines corresponding to this can be found by obtaining the stream function ψ, which obeys

$$r^2 \frac{\partial \phi}{\partial r} = -\frac{\partial \psi}{\partial x} \ , \tag{5.4}$$

$$(1 - x^2) \frac{\partial \phi}{\partial x} = \frac{\partial \psi}{\partial r} \ , \tag{5.5}$$

(*e.g.*, Batchelor 1974, Ch. 2, 4, 6). Because of reflection symmetry across the equatorial plane, we have only even n. Therefore, the leading term in equation (5.3) is proportional to r^2. On the axis, then, we must have

$$\phi = a_2 r^2 + a_4 r^4 + \cdots \ , \tag{5.6}$$

so that the flow pattern near the origin has a saddle point. Once the form of a potential is known on the axis, it can be calculated everywhere by asymptotic expansion.

These properties allow us to calculate the generic form that any flow pattern in the centre of an evaporating disc must have. This pattern can be found as follows. For $r \to \infty$, one must have $\partial\phi/\partial r \to 0$ on the axis; the simplest rational function with the required asymptotic behaviour is then

$$\phi = \phi_0 \frac{r^2}{1 + r^2} \ , \tag{5.7}$$

which has the expansion

$$\phi = \begin{cases} \phi_0 \sum_{n=1}^{\infty} (-1)^{n+1} r^{2n} \ , & \text{if } r < 1 \ ; \\ \phi_0 \sum_{n=0}^{\infty} (-1)^n r^{-2n} \ , & \text{otherwise.} \end{cases} \tag{5.8}$$

Off-axis, the equations are exactly the same, but there is an extra factor $P_{2n}(x)$ in the case $r < 1$ and a factor $P_{2n-1}(x)$ in the case $r \geq 1$. Solving equations (5.4) and (5.5) gives the corresponding stream function:

$$\psi = \begin{cases} \phi_0 \sum_{n=1}^{\infty} (-1)^{n+1} \frac{2n}{2n+1} r^{2n+1} (P_{2n-1} - x P_{2n}) \ , & \text{if } r < 1; \\ \phi_0 \sum_{n=1}^{\infty} (-1)^{n+1} r^{1-2n} (P_{2n-2} - x P_{2n-1}) \ , & \text{otherwise.} \end{cases} \tag{5.9}$$

Now lines of equal ψ are streamlines, while those of equal ϕ are perpendicular to streamlines. Because of its asymptotic properties, the flow pattern just calculated in the form of ϕ and ψ is generic for the flow in the centre of an evaporating disc. An example is shown in Fig. 5.1. The boundary conditions at $r = 0$ and at $r = \infty$ are responsible for this generic form; because these conditions are expected to be almost always valid, we expect that any evaporating disc must look very similar to this.

What if there is a central outflow source in the disc? A point source of gas at $r = 0$ is represented by the flow functions

$$\phi = -\frac{S}{r} \ ; \qquad \psi = -Sx \ , \tag{5.10}$$

where S is the source strength. Adding these to equations (5.8) and (5.9) for the disc flow, we obtain the composite flow pattern. Examples of this are shown in Fig. 5.2. Clearly, the dynamic pressure from an evaporating disc can produce considerable jet collimation.

If one wishes to calculate the exact density and velocity distributions, one must disentangle the density and the velocity from the momentum density $\rho\mathbf{v}$. This can be done by applying Bernoulli's Equation along the streamlines. For the sake of brevity, I will omit this step, but merely point out that the flow must contain a sonic transition at the point where the distance between adjacent streamlines is smallest (Icke 1983). Upstream from this point, the flow is subsonic; however, near the central source one must also have $|\rho\mathbf{v}| \propto r^{-2}$. Consequently, there must be a region around the jet source where the flow cannot smoothly connect with the downstream jet. In that region, we must have a standoff shock around the central source. For very fast jets, the size of this shock bubble is very small, and therefore will be difficult to observe.

5.3 Bending a jet

5.3.1 Pseudo-bending

A jet can assume a shape which has little, if indeed any-

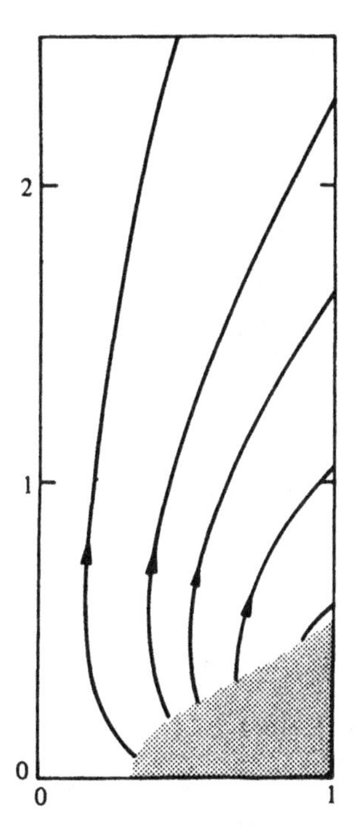

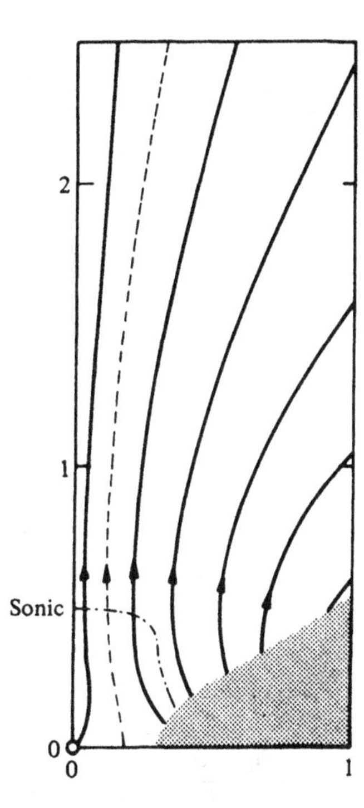

Fig. 5.1. Left: saddle point flow in the centre of an evaporating disc. The left edge of the frame is the axis of cylindrical symmetry, the bottom edge is the plane of reflection symmetry. Right: the same flow field, but with a central source added, strength $S = 0.005$. The dot-dashed line indicates the "nozzle" surface, through which a smooth transition from subsonic to supersonic flow can take place. The dashed line is the contact discontinuity between disc wind and gas from the central source. The dotted line is a reminder of the expected instability of this interface.

thing, to do with the shape of its streamlines. Several clear cases of "pseudo-bending" have been identified as being due to ballistic motion which is disturbed by nuclear precession, external gradients, acceleration by an intruder, or a combination of these (Blandford & Icke 1978; Icke 1981; Gower *et al.* 1982).

The description of the radio trails that are formed by pseudo-bending is easy; all models are basically variants of the following two possibilities. If the velocity vector $\mathbf{u}$ of a jet is inertially fixed, and if its source moves along the trajectory $\mathbf{R}(t')$, then the trail of the jet (*i.e.*, the locus of jet material at a given observing time t)

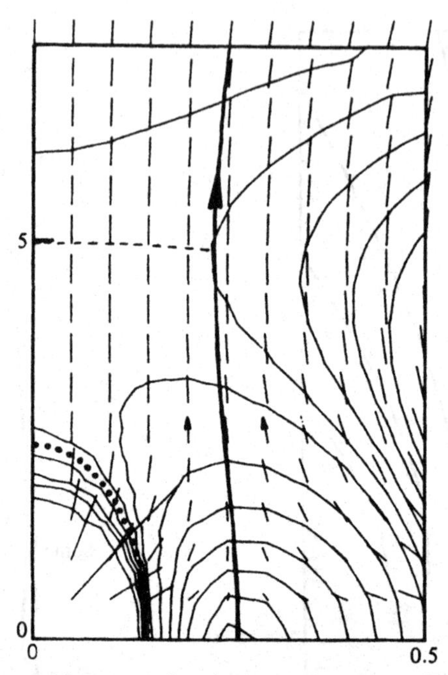

Fig. 5.2. Jet confinement by dynamic pressure from an evaporating disc. The flow field is as in Fig. 5.1, with a source strength $S = 0.02$. The vectors show the momentum density; contours are lines of equal momentum density. The heavy dotted line indicates the standoff shock around the central object, while the dashed line shows the position of the sonic transition.

is described by the parameterisation

$$\mathbf{r}(t') = \mathbf{R}(t') - (t' - t)\left(\mathbf{u} + \frac{d\mathbf{R}(t')}{dt'}\right) , \tag{5.11}$$

for $t' \leq t$; the present source position is given by $\mathbf{R}(0)$. The actual form of $\mathbf{R}(t')$ can be due to various mechanisms, the most common of which is orbital motion.

If the jet material is ballistic but not free, then the above becomes only marginally more complicated. If the cross-section of the jet increases linearly with distance, such as would be the case for a uniformly expanding jet, then the effective drag force due to the medium which surrounds the jet is

$$\mathbf{F} = \frac{\mathbf{w} - \mathbf{u}}{T} , \tag{5.12}$$

where $\mathbf{w}$ is the velocity of the jet source and $\mathbf{u}$ is the velocity of the jet material. The parameter T is an effective stopping time. The equation of motion of the jet and its solution are then

$$\frac{d\mathbf{u}}{dt} = \frac{1}{T}(\mathbf{w} - \mathbf{u}) \; , \tag{5.13}$$

$$\mathbf{u} = \mathbf{w} - (\mathbf{w} - \mathbf{u}_0)\mathrm{e}^{-t/T} \; , \tag{5.14}$$

where $\mathbf{u}_0$ is the initial velocity of the beam. One more integration produces the equation for the trail shape,

$$\mathbf{r} = \mathbf{w}t + T(\mathbf{u}_0 - \mathbf{w})(1 - \mathrm{e}^{-t/T}) \; . \tag{5.15}$$

Notice that this leaves open the possibility that the jet velocity (and therefore its direction) is a function of time; in that case, $\mathbf{u}_0(t')$ parameterises the trail $\mathbf{r}$ with the "retarded" time t'.

Models based on these approximations turn out to be extremely accurate descriptors of actually observed radio source shapes, even in the case of quite complicated trails (*e.g.*, Yokosawa & Inoue 1985). The remarkable thing about this from the point of view of the present discussion is that the good match with observations is achieved *without any hydrodynamics at all*: the ballistic models essentially describe streams of gaseous bullets. If the jet gas were dense compared with its surroundings, and if the jet were extremely cold, this would be a reasonable approximation of the hydrodynamics, because of the very large internal Mach number. However, the current consensus seems to be that jets are very tenuous and hot, which cannot be reconciled with the ballistic approximation.

5.3.2 Bending by a pressure gradient

When the surroundings of a jet contain large inhomogeneities, the above approximations of the interaction between a jet and its environment are no longer valid: we must go to the other extreme, and see what happens if a jet encounters an "obstacle". Precisely what is meant by this depends on circumstances, but real walls are presumably rare in intergalactic space, so that an obstacle is usually a density and pressure gradient.

Gradients can be encountered within galaxies or clusters, or in isolated inhomogeneities. All these, such as head-tail bending in a galaxy, are special cases of the equations below; the only quantity that plays a major role is the scale height of the gradient, ranging from intracluster scales, via galactic lengths, to interstellar cloud scales. For what follows, it is immaterial what produces the gradient. We will first consider the details of bending by pressure gradients, and thereafter consider what happens if these gradients are so strong that the jet can be thought to be deflected by a wall.

If a gas cloud causes a jet to curve, then in equilibrium the centripetal acceleration is provided by the component of the external pressure gradient that is perpendicular to the path of the jet:

$$\frac{v^2}{R} = \frac{1}{\rho}\frac{\partial P}{\partial R}\bigg|_{\perp} , \tag{5.16}$$

where v is the speed of the jet, ρ is its density, R is the radius of curvature of the central line of the jet, and P is the external pressure (cf. Jones & Owen 1979; Henriksen, Vallée & Bridle 1981; Sparke 1982a,b; Burns & Balonek 1982; Fiedler & Henriksen 1984). If we call $\rho v^2 \equiv K$, and if we use the equation for the radius of curvature of a function $y = f(x)$ which describes the jet path, then

$$\frac{1}{R} = \frac{d^2 f}{dx^2}\left(1 + \left(\frac{df}{dx}\right)^2\right)^{-3/2} = \frac{1}{K}\frac{\partial P}{\partial R}\bigg|_{\perp} . \tag{5.17}$$

For simplicity, I assume that the jet moves in a plane. The perpendicular derivative along the jet path can be written in terms of the coordinates (x, y) as

$$\partial P/\partial R\bigg|_{\perp} = \sin\alpha \frac{\partial P}{\partial x} - \cos\alpha \frac{\partial P}{\partial y} , \tag{5.18}$$

$$\tan\alpha = df/dx . \tag{5.19}$$

Using these relationships, the differential equation for the shape $f(x)$ of the jet becomes

$$\frac{d^2 f}{dx^2} = \frac{1}{K}\left(1 + \left(\frac{df}{dx}\right)^2\right)\left(\frac{df}{dx}\frac{\partial P}{\partial x} - \frac{\partial P}{\partial y}\right) . \tag{5.20}$$

A simple example of the bending in a pressure gradient is provided by the plane parallel layer $P = P(y)$. Then equation (5.20) simplifies to

$$\frac{d^2 f}{dx^2} = -\frac{1}{K}\left(1 + \left(\frac{df}{dx}\right)^2\right)\frac{dP}{dy} . \tag{5.21}$$

Since $y = f(x)$ along the jet, this has the solution

$$x = x_0 + \int_0^y \left[L e^{-2P(f)/K} - 1\right]^{-1/2} df , \tag{5.22}$$

with integration constants x_0 and L. Examples are given in Fig.

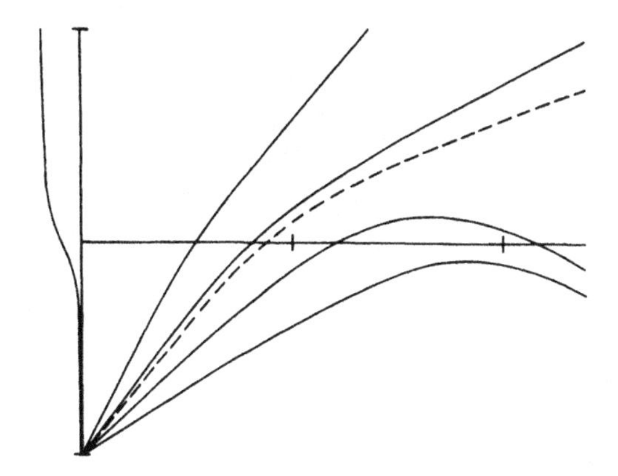

Fig. 5.3. Bending of a jet by the plane parallel pressure distribution $P = P_0(\frac{1}{2} + \frac{2}{\pi}\arctan(y/y_0))$, which is plotted along the vertical axis. Parameter $K = 2P_0$ (see text). Distance between tick marks is $5y_0$. Dashed trajectory corresponds to the critical value $L = \mathrm{e}$, below which the jet is reflected by the pressure gradient.

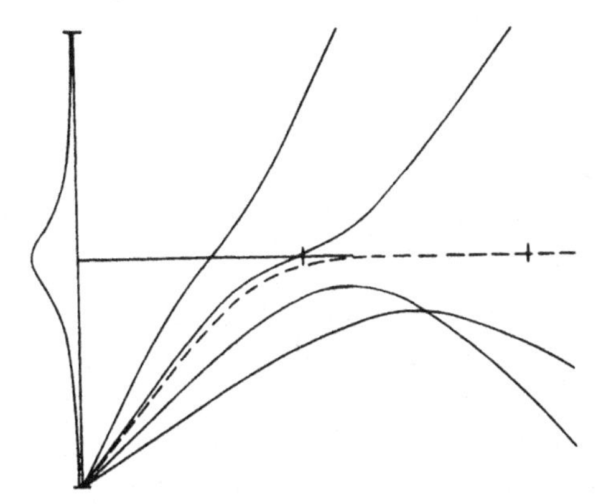

Fig. 5.4. As Fig. 5.3, with $P = P_0/(1+y^2/y_0^2)$, which is plotted along the vertical axis.

5.3 for $P = P_0(\frac{1}{2}+\frac{2}{\pi}\arctan(y/y_0))$, and in Fig. 5.4 for $P = P_0/(1+y^2/y_0^2)$.

The integration constant L is related to the angle of incidence i of the jet:

$$\tan i = (L-1)^{-1/2} , \tag{5.23}$$

provided we adopt the trivial convention $P(-\infty) = 0$. The square root in equation (5.22) must always exist, which implies that

$$L \geq \mathrm{e}^{2P/K} . \tag{5.24}$$

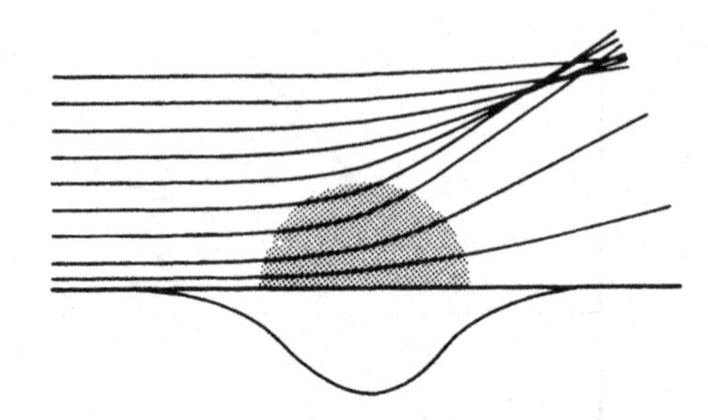

Fig. 5.5. As Fig. 5.3, for the Gaussian pressure distribution $P = P_0 \exp(-r^2/r_0^2)$. The target sphere is indicated by the hatched area, which has a radius r_0. The pressure distribution is sketched below the horizontal axis.

Accordingly, for all values of L smaller than

$$L_c = \mathrm{e}^{2P(\infty)/K} = \mathrm{e}^{2P(\infty)/\rho v^2} , \tag{5.25}$$

the jet is bounced back by the layer $P(y)$; the critical angle corresponding to this is

$$i_c = \arctan \left(\mathrm{e}^{2P(\infty)/K} - 1\right)^{-\frac{1}{2}} . \tag{5.26}$$

In Figs. 5.3 and 5.4 I have taken $K = 2P(\infty)$, so that $i_c = 37°$.

If the pressure distribution $P = P(r)$ is spherically symmetric, equation (5.20) becomes

$$\frac{d^2 f}{dx^2} = \frac{1}{K}\left(1 + \left(\frac{df}{dx}\right)^2\right)\left(x\frac{df}{dx} - y\right)\frac{1}{r}\frac{dP}{dr} . \tag{5.27}$$

An example of this is given in Fig. 5.5, where $P = P_0 \exp -(r/r_0)^2$, and $2P_0 = K$. The angle of deflection of the jet is Δ, which depends on the impact parameter b as shown in Fig. 5.6. As b increases from zero, the scattering angle goes from zero, via a maximum $\Delta_{\max}$, back to zero. The values $\Delta_{\max}$, and the corresponding $b_{\max}$ (expressed in units of r_0), are plotted in Fig. 5.7, as a function of $2P_0/\rho v^2$. The deflecting effect of the cloud drops drastically when P_0 becomes smaller than the dynamic pressure $\frac{1}{2}\rho v^2$ of the jet.

5.3.3 Cross section of the jet

What does the curvature of the jet do to its cross-section? In the previous discussion, I have assumed that the jet is infinitesimally thin, *i.e.*, the inner enthalpy gradient is negligible compared with the acceleration due to the bending. Now let us introduce a small jet enthalpy w_j, so that the jet acquires a finite thickness.

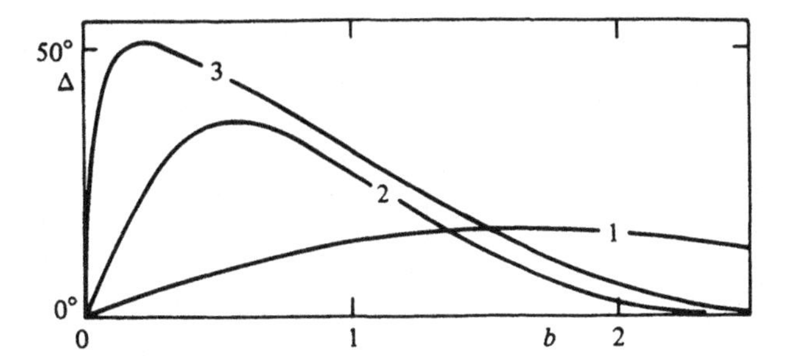

Fig. 5.6. Deflection angle Δ versus impact parameter b (in units of the core radius $r_0 = a$) for jets bent by a spherical Gaussian cloud.

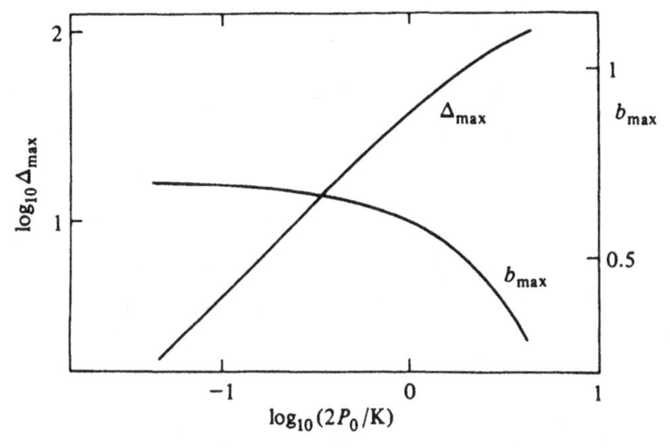

Fig. 5.7. Maximum deflection angle Δ_{max}, and corresponding impact parameter b_{max} (in units of the core radius r_0) for deflection of a jet by a spherical Gaussian cloud. Abscissa is the pressure ratio $2P_0/K = 2P_0/\rho v^2$. If this ratio is large, the dynamic pressure of the jet is weak compared with the pressure in the cloud; if it is small, the jet is strong.

Cut the jet with a plane perpendicular to its centre line; choose coordinates (x, y) such that y goes through the jet centre line, and let x be perpendicular to the plane in which the jet curves. Then the centrifugal acceleration $\mathbf{A}$ is parallel to the y-axis. If the velocity of the jet gas tangential to the jet cross-section is small, as it should be to keep the jet collimated, then the internal pressure gradient ∇P_j (and therefore also the enthalpy gradient) of the jet must be perpendicular to the cross-section $y = y(x)$, so that (see Fig. 5.8)

$$\nabla w_j = g(x, y)(-\sin\alpha, \cos\alpha) \ , \tag{5.28}$$

$$\tan\alpha \equiv dy/dx \ , \tag{5.29}$$

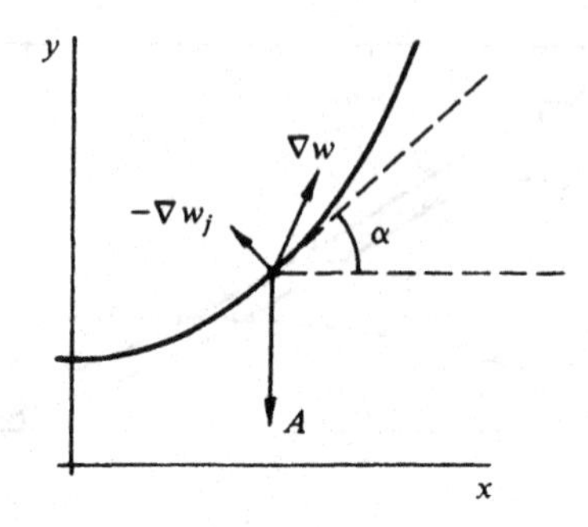

Fig. 5.8. Geometry of the cross-section of a bent jet. The plane of the drawing is perpendicular to the centre line of the jet. The centrifugal acceleration **A** is parallel to the y-axis.

where w_j is the specific enthalpy of the jet. At every point on the cross-section, the acceleration and the pressure gradients must be in balance, so that

$$\mathbf{A} = \nabla w_j + \nabla w \ , \tag{5.30}$$

where w is the specific enthalpy of the gas surrounding (and bending) the jet. Since the centrifugal acceleration $\mathbf{A} = (0, -A)$ is fixed, I get in component notation

$$0 = -g(x, y) \sin \alpha + \partial w / \partial x \ , \tag{5.31}$$

$$-A = g(x, y) \cos \alpha + \partial w / \partial y \ . \tag{5.32}$$

Elimination of g yields, using equation (5.29), the equation for the cross-section $y(x)$ of the jet:

$$\frac{dy}{dx} = -\frac{\partial w / \partial x}{A + \partial w / \partial y} \ . \tag{5.33}$$

For example, if the bending cloud is spherical, then the origin of (x, y) can be chosen such that $w = w(r)$ with $r^2 = x^2 + y^2$, and

$$\frac{dy}{dx} = -\frac{x}{Ar / \frac{\partial w}{\partial r} + y} \ . \tag{5.34}$$

In the case of the spherical Gaussian cloud used in the previous section, we have

$$w = w_0 \mathrm{e}^{-r^2 / r_0^2} \ . \tag{5.35}$$

Expressing all lengths in units of r_0 as before, I find for equation (5.34)

$$\frac{dy}{dx} = \frac{x}{Q\mathrm{e}^{r^2} - y} \ , \tag{5.36}$$

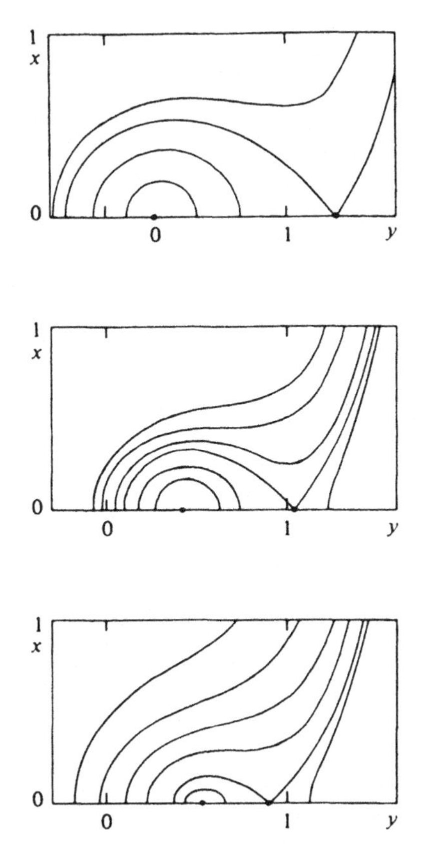

Fig. 5.9. Shapes of possible cross-sections of a jet that is bent by passage through a spherical Gaussian cloud (see Fig. 5.5). Length unit is the core radius r_0. Dots indicate possible jet locations if the internal pressure of the jet is negligible. Top: $Q = 0.2$, middle: $Q = 0.35$, bottom: $Q = 0.4$. Note that if the internal pressure of the jet exceeds a certain minimum, no closed surface is possible: the cloud pressure and the centrifugal pressure combined do not suffice to contain the jet.

$$Q \equiv \frac{Ar_0}{2w_0} = \frac{v^2}{2w_0}\frac{r_0}{R} \ . \tag{5.37}$$

Equation (5.36) was solved, by means of isoclines, for the values $Q = 0.25$, 0.35, and 0.4 (Fig. 5.9). The corresponding contours of the function $-r_0 g(x,y)/2w_0$, a dimensionless expression which is proportional to the pressure gradient at the jet surface, are shown in Fig. 5.10.

Evidently, equation (5.36) has singular points at ($x = 0$, $y =$

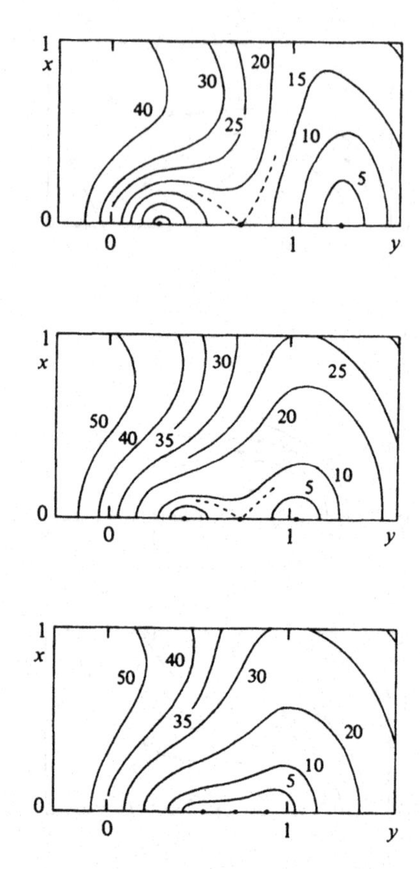

Fig. 5.10. Contours of the dimensionless function $-100r_0g(x,y)/2w_0$, which is proportional to the dimensionless pressure gradient at the surface of the jet. The values of Q are 0.25, 0.35, 0.4, and the coordinates are the same as in Fig. 5.9.

$Q \exp r^2$), *i.e.*, at those points on the y-axis where

$$y = Qe^{y^2} \ . \tag{5.38}$$

By tracing back equation (5.36) to its ancestor (equation (5.20)), it can be seen that these points correspond to the degenerate case $w_j = 0$, as it should be: if the jet enthalpy is negligible, the cross-section of the jet is reduced to zero, and equation (5.38) gives the point where the central line of the jet pierces the plane perpendicular to it.

It is immediately clear from Fig. 5.9 that (i) there is only a restricted family of possible closed jet cross-sections, (ii) closed sections appear only around the smallest of the singular points (equation (5.38)), *i.e.*, the point nearest the centre of the cloud. This is no surprise: the pressure on the periphery of a Gaussian cloud decreases so rapidly that even a small internal pressure in the jet breaks open its surface. In this region, the assumption that w_j is comparatively small does no longer apply, and the jet pops open on that side. In the direction facing the centre of the cloud, the expansion of the jet is stopped by the pressure gradient. Hence, the jet loses confinement on its concave side only. But in the inner part of the cloud, this problem does not arise; here the jet can maintain a finite cross-section.

Of course the jet would not extend to infinity if it "breaks open": all this means, in the above approximation, is that the surface on which the effective forces are in equilibrium is dish-shaped in cross-section. This is, in fact, what some numerical calculations show (Soker, O'Dea & Sarazin 1988).

Fig. 5.9 also shows the stratification that occurs in the interior of the jet because of the centrifugal acceleration. A simple dimensional analysis of equation (5.16) shows that the effective scale height of this stratification is proportional to $R/\mathcal{M}^2$, where $\mathcal{M}$ is the Mach number of the jet (*e.g.*, Sparke 1982a,b). Because of the limitations on jet bending that will presently be derived, this expression for an expected finite thickness of the jet will turn out to be important.

5.3.4 Bouncing off an unaccelerated cloud

The question how a jet can be deflected, and whether it can survive an imposed deflection, is deferred until §5.3.6; here I assume that the deflection is given, and calculate the locus of the jet gas consequent on this deflection. I will assume that the material that deflects the jet is sufficiently localized that it will not influence the jet much beyond a smallish interaction region. If the interaction is continuous, as in the case of a very tenuous jet, the assumptions of the previous sections apply, and we should more properly think of entrainment rather than deflection.

If the jet, moving with speed v, bounces off a cloud that moves with a constant velocity that has a component w in the direction of the jet, and if the deflection angle is α, then by Galilei invariance the locus of the deflected jet is still a straight line, but making an angle β with the incoming jet (see Fig. 5.11):

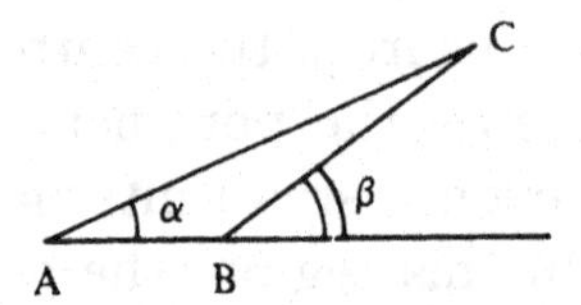

Fig. 5.11. Jet bouncing off a cloud that moves at uniform speed w. The jet comes in with speed v parallel to AB, and bounces off at A under an angle α. In a time t, we have AB $= wt$ and AC $= vt$; since the jet must overtake the cloud, $w < v$ and $\beta > \alpha$.

$$\tan \beta = \frac{\sin \alpha}{\cos \alpha - w/v} . \tag{5.39}$$

Evidently, if $w \lesssim v$, the apparent locus of the jet after deflection can make a large angle with the incoming jet, without, however, actually flowing along the direction of the locus. This leads to the simple, yet very frequently overlooked, conclusion that one should not conclude from sudden bends in radio jets that the gas necessarily turns a sharp corner.

5.3.5 Bouncing off a uniformly accelerated cloud

If the cloud moves with a velocity component $w(t)$ that increases uniformly with time, then the locus after deflection becomes a parabola, with its convex side towards the origin of the jet. Let y be perpendicular to x; the jet is deflected by an angle α in the (x, y)-plane. Let $(x_1, 0)$ be the place at which a piece of jet gas is bounced off, at time t_1. If we observe the locus of the jet at time t_2, then

$$x = x_1 + v(t_2 - t_1) \cos \alpha , \tag{5.40}$$

$$y = v(t_2 - t_1) \sin \alpha . \tag{5.41}$$

Because the deflecting cloud is accelerated with constant acceleration a, we have

$$x_1 = x_0 + w_0 t_1 + \frac{1}{2} a t_1^2 , \tag{5.42}$$

where x_0 is the place and w_0 is the speed of the cloud at $t = 0$. Substitution of equations (5.41) and (5.42) into equation (5.40) gives after some algebra

$$x = x_2 - \frac{y}{\sin \alpha}(w_0/v - \cos \alpha + a t_2/v) + \frac{a y^2}{2 v^2 \sin^2 \alpha} . \tag{5.43}$$

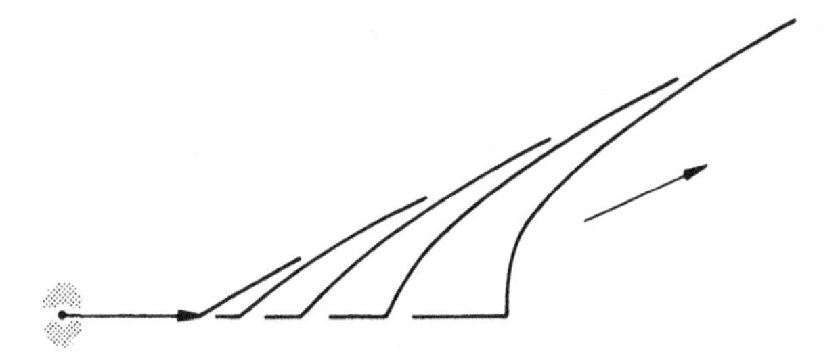

Fig. 5.12. Locus of a jet bouncing off a uniformly accelerated cloud, at various times of observation. The slanted arrow indicates the motion of the gas after deflection; this gas moves in a straight line, but the jet locus is a forward facing parabola.

Here x_2 is the position of the deflecting cloud at the time t_2 when the whole configuration is observed. Clearly, equation (5.43) describes a forward facing parabola (Fig. 5.12).

5.3.6 Deflection through a shock

When a jet is bent by an external force of whatever origin, it may develop internal shocks. At such a shock, the direction of the jet gas can change abruptly, and the jet appears to be deflected by the shock.

It is debatable whether or not a transverse shock in a jet can destroy it. Internally generated shocks typically occur at an angle to the flow that approaches the Mach angle, so these tend to be weak enough not to devastate the flow. But nearly perpendicular shocks are another matter. Internal shocks cannot cross, but must form a nearly perpendicular Mach disc, which serves to randomize the jet energy (Courant & Friedrichs 1976). At the terminal shock of a jet, the ram pressure of the external medium keeps the head of the jet more or less together (Norman *et al.* 1983), but in the middle of a jet a transverse shock could have serious consequences. And yet there are cases where transverse shocks very likely occur, such as in M 87 (Biretta, Owen & Hardee 1983), or where it can be inferred, such as in the sudden broadening of jets like 3C 31 and 3C 449 (Cornwell & Perley 1984). In these cases, the radio source remains fairly narrow and apparently quite smooth; the reason for this is not known.

If the jet passes through a flat shock, and if that shock makes an angle β with the flow velocity in the jet, then the velocity components beyond the shock, in the directions perpendicular and parallel to the plane of the shock, are

$$v_{1p} = \frac{1}{\Gamma+1}\left((\Gamma-1)v_0\sin\beta + \frac{2c_0^2}{v_0\sin\beta}\right) , \tag{5.44}$$

$$v_{1t} = v_{0t} = v_0\cos\beta , \tag{5.45}$$

in a frame that is stationary with respect to the shock (*e.g.*, Zel'dovich & Raizer 1966); here c_0 is the speed of sound in the incoming gas and Γ is Poisson's adiabatic index* of the gas. If we choose the positive x-axis along the incoming jet and the y-axis perpendicular to that, such that the (x, y)-plane is normal to the plane of the shock, then

$$\mathbf{v}_0 = (v_0, 0) , \tag{5.46}$$

$$\mathbf{v}_{0p} = v_0\sin\beta(\sin\beta,\cos\beta) ; \quad \mathbf{v}_{0t} = v_0\cos\beta(\cos\beta,-\sin\beta) , \tag{5.47}$$

$$\mathbf{v}_{1p} = \frac{v_0\sin\beta}{\Gamma+1}\left(\Gamma-1+2(\mathcal{M}_0\sin\beta)^{-2}\right)(\sin\beta,\cos\beta) , \tag{5.48}$$

where $\mathcal{M}_0$ is the Mach number, in the frame moving with the shock, of the incoming jet. Let the shock move with a velocity that has a component U in the direction of the jet (such as might be the case in the jet-accelerated clouds considered above). Then the angle χ, which a stationary observer sees between the jet before and after the shock, obeys

$$\cos\chi = \frac{(\mathbf{U}+\mathbf{v}_1)_x}{|\mathbf{U}+\mathbf{v}_1|} . \tag{5.49}$$

This can be rewritten as an equation for the y-component of the postshock speed,

$$v_{1y} = (U + v_{1x})\tan\chi , \tag{5.50}$$

which, using equations (5.46), (5.47) and (5.48), becomes

$$\frac{2}{\Gamma+1}\left(\mathcal{M}_0^{-2}-\sin^2\beta\right) = \tan\beta\,\tan\chi\left[\frac{U}{v_0}+1 + \frac{2}{\Gamma+1}\left(\mathcal{M}_0^{-2}-\sin^2\beta\right)\right] . \tag{5.51}$$

Accordingly, if we observe a jet deflected through a shock for which we can measure β and χ, and if we have a good estimate of the Mach

* My apologies to the hydrodynamics clan – we wish to avoid confusion with the Lorentz factor γ.

number, then the ratio U/v_0 of the shock speed over the jet speed can be calculated.

5.3.7 Internal shocks

If a jet is bent by an external pressure gradient, it is evidently in direct contact with its surroundings, and we must fear for the life of the jet because of the dangers of admixing external gas into it. But let us suppose that the latter difficulty can be sidestepped, for example by efficient cooling, and we have a jet with high Mach number that is bent. Then the question remains whether this can be done without destroying the jet from the inside, even if the contact with the outer world does not do any damage.

The greatest danger comes from destruction by internal shocks. Consider a parallel stream of gas with Mach number $\mathcal{M}$ which flows along a smooth flat wall. An infinitesimal bump on the wall generates a pressure wave that propagates into the gas, communicating the presence of the bump to some parts of the flow. The crest of the wave moves along a characteristic curve; locally, the angle between that curve and the direction of the flow is given by the Mach angle A:

$$\sin A = 1/\mathcal{M} \ , \tag{5.52}$$

(Courant & Friedrichs 1976, §106; because of the intricacies of characteristic theory, this chapter will lean heavily on the book, which the reader is invited to study). Now if the jet flows along a curved section of the wall, the news that the wall is no longer straight also propagates along characteristics. For simplicity, I assume that the curve is a section of a circle with radius R. If, to a first approximation, the Mach number does not change in the bend, then the characteristics emanating from the wall bunch together (Fig. 5.13, top). Where characteristics overlap, the pressure buildup would be infinite: all the infinitesimal pressure waves would be crowded together in zero space. In practice, a shock occurs at such a point (Fig. 5.13, bottom).

An internal shock is potentially disastrous for the jet, because behind it, the gas motion perpendicular to the shock is subsonic: behind a shock the jet would flare outward. However, Fig. 5.13 shows that if the jet is narrow enough, the characteristics do not intersect within the jet.

Fig. 5.13. Crowding of characteristics at a bend in a supersonic jet. Top: theoretical case for constant Mach angle. Bottom: observed case for a concave bullet (Van Dyke 1982, p. 160).

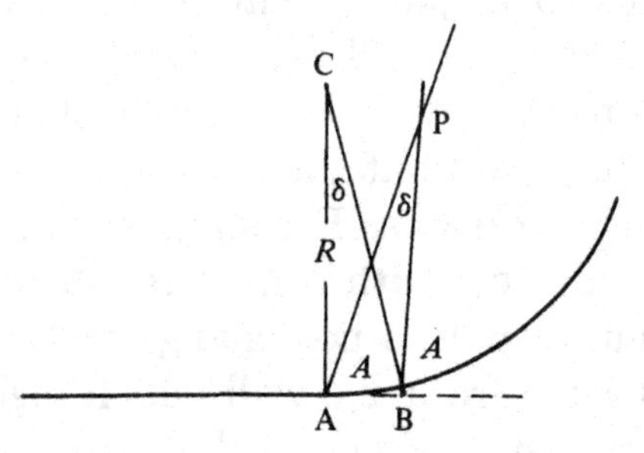

Fig. 5.14. Geometry of the intersection of the characteristics that emanate from the onset of a bend. It is assumed that the Mach angle A is constant in the bend.

Consider a small angle δ at the centre of curvature of the wall, and the distance AB it subtends at the onset of the curve (Fig. 5.14). If A is the local Mach angle, then characteristics from point A and point B intersect at P, and the angle APB $\simeq \delta$. If we call D the distance from P to the extension of the straight part of the jet wall, then $D/\mathrm{AP} = \sin A$. Moreover, we have $\mathrm{AB} \simeq R\delta$ and $\mathrm{AP} \simeq R \sin A$. Therefore,

$$D \simeq R \sin^2 A \quad ; \qquad R/D \simeq \mathcal{M}^2 , \tag{5.53}$$

by means of equation (5.52). Accordingly, if the jet is thinner than the radius of curvature divided by the square of the Mach number, it can be bent without destruction.

As we saw in the previous section, the effective scale height of the internal stratification of the jet is of the order of $R/\mathcal{M}^2$, so it might seem at first that all bent jets are safe. However, a scale height is not an actual thickness, and in §5.4 we will see that jets must be expected to have extensive entrainment cocoons around them. Therefore, although the condition (5.53) may not be quite as devastating as it seems at first, it would appear that this condition must be cleared with a good margin of safety.

The Mach angle does not remain constant in a bend. The classical theory of simple waves allows us to calculate analytically the way in which $\mathcal{M}$ changes (Courant & Friedrichs 1976, §§106-114*). In general, the Mach number decreases progressively as the bending proceeds.

From Bernoulli's Equation and from the definition (5.52) of the Mach angle A, a relationship can be found between A and the angle ϕ through which the jet is bent (Courant & Friedrichs 1976, §108):

$$\tan A = -\mu / \tan \mu(\omega - \omega_*) , \tag{5.54}$$

$$\mu^2 \equiv \frac{\Gamma - 1}{\Gamma + 1} \quad ; \qquad \omega \equiv \phi + A - \pi/2 . \tag{5.55}$$

The constant ω_* is determined by the incoming Mach angle A_0 in the direction $\phi = 0$. By inversion of equation (5.54), one finds $\phi(A)$, which is plotted in Fig. 5.15. To determine the change of A (or $\mathcal{M}$) in a bend, find the value of A of the incoming gas on the ordinate, and read off ϕ on the abscissa. Then go back along the ϕ-axis over as many degrees as the bend sweeps out, and read off the corresponding value on the A-axis: this is the outgoing Mach angle.

When A has been found, we know the corresponding characteristic, namely a straight line making an angle A with the local tangent at the bend. Since A is constant along the entire characteristic, we

* Note that in their Fig. 8, the angle labelled A is really $90° - A$

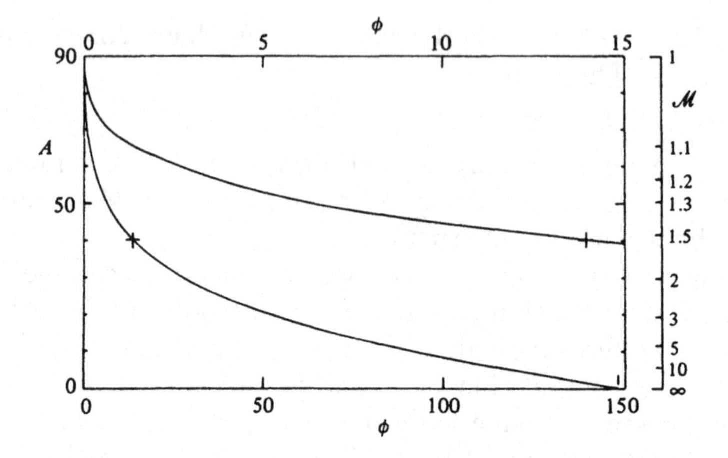

Fig. 5.15. Mach angle A and Mach number $\mathcal{M}$ versus bending angle for a jet passing along a circular bend. The adiabatic index is 4/3. The top curve corresponds to the top scale, bottom curve to bottom scale. The crosses indicate the values of the Mach number below which $dA/d\phi > 1$. The small vertical dash indicates the value of ϕ at $\mathcal{M} = \infty$.

can immediately construct the flow directions on this line. Then we can find the flowlines by graphical integration. Some results of this procedure are shown in Fig. 5.16. The jet becomes narrower in the bend, as it must because the presence of the wall curvature is not communicated instantaneously throughout the jet, but moves along the characteristic that departs from the onset of the bend.

After the bend, though, the jet can expand with a flaring angle which is at most equal to the local Mach angle. In other words, the curvature has the net effect of increasing the transverse scale height of the jet.

As we had seen in equation (5.53), if a jet is narrow enough, a bend does not necessarily create a shock that cuts into the jet when it begins to curve: the thickness can be chosen in such a way that the characteristics intersect outside the jet proper. In fact, the thickness of the jets in the figures was chosen deliberately so that the jets just avoided being shocked. But equation (5.54) shows that there is another danger, this one not at the beginning of the curve, but at the end. Differentiation of equation (5.54) with respect to the bending angle ϕ shows that

$$\frac{dA}{d\phi} = \frac{1}{2}\left(\Gamma - 1 + \frac{\Gamma + 1}{\mathcal{M}^2 - 1}\right) . \tag{5.56}$$

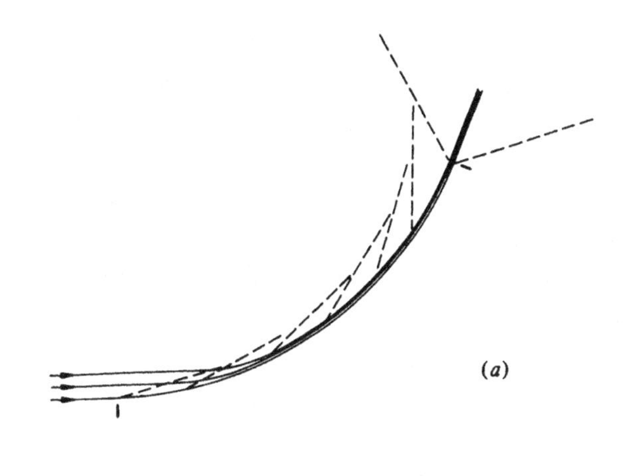

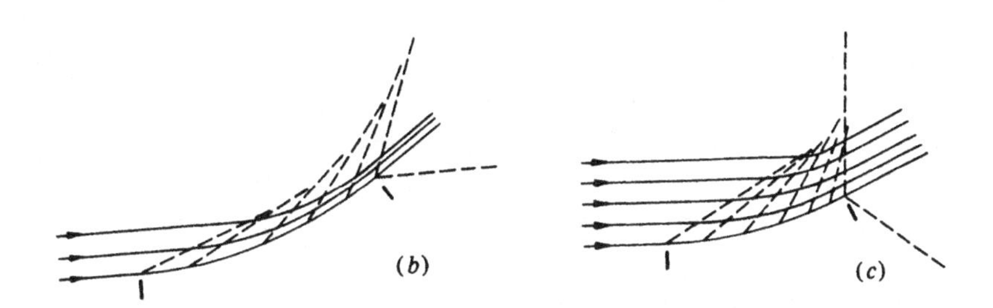

Fig. 5.16. Characteristics (dashed) and flow lines in jets passing through a circular bend. Tick marks show the beginning and the end of each circular section. The adiabatic constant is 4/3. (a) Incoming Mach number 4, outgoing $\mathcal{M} = 1.29$. (b) Incoming $\mathcal{M} = 3$, outgoing $\mathcal{M} = 1.74$. (c) Incoming $\mathcal{M} = 2$, outgoing $\mathcal{M} = 1.13$. The thickness of the jets has been adjusted so that the upper flowline just avoids being shocked; the length of the curved section has been adjusted such that a shock at the end of the jet is avoided.

Clearly, the rate of change of the Mach angle increases without bounds as $\mathcal{M} \downarrow 1$. Because $\mathcal{M}$ decreases steadily in a bend, there always comes a point at which the Mach angle increases faster than the bending angle. This means, as the bottom diagram in Fig. 5.16 shows, that a shock cuts into the jet at the end (there is only one special shape for the curve of the jet wall where this is avoided (Courant & Friedrichs 1976, §113), by matching the run of ϕ exactly with the increase in A, but that is rather contrived). Apparently a jet, whatever its thickness, cannot be bent with impunity through the point where $dA/d\phi$ exceeds the rate of change of ϕ; since $d\phi/d\phi = 1$, the ϕ-derivative of A is limited by $dA/d\phi \leq 1$. Applying this condition to equation (5.56) shows that the jet must not be bent so far that the Mach number drops below

$$\mathcal{M} = \frac{2}{\sqrt{3-\Gamma}} , \tag{5.57}$$

which value equals 1.549 for $\Gamma = 4/3$ and 1.732 for $\Gamma = 5/3$; the corresponding values of A are 40.21° and 35.26°, respectively. It is not to be expected that a violation of the limit (5.57) has consequences as serious as violating (5.53), because in the latter case the Mach number can be very high, whereas in the former case $\mathcal{M}$ is less than 2; consequently, the shock at the end of the bending is probably weak. However, that is not much comfort if one wants to keep the jet collimated after passing the bend: the steady decrease in $\mathcal{M}$ which is dictated by the behaviour of the characteristics makes the jet flare out even if the influence of a terminal shock is avoided or weak.

The condition (5.57) determines a maximum bending angle. Inversion of equation (5.54) gives, with equation (5.52),

$$\phi = \arcsin \frac{1}{\mathcal{M}} + \frac{1}{\mu} \arctan(\mu\sqrt{\mathcal{M}^2 - 1}) - \frac{\pi}{2} . \tag{5.58}$$

For $\Gamma = 4/3$, one finds $\phi = 13.95°$ for the limiting value in equation (5.57). Because $\phi = \pi/(2\mu - 2) = 148.12°$ if $\mathcal{M} = \infty$, I conclude that no jet can be bent through more than 134° (if $\Gamma = 4/3$). If we compare this with the bending by pressure gradients in Gaussian clouds (Fig. 5.7), we see that $\phi_{\rm max} \simeq \Delta_{\rm max}$ at $2P_0/\rho v^2 \sim 10$. Therefore, clouds with a central pressure less than 10 times the dynamic pressure of the jet cannot achieve bending above this maximal angle. Since it seems plausible that in general $P_0 \lesssim \frac{1}{2}\rho v^2$, bending through large angles (and its consequent problems for the jet) must be expected to be exceedingly rare.

The fact that the Mach angle changes in the bend according to equation (5.56) causes a change in the estimate (5.52) of the allowed jet thickness. If the x-direction is taken along the incoming jet,

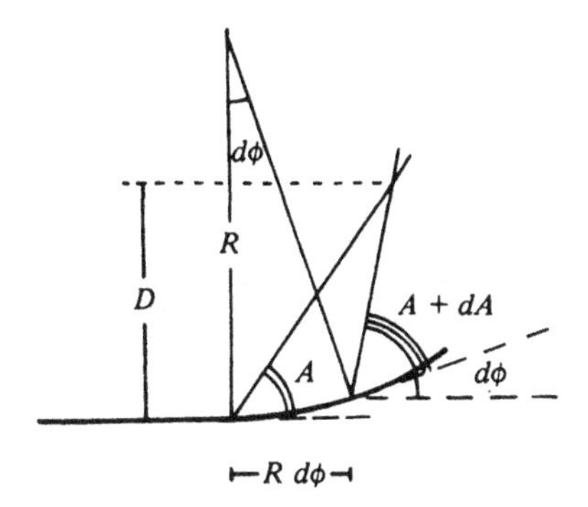

Fig. 5.17. Geometry for an improved estimate of the point where a shock occurs in a supersonic jet that is bent with a constant radius of curvature R. See Fig. 5.14 and the text.

and the y-direction is perpendicular to that and going through the centre of curvature of the jet, then the equation of the characteristic emanating from the beginning of the bend is

$$y = x \tan A \ . \tag{5.59}$$

The characteristic emanating from a point at which the bend has curved through $d\phi$ degrees (see Fig. 5.17) is

$$\begin{aligned} y &= (x - R\,d\phi)\tan(A + dA + d\phi) \\ &\simeq x \tan A + x(dA + d\phi)/\cos^2 A - R\,d\phi \tan A \ , \end{aligned} \tag{5.60}$$

so that the intersection of the two has the y-coordinate

$$D = \frac{R \sin^2 A}{1 + dA/d\phi} \ . \tag{5.61}$$

By means of equations (5.52) and (5.56) this becomes

$$\frac{D}{R} = \frac{2}{\Gamma + 1}(\mathcal{M}^2 - 1)\mathcal{M}^{-4} \ . \tag{5.62}$$

The conditions (5.62) and (5.58) allow the construction of a "safe zone" for the parameters of a supersonic jet that is bent along a circle with radius R. Jets whose thickness and bending angle lie in the region below the curve in Fig. 5.18 are safe from destruction by internal shocks. For example, a jet with $\Gamma = 4/3$ coming in at Mach 4 must have a thickness less than 5% of its radius of curvature, or else a shock occurs near the onset of the bend; it must not curve through more than 72°, or else a shock occurs near the end of the bend.

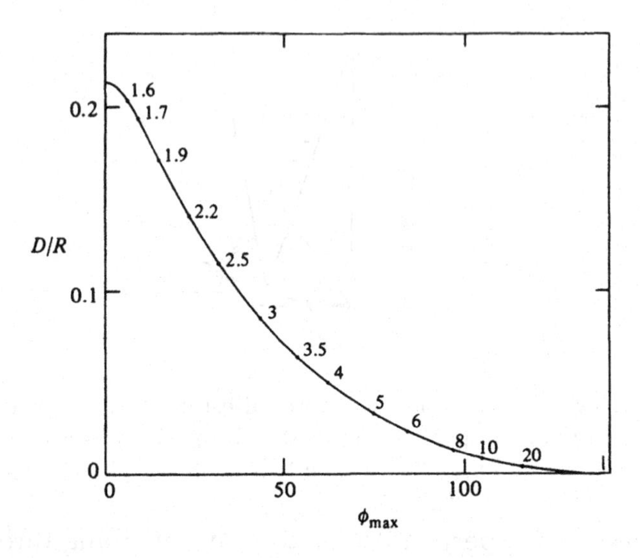

Fig. 5.18. Maximum ratio D/R of the jet thickness over the bending radius, and maximum bending angle ϕ_{max}, as a function of the initial Mach number of the jet (indicated on the curve). Jets with parameters below the curve are safe from destruction by internal shocks. If a jet with a given initial Mach number and ratio D/R is bent through such a large angle that ϕ_{max} is exceeded, the jet develops a terminal shock. If a jet with a given initial Mach number and bending angle is made thicker such that the allowed value of D/R is exceeded, the jet develops a shock at the onset of the bend.

Even though the above analysis is rather idealized, it presents serious problems for jet models because the constraints implied by equations (5.58) and (5.62) place an upper limit on the dynamic pressure a jet can exert without destroying itself. The centrifugal acceleration exerted by the jet on the deflecting medium is approximately $a = v^2/R$. The mass in a slice of jet with width D that turns through an angle ϕ is about $\rho D^2 R\phi$, so that the centrifugal force corresponding to the acceleration a is

$$F = \rho v^2 D^2 \phi \ . \tag{5.63}$$

The surface on which this force acts is ϕDR, so that the centrifugal pressure is

$$P_c = \rho v^2 D/R \ . \tag{5.64}$$

Using $\mathcal{M} = v/c_s$, and noting that $\rho c_s^2 \approx P_j$ (the gas pressure in the jet), I obtain

$$P_c = P_j \mathcal{M}^2 D/R = \frac{2}{\Gamma+1}(1 - \mathcal{M}^{-2})P_j \ . \tag{5.65}$$

Therefore, I conclude that even if the Mach number is as good as infinite, the jet cannot exert more dynamic pressure than an amount equal to its internal gas pressure without destroying itself. For small Mach numbers, the maximal possible dynamic pressure is even less.

5.4 Entrainment

5.4.1 Generic flow pattern of entrained gas

At present, there are no models of jet collimation and acceleration which are really compelling. The bulk of the numerical hydrodynamics has concentrated on the propagation of the jets, assuming their initial perfect collimation as given. Let us assume, then, that somehow the jet is collimated well enough to be approximated as a point source of linear momentum in two opposite directions. How does such a flow evolve? The details of the leading edge of the flow are discussed in Chapter 7; let us consider the generic case of viscous entrainment of the surroundings of the jet (the viscosity is due to turbulence, of course).

In the (r, θ) coordinate system used above, the stream function of a linear momentum source is

$$\psi = r\mu g(x) \ , \tag{5.66}$$

(Batchelor 1974, §4.6), where μ is the effective viscosity of the fluid, and $x \equiv \cos\theta$. The velocity has the components in the two coordinate directions

$$v_r = -\frac{\mu}{r}\frac{\partial g}{\partial x} \ , \tag{5.67}$$

$$v_\theta = -\frac{\mu}{r}\frac{g(x)}{\sqrt{1-x^2}} \ . \tag{5.68}$$

The function g can be shown to obey

$$g^2 - 2(1-x^2)\frac{dg}{dx} - 4xg = C_1 x^2 + C_2 x + C_3 \ . \tag{5.69}$$

This yields a similarity solution for the stream function and its associated flow field. The boundary conditions are given on a cone centred on $r = 0$; this cone represents the surface of the disc. The boundary conditions are, that v_θ must be zero on the axis and in the disc plane, and v_r must vanish at the surface of the disc. Therefore, if θ_0 is the opening angle of the disc surface, we require

$$\lim_{x \to 1} g(x) \propto 1 - x \ , \tag{5.70}$$

$$\lim_{x \to x_0} \frac{\partial g}{\partial x} \propto x - x_0 \ ; \qquad x_0 \equiv \cos\theta_0, \tag{5.71}$$

$$\lim_{x \to -1} g(x) \propto 1 + x \ . \tag{5.72}$$

Accordingly, the expression on the left hand side of equation (5.69) varies as $(1-x)^2$ near $x = 1$ and as $(1+x)^2$ near $x = -1$, which is impossible unless $C_1 = C_2 = C_3 = 0$. Therefore, the entrainment flow is determined by a solution of

$$g^2 - 2(1 - x^2)\frac{dg}{dx} - 4xg = 0 \ . \tag{5.73}$$

This is readily seen to have the solution

$$g(x) = \frac{2 - 2x^2}{A - x} \ , \tag{5.74}$$

in which A is a constant which is fixed by the boundary conditions, namely

$$A = \frac{1 + x_0^2}{2x_0} \ . \tag{5.75}$$

Finally, the radial and tangential velocities turn out to be

$$v_r = -\frac{2\mu}{r}\frac{1 - 2Ax + x^2}{(A - x)^2} \ , \tag{5.76}$$

$$v_\theta = -\frac{2\mu}{r}\frac{\sqrt{1 - x^2}}{A - x} \ . \tag{5.77}$$

The flow patterns that occur when the jet drags the surrounding gas along, as described by these equations, are shown in Fig. 5.19. For comparison with such flows observed in the laboratory, see Van Dyke (1982, p. 99).

The force F acting on the surrounding fluid can be found in the same formalism (Batchelor 1974, §4.6); if the fluid density is ρ, it is

$$\frac{F}{2\pi\rho\mu^2} = \frac{32A}{3A^2 - 3} + 4A^2 \log\left(\frac{A - 1}{A + 1}\right) + 8A \ . \tag{5.78}$$

The mass dragged along is approximately

$$\rho D v_r \approx \frac{F}{v_r} \ , \tag{5.79}$$

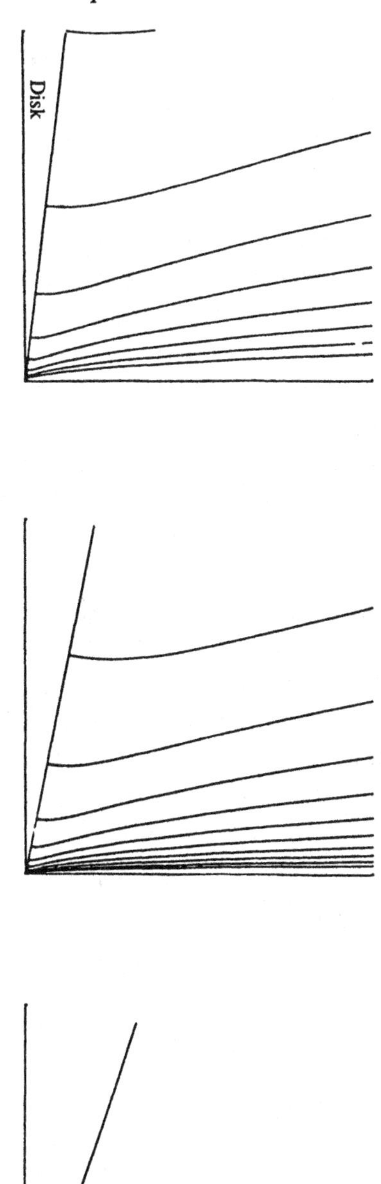

Fig. 5.19. Flow pattern of the gas of a conical disc which is dragged along by a jet. From top to bottom, disc opening angle: 83°, 79°, 71°; central source mass flux: $\dot{M} = 0.001, 0.01, 0.1$ (arbitrary units). Compare with Van Dyke (1982), p. 99, and Icke (1983), Fig. 4.

where D is the cross-section of the entrained stream. Near the axis, we have $x \simeq 1$, so that $v_r \simeq 4\mu x_0/r$.

5.4.2 Mixing extraneous gas into the jet: nonrelativistic case

Cosmic jets will mix with the gas surrounding them; the only question is, to what extent? In numerical simulations, it appears that there are corners of parameter space where the core of the jet is shielded from the outside world by a "backflow cocoon". Analytic treatment of this is very difficult. The numerical results published so far indicate that the cocoon can diminish the radial gradients around the jet, which make mixing problems less severe. In what follows, however, I will consider the case in which the jet can mix directly with the gas surrounding it.

Mixing and deflection of jets ought to occur together. The jets are not deflected by smooth metal walls, but by soft and ragged masses of gas. Therefore, one must consider what happens when matter from the outside is mixed into the jet. If a particle of the gas in the jet, having a mass m and moving with speed v, encounters a stationary extraneous parcel of gas consisting of N particles of the same mass, the centre-of-mass speed $v_{\rm cm}$ of the mass $(N+1)m$ that results when the two coalesce is

$$v_{\rm cm} = v/(1+N) \; . \tag{5.80}$$

The energy liberated in this coalescence is

$$\Delta E = \frac{mNv^2}{2(1+N)} \; . \tag{5.81}$$

The energy liberated per particle is then

$$\frac{\Delta E}{N+1} = \frac{3}{2}k_B T = \frac{1}{2}mNv^2(1+N)^{-2} \; . \tag{5.82}$$

The speed of sound in an ideal gas with adiabatic index Γ is $c_s^2 = \Gamma k_B T/m$, whence the Mach number $\mathcal{M}$ of the jet after mixing is

$$\mathcal{M} \equiv \frac{v_{\rm cm}}{c_s} = \frac{3}{\sqrt{5N}} \; . \tag{5.83}$$

If the jet and its environment are not very cold, but have temperatures T_j and T_e, respectively, a consideration of the thermal energies per particle shows that instead of equation (5.83) we get

$$\mathcal{M} = v\left[(1+N)\left(c_j^2 + Nc_e^2\right) + 5Nv^2/9\right]^{-1/2} \; , \tag{5.84}$$

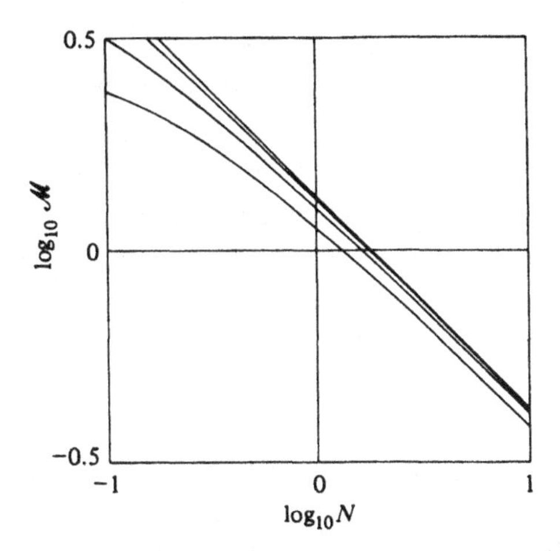

Fig. 5.20. Mach number after the mixing of extraneous matter into a nonrelativistic jet. The abscissa is the number of particles N of extraneous gas admixed per particle of the jet. The adiabatic constant is 5/3. The Mach numbers of the initial jet are: ∞ (upper curve), 10, 5, and 3 (lower curve).

where c_j is the sound speed in the jet and c_e is the sound speed in the environment. A useful alternative for equation (5.84) is

$$\mathcal{M} = \left((1+N)(1+NT_e/T_j)\mathcal{M}_j^{-2} + 5N/9\right)^{-1/2} , \qquad (5.85)$$

where $\mathcal{M}_j$ is the Mach number of the jet before mixing. Because I expect $T_j \gg T_e$, this simplifies further to

$$\mathcal{M} \simeq \left((1+N)\mathcal{M}_j^{-2} + 5N/9\right)^{-1/2} . \qquad (5.86)$$

A graph of $\mathcal{M}(N)$ is shown in Fig. 5.20. Even the admixture of 10% extraneous material makes the jet Mach number drop below 4, and consequently the Mach angle would become bigger than 15° at the slightest provocation. Since opening angles of jets are usually well below 20°, this means something of an embarrassment if external material plays a role in confining the jet. Instabilities are certain to occur on the interface between the jet and the cloud, and thorough mixing is a likely occurrence (cf. Bicknell & Melrose 1982).

The dearth of direct observational information about the mixing between a jet and its surroundings might indicate that such mixing, where it occurs, is only mild, so that it may well be that an admixture of 10% is comparatively large. In fact, one might well argue that the absence of observational evidence means that mixing is unimportant. One promising way to avoid direct contact with the environment is to have the jet core shielded from the external gas by means of its own cocoon. It is quite plausible that this works, as long as the jet propagates through a smooth background (see Chapter 7). But it is not obvious that the cocoon (which, after all, is only shocked low-density jet gas) is of much help when the outer density and pressure gradients are large.

5.4.3 Mixing extraneous gas into the jet: relativistic case

One might be tempted to think that the situation is better for relativistic jets. If a particle with mass m moves with speed v past N such particles at rest, the energy-momenta are $\gamma(mv, mc^2)$ and $(0, Nmc^2)$, respectively. The Lorentz transformation to the centre-of-momentum system gives

$$\gamma_{\rm cm} = \frac{N+\gamma}{\sqrt{N^2+2\gamma N+1}} , \tag{5.87}$$

$$\beta_{\rm cm} = -\gamma\beta/(N+\gamma) . \tag{5.88}$$

It follows that the difference between the energies before and after mixing is

$$\Delta E = mc^2\left(\sqrt{N^2+2\gamma N+1} - N - 1\right) , \tag{5.89}$$

which is the energy liberated in the centre-of-momentum system. If this energy is thermalized,* and if the original thermal content of the jet is $\frac{3}{2\Gamma}mc_i^2$ per particle, then the total energy per particle after mixing is given by

$$\begin{aligned} E &= \frac{3}{2\Gamma}(N+1)mc_s^2 \\ &= \frac{3}{2\Gamma}\frac{mc^2\beta^2}{\mathcal{M}_i^2} + mc^2\left(\sqrt{N^2+2\gamma N+1} - N - 1\right) , \end{aligned} \tag{5.90}$$

where Γ is the adiabatic index, $\mathcal{M}_i$ is the initial Mach number of the jet, and c_s is the sound speed after mixing. The Mach number $\mathcal{M} = v_{\rm cm}/c_s$ after mixing becomes, by combination of equations

* The subject of thermodynamic quantities in relativity is a little contentious, because it is not obvious how thermodynamic things like "random velocity" and "thermal equilibrium" can be made covariant. Readers interested in the fine points of relativistic kinetic theory might consult Ehlers (1971).

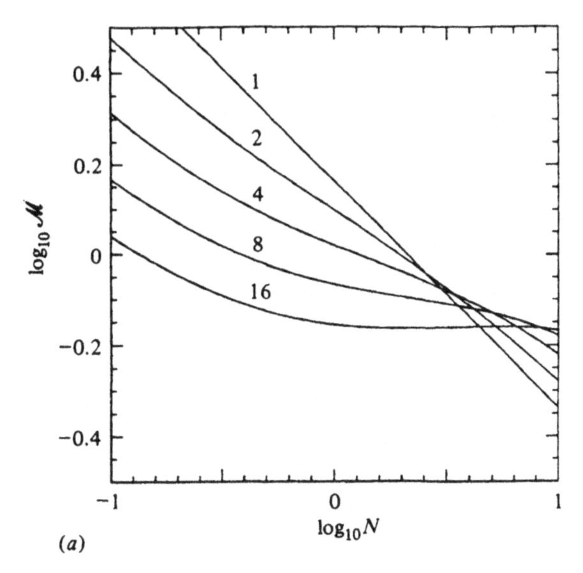

Fig. 5.21a. As Fig. 5.20, but for a relativistic jet; the adiabatic index is 4/3. The Lorentz factors of the initial jet are indicated near the curves. The initial thermal content of the jet and its surroundings is neglected, so that the Mach number of the initial jet is infinite.

(5.88), (5.89) and (5.90),

$$\mathcal{M} = \frac{\gamma\beta\sqrt{N+1}}{N+\gamma}\left[\frac{\beta^2}{\mathcal{M}_i^2} + \frac{2\Gamma}{3}\left(\sqrt{N^2+2\gamma N+1} - N - 1\right)\right]^{-1/2} . \tag{5.91}$$

Graphs of $\mathcal{M}(N)$ and of $\mathcal{M}(N)/\mathcal{M}_i$ are shown in Fig. 5.21. The situation in a relativistic jet is worse than in a classical one, which is no great surprise because the centre-of-momentum energy becomes a disproportionately large fraction of the total energy as $v \to c$. Since this is exclusively due to the Lorentz transformation, the problem is generic, and cannot be solved by invoking details of the interactions between the particles, magnetic fields, and the like.

5.5 From nucleus to hotspot

5.5.1 How do the jets manage?

How do jets manage to bridge the one-billion chasm between their point of departure and their fiery end at a hotspot? After 15 years of contemplating bulk-flow energy transport in radio galaxies, we still do not know. Leaving aside an even more important

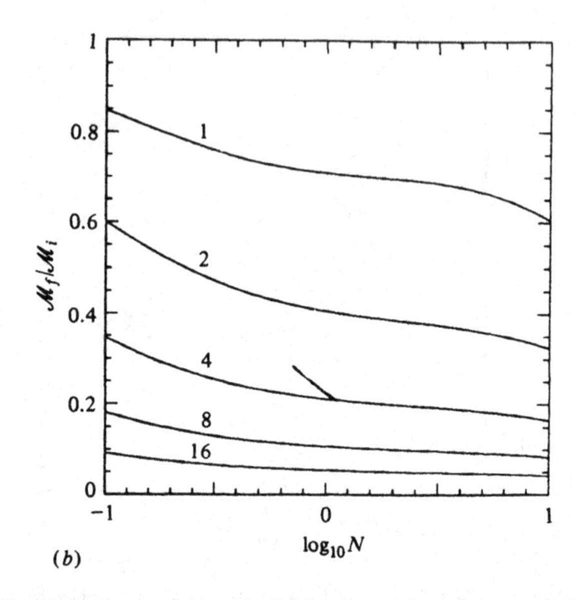

Fig. 5.21b. Ratio of the final Mach number $\mathcal{M}_f$ and the initial Mach number $\mathcal{M}_i$ for a relativistic mixing jet with $\gamma = 8$. The initial internal energy of the jet is given through $\mathcal{M}_i$, which is indicated near the curves. Unless $\mathcal{M}_i$ is very small (in which case the jet has a large opening angle to begin with), the mixing makes $\mathcal{M}_f \ll \mathcal{M}_i$.

problem – namely, why are there jets in the first place? – the most unpleasant uncertainties concern three entirely elementary parameters of moving gas: the density, the velocity, and the Mach number.

Are jets dense or tenuous? The evidence for tenuous jets is indirect. Because the energy density of a bulk flow scales as ρv^2, it is easier to make powerful active galaxies by increasing v. Moreover, the mass loss rate scales as $\dot{M} \propto \rho v$, while the luminosity of accretion sources obeys $L \propto \dot{M}$. For a fixed L, low speeds then imply high densities, high enough to begin to conflict with the observations. Thus, the current consensus is that jets are tenuous.

On the other hand, in the one source where we can observe such things directly, the famous SS 433, we see that "hot, low-density" features – the 6.7 keV Fe(K_α) line – and "cool, high-density" ones – the hydrogen lines – both occur in the jets. We know that they partake of the same motion because both systems obey the jet kinematics. Furthermore, in SS 433 as well as in the active galaxies, the best-fitting models for the jet shapes are simply trails of ballistic blobs, suggesting *high* gas densities.

Are jets slow or fast? The observations of superluminal sources say that they are fast, with $\gamma \sim 5$ or maybe larger; direct observations of SS 433 give $\beta = 0.26$. On the other hand, some of the physics discussed above has been used to estimate jet speeds from opening angles, curvature, and the like; these estimates indicate nonrelativistic speeds on length scales above a kiloparsec. Also, the same relativistic Lorentz boosting that is invoked to hide the counterjets in high luminosity sources is somehow absent when jets curve around sharp bends (but that may merely mean that jet trails are not the same as flowlines).

If we cannot decide on velocity or density separately, what of the product ρv? The finding that quasar jets are more deformed at high redshift indicates that the momentum density cannot be too high. The simplest interpretation of the deformation is that it is caused by ambient gas, which surely was much denser at high redshifts than it is today. Thus, ρv is not high enough to simply brush aside the galactic gas.

The opening angles of jets are usually small, indicating very high internal Mach numbers. At least in this respect, there is less contradiction than with the other flow parameters. But care is needed: the cross-section of well-studied jets (*e.g.*, Cygnus A) is practically independent of the distance to the nucleus. Here the concept of opening angle seems pointless; such features are more reminiscent of trails left behind by dense clouds.

5.5.2 Observation of jet processes

All of the above emphasises our profound uncertainty about the connection between flow parameters and the signal which is actually observed. Of what, precisely, is the radio brightness a measure?

Some jets, for example those in Galactic bipolar nebulae, have a morphology that strongly resembles those observed in radio galaxies; but bipolars are about a million times smaller, and do not appear to possess a nonthermal particle distribution. It would be nice if all such bisymmetric flow phenomena were hydrodynamically similar. If that is in fact the case, one should ask why Galactic bipolars are thermally emitting objects, whereas jets in radio galaxies are predominantly nonthermal.

To make an estimate of the extent to which a nonthermal particle distribution is produced, I consider the consequences of two competing effects, namely Larmor gyration and Coulomb collisions. In the canonical picture of a hydrodynamic chaotic cascade, energy is handed down to smaller and smaller length scales, until a dissipative microscale is reached. In everyday gases and fluids, this scale is the collisional mean free path. But in a very tenuous magnetoplasma, the smallest relevant length scale is the Larmor radius, which is not directly associated with a dissipative process, so that the cascade tends to pile up energy at that scale.

Thus, one may view the establishment of a nonthermal particle spectrum as a competition between thermal dissipation on the Coulomb scale, and accumulation of energy in a nonthermal distribution on the Larmor scale. In a proper theory, losses must of course be added: radiation affects both thermal and nonthermal populations, and diffusion adds spatial dimensions to the problem (*e.g.*, Hasselmann & Wibberenz 1968). In a homogeneous and stationary case, however, the primary concern is the competition between Coulomb and Larmor effects.

Because of the very large number of competing models of particle acceleration (see Chapter 9), none of which is categorically excluded by the observations, I can only try to consider the largest common denominator of all these mechanisms. The magnetoplasma is shaken around by some external agent such as shocks or turbulence; every time a particle goes through a characteristic (Larmor) collisionless time scale, its energy is amplified with a factor $1+\epsilon$ (in some models, the time scale may be longer than this, because the particle that is being accelerated must cross an accelerating region by diffusion – *e.g.*, in the form of pitch angle scattering – but the time scale for this is itself of the order of the Larmor time). This establishes a power law spectrum, because an injection energy E_0 has become $E = E_0(1+\epsilon)^N$ after N gyrations. As soon as N has increased to Ω/ν_c, *i.e.*, the ratio of the Larmor frequency and the Coulomb collision frequency, odds are better than half that the acceleration of the particle is interrupted. Therefore, it is plausible that the ratio of the energies E and E_0, between which the power law distribution is established, is given by

$$E/E_0 = (1+\epsilon)^N , \tag{5.92}$$

$$N = \Omega/\nu_c = \frac{eB}{\gamma m_e}\frac{1}{vn\sigma} , \qquad (5.93)$$

where n is the ion density, v is the speed of the electrons, and σ is the electron-ion Coulomb cross-section

$$\sigma = \pi r_e^2 (c/v)^4 , \qquad (5.94)$$

$$r_e \equiv e^2/4\pi\epsilon_0 m_e c^2 , \qquad (5.95)$$

(*e.g.*, Chen 1984). It is assumed in the following that v is not too different from the systematic speed of the jet. This appears to be a plausible assumption, whether the acceleration is due to shocks or to turbulence, because turbulent speeds depend only weakly on the length scale, and the jump in velocity across a shock is, for the strong shocks that are likely to occur, practically the same as the gas speed. From equation (5.92) we obtain

$$N = \frac{\log(E/E_0)}{\log(1+\epsilon)} , \qquad (5.96)$$

so that, using equations (5.93), (5.94) and (5.95),

$$\frac{\log(E/E_0)}{\log(1+\epsilon)} = 16\pi\epsilon_0^2 m_e (c/e)^3 \frac{B\beta^3}{\gamma n} , \qquad (5.97)$$

where $\beta \equiv v/c$ as usual. In most acceleration models, the efficiency ϵ is roughly proportional to β; for example, in Bell's (1978) shock mechanism, one finds $\epsilon \approx 4\Delta v/3c$, where Δv is the shock jump; in a strong shock, $\Delta v \approx 3v$, so that $\epsilon \approx 4\beta$ (although we will see below that the value of the efficiency hardly matters). Thus, equation (5.97) becomes

$$n = \frac{16\pi\epsilon_0^2 m_e (c/e)^3}{\log(E/E_0)} \frac{B\beta^3}{\gamma} \log(1+4\beta) . \qquad (5.98)$$

I expect that sources with a density higher than this cannot maintain a nonthermal spectrum over more than the range $E_0 \to E$. Assuming that this range is about five decades, which is about midway between galactic thermal and extragalactic nonthermal sources, I obtain

$$n \approx 2.042 \times 10^{21} B_9 \beta^3 \sqrt{1-\beta^2} \log(1+4\beta) \quad \mathrm{m}^{-3} , \qquad (5.99)$$

where B_9 is the magnetic induction in nanotesla. This curve, for $B_9 = 1$, has been drawn in Fig. 5.22, as well as a curve for which $E/E_0 = 10^{10}$ and $B_9 = 0.1$. It may be seen that the band spanned by these curves divides the (β, n)-plane into two halves; in the upper

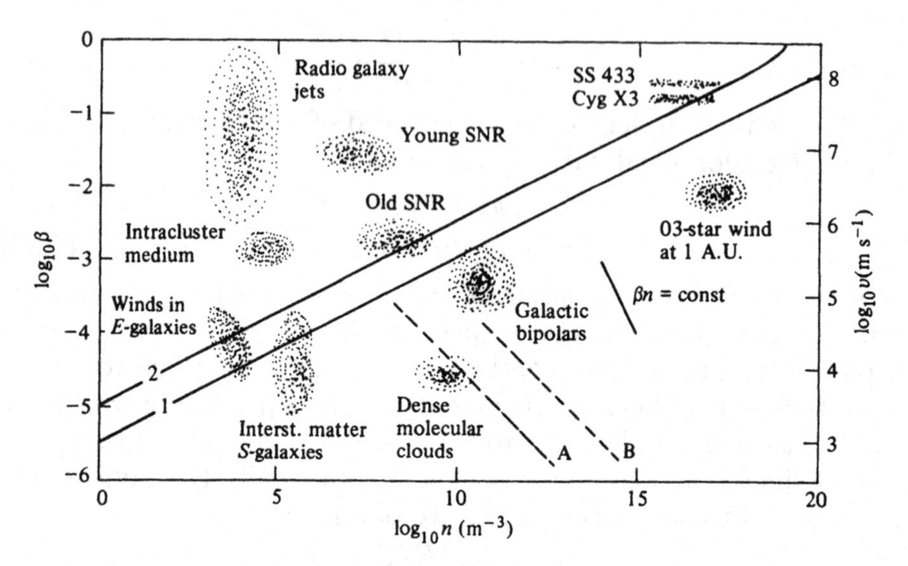

Fig. 5.22. Diagram indicating which combinations of flow parameters allow the production of a nonthermal particle spectrum. If a combination of density n and speed v (or $\beta \equiv v/c$) falls above the diagonal band, nonthermal behaviour is to be expected. Below that band, it ought not to occur. The line marked "1" is a solution of equation (5.99) with a nonthermal energy range $E/E_0 = 10^5$, and a magnetic induction of 1 nT. The line marked "βn =const" indicates the slope of lines of equal momentum density. The dashed lines have a slope on which the energy density is constant, assuming that the ions are protons. On line "A" the energy density is 10^{-9} J m^{-3}, on line "B" it is 10^{-7} J m^{-3}.

half I expect to find the nonthermal sources. Known classes of high energy objects seem to conform reasonably well to this expectation.

What does the above estimate imply for the appearance of the working surface of extragalactic jets? Ignoring the ultrarelativistic case $\beta \lesssim 1$, which is only relevant in superluminal nuclei, equation (5.98) can be written as

$$E = E_0(1+4\beta)\exp\left(16\pi\epsilon_0^2 m_e (c/e)^3 \frac{B\beta^3}{n}\right)$$
$$= E_0(1+4\beta)\exp(2.351\times 10^{22} B_9\beta^3/n) \ . \qquad (5.100)$$

The exponential dependence emphasizes that the occurrence of a nonthermal distribution depends very sensitively on β^3/n. The working surface is in approximate equilibrium between the dynamic pressure of the jet and the ram pressure of the medium ahead (Blandford & Rees 1974). Thus, if n_j is the density of the jet,

n_{ext} the density of the medium outside the jet, β_j the jet speed, and β_b the speed with which the bow shock moves through the gas outside the jet, we have

$$n_{\text{ext}}\beta_b^2 \approx n_j\beta_j^2 \ . \tag{5.101}$$

Thus, the ratio of the relevant variable combination β^3/n in the exponent of the energy equation (5.100) becomes

$$\frac{\beta_b^3}{n_{\text{ext}}}\frac{n_j}{\beta_j^3} = \left(\frac{n_j}{n_{\text{ext}}}\right)^{5/2} \ . \tag{5.102}$$

This is quite a steep dependence on the ratio of the jet density and the density of the ambient medium. Therefore, shocks in the lowest density medium should completely dominate the nonthermal emission.

In those cases where there is a suspicion that the jet hits dense interstellar matter, we must expect that the particle distribution properties lie somewhere between those of low density jets and dense clouds of interstellar matter. As can be seen in Fig. 5.22, that means that such an interaction region most likely ends up on the borderline between thermal and nonthermal. Thus, it is plausible that such regions show evidence of intense jet-associated thermal gas, such as emission lines with peculiar intensity ratios. In a more general sense, Fig. 5.22 shows that thermal and nonthermal emission is expected to be *anticoincident.*

It would be interesting to see what would be the properties of sources which straddle the thermal/nonthermal line. One relevant group which comes to mind are galaxies at redshifts above 2 with very steep spectra: these show clear alignment between radio jets and extended optical emission.

5.6 Concluding remarks

From simple estimates of the physics of the propagation of supersonic jets, I conclude that:

1. When a jet bounces off a cloud, the observed jet locus can be a straight line, a forward facing parabolic section, or a backswept curve. This kind of splashing may account for wide angle tails.
2. Mixing extraneous matter into a supersonic jet is rather bad for the jet collimation, unless the mixing energy can be re-

moved very efficiently. Even as little as 10% admixing is quite serious.

3. Mixing extraneous matter into a relativistic jet is disastrous, even if as little as 1% is admixed, unless energy losses are both enormously efficient and very secretive, so that the energy does not appear in a readily observed form.
4. Jets can be bent smoothly by the pressure gradient in a gas cloud, which to a certain degree justifies the assumption that jets can bounce off clouds. The bending angle is a simple function of the ratio of the peak pressure in the cloud divided by the dynamic pressure of the jet.
5. If a bent jet is thicker than the radius of curvature divided by the square of the Mach number, it will be decollimated by the heat liberated in an internal shock wave.
6. Jets with a low Mach number (below about 2) cannot be bent without developing an internal shock wave.
7. The maximum bending angle for any supersonic gaseous jet is about 130°.
8. A jet cannot exert more dynamic pressure than an amount about equal to its internal gas pressure without destroying itself.
9. It is to be expected that extragalactic jets emit mostly a nonthermal radio spectrum, whereas Galactic bipolar jets ought to be predominantly thermal. If jet heads are dense gasballs, these also should produce mostly thermal emission.
10. The thermal emission and the nonthermal emission are anticoincident.

Symbols used in Chapter 5

Symbol	Meaning
a	absolute value of centrifugal acceleration
A	Mach angle
$\mathbf{A}$	centrifugal acceleration
b	jet impact parameter
B	magnetic induction
c	speed of light

c_s	sound speed
D	jet width
e	electron charge
e	base of natural logarithms
E	kinetic energy
E	electron energy
f	jet path
F	centrifugal force
$\mathbf{F}$	drag force
g	absolute value of enthalpy gradient
g	angular factor of stream function
k_B	Boltzmann's constant
$\frac{1}{2}K$	kinetic energy density
L	luminosity
L	jet path integration constant
m	particle mass
m_e	electron mass
$\dot{M}$	mass transport rate
$\mathcal{M}$	Mach number
n	particle density
N	particle number
P	gas pressure
P_n	n-th Legendre polynomial
Q	dimensionless bending constant
r	radial position
r_e	classical electron radius
R	radius of curvature
$\mathbf{R}$	jet source trajectory
S	jet source strength
t	time
T	temperature
T	effective stopping time
U	shock velocity component in jet direction
$\mathbf{u}$	jet gas velocity
$\mathbf{v}$	gas velocity
w	specific enthalpy
w	jet speed after collision
$\mathbf{w}$	jet source velocity
x	cosine of polar angle θ

x	abscissa of jet path
y	ordinate of jet path
α	direction of jet
β	jet direction after collision
β	speed in units of c
δ	increment of jet deflection
Δ	deflection angle
γ	Lorentz factor
Γ	Poisson's adiabatic index
ϵ	electron energy amplification
ϵ_0	dielectric constant of the vacuum
μ	effective viscosity
μ	square root of asymptotic shock jump
ν_c	Coulomb collision frequency
θ	polar angle in spherical coordinates
ρ	gas density
σ	electron-ion Coulomb cross-section
ϕ	momentum-density potential
ϕ	jet bending angle
χ	deflection angle through a shock
ψ	stream function
ω	jet bending parameterization angle
Ω	Larmor frequency

References

Batchelor, G. K., 1974, *An Introduction to Fluid Dynamics*, (Cambridge University Press: Cambridge).

Begelman, M. C., Blandford, R. D. & Rees, M. J., 1984, *Rev. Mod. Phys.*, **56**, 255.

Bell, A. R., 1978, *M. N. R. A. S.*, **182**, 147.

Bicknell, G. V. & Melrose, D. B., 1982, *Ap. J.*, **262**, 511.

Biretta, J. A., Owen, F. N. & Hardee, P. E., 1983, *Ap. J. Lett.*, **274**, L27.

Blandford, R. D. & Icke, V., 1978, *M. N. R. A. S.*, **185**, 527.

Blandford, R. D. & Rees, M. J., 1974, *M. N. R. A. S.*, **169**, 395.

Bridle, A. H. & Eilek, J. A. (eds.), 1984, *Physics of Energy Transport in Extragalactic Radio Sources*, NRAO Workshop No. 9, (NRAO: Green Bank, WV).

Bridle, A. H. & Perley, R. A., 1984, *Ann. Rev. Astr. Ap.*, **22**, 319.

Burns, J. O. & Balonek, T. J., 1982, *Ap. J.*, **263**, 546.

Chen, F. F., 1984, *Introduction to Plasma Physics and Controlled Fusion*, Vol. 1, (Plenum: New York).

Cornwell, T. & Perley, R. A., 1984, In *Physics of Energy Transport in Extragalactic*

Radio Sources, NRAO Workshop No. 9, eds. A. H. Bridle & J. A. Eilek (NRAO: Green Bank, WV), p. 39.

Courant, R. & Friedrichs, K. O., 1976, *Supersonic Flow and Shock Waves*, (Springer: Berlin).

Ehlers, J., 1971, In *General Relativity and Cosmology*, Int. Sch. "Enrico Fermi" Course XLVII, ed. R. K. Sachs, (Academic Press: New York), p. 1.

Fiedler, R. & Henriksen, R. N., 1984, *Ap. J.*, **281**, 554.

Forman, W., Jones, C. & Tucker, W., 1985, *Ap. J.*, **293**, 102.

Gower, A. C., Gregory, P. C., Hutchings, J. B. & Unruh, W. G., 1982, *Ap. J.*, **262**, 478.

Hasselmann, K. & Wibberenz, G., 1968, *Zeitschr. F. Geophys.*, **34**, 353.

Henriksen, R. N. (ed.), 1986, *Can. J. Phys.*, **64**, 351.

Henriksen, R. N., Vallée, J. P. & Bridle, A. H., 1981, *Ap. J.*, **249**, 40.

Icke, V., 1976, In *Structure and Evolution of Close Binary Systems*, IUA Symposium 73, eds. P. Eggleton, S. Mitton & J. Whelan (D. Reidel: Dordrecht, Netherlands), p. 267.

Icke, V., 1981, *Ap. J. Lett.*, **246**, L65.

Icke, V., 1983, *Ap. J.*, **265**, 648.

Icke, V. & Choe, S.-U., 1983, *Confinement of Jets by Dynamic Pressure from Evaporating Disks*, unpublished.

Jones, T. W. & Owen, F. N., 1979, *Ap. J.*, **234**, 818.

Lind, K. R., Payne, D. G., Meier, D. L. & Blandford, R. D., 1989, *Ap. J.*, **344**, 89.

Mobasher, B. & Raine, D. J., 1989, *M. N. R. A. S.*, **237**, 979.

Norman, M. L., Winkler, K.-H. A. & Smarr, L. L., 1983, In *Astrophysical Jets*, eds. A. Ferrari & A. G. Pacholczyk (D. Reidel: Dordrecht, Netherlands), p. 227.

Soker, N., O'Dea, C. P. & Sarazin, C. L., 1988, *Ap. J.*, **327**, 627.

Sparke, L. S., 1982a, *Observatory*, **102**, 129.

Sparke, L. S., 1982b, *Ap. J.*, **254**, 456.

Van Dyke, M., 1982, *An Album of Fluid Motion*, (Parabolic Press: Stanford).

Yokosawa, M. & Inoue, M., 1985, *Publ. Astron. Soc. Japan*, **37**, 655.

Zel'dovich, Ya. B. & Raizer, Yu. P., 1966, *Physics of Shock Waves and High Temperature Hydrodynamic Phenomena*, (Academic Press: New York).

6

The Stability of Jets

MARK BIRKINSHAW*
Department of Astronomy, Harvard University, Cambridge, MA 02138 , USA.

6.1 Introduction

The attractive features of the 'beam' model for galactic and extragalactic radio sources are predicated on the contention that astrophysical processes permit the formation of high-power collimated flows, and that such flows survive their passage from the generating engine to the outer parts of the source without losing most of their energy. In other words, astrophysical plasma beams must be *capable* of exceptional stability, although there are sources for which less stability is necessary. This may be seen in the 'P-D' diagram of Baldwin (1982) (cf. Chapter 2) – for a given radio power (say 10^{27} W Hz^{-1} sr^{-1}), radio sources spanning a wide range of linear sizes (about 1 to 1000 kpc) are found, and hence the beams driving these sources must be stable over distances exceeding 1 Mpc in the largest objects, but need be stable for only 1 kpc in the smallest. The purpose of this chapter is to discuss the physical mechanisms that are effective in stabilising and destabilising beam flows, and to calculate the stability properties of some beams that might be components of extragalactic radio sources. Much of the physics discussed here can be applied to Galactic jets (Chapter 10), with some modification for the difference in physical parameters.

The formalism within which most of this discussion takes place is that of (analytical) linear instability theory (*e.g.*, Drazin & Reid 1981), which has been shown to be useful and relevant to the calculation of the stability of free shear flows observed in the laboratory (see the review of Ho & Huerre 1984). The principal differences between astrophysical investigations of beams and the calculations used to interpret laboratory results are the importance of the compressibility of the beam fluid, the possible dynamical relevance of

*Alfred P. Sloan Foundation Research Fellow.

magnetic fields, and the neglect of viscosity. In the present theoretical environment where numerical hydrodynamics plays such an important role in the discussion of jets (see Chapter 7), it is worth explaining that the place of analytical calculations has not been usurped entirely by large numerical codes. Although analytical instability calculations are limited to the earliest phase of instability in simple beam structures (generally the phase of exponential growth only, although extensions to the study of growth at larger amplitudes can be made), *only* such calculations can address a wide range of physical parameters and scales of stability, and *only* such calculations are presently capable of investigating the full three-dimensional stability of beams (and many of the interesting instabilities are intrinsically three-dimensional). Furthermore, where numerical experiments can be performed, the existence of approximate analytical solutions permits a greater understanding of the origins of the instability, and can provide a check of the code.

§6.2 of this chapter contains a brief review of the methods of linear stability analysis, and the types of instability that might occur in laboratory or astrophysical environments. A summary of the important general conclusions about jet stability provided by theoretical and experimental investigations of laboratory jets is given in §6.3. §6.4 then examines the properties of the beams that are inferred to underlie astrophysical jets (see also Chapters 3, 5, and 10), and sets up the equations of motion of beams that are applicable to these situations. §6.5 discusses the properties of the Kelvin-Helmholtz instability in the astrophysical environment. §6.6 compares some observed jet flows of extragalactic radio-astronomy (see Chapter 2) with the shapes into which the Kelvin-Helmholtz instability should deform beams. Finally, §6.7 summarizes the results obtained from instability analyses of astrophysical beams.

6.2 Instabilities in plasmas

The statement that a plasma or fluid system is stable or unstable is usually a statement about the behaviour of small perturbations of that system. If infinitesimal disturbances of a system tend to relax and decrease in size then that system is said to be *stable* (or, strictly, conditionally stable, since larger perturbations might grow). If, on the other hand, the perturbations tend to increase in size and produce a change in the nature of the flow, the system is said to be *unstable.* The results of the operation of an instability may be to disrupt the flow entirely, to produce a chaotic and turbulent system, or (if the growth of the perturbations is limited by some non-linear process) to establish a new steady flow. That new flow may also be stable or unstable to the action of perturbations, and, in general, a flow may pass through a number of intermediate, metastable states before reaching a final stable or turbulent state.

Analytical examination of the stability of a fluid is normally undertaken through linear stability analysis, although the method can be extended to non-linear orders if desired. This technique is described in most textbooks on fluid dynamics, and an exhaustive description of its use can be found in Chandrasekhar (1961) or Drazin & Reid (1981). In its simplest form, the method of linear instability analysis revolves around the derivation of the *dispersion relation*

$$\mathcal{D}(\omega, \mathbf{k}) = 0 \tag{6.1}$$

where perturbations $V^{(1)}$ of the fluid variable V are assumed to be of the form

$$V^{(1)} \approx e^{i(\mathbf{k}.\mathbf{r}-\omega t)}. \tag{6.2}$$

The stability of the flow is then inferred from the behaviour of the *complex* angular frequency (ω) or wavevector ($\mathbf{k}$) roots of equation (6.1). If ω is assumed to be real, then roots of (6.1) with a positive imaginary part of $\mathbf{k}$ correspond to (spatially) decaying perturbations of the system, and roots with a negative imaginary part of $\mathbf{k}$ correspond to (spatially) growing perturbations of growth length $\lambda_e = -1/\mathrm{Im}(\mathbf{k})$. The presence of roots of the latter type is taken as an indication of *spatial instability.* Alternatively, $\mathbf{k}$ can be assumed to be real, when the presence of roots of (6.1) with a positive imaginary part of ω indicates *temporal instability* with a growth time $\tau_e = 1/\mathrm{Im}(\omega)$, and roots with negative imaginary parts of ω correspond to decaying perturbations.

Plasma instabilities can be divided into the categories of *microinstabilities*, whose primary effect is to change the distribution functions of particle species in the plasma and where the details of these distribution functions are essential to their understanding, and *macroinstabilities*, where only the bulk equations of motion of the fluid are necessary to describe the instability. Microinstabilities tend to appear on small scales and have little effect on the gross structure of a flow. By contrast, the length scales on which macroinstabilities operate are expected to reflect the physical scales of the system (such as the radius of a beam). The classic types of macroscopic instability in fluids are exemplified by the Rayleigh-Taylor and Kelvin-Helmholtz instabilities (the latter instability is the central subject of this chapter); and macroscopic instabilities in plasmas may be represented by the kink and sausage instabilities in a beam with an azimuthal magnetic field as well as the Rayleigh-Taylor and Kelvin-Helmholtz instabilities of magnetohydrodynamics. The Rayleigh-Taylor instability arises when a dense fluid rests upon a light fluid in an accelerating frame or a gravitational field, and its effect is to cause 'drips' of denser fluid to penetrate into the less dense medium. Kelvin-Helmholtz instabilities arise when two fluids are in relative motion on either side of some common boundary. Their origin may be traced to the Bernouilli equation – if a ripple develops at the interface, then fluid flowing faster to pass over that ripple exerts less pressure, and the ripple tends to grow. Kelvin-Helmholtz modes grow into striking wave structures (*e.g.*, Roberts *et al.* 1982) and cause a transfer of material across the boundary. Kink and sausage instabilities deform a plasma beam to reduce its magnetic energy density – in the sausage (or pinch) instability this occurs through the growth of periodic pinches in the diameter of the beam, and in the kink instability the beam adopts a helical shape. Theoretical discussions of these instabilities for incompressible flows are given by Chandrasekhar (1961).

The survival of astrophysical beams is largely determined by their response to the Kelvin-Helmholtz instability. This instability was invoked as a possible cause of the distortion of the lobes of radio sources by Blake (1972) and of the tails of comets by Ershkovich (1980) and Kochhar & Trehan (1988). The study of Kelvin-Helmholtz instabilities on radio jets was introduced by Turland & Scheuer (1976) and Blandford & Pringle (1976) to test the feasibility of the transport of energy to the outer parts of radio galaxies required by the beam model. In this context, the presence of instabilities is *undesirable*, because they might prevent effective transport of energy. The possibility that instabilities on the jet of M 87 might cause the optical knots was discussed by Stewart (1971), and Hardee (1979) has emphasized the possibility that instabilities on radio jets might cause knots and bends which resemble observed jet structures (Chapter 2). Instabilities have also been invoked as a source of energy that might cause *in situ* particle acceleration (Ferrari *et al.* 1979). In this context, the presence of instabilities is seen as *desirable.* Overall the study of beam stability aims

(1) to discover whether there are beam structures capable of transporting energy to radio lobes without being wholly disrupted; and
(2) to determine whether particle acceleration and observable structural distortions may arise through the operation of instabilities which do not disrupt beams.

The analysis of the stability of a beam flow through the derivation of a dispersion relation such as (6.1) involves the assumption that the perturbations are sufficiently small that the fluid/plasma equations are linear in the perturbation variables. This assumption ceases to be true when the perturbations grow to observable sizes, so that linear analyses can predict *at most* the scales and types of structure that *might* develop when the instability *saturates.* Non-linear analyses might predict the final outcome of the operation of an instability, but plausibility arguments or representative numerical calculations are commonly preferred since non-linear analyses are algebraically burdensome. The structures produced by an instability may develop shocks within (or outside, or both) a beam if the flow is supersonic, and the flow is likely to deviate strongly from the smooth pattern that is assumed as a basis for the calculation. If shocks are formed, they provide an opportunity for particle acceleration (Chapter 9), and can cause abrupt changes in the direction or the physical properties of the beams. Alternatively, the operation of the instability might lead to the formation of *solitons*, non-linear disturbances that propagate with no (first order) change of shape, and which retain their identity in collisions with other solitons. The vortex rings developed around jets observed in the laboratory (§6.3) may be disturbances of this type. Lerche & Wiita (1980), Fiedler (1986), and Roberts (1987) have investigated the possibility of solitons in astrophysical situations. Instabilities are likely to cause the beam flow to become turbulent if it is initially laminar (see §6.4), as is seen in laboratory jets, and the energy input that they provide may drive turbulent particle acceleration (Ferrari *et al.* 1979) and heat the beam material.

6.3 Laboratory jets

Laboratory investigations of the stability of free shear flows, including jets and wakes, have lead to theoretical insights into the development of these flows under the influence of forced and random perturbations. Such insights are directly applicable to investigations of *astrophysical* flows insofar as the dimensionless numbers characterising the flow (*e.g.*, the Mach number, $\mathcal{M}$) are comparable in the two situations. More importantly, the *general* insights into the relevance of possible approaches to the simplification and analysis of the flow equations are of considerable help in understanding the flow. The most important of these insights are discussed below in notes based on reviews of the stability of free shear layers by Maslowe (1981) and Ho & Huerre (1984).

6.3.1 The steady flow

The discussion of the stability of a jet begins with the statement of its steady-state structure, and analytical studies are restricted to flows which are slowly-varying and near pressure equilibrium, so that the steady-state flow is relatively simple. The flow may be laminar or turbulent, but the lack of a convincing theoretical description of turbulence leads us to consider the development of instabilities on a laminar jet (see §6.4). Jets of circular section (and constant or slowly-changing radius) are of greatest interest for astrophysicists, and the typical structure of such a flow is sketched in Fig. 6.1, where the velocity profile $v_z^{(0)}(r)$ and the density profile $\rho^{(0)}(r)$ of the unperturbed flow are defined.

6.3.2 Observed pattern of instability

The *shear layer* between a jet flow and the ambient medium is unstable, and the jet distorts to become non-cylindrical through the growth of *Kelvin-Helmholtz waves* downstream of the nozzle where the jet emerges into the ambient medium (see the experiments of Sato 1956, 1959; Browand 1966; Freymuth 1966; Miksad 1972; Mattingly & Chang 1974). The jet displays axisymmetric distortions at first, but helical and other non-axisymmetric structures dominate further downstream (Browand & Laufer 1975). Axisym-

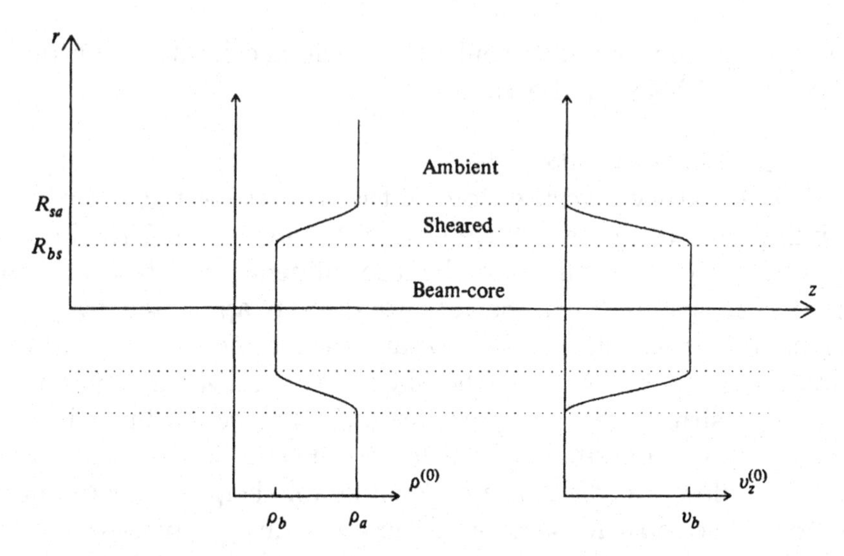

Fig. 6.1. A schematic diagram of the assumed structure of a cylindrical beam. The density and velocity profiles of the beam and the ambient medium are specified by $\rho^{(0)}(r)$ and $v_z^{(0)}(r)$ respectively. These functions are constant except in a sheared layer, $R_{bs} < r < R_{sa}$, shown shaded in the figure. Magnetic fields may also be included in the flow (see §6.4). The ambient medium is assumed to be infinite and uniform. The beam is assumed to be long, so that there is no feedback from pressure waves generated at the head of the beam, and in pressure equilibrium in the steady state.

metric disturbances are less important at higher jet speeds, and helical distortions of the jet are more dominant. The initial growth of the dominant modes is exponential in time (as expected from linear instability theory), until the fastest-growing mode reaches non-linear amplitude. The later development of the flow involves the passage of energy from some modes into others in a cross-talk that is intrinsically non-linear and is said to reflect *mode competition* (Freymuth 1966; Miksad 1972, 1973). Modes which survive this non-linear phase display large growth rates and a favourable balance between their gains and losses of energy from interaction with other waves. The surviving surface waves 'roll up' to produce large-scale vortical structures which form a quasi-periodic pattern within the mixing layer (Freymuth 1966; Miksad 1972; Ho & Huang 1982). In 2-D shear layers (and, presumably, jet flows) the nature of the large-amplitude flow depends sensitively upon the structure of the *critical layer* where the phase speed of surface waves equals the flow velocity (Stuart 1971). Eventually the large vortices merge to form larger structures, which broaden the (turbulent) mixing layer between the (laminar) core of the jet and the (laminar) ambient medium (Winant & Browand 1974). In unforced shear layers, where the locations of the vortex mergers are random, the width of the mixing layer increases linearly with distance from the nozzle. Instability waves on jets are less unstable than those on two-dimensional shear layers because the production and merging of vortices is predominantly a two-dimensional process (Michalke 1969).

6.3.3 Small-scale transition

The cores of the large vortices tend to become turbulent after a few turning times, and flow within the mixing layer undergoes the *small-scale transition* to turbulence. Even when turbulence is fully established and the mixing layer has become broad (a large fraction of the jet radius) Kelvin-Helmholtz waves with wavelengths much larger than the width of the layer can still be supported, and large-scale structures can still be seen within the flow. This result is only well documented for two-dimensional shear layers, but is likely to remain true for jets (Brown & Roshko 1974). The growth of the turbulent mixing layer under the influence of long-wavelength surface waves and large vortices eventually leads to the disappearance of the *potential core* of laminar flow at the centre of the jet, but large-scale structures may still be produced and propagate after the entire flow appears turbulent.

6.3.4 Calculation of instabilities

Laboratory experiments are particularly important for their elucidation of the applicability of linear instability theory to the stability of jets. Since the dominant structures developed on a jet have non-linear amplitudes and sizes which are a large fraction of the jet radius, it might be thought that linear theory cannot provide useful predictions of the form of the instability. Fortunately this is not the case, and the scale on which a jet begins to distort, the wavelengths of the modes, and the wavelength of the dominant instability of the jet are all well described by linear instability theory (Gerwin 1968; Miksad 1972; Mattingly & Chang 1974; Ho & Huang 1982; Ho & Huerre 1984), although no information on the amplitudes of the fully-developed vortex rolls can be obtained in this way.

It is not clear mathematically whether Kelvin-Helmholtz instabilities are *absolute* (grow at fixed space points) or *convective* (grow in the frame in which the fluid is at rest, but decrease at fixed space points because the mean flow carries the instability downstream), and hence whether a temporal or a spatial analysis of the dispersion relation is to be preferred. However, experimental data show clearly that jet instabilities are described better by a spatial than a temporal approach (Mattingly & Criminale 1972). This distinction is important because although the growth time, τ_e, and the growth length, λ_e, of a mode of wavelength λ_0 can be related by the group velocity if the mode is only slightly unstable, no simple relation can be found for the most unstable modes, which are precisely those of most interest (Gaster 1962). A logical inconsistency in the spatial treatment of instability is that the modes become of infinite size at downstream infinity, and a counter-example to the use of spatial analysis of a dispersion relation has been given by Drazin & Reid (1981). This inconsistency is generally unimportant, since it can be shown that the divergence at downstream infinity arises after the mode becomes large at any given fixed point on the jet. Spatial analysis provides a consistent description of instabilities in situations in which the dispersion relation makes a smooth mapping between the Re($\mathbf{k}$) and Re(ω) axes, which seems to be the case for jet flows.

Linear instability theory successfully reproduces the dominance of axisymmetric instabilities close to the nozzle of a jet, and of helical instabilities further downstream, and shows that this phenomenon is related to the disappearance of a potential core within the jet. Batchelor & Gill (1962) showed that only helical instability waves can be supported by the type of smooth velocity profile adopted by jets well downstream from their nozzles. This work has been extended to slowly diverging jets by Plashko (1979). Axisymmetric vortex rings produced in the early part of the evolution of the jet can also be shown to be linearly unstable to non-axisymmetric modes (Widnall *et al.* 1974). The investigation of the later development of instabilities can be made through numerical calculations (*e.g.*, Corcos & Sherman 1983; Corcos & Lin 1983), or through the analytical approaches reviewed by Stuart (1971) and Maslowe (1981).

6.3.5 Preferred mode

It is generally more convenient in the laboratory to study the response of a jet to *forcing* by an applied sinusoidal perturbation of frequency ν_{f}, rather than allowing ambient noise to perturb the jet. Crow & Champagne (1971) showed that jets have a maximum response to forcing when $\nu_{\mathrm{f}} = \nu_{\mathrm{p}}$, the *preferred frequency*, given by $\nu_{\mathrm{p}} \approx 0.3v_j/D$, where D is the jet diameter and v_j is the jet speed. The preferred mode (or *jet-column mode*) is a global, axisymmetric distortion of vortex type, and is not the same as the shear-layer mode of greatest instability, the natural mode of the original velocity profile of the jet (of frequency ν_{n}). Instead, the preferred mode develops on the changed velocity profile that results from the spreading of the mixing layer under the influence of the original instabilities.

Experimental studies of the natural and preferred modes of jets show that ν_{n} and ν_{p} are strongly related. The form of the relationship is unexplained, although the analysis of Crighton & Gaster (1976) does predict the preferred mode frequency for a jet. This calculation involved a linear instability analysis based on the observed velocity profile of the jet in the region where the preferred mode is strong, rather than the original jet velocity profile, and hence may be regarded as describing the non-linear evolution of structures on the original jet profile.

6.3.6 Global feedback

It is generally supposed that the development of instabilities and mixing in the shear layer of a jet is a local phenomenon which arises through the conditions obtaining at each point on the jet. Dimotakis & Brown (1976) suggested that this may not be the case in many laboratory experiments, and that the development of large-scale vortices of large amplitude far downstream may have a significant feedback effect on the jet. This feedback from the non-linear products of instability to the generation of instabilities near the nozzle may be important in astrophysics as well as in the laboratory – there is evidence from numerical codes that this type of feedback may occur (Norman *et al.* 1981, 1983). This effect will be largest in cases where the jet is short.

6.3.7 Electromagnetically-active jets

The differences between laboratory and astrophysical flows are far more pronounced when jets containing magnetic fields and ionised particles are studied. Most laboratory jets do not resemble astrophysical jets in the sizes of key physical parameters, such as the ratio of the electron gyroradius or the Debye length to the radius of the jet (small in astrophysical jets, but close to unity in most laboratory experiments). Thus laboratory jets are more particle-like than fluid-like (Spulak & Burns 1984), and are not necessarily useful indicators of the properties of astrophysical flows except for purely electrodynamical effects. It is known, for example, that laboratory jets can carry currents of dynamical importance (*e.g.*, Ekdahl *et al.* 1974; Benford & Smith 1982), and that self-confinement may arise from the magnetic fields driven by these currents (Benford 1983), but the importance of currents in astrophysical situations is unclear.

6.4 The steady flow in astrophysical jets

A correct analysis of the stability of a beam depends on the nature of the flow, the type of material participating in the flow, the width of the flow, the density and velocity of the flow, and the boundary conditions imposed by the environment of the flow. The extensive observational data on jets (Chapters 2, 3 and 10) do not provide unambiguous estimates of the flow parameters, and it is even unclear how the observed emissions relate to the flow itself (Chapters 5 and 9). Indeed, one rationale for the study of the instabilities of beams is that a comparison of the predicted stability properties with the observed structures might reduce the parameter ranges by eliminating beams which are too unstable to be observed. In this section, the range of *possible* physical parameters of jets are collected, and the equations of motion describing beams are developed.

6.4.1 Physical conditions

Table 6.1 collects the ranges of the important physical parameters that characterise jets.

Table 6.1. *Physical parameters characterising jets and the ambient medium.*

Quantity		Value
Ambient medium sound speed	c_{Sa}	$(300 - 1000)$ km s^{-1}
Ambient medium density	ρ_a	$(10^{-25} - 10^{-22})$ kg m^{-3}
Ambient medium magnetic field	B_a	$(0 - 1000)$ nT
Ambient medium temperature	T_a	$(10^6 - 10^8)$ K
Ambient medium gas pressure	P_a	$(10^{-15} - 10^{-10})$ Pa
Ambient medium polytropic index	Γ_a	$\frac{5}{3}$
Beam radius	R	$(0.1 - 1000)$ pc
Beam speed	v_b	$c_{Sa} - 0.95c$
Beam sound speed	c_{Sb}	$(0.1 - 1000)\ c_{Sa}$
Beam density	ρ_b	$(10^{-6} - 10^2)\ \rho_a$
Beam magnetic field	B_b	$(0 - 1000)$ nT
Beam fluid polytropic index	Γ_b	$\frac{4}{3} - \frac{5}{3}$
Sheared region width	δ	$0 - 2$
Sheared fluid polytropic index	Γ_s	$\frac{4}{3} - \frac{5}{3}$

The beam flows underlying astrophysical jets have been divided into the three segments shown in Fig. 6.1 – a beam core (subscript b), the ambient medium (subscript a), and a sheared region between them (subscript s). Few of the physical parameters contained in Table 6.1 can be considered to be well-determined, and it is clear that many structures may be supported without contradiction with observational data.

The parameters for the medium surrounding radio jets are relatively well determined by virtue of X-ray data obtained by satellites such as *Einstein*, and the values given in Table 6.1 reflect the range of properties of the interstellar medium in galaxies and the intergalactic medium in which they are embedded (see the reviews by Forman & Jones 1982 and Sarazin 1986): this range should be extended to lower temperatures ($T_a \sim 10^4$ K) and higher densities ($\rho_a \sim 10^{-12}$ kg m^{-3}) if the stability of the jet in the broad-line and narrow-line emission regions is to be discussed. Less evidence is available on the magnetic environment of radio jets, but if the field in the magneto-ionic halo around M 84 is typical, magnetic fields $B_a \sim 0.1$ nT which are ordered on scales of a few kpc will be common (Laing & Bridle 1987). Much stronger external fields are necessary if jets are magnetically confined, and radial gradients of azimuthal field intensity are expected: the upper limit to B_a in Table 6.1 reflects this possibility. The pressure exerted by a confining field may be as large as 10^{-6} Pa, which may greatly exceed the ambient gas pressure, and the Alfvén speed

$$c_A = \frac{B}{(\mu_0 \rho)^{1/2}} \tag{6.3}$$

in the ambient medium may approach c (*i.e.*, $c_{Aa} \gg c_{Sa}$). If the magnetic field in the ambient medium is low, $c_{Aa} \ll c_{Sa}$ and magnetic pressure is dynamically unimportant.

The ambient medium will be assumed to be uniform, with no density, temperature, or magnetic field gradients that might cause the steady-state flow to be non-cylindrical. Where this is not the case, the equations of motion must take explicit account of structures in the ambient medium and their effect on the steady-state flow. The problem of the stability of a beam is rendered intractable if rapid variations of the properties of the ambient medium are taken into account. For long-wavelength modes the assumption of uniformity is unlikely to be valid, and more complicated calculations will be necessary.

§6.3.1 has noted that analytical instability studies are restricted to flows near dynamical equilibrium, where the internal pressure in the beam (magnetic, thermal, and relativistic-particle) balances the external pressure of the ambient medium. For weak radio sources such as Cen A, 3C 449, and NGC 6251 (Chapter 2), the (minimum) energy contents of the jets are low, and the minimum pressures of the radio-emitting electrons (10^{-14} to 10^{-11} Pa) indicate that the thermal pressure of the ambient medium (up to 10^{-10} Pa) can confine the jets. For the most powerful jets in quasars and high-power radio galaxies, on the other hand, the assumption that the jets are confined implies that the external magnetic field, B_a, is making a significant contribution to the pressure (perhaps by adopting a 'pinch' configuration). If the jets are *free*, and not in effective contact with the ambient medium, then the instability analyses described here are inapplicable.

The opening angles of radio jets are small, from less than about 3° in high-power jets (such as 3C 111, Linfield & Perley 1984) to about 15° (*e.g.*, in the rapidly-expanding parts of NGC 315, Bridle *et al.* 1979). In the simplest representation of the flow, the beam will be assumed to be cylindrical (Fig. 6.1), although conical beams with small opening angles have also been discussed (Hardee 1984). The radii of radio jets, R, vary from less than 1 parsec in VLBI jets to more than 1 kpc in low-power radio sources (Chapter 5), and this large range of R is used in Table 6.1. In cases where the model beam contains a core surrounded by a shear layer (Fig. 6.1), R will be taken as the mid-point of the layer, $R = \frac{1}{2}(R_{sa} + R_{bs})$. The width of the layer will be described by the dimensionless constant

$$\delta = 2\,\frac{R_{sa} - R_{bs}}{R_{sa} + R_{bs}} \tag{6.4}$$

and all variations of $v_z^{(0)}$ and $\rho^{(0)}$ will be assumed to be confined within the shear layer.

The major dynamical constituents of the beam fluid are, presumably, thermal gas and relativistic particles (with an effective polytropic index, Γ_b, in the range 4/3 to 5/3), and magnetic fields. The mixture of relativistic gas and thermal particles is treated as a single fluid, with no explicit account being taken of the transfer of energy between the constituents of the fluid. It is generally assumed that the beam fluid is a neutral mixture of electrons and protons, and this assumption will be made here, but the beams may consist of an electron-positron fluid (see, for example, Burns & Lovelace 1982 and Chapter 7) – if this is so, it would require a substantial re-evaluation of the parameters of Table 6.1.

Arguments about the speed of the beam fluid are reviewed in Chapters 3 and 5 – in view of the uncertainty in the value of v_b, a wide range is given in Table 6.1. We are also ignorant of the velocity structure across a beam, so that it is of interest to treat a range of velocity profiles. The steady-state velocity profile of the beams will be treated as an undetermined function, $v_z^{(0)}(r)$, which decreases towards zero (not necessarily monotonically) from its central value, $v_z^{(0)}(0) = v_b$, as r increases from R_{bs} to R_{sa}.

The above discussion has referred to the *mean flow velocity*, which is an incomplete representation of the flow field if the beam is turbulent. The Reynolds number determines whether an astrophysical beam can support turbulence: the relevant Reynolds numbers for magnetically-active plasmas are the usual fluid Reynolds number

$$Re = \frac{\rho\, v\, l}{\eta}, \tag{6.5}$$

and the magnetic Reynolds number

$$Re_m = v\, l\, \sigma\, \mu_0, \tag{6.6}$$

where v is the velocity scale and l is the length scale of the flow ($v \approx v_b$ and $l \approx R$). ρ, η, and σ are the density, viscosity, and electric conductivity of the fluid, respectively, and μ_0 is the permeability of free space. Using representative values for the temperature, magnetic field, density, velocity, and radius of a jet (from Table 6.1) and the expressions for σ and η given by Spitzer (1962), these Reynolds numbers may be calculated to be

$$Re \sim 4 \times 10^{28} \tag{6.7}$$

and

$$Re_m \sim 10^{28} \tag{6.8}$$

unless the velocity and velocity gradient in the flow are accurately parallel to the magnetic field. This large value of Re_m indicates that magnetic field diffusion is much less important than the advection of the field by the flow (but note that this may not be true in thin *current sheets*, where collective plasma processes may produce anomalous transport coefficients, cf. Chapter 9). The large value of Re implies that turbulent motions are almost undamped (the *inertial range* of the turbulent spectrum is large), and astrophysical beams should be highly turbulent (De Young 1984), or should rapidly undergo the transition to turbulence under the influence of instabilities if the beams are injected into the ambient medium as laminar flows. The principal emphasis of this chapter, and most published work, is on the stability of laminar flows. Studies of the mean flow and stability of turbulent beams are handicapped by the lack of proven methods for calculating turbulent shear stresses, although some calculations have been made under particular assumptions about the physics of the processes involved (Bicknell 1984; Henriksen 1987).

Internal magnetic fields in jets are likely to be dynamically important in powerful sources. Although at least partially ordered in two dimensions, they need not be ordered in three dimensions (Laing

1981, Chapter 3). However, the instability analyses performed here assume that any dynamically-important field is well-ordered (to make the problem tractable), and dominated by azimuthal and axial components.

The procedures outlined in Chapter 3 for estimating the jet density are fraught with interpretational difficulties, and an alternative approach is to compare the results of numerical simulations of beam flows with the observed jet morphologies. High Mach numbers ($\mathcal{M}_b > 10$) and low densities ($\rho_b/\rho_a < 0.1$) are generally deduced (Chapter 7). The calculations also suggest that low-density beams develop hot cocoons where back-flow from the head of the beam is important, but that high-density beams do not. Such temperature and density structures transverse to the beam axis will be parameterized by a model function, $\rho^{(0)}(r)$, which varies from ρ_a to ρ_b within the sheared layer, $R_{sa} > r > R_{bs}$ (Fig. 6.1).

Finally, we should examine the continuity of the flows. Stability analyses generally begin from the assumption that the flow velocity and density are steady, but if the observations of one-sided jets are interpreted as evidence for intrinsically one-sided ejection rather than Doppler preference (Burns 1984, Chapter 8) then the directions of jet ejection must vary on time scales much less than the flow time from the core to the lobes. Most stability analyses are incapable of handling the drastic change in flow velocity implied if significant velocity changes occur within an instability growth length.

6.4.2 Equations of motion of beams

In accordance with the arguments given above, we model the beam fluid flows as an inviscid, electrically-conducting (but not heat-conducting), compressible, relativistic gas moving in the absence of any gravitational field. The equations describing this motion are the *equation of conservation of energy and momentum*

$$T^{\mu\nu}{}_{,\nu} = 0 \tag{6.9}$$

where the energy-momentum tensor $T^{\mu\nu}$ contains the energy and momentum of the fluid and the electromagnetic field

$$T^{\mu\nu} = T^{\mu\nu}_{\text{fluid}} + T^{\mu\nu}_{EM} \tag{6.10}$$

$$T^{\mu\nu}_{\text{fluid}} = (\rho + \frac{P}{c^2})\, u^\mu u^\nu + P\, \eta^{\mu\nu} \tag{6.11}$$

$$T^{\mu\nu}_{\rm EM} = \frac{1}{\mu_0}\left(F^{\mu\alpha}\, F^{\nu}{}_{\alpha} - \frac{1}{4} F^{\alpha\beta}\, F_{\alpha\beta}\, \eta^{\mu\nu}\right). \tag{6.12}$$

The density of the fluid, ρ, includes both the rest-mass and the mass equivalent of the energy density of the fluid, P is the pressure of the fluid, and u^μ is the usual velocity four-vector. Greek indices are assumed to run from 0 to 3, and $\eta^{\mu\nu}$, the Minkowski metric tensor, is defined according to a spacelike convention. $F^{\mu\nu}$ is the electromagnetic field tensor. μ_0 is the permeability of free space, and the relative permittivity and permeability of the fluid, ϵ and μ, are assumed to be unity. In addition to the equations of motion, an equation is needed to describe the presence of a gradient in the entropy due to the density and velocity shears and the generation of entropy by Joule heating. This *entropy equation* is

$$(n\, s\, u^\mu)_{,\mu} = -\frac{1}{T}\, u_\mu\, F^{\mu\nu}\, J_\nu, \tag{6.13}$$

where n is the baryon number density, s is the entropy per baryon in the fluid, T is the temperature of the fluid and J^μ is the charge-current four-vector. The right hand side of this equation is the entropy generation rate per unit volume. The electromagnetic field tensor $F^{\mu\nu}$ is constrained by the *Maxwell equations*

$$F_{\alpha\beta,\gamma} + F_{\beta\gamma,\alpha} + F_{\gamma\alpha,\beta} = 0 \tag{6.14}$$

$$F^{\alpha\beta}{}_{,\beta} = \mu_0\, J^\alpha \tag{6.15}$$

which contain within them the equation of conservation of electric charge, $J^\mu{}_{,\mu} = 0$, as can be seen by differentiating (6.15) and using the anti-symmetry of $F^{\mu\nu}$. We also require a constitutive equation describing the relation between the charge current and the fields (the *conduction equation*),

$$J^\mu + \frac{u^\mu u_\nu}{c^2}\, J^\nu = \sigma\, F^{\mu\nu}\, u_\nu \tag{6.16}$$

where σ is the electrical conductivity, and equations describing the interrelation of the thermodynamic parameters n, ρ, P, s, and T. These equations are the *equation of baryon conservation*

$$(n\, u^\mu)_{,\mu} = 0, \tag{6.17}$$

two *equations of state* (whose consistency with the Maxwell relations is required),

$$P = P(n, s) \tag{6.18}$$

$$T = T(n, s), \tag{6.19}$$

and the *first law of thermodynamics*

$$d\rho = (\rho + \frac{P}{c^2})\frac{dn}{n} + \frac{nT}{c^2}\, ds \qquad (6.20)$$

(Landau & Lifshitz 1959; Misner *et al.* 1973; Königl 1980).

Equations (6.9 – 6.20) contain twenty-one relations (three being redundant) between the eighteen functions of position and time $\mathbf{E}$, $\mathbf{B}$, $\mathbf{j}$, ρ_c, ρ, P, $\mathbf{v}$, n, s and T, and (6.9 – 6.17) can be written in terms of these variables as

$$\frac{\partial}{\partial t}\left(\gamma^2(\rho + \frac{P}{c^2}) - \frac{P}{c^2}\right) + \nabla.\left(\gamma^2(\rho + \frac{P}{c^2})\mathbf{v}\right) = \frac{\mathbf{E}.\mathbf{j}}{c^2} \qquad (6.21)$$

$$\gamma^2(\rho + \frac{P}{c^2})\left(\frac{\partial \mathbf{v}}{\partial t} + (\mathbf{v}.\nabla)\mathbf{v}\right) + \nabla P + \frac{\mathbf{v}}{c^2}\frac{\partial P}{\partial t} = \rho_c \mathbf{E} + \mathbf{j} \times \mathbf{B} - \frac{\mathbf{v}}{c^2}(\mathbf{E}.\mathbf{j}) \qquad (6.22)$$

$$nT\left(\frac{\partial s}{\partial t} + \mathbf{v}.\nabla s\right) = (\mathbf{E}.\mathbf{j} - \mathbf{v}.\mathbf{j} \times \mathbf{B} - \rho_c \mathbf{v}.\mathbf{E}) \qquad (6.23)$$

$$\frac{\partial(\gamma n)}{\partial t} + \nabla.(\gamma n \mathbf{v}) = 0 \qquad (6.24)$$

$$\nabla.\mathbf{E} = \frac{\rho_c}{\epsilon_0} \qquad (6.25)$$

$$\nabla.\mathbf{B} = 0 \qquad (6.26)$$

$$\nabla \times \mathbf{E} = -\frac{\partial \mathbf{B}}{\partial t} \qquad (6.27)$$

$$\nabla \times \mathbf{B} = \mu_0 \mathbf{j} + \frac{1}{c^2}\frac{\partial \mathbf{E}}{\partial t} \qquad (6.28)$$

$$\gamma(\mathbf{v}.\mathbf{j} - \rho_c \mathbf{v}.\mathbf{v}) = \sigma\, \mathbf{v}.\mathbf{E} \qquad (6.29)$$

$$\mathbf{j} + \frac{\gamma^2}{c^2}\mathbf{v}(\mathbf{v}.\mathbf{j} - \rho_c c^2) = \gamma\sigma\,(\mathbf{E} + \mathbf{v} \times \mathbf{B}) \qquad (6.30)$$

where

$$\gamma = \left(1 - \frac{\mathbf{v}.\mathbf{v}}{c^2}\right)^{-1/2} \qquad (6.31)$$

These equations may be simplified for the highly-conducting plasmas of astrophysical beams, since the large value of the magnetic Reynolds number Re_m (6.8) causes $F^{\mu\nu}u_\nu = 0$, or equivalently

$$\mathbf{E} = -\mathbf{v} \times \mathbf{B} \qquad (6.32)$$

to an accuracy $O(Re_m^{-1})$, or about one part in 10^{28} (see equations 6.16 and 6.30). This is the same basic simplification that is used in non-relativistic magnetohydrodynamics. Equations (6.18 – 6.32) provide a consistent description of the flow of a plasma if the fields change slowly compared with the time on which the plasma re-adjusts to charge neutrality. That is, the equations apply only when the frequencies of interest are much lower than the plasma frequency, the ion gyrofrequency, and so on. For this reason, the equations describe only macroscopic fluid properties, without charge separation. Ferrari *et al.* (1980) have discussed the internal consistency of the equations of relativistic magnetohydrodynamics, and note that the displacement currents cannot be ignored if $c \gg c_A$, where c_A is the Alfvén velocity (6.3), and that the equations do not consistently describe the physics of the plasma if $c < c_A$, since in this limit the Debye length, $\lambda_D = (\epsilon_0 k_B T/ne^2)^{1/2}$, exceeds $(r_{ge} r_{gi})^{1/2}$, where r_{ge} and r_{gi} are the electron and ion gyroradii, and significant charge separation can arise.

The set of equations (6.18 – 6.32) is sufficiently rich that no general solution is known. The first step in performing an analysis of beam stability involves finding an analytically-simple solution of (6.18 – 6.32) which describes an unperturbed beam. It is an advantage if this 'zero-order' solution contains sufficient functional flexibility that realistic flow structures can be described without causing excessive difficulties in the stability analyses that will follow. For a cylindrical beam flow it is convenient to work in cylindrical coordinates (r, θ, z), and a useful simple solution is provided by the steady, non-rotating, cylindrical beam flow with

$$\rho = \rho^{(0)}(r) \tag{6.33}$$

$$\mathbf{v} = (0, 0, v_z^{(0)}(r)) \tag{6.34}$$

$$\mathbf{B} = (0, B_\theta^{(0)}(r), B_z^{(0)}(r)) \tag{6.35}$$

where $\rho^{(0)}(r)$, $v_z^{(0)}(r)$, $B_\theta^{(0)}(r)$ and $B_z^{(0)}(r)$ are arbitrary functions of the distance from the beam axis, r. The pressure, $P = P^{(0)}(r)$, must then obey the equation of transverse force equilibrium

$$\frac{dP^{(0)}}{dr} = \frac{1}{\mu_0}\left(B_\theta^{(0)2}\frac{v_z^{(0)}}{c^2}\frac{dv_z^{(0)}}{dr} - \frac{1}{\gamma^{(0)2}}\frac{B_\theta^{(0)}}{r}\frac{d}{dr}(rB_\theta^{(0)}) - B_z^{(0)}\frac{dB_z^{(0)}}{dr}\right) \tag{6.36}$$

where $\gamma^{(0)}$ is the value of γ with $\mathbf{v}$ given by (6.34). The remaining state variables (s, T, and n) are obtained from equations (6.18 – 6.20).

Formally, (6.18 – 6.32) and the steady-state flow (6.33 – 6.36) describe the beam and the ambient medium together by a single set of velocity, density, pressure, and magnetic field functions. However, the problem is simplified by defining boundaries between regions of the flow which are described by different functions. One convenient choice, shown in Fig. 6.1, is to refer to the flow as three distinct regions (the beam core, the sheared flow, and the ambient medium), where only one of these regions (the sheared flow) contains gradients in the velocity or density. The functions describing the flow are then matched at the interfaces between these artificial domains.

The steady-state solutions of the fundamental equations will be denoted by the suffix $^{(0)}$, as in the unperturbed pressure, $P^{(0)}$. These suffices should not be confused with the contravariant 'time' component of the four-vectors, written as u^0, without the brackets. To investigate the stability of these $^{(0)}$ solutions, the physical variables are expanded as sums of the unperturbed $^{(0)}$ parts and small $^{(1)}$ perturbations (for example, $P = P^{(0)} + P^{(1)}$). These forms for the variables are then substituted into equations (6.18 – 6.32), and the equations are linearized (expanded to first order in the perturbed quantities) by requiring that the perturbed flow variables are much smaller than the unperturbed variables (for example, $v_r^{(1)} \ll v_z^{(0)}$). The resulting equations are homogeneous in the perturbed quantities, and are to be solved in the instability analysis. Unfortunately the set of equations that is produced is too complicated for most purposes, and analyses of the stability of beam flows are based on simplifying assumptions about the functions $\rho^{(0)}$, $v_z^{(0)}$, $B_\theta^{(0)}$, and $B_z^{(0)}$.

For a cylindrical flow geometry, and a steady-state flow of the form (6.33 – 6.36), the homogeneity of the linearized equations allows the perturbations to be decomposed into wave-like eigenfunctions

$$P^{(1)} = P^{(1)}(r)\, e^{i(kz+n\theta-\omega t)} \tag{6.37}$$

$$\rho^{(1)} = \rho^{(1)}(r)\, e^{i(kz+n\theta-\omega t)} \tag{6.38}$$

$$\mathbf{v}^{(1)} = \mathbf{v}^{(1)}(r)\, e^{i(kz+n\theta-\omega t)} \tag{6.39}$$

$$\mathbf{B}^{(1)} = \mathbf{B}^{(1)}(r)\, e^{i(kz+n\theta-\omega t)}. \tag{6.40}$$

The stability of the flow is described by the properties of the complex wave-number k, or angular frequency ω, of the Kelvin-Helmholtz *normal modes* (6.37 – 6.40), and the corresponding radial structures imposed on the beam. k and ω appear as eigenvalues of the solutions of the flow equations subject to the boundary conditions on the flow. General perturbations of the flow can be expressed as a superposition of these wavelike solutions (a Fourier decomposition) by virtue of the homogeneity of the linearized flow equations.

(6.37 – 6.40) imply that the Kelvin-Helmholtz normal modes are wavelike in z with wavenumber k, wavelike in time with angular frequency ω, and develop $|n|$ oscillations around the circumference of the beam ('wavelike in θ'). The radial structure, specified by (for example) $P^{(1)}(r)$, dictates the extent to which these waves are localised near the beam. The azimuthal mode number, n, is a positive or negative integer, so that the eigenfunctions are single-valued. For any given n, we expect the dispersion relation (6.1) to describe a solution $k(\omega)$ (or a family of solutions) with similar azimuthal characteristics. $n = 0$ corresponds to a 'breathing' or 'pinching' mode of the beam, where the beam expands or contracts coherently at any z. $n = \pm 1$ modes display sideways displacements of the beam with a helical pattern in z (these are the 'helical' or 'kink' modes). Modes with $|n| > 1$ (referred to as 'fluting' modes) produce distortions with ripples around the circumference of the beam – for $n = \pm 2$ there are two nodes on the beam surface. These beam distortions are illustrated in Fig. 6.2 for $n = 0$, 1, 2, and 3. Note that the beam area changes only for the pinching ($n = 0$) modes, and that the centre of the beam is displaced only by the helical ($|n| = 1$) modes. The eigenfrequencies for positive and negative n are degenerate for the simplest beam configurations (non-rotating beams with only axial magnetic fields).

The boundary conditions on the flow variables are imposed at $r \to \infty$ in the ambient medium, and at the boundaries between the ambient and sheared fluids (at $r = R_{sa}$ before perturbation) and the sheared and beam fluids (at $r = R_{bs}$ before perturbation). These conditions are

(a) all the flow variables should be finite at all r at finite t (in-

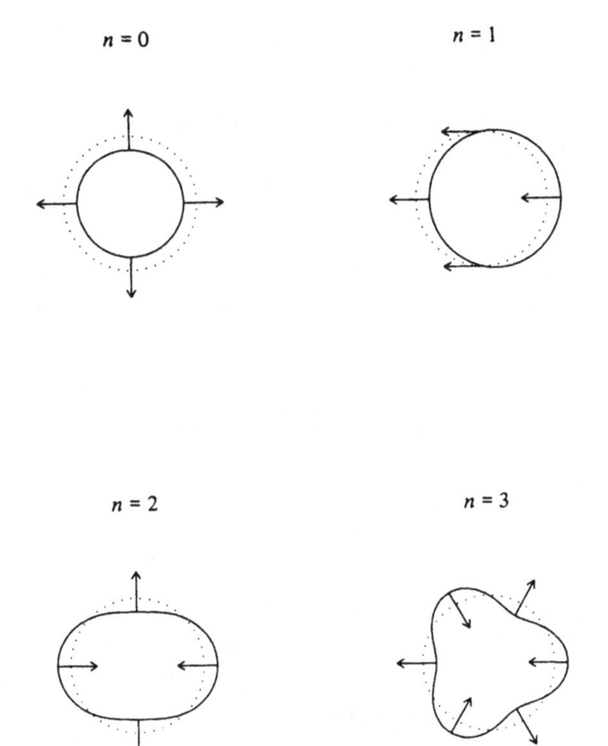

Fig. 6.2. A schematic diagram of the transverse beam deformations caused by Kelvin-Helmholtz modes with azimuthal mode numbers $n = 0$, 1, 2, and 3. The dotted lines indicate the location of the boundary of the unperturbed beam, and the full lines indicate the distortion of the boundary caused by the modes.

cluding downstream from any z – see the discussion of spatial and temporal instabilities in §6.3);

(b) all the flow variables should $\to 0$ as $r \to \infty$ at all finite time;

(c) the flow variables should be continuous functions of r;

(d) the radial phase velocity of the perturbations should be positive as $r \to \infty$ (so that perturbations are radiated from the beam);

(e) the radial displacements of the beam fluid and the ambient medium, $\xi_r^{(1)}$, where

$$\frac{d\xi^{(1)}}{dt} = v_r^{(1)}, \tag{6.41}$$

should match at $r = R_{bs}$ and R_{sa};

(f) the total (fluid plus magnetic) pressures should match at $r = R_{bs}$ and R_{sa};

(g) the perpendicular (radial) component of the magnetic field should match at the boundaries; and

(h) the parallel component of the electric field should match at the boundaries.

If the magnetic and electric fields are related by (6.32), then (g) and (h) imply that the magnetic field is continuous at the boundary. Note the use of the Sommerfeld finiteness and radiation conditions (a, b, and d) for the perturbations, and the relevance of the problem of the divergence of the flow at downstream infinity (condition b; §6.3.4).

6.5 The Kelvin-Helmholtz instability in astrophysics

This section applies the instability theory outlined above to two types of steady flow of the form (6.33 – 6.36): non-magnetic, relativistic flows, which provide simple illustrations of the conceptual difficulties of applying instability theory to beam flows; and non-relativistic, magnetic flows, which will be given a briefer treatment.

6.5.1 Kelvin-Helmholtz instabilities in non-magnetic beams

The simplest sub-case of equations (6.18 – 6.32) to have been subjected to theoretical analysis is that of non-magnetic, non-relativistic beam flows. The general forms for the unperturbed pressure, density, and velocity are given in (6.33 – 6.36). If no steady-state magnetic field is imposed, linear perturbations cannot generate a field, since the generation of magnetic field is quadratic in the field intensity. Arbitrary forms for the density gradient, $\rho^{(0)}(r)$, and the velocity field, $v_z^{(0)}(r)$, may be independently chosen for this flow.

The flow variables obey the general set of equations

$$\frac{\partial}{\partial t}\left(\gamma^2(\rho + \frac{P}{c^2}) - \frac{P}{c^2}\right) + \nabla.\left(\gamma^2(\rho + \frac{P}{c^2})\mathbf{v}\right) = 0 \qquad (6.42)$$

$$\gamma^2(\rho + \frac{P}{c^2})\left(\frac{\partial \mathbf{v}}{\partial t} + (\mathbf{v}.\nabla)\mathbf{v}\right) + \frac{\mathbf{v}}{c^2}\frac{\partial P}{\partial t} + \nabla P = 0 \qquad (6.43)$$

$$\frac{\partial s}{\partial t} + \mathbf{v}.\nabla s = 0 \tag{6.44}$$

$$s = s(P, \rho) \tag{6.45}$$

which may be recognised as the usual equations of relativistic fluid flow. If these equations are taken as describing a superposition of the cylindrical, steady-state equilibria defined by (6.33 – 6.36) and perturbations of the form (6.37 – 6.40), they may be manipulated to a set of equations in the perturbations

$$\begin{aligned}
&-i\omega\Big(\gamma^{(0)^2}(\rho^{(1)} + \frac{P^{(1)}}{c^2}) - \frac{P^{(1)}}{c^2} + \\
&2\gamma^{(0)^4}\frac{v_z^{(0)}v_z^{(1)}}{c^2}(\rho^{(0)} + \frac{P^{(0)}}{c^2})\Big) \\
&+\frac{1}{r}\frac{d}{dr}\left(r\gamma^{(0)^2}(\rho^{(0)} + \frac{P^{(0)}}{c^2})v_r^{(1)}\right) \\
&+\frac{in}{r}\gamma^{(0)^2}(\rho^{(0)} + \frac{P^{(0)}}{c^2})v_\theta^{(1)} \\
&+ik\Big(\gamma^{(0)^2}v_z^{(0)}(\rho^{(1)} + \frac{P^{(1)}}{c^2}) \\
&+\gamma^{(0)^4}(\rho^{(0)} + \frac{P^{(0)}}{c^2})(1 + \frac{v_z^{(0)^2}}{c^2})v_z^{(1)}\Big) = 0
\end{aligned} \tag{6.46}$$

$$i\gamma^{(0)^2}(\rho^{(0)} + \frac{P^{(0)}}{c^2})(kv_z^{(0)} - \omega)v_r^{(1)} + \frac{dP^{(1)}}{dr} = 0 \tag{6.47}$$

$$i\gamma^{(0)^2}(\rho^{(0)} + \frac{P^{(0)}}{c^2})(kv_z^{(0)} - \omega)v_\theta^{(1)} + \frac{in}{r}P^{(1)} = 0 \tag{6.48}$$

$$\begin{aligned}
&i\gamma^{(0)^2}(\rho^{(0)} + \frac{P^{(0)}}{c^2})(kv_z^{(0)} - \omega)v_z^{(1)} \\
&+\gamma^{(0)^2}(\rho^{(0)} + \frac{P^{(0)}}{c^2})v_r^{(1)}\frac{dv_z^{(0)}}{dr} + i(k - \frac{\omega v_z^{(0)}}{c^2})P^{(1)} = 0
\end{aligned} \tag{6.49}$$

$$i(kv_z^{(0)} - \omega)\left(\frac{P^{(1)}}{P^{(0)}} - \Gamma\frac{\rho^{(1)}}{\rho^{(0)}}\right) - \frac{\Gamma}{\rho^{(0)}}\frac{d\rho^{(0)}}{dr}v_r^{(1)} = 0 \tag{6.50}$$

where the final equation results from a model for the beam fluid in which the entropy is given by

$$s = s_0 + \frac{k_B}{\Gamma - 1}\ln P\rho^{-\Gamma} \tag{6.51}$$

with s_0 and Γ, the polytropic index, being constants. Few investigations of the effects of variable polytropic index have been made, but see Turland & Scheuer (1976). It should be noted that although (6.51) is commonly used for $s(P, \rho)$, and is identical with the familiar non-relativistic form (and hence should be adequate when $nk_BT < \rho c^2$), it is, nevertheless, not consistent with the thermodynamics of relativistic fluids unless those fluids have unusual equations of state. See Synge (1957) for a further discussion of the equation of state of a relativistic gas.

Equations (6.46 – 6.51) can be manipulated to yield a single, second-order, differential equation for the pressure perturbation,

$$\frac{d^2P^{(1)}}{dr^2} + \frac{dP^{(1)}}{dr}\left(\frac{1}{r} + 2\,\gamma^{(0)2}\frac{\left(k - \frac{\omega v_z^{(0)}}{c^2}\right)}{\left(\omega - kv_z^{(0)}\right)}\frac{dv_z^{(0)}}{dr} - \frac{\frac{d\rho^{(0)}}{dr}}{\rho^{(0)} + \frac{P^{(0)}}{c^2}}\right)$$
$$+ P^{(1)}\left(\gamma^{(0)2}\left(\frac{(\omega - kv_z^{(0)})^2}{c_S^2} - \left(k - \frac{\omega v_z^{(0)}}{c^2}\right)^2\right) - \frac{n^2}{r^2}\right) = 0 \quad (6.52)$$

where c_S is the sound speed,

$$c_S = \left(\frac{\Gamma P^{(0)}}{\rho^{(0)}}\right)^{1/2} \quad (6.53)$$

(Birkinshaw 1984). (6.52) represents a generalisation of the results of Ferrari *et al.* (1978) and Hardee (1979) for cylindrical beams bounded by vortex layers. It is the cylindrical, relativistic, counterpart of the pressure perturbation equation derived by Ferrari *et al.* (1982) for a sheared box-shaped beam, and by Ray (1982) for a slab beam. A similar equation may be found for the pressure perturbation on an expanding beam (using spherical coordinates), see Hardee (1984). The solutions for the velocity perturbation and other physical variables can be obtained from the solutions for $P^{(1)}$ and (6.46 – 6.51).

In general the solutions $P^{(1)}$ of (6.52) cannot be expressed in terms of the standard 'special' functions of mathematical physics (*e.g.*, the hypergeometric function) because (6.52) allows general forms for $v_z^{(0)}$ and $\rho^{(0)}$. Simple solutions do arise in those parts of the flow where the density and velocity gradients can be ignored (the ambient medium and the beam core) since (6.61) then simplifies to Bessel's equation in αr, where the radial wavenumber

$$\alpha = \gamma^{(0)}\left(\frac{(\omega - kv_z^{(0)})^2}{c_S^2} - \left(k - \frac{\omega v_z^{(0)}}{c^2}\right)^2\right)^{1/2}. \quad (6.54)$$

Since the ambient medium is taken to be uniform and at rest, the pressure perturbation in $r > R_{sa}$ is

$$P^{(1)}(r) = \epsilon_a P^{(0)} \frac{H_n^{(1)}(\alpha_a r)}{H_n^{(1)}(\alpha_a R_{sa})} \qquad r > R_{sa} \tag{6.55}$$

where $H_n^{(1)}(z)$ is a Hankel function of the first kind with order n. $\epsilon_a \ll 1$ is a measure of the smallness of the pressure perturbation relative to $P^{(0)}$. R_{sa} is the location of the interface between the sheared region and the ambient medium, α_a is given by (6.54) with $v_z^{(0)} = 0$, $\rho^{(0)} = \rho_a$, and $c_S = c_{Sa}$, the sound speed in the ambient medium (6.53). This solution represents an outward-going, decaying mode (a radiated sound wave) as $r \to \infty$, as required by boundary conditions (a – d) of §6.4.2.

In the unsheared beam core, the solution of (6.52) which is regular at $r = 0$ is

$$P^{(1)}(r) = \epsilon_b P^{(0)} \frac{J_n(\alpha_b r)}{J_n(\alpha_b R_{bs})} \qquad r < R_{bs} \tag{6.56}$$

where $J_n(z)$ is a Bessel function of the first kind with order n. $\epsilon_b \ll 1$ is a constant defining the smallness of the internal pressure perturbation. R_{bs} locates the interface between the beam core and the sheared part of the flow. The internal radial wavenumber, α_b, is given by (6.54) with $\rho^{(0)} = \rho_b$, $v_z^{(0)} = v_b$, and $c_S = c_{Sb}$, the sound speed in the core (6.53 with $\rho^{(0)} = \rho_b$). (6.56) is a standing pressure wave (contrast the radiated wave of 6.55), and may support pressure nodes within the beam.

Finally, the pressure perturbation must be found in the sheared region between the beam core and the ambient medium ($R_{bs} < r < R_{sa}$). No general solution is available in this case, but particular solutions may be found for any specified velocity or density structure. The simplest structure that might be used to link the beam core and the ambient medium is *a vortex sheet*, where the width of the sheared layer is zero and $R_{bs} = R_{sa} = R$. In this case, Ferrari *et al.* (1978), Hardee (1979), and others match the radial fluid displacements

$$\xi_r^{(1)} = \frac{\frac{dP^{(1)}}{dr}}{\gamma^{(0)^2}(\rho^{(0)} + \frac{P^{(0)}}{c^2})(\omega - k v_z^{(0)})^2} \tag{6.57}$$

and the perturbed pressures, $P^{(1)}$, at the interface (as required by the boundary conditions of §6.4.2) to obtain the dispersion relation

$$\mathcal{D}(k,\omega) = \frac{H_n^{(1)\prime}(\alpha_a R)}{H_n^{(1)}(\alpha_a R)} \cdot \frac{\alpha_a}{\omega^2 \left(\rho_a + \frac{P^{(0)}}{c^2}\right)} - \frac{J_n'(\alpha_b R)}{J_n(\alpha_b R)} \cdot \frac{\alpha_b}{\gamma_b^2(\omega - kv_b)^2 \left(\rho_b + \frac{P^{(0)}}{c^2}\right)} = 0 \tag{6.58}$$

where the primes (′) denote differentiation with respect to argument, and γ_b is the beam Lorentz factor (given by (6.31) with $v_z^{(0)} = v_b$). (6.58) codifies the first order stability of a cylindrical flow bounded by a vortex sheet, and its interpretation will be addressed in §6.5.3.

A dispersion relation based upon similar physics, but for a conical beam with small opening angle (so that the gradients in the physical variables are small), and bounded by a vortex sheet, has been derived by Hardee (1984) – the essential character of the dispersion relation is the same, but it must be regarded as describing the local stability of the beam (although the dispersion relation may provide a global description of the stability of the flow if it is analysed from a spatial viewpoint; Hardee 1986).

The use of a vortex sheet as the interface between the beam core and the ambient medium is beguilingly simple, and provides a convenient dispersion relation that is susceptible to detailed mathematical analysis. However, this approach obscures several important points, and a sheared matching layer between the beam-core and the ambient medium should be included to make the problem more realistic. First, note that the shear layer around a supersonic beam contains the *critical surface* at which the speed of a Kelvin-Helmholtz mode equals the speed of the fluid flow. It is known from laboratory experiments (§6.3) that the nature and location of this critical surface plays an important role in the development of the instability (equation 6.52 demonstrates that the critical surface represents the location of a regular singularity in the differential equation for $P^{(1)}$), and the use of the vortex layer approximation conceals this phenomenon within the matching condition at the interface. A second rationale for a study of a shear layer is as a test of the assertion that shear layers are transparent to long-wavelength modes, but suppress short-wavelength instabilities. This view may be an oversimplification, since the presence of shear may *destabilise* some modes at sufficiently high Mach number $\mathcal{M}_b = v_b/c_{Sb}$ for shear layers which are not too wide.

The pressure perturbation in a linearly-sheared sheathing layer around a beam is

$$P^{(1)}(r) = \epsilon_{s1}\, P^{(0)} \frac{F_n^{(1)}(r)}{F_n^{(1)}(R_{bs})} + \epsilon_{s2}\, P^{(0)} \frac{F_n^{(2)}(r)}{F_n^{(2)}(R_{sa})} \qquad R_{sa} > r > R_{bs} \tag{6.59}$$

where $F^{(1)}$ and $F^{(2)}$ are independent solutions of (6.52) in the sheared layer, and can be expressed as infinite, convergent, series of terms or pairs of series (Birkinshaw 1990). A dispersion relation $\mathcal{D}(\omega, k)$ similar to (6.58), but involving $F^{(1)}$ and $F^{(2)}$ as well as J_n and $H_n^{(1)}$, can be deduced by matching the pressures and fluid displacements at $r = R_{sa}$ and $r = R_{bs}$.

Bodo *et al.* (1989) have made a further extension of the theory of non-magnetic Kelvin-Helmholtz instabilities by including the effects of steady-state, uniform rotation of a beam on the growth of the modes. This involves the modification of (6.33 – 6.36) by adding a velocity component $v_\theta^{(0)} \propto r$. Their work has discussed only non-relativistic, $n = 0$ modes, and indicates the existence of a new channel of instability.

It is clear from (6.52) that Kelvin-Helmholtz modes are sound waves associated with the beam/ambient medium interface. The growth rate of the instability is related to the extent to which these waves are localised in the beam or efficiently radiated. That is, features in curves of the growth length of the instability as a function of wavelength can be related to the match between the wave impedances of the beam and ambient medium fluids as a function of frequency (Payne & Cohn 1985).

6.5.2 Non-relativistic, magnetic beams

The basic equations of fluid motion for an ideal, compressible, magnetic fluid in non-relativistic motion ($v \ll c$) are

$$\frac{\partial \rho}{\partial t} + \nabla.(\rho \mathbf{v}) = 0 \tag{6.60}$$

$$\rho\left(\frac{\partial \mathbf{v}}{\partial t} + (\mathbf{v}.\nabla)\mathbf{v}\right) = -\nabla P + \frac{1}{\mu_0}(\nabla \times \mathbf{B}) \times \mathbf{B} \tag{6.61}$$

$$\frac{\partial s}{\partial t} + \mathbf{v}.\nabla s = 0 \tag{6.62}$$

$$s = s(P, \rho) \tag{6.63}$$

$$\nabla.\mathbf{B} = 0 \tag{6.64}$$

$$\frac{\partial \mathbf{B}}{\partial t} = \nabla \times (\mathbf{v} \times \mathbf{B}) \tag{6.65}$$

(derived from 6.18 – 6.32 for large Re_m). These equations represent a considerable simplification of (6.18 – 6.32), but the presence of the magnetic field causes additional difficulties over the non-magnetic case, although all relativistic effects have been ignored. If these equations are interpreted as a superposition of the steady-state equilibria defined by (6.33 – 6.36) and perturbations of the form (6.37 – 6.40), they may be reduced to a set of eight non-redundant equations in the perturbations, but these eight equations are too complicated for useful manipulation if general $v_z^{(0)}(r)$, $\rho^{(0)}(r)$, $B_\theta^{(0)}(r)$ and $B_z^{(0)}(r)$ are assumed.

A variety of treatments of these equations in different limits, or with different forms for the magnetic field distributions have been discussed. Cohn (1983) considered the stability of a magnetically-pinched beam of this type, which had a vortex boundary layer at $r = R$, no density or velocity gradients in the ambient medium or the beam, and an azimuthal field $B_\theta^{(0)} \propto \frac{1}{r}$ for $n = 0$ modes only. In this limit (6.60 – 6.65) can be reduced to a form of the confluent hypergeometric equation, and an analytical discussion of the stability of the flow can be performed using the known properties of the confluent hypergeometric functions. Ray (1981) studied a cylindrical plasma beam with an incompressible equation of state, and with uniform internal magnetic field parallel to the axis of the beam – again, no gradients of velocity or density were considered. Finally, Ferrari *et al.* (1981) have discussed the stability of cylindrical flows without gradients of velocity, density, or the axial magnetic field, but with a relativistic beam velocity. A non-relativistic analogue of their results may be found by reducing (6.60 – 6.65) to a single governing equation for the pressure perturbation,

$$\frac{d^2P^{(1)}}{dr^2} + \frac{1}{r}\frac{dP^{(1)}}{dr}$$

$$+P^{(1)}\left(\frac{\left(k^2-\frac{(kv_z^{(0)}-\omega)^2}{c_S^2}\right)\left(\frac{(kv_z^{(0)}-\omega)^2}{c_A^2}-k^2\right)}{\left(k^2-(kv_z^{(0)}-\omega)^2\left(\frac{1}{c_A^2}+\frac{1}{c_S^2}\right)\right)}-\frac{n^2}{r^2}\right)=0 \qquad (6.66)$$

which is a form of Bessel's equation. The solution for $P^{(1)}$ can be written

$$P^{(1)}(r)=\epsilon_a P_a\frac{H_n^{(1)}(\alpha_a r)}{H_n^{(1)}(\alpha_a R_{sa})} \qquad r>R_{sa} \qquad (6.67)$$

in the ambient medium and

$$p^{(1)}(r)=\epsilon_b P_b\frac{J_n(\alpha_b r)}{J_n(\alpha_b R_{bs})} \qquad r<R_{bs} \qquad (6.68)$$

in the beam, where the radial wavenumbers are given by

$$\alpha=k\left(\frac{\left(1-\left(\frac{\omega-kv_z^{(0)}}{kc_S}\right)^2\right)\left(\left(\frac{\omega-kv_z^{(0)}}{kc_A}\right)^2-1\right)}{1-\left(\frac{\omega-kv_z^{(0)}}{kc_A}\right)^2\left(1+\left(\frac{c_A}{c_S}\right)^2\right)}\right)^{1/2} \qquad (6.69)$$

with sound speed $c_S=c_{Sa}$ or c_{Sb} (6.62), Alfvén velocity $c_A=c_{Aa}$ or c_{Ab} (6.3), density $\rho^{(0)}=\rho_a$ or ρ_b, flow velocity $v_z^{(0)}=0$ or v_b, and pressure $P^{(0)}=P_a$ or P_b for the ambient and beam wavenumbers, α_a and α_b, respectively. (6.36), the condition of transverse pressure equilibrium, requires

$$P_a+\frac{B_{za}^2}{2\mu_0}=P_b+\frac{B_{zb}^2}{2\mu_0}. \qquad (6.70)$$

(6.66) specifically excludes radial gradients in the velocity and density, and the dispersion relation may be derived only for a vortex layer boundary (associated with a current sheet) by matching the displacement of the boundary and the total (fluid plus magnetic) pressure at $r=R$ to obtain

$$\mathcal{D}(k,\omega)=\frac{H_n^{(1)\prime}(\alpha_a R)}{H_n^{(1)}(\alpha_a R)}\cdot\frac{\alpha_a}{\rho_a}\cdot\frac{1}{\omega^2-k^2c_{Aa}^2}-\frac{J_n'(\alpha_b R)}{J_n(\alpha_b R)}\cdot\frac{\alpha_b}{\rho_b}\cdot\frac{1}{(\omega-kv_b)^2-k^2c_{Ab}^2}=0 \quad (6.71)$$

The Kelvin-Helmholtz modes are magnetosonic waves which are partially trapped in the beam cavity. The growth rates of the modes depend on the ability of the system to radiate these waves (*i.e.*, on the impedance match between the beam-fluid and the ambient medium).

Ferrari *et al.*'s (1981) treatment is more general than that given above, since it includes the effects of relativistic beam motion. Another version of (6.71), but incorporating the effects of uniform beam rotation, was derived by Bodo *et al.* (1989). A further refinement was made by Fiedler & Jones (1984), who considered a beam with both axial and azimuthal fields, bounded by a vortex sheet, and with no gradients in density or velocity. The internal and external azimuthal fields were chosen to be appropriate for a beam carrying a uniform current density. For this case, no simple analytical dispersion relation resembling (6.71) can be derived, and the properties of the unstable modes must be derived numerically. A further modification of the theory by Trussoni *et al.* (1988) considers the effect of allowing the fluid pressure to be anisotropic, with independent components parallel and perpendicular to an axial magnetic field inside the beam. No qualitatively-new features appear for fast beams, but flows with low Mach numbers tend to become unstable.

Dispersion relations (6.58) and (6.71) which describe the first-order stability of vortex-sheet bounded fluid and (axial-field) magnetic beams are adequate for most of the discussion in this chapter, but the differences arising in more general flows (such as those calculated by Fiedler & Jones 1984 for magnetic flows, or Ray 1982 for sheared flows) will be noted where appropriate. It should be clear that only very restricted beam models are amenable to this type of stability analysis. General beam flows are too complicated for analytical dispersion relations (which allow extensive mathematical analysis) to be derived, but numerical analyses do not allow such exhaustive investigations of the stability of a flow.

6.5.3 Describing the instability

The analysis of a dispersion relation such as (6.58) or (6.71) is performed in terms of parameters which describe the properties of the modes that are supported: λ_0, the wavelength of a mode, τ_e, the time scale on which that mode grows by a factor of e, and λ_e, the distance down the beam on which that mode grows by a factor e. τ_e is likely to be referred to the frame of the beam, where the rate of growth of the instability may also be the time scale on which energy is fed into a turbulent spectrum of magnetohydrodynamical waves (*e.g.*, Ferrari *et al.* 1979). λ_0 and λ_e are more informative in the frame of the ambient medium (the frame of the observer), since they describe the large-scale configuration taken up by the beam under the influence of the Kelvin-Helmholtz instability (this constitutes an adoption of the spatial viewpoint, see §6.3.4).

The growth length of a perturbation of a particular angular frequency is found by solving the dispersion relation for complex k as a function of real ω. The wavelength and growth length of instabilities are

$$\lambda_0 = \frac{2\pi}{\mathrm{Re}(k)} \tag{6.72}$$

$$\lambda_e = -\frac{1}{\mathrm{Im}(k)} \tag{6.73}$$

and the pressure perturbation grows with distance down the beam, z, as

$$|P^{(1)}| = |P^{(1)}(r)e^{i(kz+n\theta-\omega t)}| \tag{6.74}$$

$$\propto e^{z/\lambda_e}. \tag{6.75}$$

The operation of an instability is generally said to be that the fastest-growing mode (of wavelength λ_0^* and growth length λ_e^*) comes to dominate the flow after a short incubation period (of length a few λ_e^*). Thus the dispersion relation is analyzed to search for the values of ω and n that produce the minimum value for λ_e for a beam, when the stability of the beam is described by λ_e^*, and the shape to which the beam deforms during the linear phase of the instability is described by n, λ_0^*, and λ_e^*.

The results of a spatial analysis of (6.58) have been described by Birkinshaw (1984), and curves of $\lambda_e(\lambda_0)$ for several beams are shown in Fig. 6.3. These instability curves are drawn for azimuthal mode numbers $n = 0$, 1, and 2 (6.58 indicates that positive and negative n modes are degenerate), and it can be seen that for each value of n the curve may show many branches. These branches have been labelled by N, the radial mode number, which represents the number of nulls in the pressure perturbation in $0 < r < R$. $N = 0$ modes are the 'Ordinary Modes' and $N > 0$ modes are the 'Reflection Modes' of Gill (1965) and Ferrari *et al* (1981). Reflection modes appear only for sufficiently large beam velocity (if $v_b > c_{Sa} + c_{Sb}$; Payne & Cohn 1985); at smaller v_b only the ordinary mode is

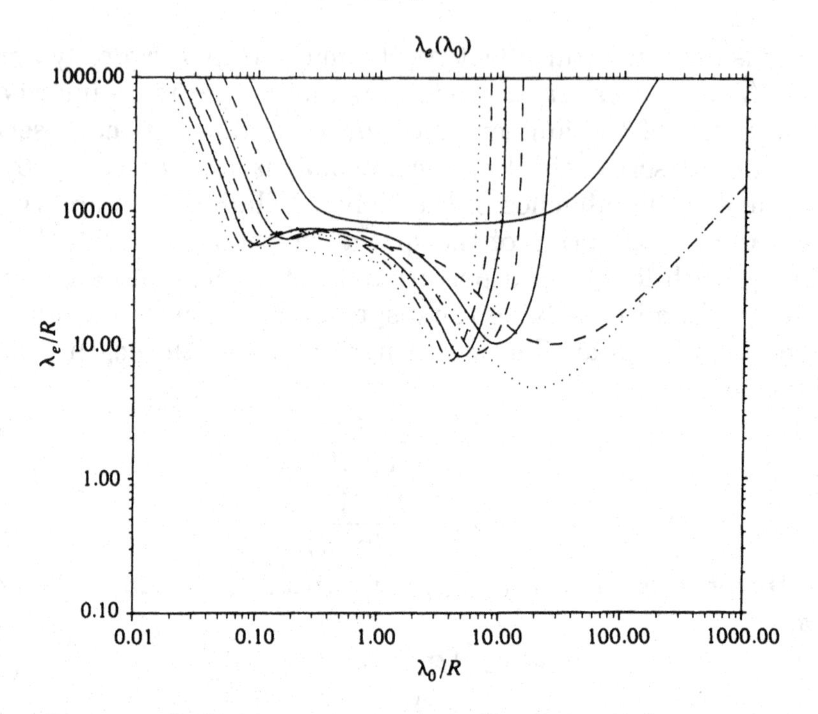

Fig. 6.3. Spatial instability growth length, λ_e, plotted against wavelength, λ_0, for several field-free beams bounded by vortex layers. λ_e and λ_0 are normalised by the radius of the beam, R. 'Pinching' ($n = 0$) modes are drawn as solid lines, 'helical' ($n = 1$) modes are dashed lines, and the first 'fluting' ($n = 2$) modes are dotted lines. The ordinary ($N = 0$) modes are those present at $\lambda_0 \to \infty$, and the reflection ($N > 0$) modes appear at regular intervals of $\lambda_0 < \lambda_0^{\rm crit}(N)$. Only the first two reflection modes ($N = 1, 2$) are shown for each n. Note that the beams (d), (e), and (f) display only ordinary modes. All figures are drawn on the same scale to illustrate the relative stability of the beams. Only positive n need be considered since modes with $n = \pm |n|$ are degenerate for these beams.
a. $v_b = 3000$ km s^{-1}, $c_{Sa} = 300$ km s^{-1}, $\rho_b = \rho_a$ ($\mathcal{M}_a = \mathcal{M}_b = 10$).

present. Analytical approximations for the values of λ_0^* and λ_e^* are given by Hardee (1987a, b) for low-n and low-N modes, and by Zaninetti (1986a, b) for short-wavelength modes in magnetic beams.

For beams which support reflection modes, Fig. 6.3 shows that many modes can be excited at sufficiently small wavelength, λ_0. For example, at $\lambda_0 = 10R$ in a beam with Mach number $\mathcal{M}_b = 10$ and $\rho_a = \rho_b$ (Fig. 6.3a), modes with $n = 0$, $N = 0, 1, 2$, with $n = 1$,

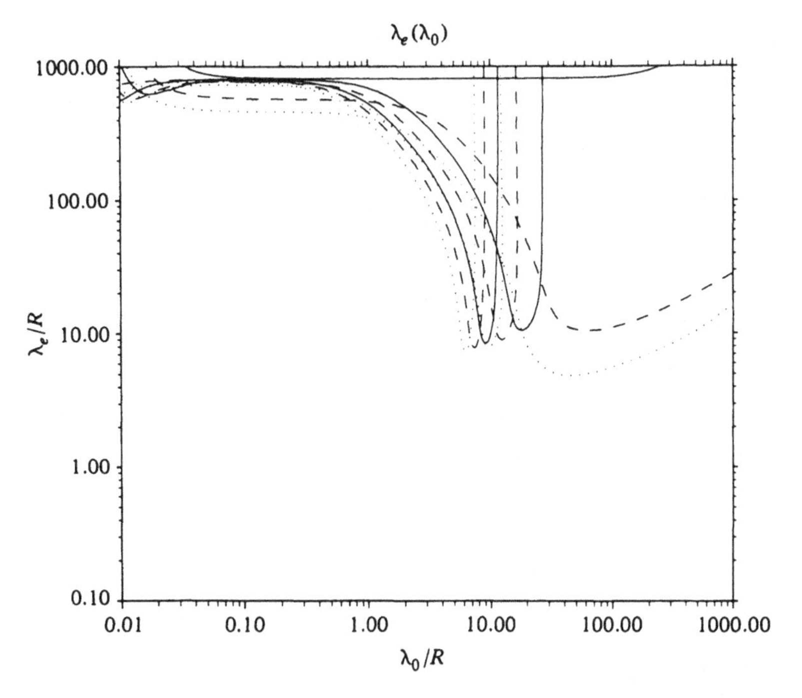

Fig. 6.3b. As Fig. 6.3a, with $v_b = 3000\ \mathrm{km\,s^{-1}}$, $c_{Sa} = 30\ \mathrm{km\,s^{-1}}$, $\rho_b = 10^{-2}\rho_a$ ($\mathcal{M}_a = 100$, $\mathcal{M}_b = 10$).

$N = 0, 1$, and with $n = 2$, $N = 0, 1$ are excited (as well as all modes with $n > 2$ and $N = 0$). If only modes of this wavelength and $n < 3$ are accessible to the beam then the $n = 2$, $N = 0$ mode, which has the smallest growth length, $\lambda_e \approx 7R$, would dominate the structure of the beam and should be visible to observers. Since there is no reason to suspect that $\lambda_0 = 10R$ is the only wavelength that is excited on the beam, Fig. 6.3a suggests that the fastest-growing mode tends to be that of smallest wavelength and largest n (the cylindrical analogue of the result obtained for a planar shear layer by Turland & Scheuer 1976). Thus vortex layer theory does not predict an unambiguous mode which dominates the instability of a beam – rather we must appeal to other physical principles to set the ranges of ω and n which are accessible to instabilities, and then locate the values of λ_0^* and λ_e^* given this range. However, it can be seen that the ordinary ($N = 0$) mode has a smaller growth length than an increasing number of reflection ($N > 0$) modes as n increases. Hence high-n and N modes are unlikely to have a signifi-

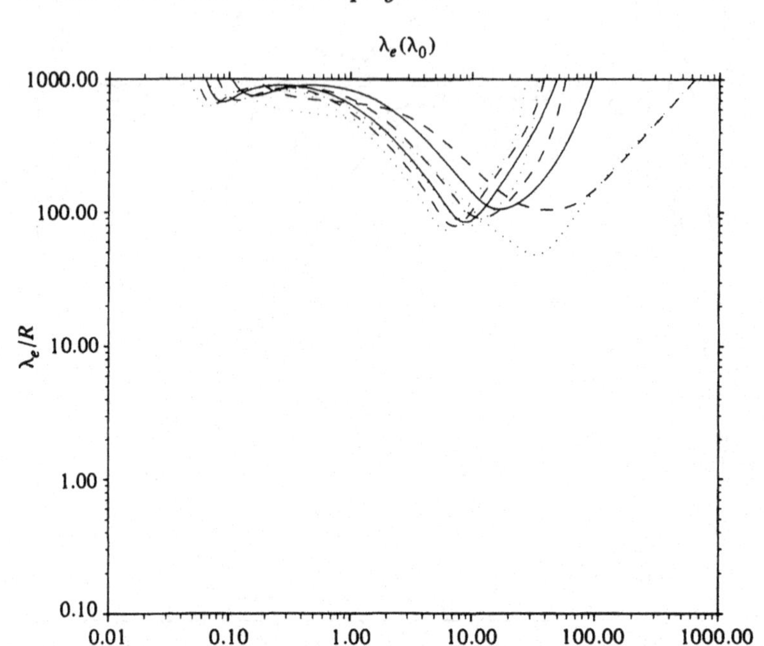

Fig. 6.3c. As Fig. 6.3a, with $v_b = 3000$ km s^{-1}, $c_{Sa} = 300$ km s^{-1}, $\rho_b = 10^2 \rho_a$ ($\mathcal{M}_a = 10$, $\mathcal{M}_b = 100$).

cant individual effect on a beam, although their effective *continuum* at small wavelengths may cause a smooth beam profile to develop by broadening the vortex layer to create a hot, sheared, sheathing layer that will inhibit the further growth of small-λ_0 instabilities.

Fig. 6.3 demonstrates that the spatial instability of low-density beams is greater than that of high-density beams with the same velocity. This result is different from the result for the temporal instability of beams: Ferrari *et al.* (1981) found that the temporal growth of a Kelvin-Helmholtz instability is fastest if the beam and ambient medium densities are equal. The difference presumably arises from the strong dependence of the mode velocity on the properties of both the beam and the ambient medium. If the beam is dense, then the mode speed is close to the beam speed. If the beam is light, then the inertia of the ambient medium has a significant effect on the mode speed. Thus although the growth times increase for light beams, the growth lengths decrease because the mode velocity decreases.

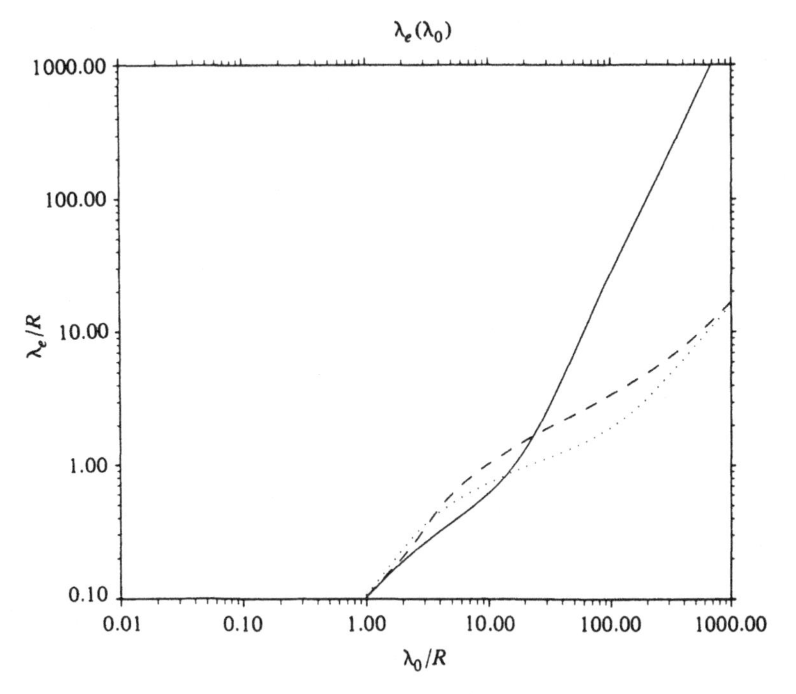

Fig. 6.3d. As Fig. 6.3a, with $v_b = 3000\ \mathrm{km\,s^{-1}}$, $c_{Sa} = 300\ \mathrm{km\,s^{-1}}$, $\rho_b = 10^{-2}\rho_a$ ($\mathcal{M}_a = 10$, $\mathcal{M}_b = 1$).

The overall stability of a beam is increased as the Mach number $\mathcal{M}_b$ is increased, but the growth lengths of pinching ($n = 0$) modes increase faster then the growth lengths of helical ($n = 1$) modes (Ray 1981). This suggests that beams tend to display knots at small Mach numbers and tend to twist at high Mach numbers where the flows are more stable. Relativistic effects do not have a strong influence on the stability of the beam for moderate γ_b (Turland & Scheuer 1976).

If the density of a beam is increased whilst the kinetic energy flux and the internal pressure are held constant, the velocity, v_b, and the internal sound speed, c_b, of the flow decrease. The combined effect of these changes is that the growth lengths λ_e^* of the modes increase slightly, and that the wavelengths λ_0^* increase faster (if $\rho_b > \rho_a$) as functions of ρ_b. An exception arises for the $n = 0$, $N = 0$ mode, which is destabilised. If the density of the beam is increased whilst the momentum flux is held constant, only small changes in the values of λ_e^* and λ_0^* result, except that the $n = 0$, $N = 0$ mode

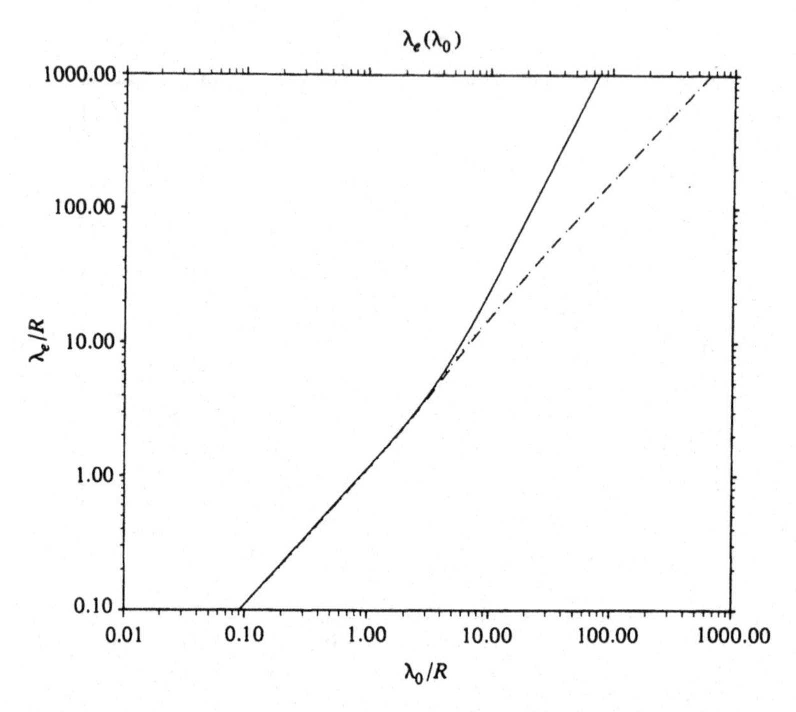

Fig. 6.3e. As Fig. 6.3a, with $v_b = 3000\ \mathrm{km\,s^{-1}}$, $c_{Sa} = 3000\ \mathrm{km\,s^{-1}}$, $\rho_b = 10^2\rho_a$ ($\mathcal{M}_a = 1$, $\mathcal{M}_b = 10$).

is destabilised again. A sufficiently large increase or decrease of ρ_b subject to either a fixed kinetic energy flux or momentum flux may cause v_b to decrease below $c_{Sa} + c_{Sb}$, so that the reflection modes no longer appear.

The pressure perturbations $P^{(1)}(r)$ for the $n = 0$, $N = 0$ and $n = 1$, $N = 2$ modes of Fig. 6.3a near maximum instability are illustrated in Fig. 6.4. These figures illustrate some general properties of the solutions (6.55 – 6.56). The pressure perturbation has an appreciable amplitude over a significant fraction of the beam radius, although $P^{(1)}$ tends to zero as $r \to 0$ for $n > 0$. Modes with $N > 0$ exhibit N pressure nulls within the beam. The ordinary mode ($N = 0$) is a consequence of the existence of any boundary between a flowing and a static region of a fluid – it arises for planar flows (*e.g.*, Blake 1972; Turland & Scheuer 1976) as well as in other geometries, and is unstable for all wavelengths. Reflection ($N > 0$) modes, on the other hand, require the fluid to contain more than a single planar surface, and are particularly strong where the

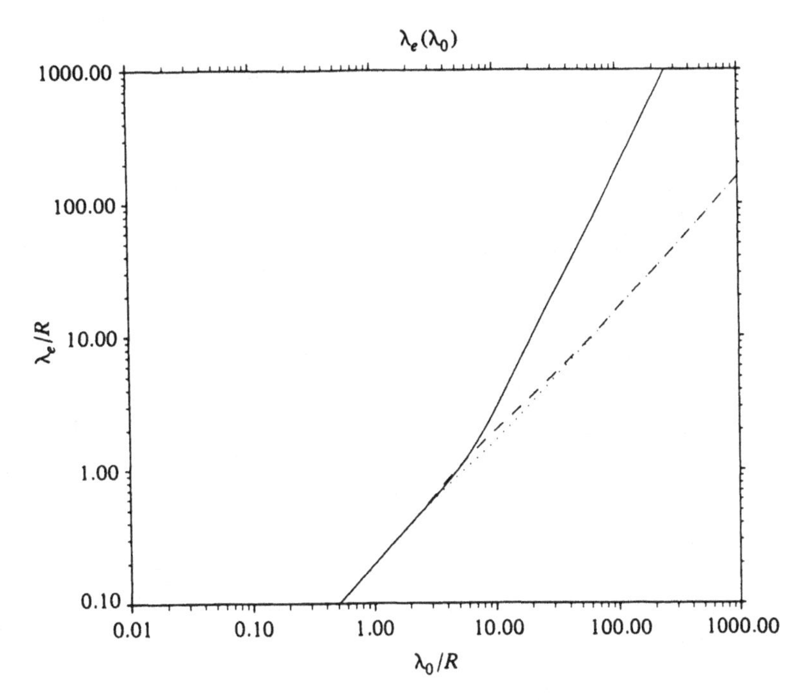

Fig. 6.3f. As Fig. 6.3a, with $v_b = 300\ \mathrm{km\,s^{-1}}$, $c_{Sa} = 300\ \mathrm{km\,s^{-1}}$, $\rho_b = \rho_a$ ($\mathcal{M}_a = \mathcal{M}_b = 1$).

fluid contains an enclosed region whose walls can vibrate coherently (*e.g.*, the beam-cavity). In that case, instabilities occur as sound waves radiated from one surface are reflected back from the other surface (with some phase delay) to interfere constructively. Thus the beam acts as an acoustic waveguide: standing waves across the waveguide allow co-operative modes of the walls to be excited, and growing perturbations can be produced if the reflection coefficient for a wave striking a wall is greater than unity (over-reflection; Payne & Cohn 1985; Bodo *et al.* 1989). The appearance of reflection modes at regular intervals of λ_0 in Fig. 6.3 can be interpreted in terms of this frequency-dependent coupling of internal modes of the beam to radiated magnetosonic waves (Bodo *et al.* 1989). A regular pattern of appearance of such modes as λ_0 is decreased corresponds to the 'fitting in' of increasing numbers of nulls of the pressure perturbation within the beam.

The effect of a velocity shear layer on the ordinary mode has been discussed by Ray (1982) for a two-boundary planar (moving

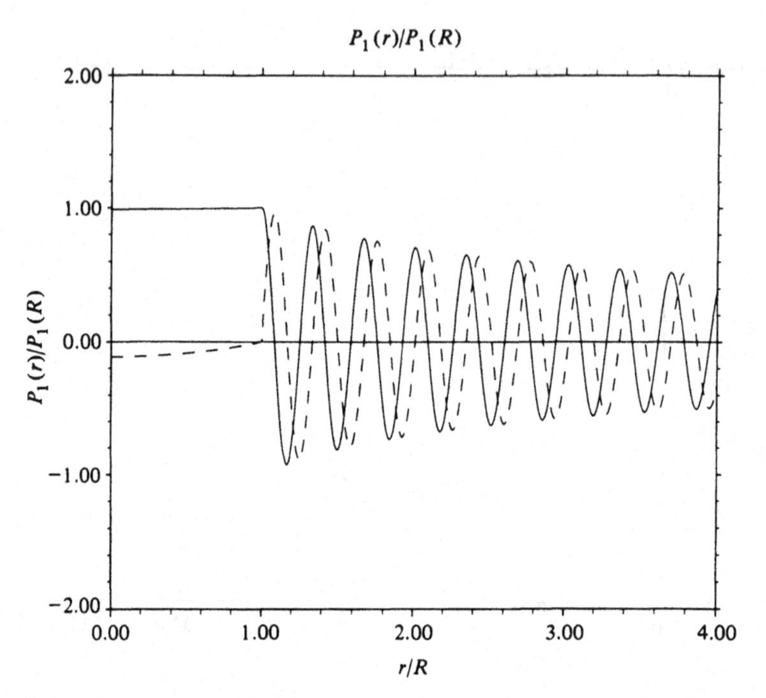

Fig. 6.4a. Representative solutions for the pressure perturbations as a function of radius at maximum instability $(\lambda_0^*, \lambda_e^*)$ for the beam with $v_b = 3000\ \mathrm{km\,s^{-1}}$, $c_{Sa} = 300\ \mathrm{km\,s^{-1}}$, and $\rho_b = \rho_a$ $(\mathcal{M}_a = \mathcal{M}_b = 10)$ whose modes are shown in Fig. 6.3a. The real and imaginary parts of the eigenfunctions $P^{(1)}$ are indicated by solid and dashed curves, and the solutions are normalised to their value at $r = R$. The mode is $n = 0$, $N = 0$.

slab) geometry, and for a somewhat different velocity profile by Ferrari *et al.* (1982). Further investigations have been made by Roy Choudhury & Lovelace (1984). Earlier work on the structures of instabilities in shear layers has been done for laboratory jets by Blumen *et al.* (1975) and Drazin & Davey (1977). The results of Ferrari *et al.* (1982) suggest that the fastest instabilities arise for wavelengths similar to the slab width (or the beam radius), and that the reflection modes cease to be important at wavelengths smaller than the thickness of the shear layer. They find that the sharper the velocity shear, the larger the growth rate for all wavelengths, and that the effect of density gradients is much smaller than that of velocity gradients. Finally, they find that it is possible to suppress the instability at short wavelengths by choosing an ar-

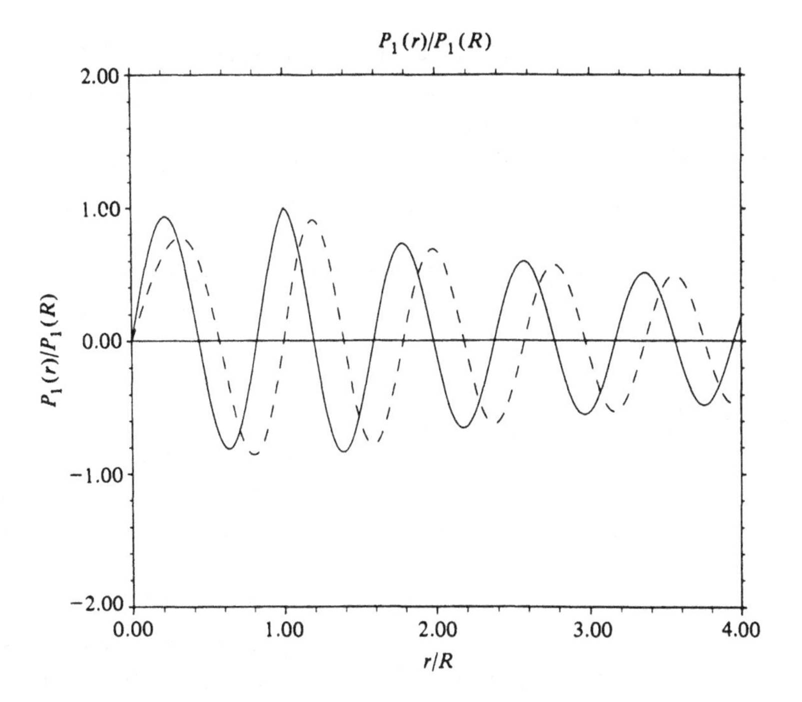

Fig. 6.4b. As Fig. 6.4a, for the mode $n = 1$, $N = 2$.

bitrarily smooth velocity shear. Hardee (1982a, 1983, 1984) found that a sheared beam is further stabilised by its expansion.

The stability of a cylindrical beam bounded by linear shear layers has been investigated by Birkinshaw (1990). As the dimensionless thickness of the shear layer, δ (6.4), is increased, the ordinary modes are slightly stabilized at long wavelengths. For wavelengths near the beam radius, R, and at moderate δ, mode mixing causes slightly *greater* ordinary mode instability than in a vortex layer. The reflection modes are stabilised by the effects of the shear, and the wavelengths at which they appear, $\lambda_0^{\mathrm{crit}}$, change slightly. Large values of δ cause increasing stabilisation of the flow, and the reflection modes become less important as the smoothness of the velocity profile increases. The increases in the growth lengths for ordinary modes are less than for reflection modes. These results confirm those of Ferrari *et al.* (1982), and suggest that a shear layer effectively provides a wavelength cutoff below which no Kelvin-Helmholtz modes appear, and that at $\lambda_0 \gg R$ the shear layer has little effect on the stability of a beam (as for laboratory jets). The calculations performed so far have considered only shear layers in which the velocity or density are monotonic functions of radius. This may not be appropriate, since numerical calculations (Chapter 7) suggest that beams are surrounded by thin, hot sheathing layers which may be developed by the growth of short-λ_0 instabilities or backflow from the head of the beam.

An interesting analysis of the instabilities of sheared slab flows has been given by Glatzel (1988), who finds a close analogy between Kelvin-Helmholtz modes and the instabilities in accretion tori, and that the density structure of the shear layer has a strong, qualitative, effect on the modes of instability.

Bodo *et al.* (1989) found that uniform rotation of a beam permits additional destabilising $n = 0$ modes for low Mach number flows, and that these modes remain active even when the magnetic field suppresses other $n = 0$ modes. Rotation does not greatly affect the stability properties of the modes of Fig. 6.3, however. Since modes with $n \neq 0$ may be described as rotating, it is to be expected that the analysis of these modes for rotating beams will show stronger effects on the pattern of instability, with several branches of these modes appearing and the degeneracy of modes with positive and negative n being broken.

Magnetic effects for Kelvin-Helmholtz instabilities have been discussed by Cohn (1983), Ray (1981), Fiedler & Jones (1984), Ferrari *et al.* (1980, 1981) and by Benford (1981) for a cold beam (with no internal dynamics). Bodo *et al.* (1989) have discussed the stability of a rotating beam to the pinching ($n = 0$) mode. Roy Choudhury & Lovelace (1986) and Ray & Ershkovich (1983) have discussed the combined effects of magnetic field and shear on a planar flow. The results of these investigations are that all modes are stable for sub-Alfvénic flows ($v_b < c_{Ab}$) and axial magnetic fields (Ray 1981; Ferrari *et al.* 1981; Fiedler & Jones 1984). $n = 1$ modes stabilise at a lower axial field ($c_{Ab} > 0.5v_b$). By contrast, azimuthal fields are destabilising (as shown for the $n = 1$ modes by Ferrari *et al.* 1981; for the $n = 0$ and 1 modes by Fiedler & Jones 1984; and for $n = 0$ modes on magnetically-confined beams by Cohn 1983). In all cases, the nature of the instability is similar to that of a non-magnetic flow, and the rates of growth of instabilities are not much different from those of non-magnetic flows for axial fields less than the critical values for stabilisation.

None of the above discussions of the stability of magnetic beams identified the presence of slow magnetosonic reflection modes in the dispersion relation, but discussed only the ordinary modes and fast magnetosonic reflection modes. Bodo *et al.* (1989) have described the importance of the slow magnetosonic modes. Only ordinary modes are present at low (subsonic) beam speeds, the slow magnetosonic reflection modes dominate the stability next, and at high Mach numbers it is the fast magnetosonic reflection modes that determine the stability of the flow.

The effect of Kelvin-Helmholtz modes on the *survival* of the beam cannot be predicted with certainty from the simple linear instability results – since small-λ_0 modes are the most unstable it is likely that the structure of the beam will adjust to suppress these modes before growth of the longer (λ_0, λ_e) modes (which may produce observable structures) can occur. The rapid adjustment of the structure of the beam to small-wavelength modes may invalidate the assumed z-independence of the flow used to derive the dispersion relations (6.58) and (6.71). Where longer-wavelength modes occur, they should grow on a severely modified beam structure. Numerical work, which suppresses the shortest-scale structures by averaging over the resolution element of the code, is not presently useful in predicting the outcome of the operation of the full spectrum of Kelvin-Helmholtz instabilities. However, the results of simple shear-layer calculations tend to suggest that the modification of the structure of the beam can be ignored in predicting the instability at long wavelengths. If we assume that the shorter-wavelength modes saturate by establishing a thin shear layer, then the shear-layer calculations of Ray (1981), Ferrari *et al.* (1982) and Birkinshaw (1990) suggest that the dominant instabilities will always be those with wavelengths comparable with the width of the shear layer. Since the width of the shear layer that is produced is not known, it is not possible to predict which wavelengths will dominate the stability of a given beam. The best use of the theory in its present state is, therefore, to compare the wavelengths and growth lengths for modes which are inferred from maps of sources with the values that are predicted using jet parameters determined in other ways, as a consistency check.

The linear calculations described above contain no information on the limits to the amplitudes of instability. The method that was used to derive the dispersion relations (6.58) and (6.71) can be extended to second order, by writing the pressure (for example) as

$$P = P^{(0)}(\mathbf{r}, t) + P^{(1)}(\mathbf{r}, t) + P^{(2)}(\mathbf{r}, t), \qquad (6.76)$$

and carrying the expansion of (6.18 – 6.32) to second order to obtain equations linear in $P^{(2)}$ (but quadratic in $P^{(1)}$), and then the behaviour of $P^{(2)}$ might yield limits to the growth of the instability. The analytical difficulties in performing such a calculation are severe, and no direct analytical discussion of the limits to growth has been made. Benford (1981) argues that the growth of the perturbations will cease when the velocity excursions produced by the Kelvin-Helmholtz modes become sonic and develop shocks – at $|\mathbf{v}^{(1)}| = c_S$. This may be regarded as a strong upper limit on the linear phase of growth, but it is possible that the (weak) shocks produced in this way might intensify under further non-linear growth, and further development of the theory is necessary to discover whether there are conditions under which instability growth terminates earlier.

An alternative approach to the determination of the end-points of the operation of the Kelvin-Helmholtz instability is to use numerical simulations of beams (*e.g.*, Nepveu 1982a, b; Norman *et al.* 1982, 1983, 1984; Williams & Gull 1984; Chapter 7) to discover the results of the growth of the instability in axisymmetrical cases, and then to extend these results (by inspired guesswork) to non-axisymmetrical instabilities (which are stronger, according to linear theory). Norman *et al.* (1984) have examined their results in the light of analytical instability theory, and conclude that pinch reflection modes (with $n = 0$ and $N > 0$) do not have a significant effect in disrupting beams since their non-linear growth saturates with a series of weak conical shocks which cause alternate divergence and refocussing of the flow (that is, the growth of the pinch reflection modes terminates by Benford's (1981) sonic criterion). Recent calculations on slab beams (which can model the $n = 0$ and $n = 1$ modes of cylindrical beams) by Hardee & Norman (1988) and Norman & Hardee (1988) indicate that the reflection modes saturate with the formation of weak oblique shocks, whilst the ordinary modes are unsaturated and can cause large-scale distortions and disruption of the flow. Since the ordinary pinching mode is only important for trans-sonic flows, this suggests that the ordinary helical mode may be the most important long-wavelength instability channel for supersonic beams. The numerical simulations found growth lengths $\lambda_e^* = (2-4)\mathcal{M}_b R$ and wavelengths $\lambda_0^* \approx 5R$ for Mach numbers $\mathcal{M}_b = 3$ to 6, in fair agreement with linear instability theory (Fig. 6.3).

It is easy to see from Fig. 6.3 that low-density beams are expected to make a rapid transition to turbulence since their short-wavelength modes have growth lengths less than R. Higher sound-speeds in the ambient medium and the beam ameliorate this situation as far as the low-order reflection modes are concerned, but the higher-order modes will still have small values of λ_e^*. The transition to turbulence can only be prevented if the beam is well protected from the influence of modes with wavelengths less than a few R – such protection might be provided by a hot sheathing layer for which waves with $\lambda_0 < \delta R$ will propagate poorly. Beams which are more dense than the ambient medium are rather stable and only slightly affected by short-wavelength modes (until the frequency of the driving perturbation becomes very high – that is, until the wavelength of the driving mode becomes much less than the beam radius). For simple beam structures, therefore, the single clear prediction of Kelvin-Helmholtz theory is that light beams will make a fast transition from laminar to turbulent flow, whilst denser beams can remain laminar longer.

6.6 The Kelvin-Helmholtz model for structures in jets

The growth of the modes described above may lead to observable distortions on a beam if the wavelengths and growth lengths are not severely modified in their transition from the linear to the non-linear phase of growth, and if the beams are not first disrupted by the influence of the shortest wavelengths of instability. In this section, the results of §6.5 will be applied to models of some well-known jets to calculate their stability. The observed jet distortions and these predictions will be tested for consistency, and some general conclusions about the applicability of the Kelvin-Helmholtz model for jet structures will be reached.

6.6.1 The structure of a distorted beam

The linear theory of §6.5 suggests that instabilities with a range of wavelengths will grow on a beam unless the beam contains an axial magnetic field such that $c_{Ab} > v_b$, or

$$B_{zb} > (n/\mathrm{m}^{-3})^{1/2}\,(v_b/c) \quad \mathrm{nT}. \tag{6.77}$$

At lower (axial or otherwise) field strengths the instabilities will not be significantly different from those of a fluid with no magnetic field. In what follows, a low-field flow will be considered unless explicitly stated otherwise, so that dispersion relation (6.58) is relevant, and only positive azimuthal mode numbers, n, need to be considered.

Fig. 6.3 illustrates the results that the operation of a Kelvin-Helmholtz instability will be dominated by the mode with the smallest λ_0 and the largest n that is presumed to be active, and that the growth lengths of the instabilities are small unless the beam is supersonic both internally and externally (*i.e.*, $v_b > c_{Sa}$ and $v_b > c_{Sb}$). Only the ordinary mode will be active unless $v_b > c_{Sa} + c_{Sb}$ (Payne & Cohn 1985). Thus the condition that the instability growth lengths $\lambda_e > R$ for wavelengths $\lambda_0 > R$ is just the condition that the beam is supersonic, and hence that its stability is dominated by reflection modes at short wavelengths and ordinary modes at long wavelengths (as in Fig. 6.3a). It may be concluded that if the observed structures in radio jets are caused by the Kelvin-Helmholtz instability, then the beams are supersonic both internally and externally, and that the influence of the shortest-wavelength modes on the stability of the beam has been slight.

Since the result of the operation of Kelvin-Helmholtz modes, as observed in the laboratory, is the transition to turbulence of the beam flow, it is logical to interpret the greater or lesser instability of astrophysical beams as an indication of their susceptibility to this transition. Ferrari *et al.* (1979) suggested that the radio-bright jets that are common in low-power radio galaxies and high-power quasars are indicative of turbulent particle acceleration occurring in beams which are very unstable, and that the 'invisible' jets of the higher-power radio galaxies should be interpreted as beams which are more stable, and hence have not developed strong turbulence and its consequent particle acceleration. Since light beams are more unstable than dense beams, it might be suggested that this tendency to bright radio emission might also be taken as a tracer of beam density, but there is no independent evidence for this hypothesis.

The wavelength of fastest growth is always set, essentially, by the minimum allowed wavelength and the maximum allowed azimuthal mode number, so that a critical parameter in the stability of a beam is the value of that minimum wavelength (or maximum frequency). In the absence of direct evidence for the consequences of the operation of small-λ_0 modes, it will also be assumed that these modes operate to cause Kelvin-Helmholtz heating of the beam fluid, which results only in the creation of a self-consistent density, velocity, and temperature profile across the beam (§6.5.3). The predictive power of Kelvin-Helmholtz calculations is poor, and the theory will be applied here merely to check that the parameters suggested for a particular jet are consistent with the observed structure arising from one of the Kelvin-Helmholtz modes. In this model, knots are interpreted as the non-linear limits of $n = 0$ modes, and twists and bends are ascribed to $n = 1$ modes. No morphological counterparts to higher-n modes are required ($n = 2$ modes would cause beam bifurcation, $n = 3$ modes would cause trifurcation, etc.).

It can be seen from Fig. 6.4 that both the ordinary and the reflection modes involve the transmission of wave energy across the diameter of the beam, and hence that the pressure perturbations do not decrease rapidly away from the beam boundary (as noted by Ferrari *et al.* 1981). If the amplitude of the pressure perturbation is taken to reflect the radio brightness of a beam, then we would deduce that beams affected by Kelvin-Helmholtz instabilities should not display appreciable limb brightening, but only low-contrast radial structures. The contrast may be further reduced by the diffusion of relativistic particles across the beam. Observationally, there is little evidence for the presence of limb-brightening, except in circumstances (like the first knot in the Cen A radio jet; Clarke *et al.* 1986) where it is clear that complicated flow phenomena are occurring. Although the result that distorted beams show no limb-brightening is in accord with the predictions of the Kelvin-Helmholtz model, this is also a feature of other models for beam-bending, and cannot be taken as support for a Kelvin-Helmholtz origin for bent jets. It should be stressed that the effects of projection can cause complicated patterns to be seen from simple beam structures (Zaninetti & Van Horn 1988), so that the relative orientation of the beam and the line of sight is of critical importance in interpreting the observed jet structures.

6.6.2 **3C 449**

The archetypal distorted radio jet with which most comparisons are made is 3C 449 (see Chapter 2, Perley *et al.* 1979, and Birkinshaw *et al.* 1981). Hardee (1981) has interpreted the distortions of this two-sided jet in terms of an $n = 1$ Kelvin-Helmholtz mode with $\lambda_0 \approx 16R$ and $\lambda_e \approx 6R$. Hardee's fit of this structure as a growing $n = 1$ mode (on the basis of a temporal instability calculation) involved a beam flow with Mach number $\mathcal{M}_b = 10$, equal density in the beam and the ambient medium, and an ambient medium sound speed $c_{Sa} = 150 \text{ km s}^{-1}$ $(= c_{Sb})$. The density in the beam fluid was taken as $n \sim 2 \times 10^4 \text{ m}^{-3}$ on the basis of depolarization data (but note that this is likely to be a significant over-estimate: Laing 1984, Chapter 3). The observed structures are strongly non-linear ($\xi_r^{(1)} > R$) in this interpretation. Equation (6.77) indicates that the flow in 3C 449 will be strongly stabilised if $B_{zb} > 100$ nT. The equipartition pressure of the fields in the jet is about 2×10^{-13} Pa, so that the equipartition field is about 1 nT and the magnetic field should not be dynamically important.

With these parameters for the beam in 3C 449, Birkinshaw (1984) demonstrated that a spatial analysis of (6.58) leads to significantly different results than Hardee's temporal calculation, and found three $n = 2$ modes with $\lambda_0 < 2R$ which had shorter growth lengths than the $n = 1$ mode that Hardee interpreted as the cause of the structure of 3C 449. Several $n = 1$ reflection modes also are more unstable than the mode Hardee interpreted as responsible for the source configuration. On the basis of these results, it can be concluded that unless a mechanism operates that forces $\lambda_0 > 10R$ and $n < 2$, it is not likely that the structure of 3C 449 can be generated by a helical Kelvin-Helmholtz mode. Small changes in $\mathcal{M}_b$ and ρ_b/ρ_a match the wavelength and growth length of the $n = 1$, $N = 0$ mode to the wavelength and growth length fitted by Hardee (1981). Since the aspect of 3C 449 on the sky is unknown, fits of λ_0 and λ_e are intrinsically uncertain, and all models of the beam that reproduce the observables λ_e/λ_0 and a minimum value of λ_0/R are acceptable. Since it is not known which (if any) of the $n = 1$ modes is responsible for the structure of 3C 449, a variety of reflection modes may be fitted to these values of λ_e/λ_0 and λ_0/R, and a wide range of physical parameters are consistent with a Kelvin-Helmholtz model.

A simple argument against the interpretation of the structure of 3C 449 as a Kelvin-Helmholtz instability is the strong two-sided symmetry of the jets. Unless the jet perturbation is applied to the beam-generator and the conditions for instability growth are closely similar on the two sides of the source, an essentially stochastic instability is unlikely to achieve such symmetry. Further, since only a few growth lengths of the instability have occurred from the centre of the source, it is difficult to see how the structure can have grown to non-linear amplitude without a perturbation of non-linear amplitude having been applied.

To the extent that a single mode dominates the structure of 3C 449, the purity of that mode and the crispness of the structure that it produces depends on the sharpness of its $\lambda_e(\lambda_0)$ minimum (Birkinshaw 1984). At a given distance z from the inception of a given instability, modes with λ_e within about λ_e^{*2}/z of λ_e^* have amplified to within a factor $1/e$ of the maximum growth rate. For some modes (such as the $n = 0$, $N = 0$ mode in Fig. 6.3a) the corresponding range of wavelengths will be large until z is many λ_0, so that only chaotic distortions of the radio jet should arise unless strong mode coupling operates.

There are a variety of alternative explanations for the bending of 3C 449 (see also Chapter 5). Blandford & Icke (1978) have noted that the 'C-type' symmetry of 3C 449 might result from the orbital motion of UGC 12066 (the host galaxy of the radio source) in its group of galaxies if the beam of 3C 449 is slow and heavy. This model may not be consistent with the low radio depolarization of the jets. A similar model involves orbital motion of the beam-generator about a massive body in the nucleus of UGC 12066 – this permits a faster flow, since the orbital motion is faster, and may alleviate the difficulty with the density of the beam. Possible interactions between the beam and a moving ambient medium have also been discussed: if the beams of 3C 449 are subsonic (and turbulent), as suggested by the appearance of the jets, then they may be strongly affected by buoyancy forces, interactions with galactic wakes and other inhomogeneities in the intergalactic medium, and turbulent entrainment of the external medium into the flow.

6.6.3 M 87

Hardee (1982b) has also applied Kelvin-Helmholtz instabil-

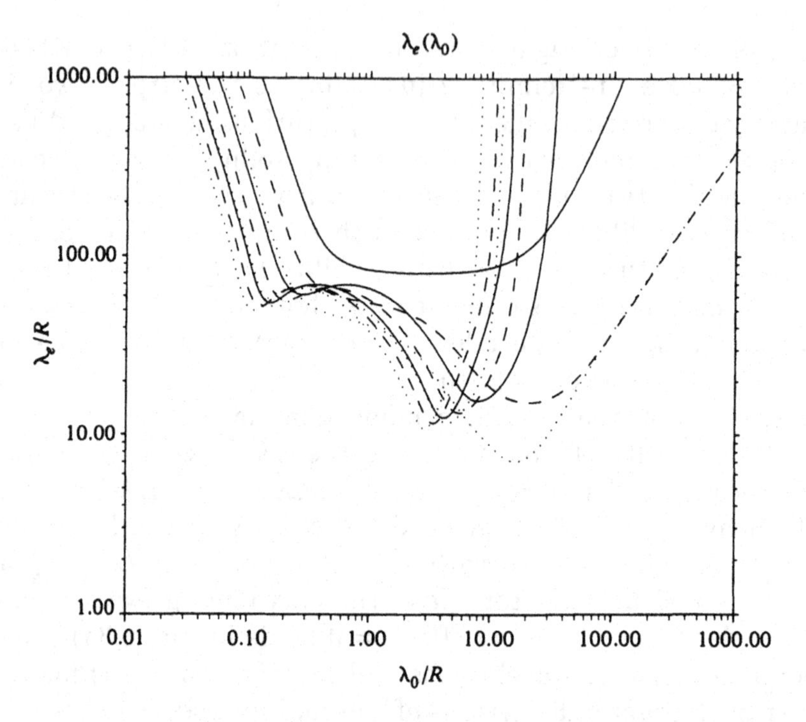

Fig. 6.5. Stability curves $\lambda_e(\lambda_0)$ for the model of M 87 described in §6.6.3. Only modes with $n = 0$ (solid lines), 1 (dashed line), and 2 (dotted lines) and $N = 0$, 1, and 2 are illustrated.

ity theory to describe the structure of the radio jet of M 87 (mapped by Biretta *et al.* 1983, cf. Chapter 7) – the optical knots were first interpreted in terms of Kelvin-Helmholtz modes by Stewart (1971). Hardee fitted the deviations of the centroid of the radio emission as a growing instability of wavelength $\lambda_0 \approx 13R$ and growth length $\lambda_e \approx 10R$, where $R \sim 0.65$ arcsec and the jet has been assumed to lie in the plane of the sky. Using approximate results from the temporal analysis of (6.58), this fit implies that the beam fluid is about 6 times denser than the external medium, and that the Mach number of the flow is about 15 (Hardee 1982b). From the depolarization of knot 4, Hardee deduced that the sound speed in the beam was $c_{Sb} \sim 2000\ \mathrm{km\,s^{-1}}$, which may be used with the Mach number to give a flow speed $v_b \sim 0.1c$.

These flow parameters have been used as the basis for a spatial instability analysis of dispersion relation (6.58) whose results $\lambda_e(\lambda_0)$ are shown in Fig. 6.5. It can be seen that a number of modes appear with $\lambda_0^* \approx 10R$, and it is clear that the instability will, once again, be dominated by modes with the largest n and N and smallest λ_0 that are permitted to disturb the beam. If, for example, attention is restricted to modes with $\lambda_0^* > 10R$, then the most unstable modes are the $n = 1$ and $n = 2$ (and $n = 3$, not shown on Fig. 6.5) ordinary modes. The helical structure of M 87 can consistently be described in terms of the $n = 1$, $N = 0$ mode if either the $n > 1$ modes are strongly suppressed or if (equivalently) the limiting amplitudes of modes with $n \leq 1$ are much larger than modes with $n > 1$.

Hardee (1982b) interprets the helical structure in this way, and notes that the knots in the M 87 jet cannot be produced by pinching ($n = 0$) Kelvin-Helmholtz modes because the growth rate of the pinching mode is much less than that of the helical mode. Fig. 6.5 shows, however, that the $n = 0$, $N = 1$ mode has a very similar growth length to the $n = 1$, $N = 0$ mode, so that helical and pinching Kelvin-Helmholtz modes should occur simultaneously, with $\lambda_0^*(n = 1, N = 0)/\lambda_0^*(n = 0, N = 1) \sim 2.5$ times as many pinches (and knots) as helical waves. This is approximately the ratio observed, and the agreement of the predicted and observed ratio constitutes supporting evidence for a Kelvin-Helmholtz view of the structure of the M 87 jet. Difficulties with this interpretation are caused by the absence of the nearby $n = 1, N = 1$ mode and the other high-n, high-N modes with λ_0^* close to $\lambda_0^*(n = 0, N = 1)$ and $\lambda_0^*(n = 1, N = 0)$ but with smaller λ_e^*. The consistency of the Kelvin-Helmholtz model is only impressive if it can be shown how these modes can be suppressed.

A further difficulty with this prediction is that it has neglected the magnetic field of the jet, but the equipartition field of about 20 nT is only slightly smaller than the field needed to stabilise the Kelvin-Helmholtz modes (30 nT; 6.77), so that magnetic stabilisation may be important if the field is predominantly axial. If the depolarization estimate of the thermal particle density is too high, then fields less than 20 nT may be sufficient to stabilise the Kelvin-Helmholtz modes. On the other hand, much of the field in M 87 may be azimuthal, and hence destabilising, since it is possible that the jet is magnetically confined (Biretta *et al.* 1983).

An alternative to Kelvin-Helmholtz instabilities as the causes of the structure of the M 87 jet is that the properties of the beam-fluid vary in a quasi-periodic fashion at injection (for example, involving changes in the beam speed and direction). Other mechanisms (Chapter 5) might act to generate either the knots or the twists of the M 87 jet, but both knots and twists can be caused by the action of Kelvin-Helmholtz instabilities. The usual problem of the suppression of short-wavelength and high-n modes in favour of longer-wavelength modes remains.

6.6.4 NGC 6251

The spectacular jet of NGC 6251 has been mapped by Waggett *et al.* (1977), Saunders *et al.* (1981), and Perley *et al.* (1984a), (cf. Chapter 7), and appears to support a number of instability-like structures. Within the first 40 arcsec, slowly-growing structures with apparent wavelength 5.7 and 9 arcsec can be identified although it is not known whether these correspond to sideways displacements of the jet (n = 1-like modes) or merely brightness enhancements and pinches (n = 0-like modes). At intermediate angular scales (120 to 240 arcsec), a rapidly-growing 31-arcsec oscillation which produces knots in the jet can be identified. Finally, on the largest scales (beyond 240 arcsec), the structure is dominated by a 143-arcsec oscillation which produces bulk displacements of the jet, and which has an almost linear growth with z.

Non-exponential growth of Kelvin-Helmholtz instabilities with z was predicted for expanding beams by Hardee (1982a). The expansion rate of the jet in NGC 6251 may be taken as linear to sufficient accuracy (Perley *et al.* 1984a). Thus the *local* growth lengths and wavelengths may be estimated from dispersion relation (6.58) or (6.71), using the local mean radius of the jet and the local physical parameters, and then Hardee's (1984) results indicate that those growth lengths and wavelengths are *global*, if the spatial instability view is valid. The best-determined structural distortions in the NGC 6251 jet are the 143-arcsec oscillations which dominate the flow more than 240 arcsec from the nucleus of the source. In this region, Perley *et al.* (1984a) estimate the Mach number of the flow to be $\mathcal{M}_b$ = 13, with external density and temperature $\rho_a \sim 10^{-24}$ kg m^{-3} and $T_a \sim 3 \times 10^7$ K, and with the beam less dense than the external medium ($\rho_b \sim 3 \times 10^{-25}$ kg m^{-3} is consistent with the radio data). In the region beyond 240 arcsec from the nucleus, the radius of the jet is about 10 arcsec. Then if the structural distortions are due to the effect of Kelvin-Helmholtz in-

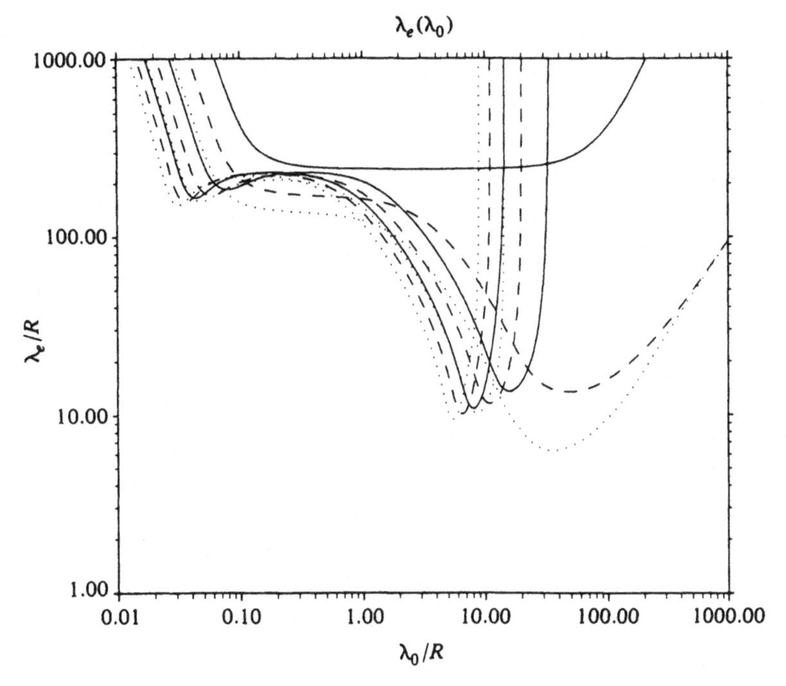

Fig. 6.6. Stability curves $\lambda_e(\lambda_0)$ for the model of NGC 6251 described in §6.6.4. Only modes with $n = 0$ (solid lines), 1 (dashed lines), and 2 (dotted lines) and $N = 0$, 1, and 2 are illustrated.

stabilities, a dominant $n = 1$ mode should appear with $\lambda_0^*/R \sim 15$. The results of a spatial instability analysis for these parameters are given in Fig. 6.6, and indicate that there are many reflection modes with $\lambda_0^* < 15R$, as well as ordinary modes with $n = 1$ and $n = 2$, which should have a strong effect on the flow.

It is difficult to see, from this figure, how a single $n = 1$ mode could dominate the stability of the beam at $\lambda_0 \approx 15R$ unless the velocity has been significantly over-estimated. Perley *et al.* (1984a) make the same point, and estimate that the beam oscillations can be important if the Mach number of the flow $\mathcal{M}_b \sim 8$, or if the jet is projected at a large angle to the sky, and the observed structures are seen strongly foreshortened. Again, if the 143-arcsec oscillation is ascribed to a Kelvin-Helmholtz mode, it is necessary that the $n = 2$ mode, and all the reflection modes, are strongly suppressed.

Similar arguments about the origin of the other instability-like structures in the NGC 6251 jet might be made, and, in general, the agreement between the flow parameters estimated by Perley *et al.* (1984a) and the observed structures is not impressive. Since NGC 6251 has only a faint counterjet, whose symmetry is not clear, it is not possible to use the symmetry of the flow from the nucleus of NGC 6251 to rule out possible precessional origins for oscillations in the outer jet. Alternative suggestions for the structures include the effects of orbital motion (Blandford & Icke 1978), and the torquing of the interstellar medium in NGC 6251 by galaxy-galaxy interactions (Wirth *et al.* 1982). The abundance of scales of structure in this jet suggests that a number of other processes of this type (including perhaps variations in the injection properties of the beam) are needed to explain them all: a Kelvin-Helmholtz view *might* provide a single-mechanism description of the instabilities if it could be shown that a realistic model for the variation of the parameters of the flow down the jet leads to the different instability wavelengths. Unfortunately, the estimates of the jet properties that are available remain too crude for this test to be made, and the physical uncertainties in the application of the Kelvin-Helmholtz model are too large for the problem to be inverted.

6.6.5 Cygnus A

The Cygnus A jet (Perley *et al.* 1984b, cf. Chapter 2) bends several times in its passage from the nucleus into the bright parts of the NW radio lobe. The minimum internal pressure of this jet is larger than the external pressure, and it is likely that the jet is confined by azimuthal magnetic fields outside the beam. Such fields cause some destabilisation of the flow, but, using the result of Cohn (1983) that the effect of an azimuthal field is similar to that of external pressure as far as the growth of Kelvin-Helmholtz modes is concerned, we can make a rough assessment of the stability of the beam ignoring the external field. Perley *et al.* use a thrust argument to suggest that the velocity of the beam flow $v_b > 0.8c$, with a density $\rho_b < 2 \times 10^{-26}$ kg m^{-3}. If the medium through which the beam propagates is taken as a typical intracluster medium, with $\rho_a \sim 10^{-23}$ kg m^{-3}, then the density of the beam is much less than the density of the ambient medium. A rough estimate of the stability of the beam to Kelvin-Helmholtz modes can be obtained by examining the properties of a fluid beam with $\rho_b \sim 10^{-26}$ kg m^{-3} and an external sound speed $c_{Sa} \sim 1000$ km s^{-1}. The results of this analysis are shown in Fig. 6.7, where it can be seen that growth

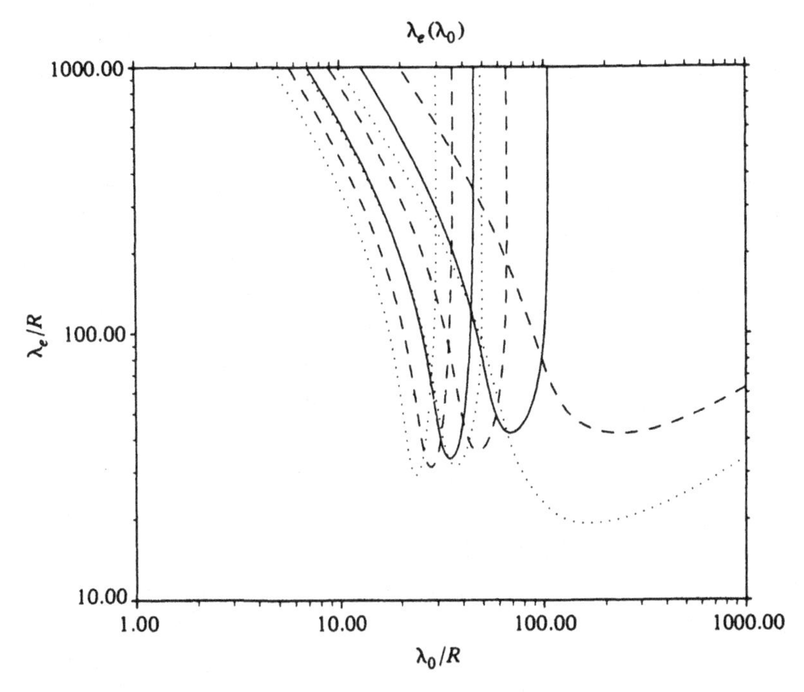

Fig. 6.7. Stability curves $\lambda_e(\lambda_0)$ for the model of Cygnus A described in §6.6.5. Only modes with $n = 0$ (solid lines), 1 (dashed lines), and 2 (dotted lines) and $N = 0$, 1, and 2 are illustrated. The $n = 0$, $N = 0$ mode has $\lambda_e > 10^3 R$ for all λ_0.

lengths less than about $40R$ at wavelengths $20 - 80R$ occur for the reflection modes, and that the $n = 1$ and $n = 2$ ordinary modes, with wavelengths greater than $100R$, are also important. The $n = 0$ ordinary mode, by comparison, has $\lambda_e > 10^3 R$ and does not appear on Fig. 6.7. If the mean radius of the jet is about 2 arcsec, and the total length of the jet is less than 60 arcsec, none of the modes in Fig. 6.7 can have a strong effect on the structure of the beam and none should display more than about one wavelength. Under these circumstances, the assumption of z-independent flow used in deriving the dispersion relations (6.58) and (6.71) (or to discuss the stability of an expanding beam; Hardee 1984) are violated, and this type of analysis cannot be valid.

There are many alternative explanations for the structures present in the jet of Cygnus A. The simplest of these is that the flow is interacting with the material involved in the radio lobes as it flows back from the heads of the source. This medium is highly structured (as can be seen in the maps of Perley *et al.* 1984b), and the high local energy densities in the filamentary loops suggest that strong interactions with the jet are possible. Difficulties arise in bending the jet if its velocity is as high as suggested here, since bending a fast beam through a large angle is likely to cause disruption of the flow.

6.7 Summary and final considerations

6.7.1 Deficiencies in the present theory

The major theoretical difficulty in discussing astrophysical beams is that they are likely to be strongly turbulent – the Reynolds numbers of radio jets are $\sim 10^{28}$ (§6.4.1). An accurate description of jet structures, including bending and the propagation of large-scale waves, requires the use of a physically-consistent model for turbulence that permits reliable calculation of the Reynolds stresses.

A further point of neglect is the effects of the current sheets and field loops, where strong energy releases and anomalous transport coefficients can arise. Other entropy-generating processes have also been ignored. Chief among these is the transfer of energy from the flow to radiating particles, and hence the loss of energy through radiation. The thermal conductivity of the beam fluid may also be relevant if there is a significant transfer of heat by diffusive conduction.

Since the observed structures in jets are certainly of large amplitude, the theory of Kelvin-Helmholtz modes needs to be extended to describe their non-linear structures and the limits to their growth. Only rudimentary ideas are available at present: numerical results only apply to the longer-wavelength modes, since only these modes are represented by the two-dimensional codes that have been used to simulate beams. These numerical simulations also suggest that the jet becomes sheathed by a hot, sheared cocoon which might protect the beam against Kelvin-Helmholtz instabilities. A fundamental conceptual improvement in the theory would be achieved by a calculation of the (turbulent) flow pattern and sheathing layer that should be produced by the operation of short-wavelength instabilities on the beam structure, and by the derivation of results on the extent of the protection effected by the presence of the cocoon. Substantial differences from the results for simpler beam models might be expected.

For large-scale coherent structures to develop on the beam, the perturbing force must be coherent for times greater than the period of the structure. Since the dominant perturbations are most likely to be those encountered by the beam as it leaves a nozzle and is injected into the ambient medium, either a coherent perturbation is applied there or (non-linear) mode competition between a wide range of co-existing modes on the beam operates to set the dominant structure (perhaps a soliton). The steadiness of the beam as it leaves the nozzle is also an issue, since theoretical investigations of the Kelvin-Helmholtz modes rely on the absence of significant variations in the injected flow. If such variations do exist, then it is likely that only full, three-dimensional numerical simulations will be capable of useful investigations of beam stability.

None of the instability calculations so far have taken account of the effects of structure in the external medium, except insofar as beam expansion has been included in Hardee's (1982a, 1984) treatments. If the scale of variations in the beam parameters is much larger than the wavelengths and growth lengths of the instabilities, then the theory is changed only by the need to apply the dispersion relation locally. If the gradients of the beam parameters are large, then the effects of Kelvin-Helmholtz waves are unlikely to be important compared with the effects of shocks and other structures produced as the beam adjusts to its environment. Buoyancy effects on the dynamics of the beam in the stratified ambient atmosphere may also be important (Achterberg 1982).

6.7.2 General conclusions and final remarks

The present state of the theory of the Kelvin-Helmholtz instability for beams is that laminar, cylindrical, fluid flows are unstable for all beam velocities and densities unless the beam contains a strong, axial magnetic field, and that high-density flows are less unstable than low-density flows of the same beam velocity. Azimuthal magnetic fields, which can help to confine the beam, tend to destabilise the flow. The growth lengths of the instability are short, and the instability is likely to produce beam heating and turbulence on small scales (perhaps in a beam-sheath). On larger scales, where observable structural distortions of the flow may result, it has not yet been possible to find unambiguous predictions of the dominant modes because of uncertainties in the beam structure and the lack of information on the non-linear limit of the modes. A number of types of Kelvin-Helmholtz mode occur: the ordinary mode is the simple analogue of the instability as seen on a semi-infinite planar surface, while reflection modes arise from the partial trapping of magnetosonic waves inside the beam and resonances of these waves with the beam/ambient medium interface (over-reflection). The ordinary mode dominates the stability of a beam for low Mach number flows, whilst the reflection modes dominate for fast flows.

In summary, although the Kelvin-Helmholtz model for the origin of structures in beams has achieved some successes, for example in 'predicting' the correct ratio of the number of knots and helical structures in the jet of M 87, and it is certain that such instabilities must influence the development of laminar beam flows, at present little can be learned about any given flow through the application of the theory. If self-consistent results for the profiles of the jets and better estimates of the flow parameters were available, slight developments of the present theory would allow strong statements about the stability of laminar flows to be made. Little work has been done so far on the development of coherent structures in turbulent flows, and the greatest single improvement that could occur in the theory of astrophysical jets would arise through the application of physically-consistent models for the origin and structure of turbulence in general sheared flows to calculations of this phenomenon.

Symbols used in Chapter 6

Symbol	Meaning
$_a$	ambient medium (subscript)
$_b$	beam core (subscript)
B	magnetic field
$\mathbf{B}$	magnetic field vector
$\mathbf{B}^{(1)}$	perturbed magnetic field vector
$B_\theta^{(0)}$	azimuthal steady-state magnetic field
$B_z^{(0)}$	axial steady-state magnetic field
B_a	magnetic field in ambient medium
B_{za}	axial magnetic field in ambient medium

B_b	magnetic field in beam core
B_{zb}	axial magnetic field in beam core
c_A	Alfvén speed
c_{Aa}	Alfvén speed in ambient medium
c_{Ab}	Alfvén speed in beam core
c	speed of light *in vacuo*
c_S	sound speed
c_{Sa}	sound speed in ambient medium
c_{Sb}	sound speed in beam core
D	jet diameter
$\mathcal{D}(\omega, \mathbf{k})$	dispersion relation
e	base of natural logarithms
e	charge on the electron
$\mathbf{E}$	electric field vector
$F^{\mu\nu}$	electromagnetic field tensor
$F_n^{(1)}$	linear shear equation solution; first kind, order n
$F_n^{(2)}$	linear shear equation solution; second kind, order n
$H_n^{(1)}(z)$	Hankel function of the first kind and order n
i	unit imaginary number
$\mathrm{Im}(z)$	imaginary part of complex number z
$\mathbf{j}$	charge-current vector
J^μ	charge-current four-vector
$J_n(z)$	Bessel function of the first kind and order n
k	wave-number
k_B	Boltzmann constant
$\mathbf{k}$	wave-vector
l	length scale
$\mathcal{M}$	Mach number
$\mathcal{M}_b$	Mach number of beam v_b/c_{Sb}
n	baryon number density
n	azimuthal mode number
N	radial mode number
P	fluid pressure
P_a	fluid pressure in ambient medium
P_b	fluid pressure in beam core
$P^{(0)}(r)$	steady-state pressure
$P^{(1)}(r)$	perturbed pressure
$P^{(2)}(r)$	second-order pressure perturbation
r	radius variable (off-axis distance)

r_{ge}	electron gyroradius
r_{gi}	ion gyroradius
$\mathbf{r}$	position vector
R	jet radius (middle of shear layer)
R_{sa}	boundary between ambient and sheared fluids
R_{bs}	boundary between beam-core and sheared fluids
$\mathrm{Re}(z)$	real part of complex number z
Re	Reynolds number
Re_m	magnetic Reynolds number
$_s$	sheared fluid (subscript)
s	entropy per baryon
s_0	constant entropy term
St	Strouhal number
t	time variable
T	temperature
T_a	temperature of ambient medium
$T^{\mu\nu}$	energy-momentum tensor
$T^{\mu\nu}_{\mathrm{fluid}}$	energy-momentum tensor of a fluid
$T^{\mu\nu}_{\mathrm{EM}}$	electromagnetic energy-momentum tensor
u^μ	velocity four-vector
u^0	'time' component of u^μ
v	fluid speed
$\mathbf{v}$	fluid velocity
$v_\theta^{(0)}(r)$	steady-state azimuthal velocity
$v_z^{(0)}(r)$	steady-state axial velocity
$\mathbf{v}^{(1)}(r)$	perturbed velocity
$v_r^{(1)}(r)$	perturbed radial velocity
$v_\theta^{(1)}(r)$	perturbed azimuthal velocity
$v_z^{(1)}(r)$	perturbed axial velocity
v_b	speed of beam-core
v_j	jet speed
V	fluid variable
$V^{(1)}$	perturbed fluid variable
z	axial distance variable
z	complex number
α	counting index; $\alpha = 0, 1, 2, 3$
α	radial wavenumber
α_a	radial wavenumber in ambient medium

α_b	radial wavenumber in beam core
β	counting index; $\beta = 0, 1, 2, 3$
γ	Lorentz factor
γ	counting index; $\gamma = 0, 1, 2, 3$
γ_b	Lorentz factor for beam-core
$\gamma^{(0)}$	steady-state Lorentz factor
Γ	polytropic index
Γ_a	polytropic index of ambient fluid
Γ_b	polytropic index of beam fluid
Γ_s	polytropic index of sheared fluid
δ	dimensionless shear layer width (6.12)
ϵ	relative permittivity of fluid
ϵ_0	permittivity of free space
ϵ_a	dimensionless amplitude of $P^{(1)}$ in ambient medium
ϵ_b	dimensionless amplitude of $P^{(1)}$ in beam core
ϵ_{s1}	dimensionless amplitude of $P^{(1)}$ in shear layer
ϵ_{s2}	dimensionless amplitude of $P^{(1)}$ in shear layer
η	coefficient of viscosity
$\eta^{\mu\nu}$	Minkowski metric tensor
θ	azimuthal angle variable
λ_D	Debye length
λ_e	*e*-folding growth length of mode
λ_e^*	*e*-folding growth length of fastest-growing mode
λ_0	wavelength of mode
$\lambda_0^{\mathrm{crit}}$	critical wavelength for a mode
λ_0^*	wavelength of fastest-growing mode
μ	counting index; $\mu = 0, 1, 2, 3$
μ	relative permeability of fluid
μ_0	permeability of free space
ν	counting index; $\nu = 0, 1, 2, 3$
ν_f	forcing frequency
ν_n	natural frequency of jet
ν_p	preferred frequency of jet
$\xi_r^{(1)}$	radial displacement of fluid in perturbation
π	mathematical constant
ρ	fluid density
ρ_a	density of ambient medium
ρ_b	density of beam core
ρ_c	charge density

$\rho^{(0)}(r)$	steady-state density profile of beam
$\rho^{(1)}(r)$	perturbation of density
σ	electrical conductivity
τ_e	e-folding growth time of mode
ω	angular frequency
∇	vector differential operator
$\times$	vector product operator
$\cdot$	scalar product operator
d	differential operator
∂	partial differential operator
$\frac{d}{dt}$	convective derivative operator
$'$	differential with respect to argument

References

Achterberg, A., 1982, *Astr. Ap.*, **114**, 233.

Baldwin, J. E., 1982, In *Extragalactic Radio Sources, IAU Symposium* **97**, eds. Heeschen, D. S. & Wade, C. M., (Reidel: Dordrecht, Netherlands), p. 21.

Batchelor, G. K. & Gill, A. E., 1962, *J. Fluid Mech.*, **14**, 529.

Benford, G., 1981, *Ap. J.*, **247**, 792.

Benford, G., 1983, In *Astrophysical Jets*, eds. Ferrari, A. & Pacholczyk, A. G. (Reidel: Dordrecht, Netherlands), p. 271.

Benford, G. & Smith, D., 1982, *Phys. Fluids*, **25**, 1450.

Bicknell, G. V., 1984, In *Physics of Energy Transport in Extragalactic Radio Sources*, eds. Bridle, A. H. & Eilek, J. A. (NRAO: Green Bank, WV), p. 229.

Biretta, J. A., Owen, F. N. & Hardee, P. E., 1983, *Ap. J.*, **274**, L27.

Birkinshaw, M., 1984, *M. N. R. A. S.*, **208**, 887.

Birkinshaw, M., 1990, *M. N. R. A. S.*, in preparation.

Birkinshaw, M., Laing, R. A. & Peacock, J. A., 1981, *M. N. R. A. S.*, **197**, 253.

Blake, G. M., 1972, *M. N. R. A. S.*, **156**, 67.

Blandford, R. D. & Icke, V., 1978, *M. N. R. A. S.*, **185**, 527.

Blandford, R. D. & Pringle, J. E., 1976, *M. N. R. A. S.*, **176**, 443.

Blumen, W., Drazin, P. G. & Billings, D. F., 1975, *J. Fluid Mech.*, **71**, 305.

Bodó, G., Rosner, R., Ferrari, A. & Knobloch, E., 1989, *Ap. J.*, **341**, 631.

Bridle, A. H., Davis, M. M., Fomalont, E. B., Willis, A. G. & Strom, R. G., 1979, *Ap. J.*, **228**, L9.

Browand, F. K., 1966, *J. Fluid Mech.*, **26**, 281.

Browand, F. K. & Laufer, J., 1975, In *Turbulence in Liquids*, eds. J. L. Zakin & G. K. Patterson (Science: Princeton, NJ), p. 33.

Brown, G. L. & Roshko, A., 1974, *J. Fluid Mech.*, **64**, 775.

Burns, J. O., 1984, In *Physics of Energy Transport in Extragalactic Radio Sources*, eds. Bridle, A. H. & Eilek, J. A. (NRAO: Green Bank, WV), p. 25.

Burns, R. L. & Lovelace, R. V. E., 1982, *Ap. J.*, **262**, 87.

Chandrasekhar, S., 1961, *Hydrodynamic and Hydromagnetic Stability*, (Oxford University Press: Oxford).
Clarke, D. A., Burns, J. O. & Feigelson, E. D., 1986, *Ap. J.*, **300**, L41.
Cohn, H., 1983, *Ap. J.*, **269**, 500.
Corcos, G. M. & Lin, S. J., 1983, *J. Fluid Mech.*, **139**, 67.
Corcos, G. M. & Sherman, F. S., 1983, *J. Fluid Mech.*, **139**, 29.
Crighton, D. G. & Gaster, M., 1976, *J. Fluid Mech.*, **77**, 397.
Crow, S. C. & Champagne, F. H., 1971, *J. Fluid Mech.*, **48**, 547.
De Young, D. S., 1984, In *Physics of Energy Transport in Extragalactic Radio Sources*, eds. Bridle, A. H. & Eilek, J. A. (NRAO: Green Bank, WV), p. 202.
Dimotakis, P. E. & Brown, G. L., 1976, *J. Fluid Mech.*, **78**, 535.
Drazin, P. G. & Davey, H., 1977, *J. Fluid Mech.*, **82**, 255.
Drazin, P. G. & Reid, W. H., 1981, *Hydrodynamic Stability*, (Cambridge University Press: Cambridge).
Ekdahl, C., Greenspan, M., Kribel, R. E., Sethian, J. & Wharton, C. B., 1974, *Phys. Rev. Lett.*, **33**, 346.
Ershkovich, A. I., 1980, *Sp. Sci. Rev.*, **25**, 3.
Ferrari, A., Massaglia, S. & Trussoni, E., 1982, *M. N. R. A. S.*, **198**, 1065.
Ferrari, A., Trussoni, E. & Zaninetti, L., 1978, *Astr. Ap.*, **64**, 43.
Ferrari, A., Trussoni, E. & Zaninetti, L., 1979, *Astr. Ap.*, **79**, 190.
Ferrari, A., Trussoni, E. & Zaninetti, L., 1980, *M. N. R. A. S.*, **193**, 469.
Ferrari, A., Trussoni, E. & Zaninetti, L., 1981, *M. N. R. A. S.*, **196**, 1051.
Fiedler, R. L., 1986, *Ap. J.*, **305**, 100.
Fiedler, R. L. & Jones, T., 1984, *Ap. J.*, **283**, 532.
Forman, W. & Jones, C., 1982, *Ann. Rev. Astr. Ap.*, **20**, 547.
Freymuth, P., 1966, *J. Fluid Mech.*, **25**, 683.
Gaster, M., 1962, *J. Fluid Mech.*, **14**, 222.
Gerwin, R. A., 1968, *Rev. Mod. Phys.*, **40**, 652.
Gill, A. E., 1965, *Phys. Fluids*, **8**, 1428.
Glatzel, W., 1988, *M. N. R. A. S.*, **231**, 795.
Hardee, P. E., 1979, *Ap. J.*, **234**, 47.
Hardee, P. E., 1981, *Ap. J.*, **250**, L9.
Hardee, P. E., 1982a, *Ap. J.*, **257**, 509.
Hardee, P. E., 1982b, *Ap. J.*, **261**, 457.
Hardee, P. E., 1983, *Ap. J.*, **269**, 94.
Hardee, P. E., 1984, *Ap. J.*, **287**, 523.
Hardee, P. E., 1986, *Ap. J.*, **303**, 111.
Hardee, P. E., 1987a, *Ap. J.*, **313**, 607.
Hardee, P. E., 1987b, *Ap. J.*, **318**, 78.
Hardee, P. E. & Norman, M. L., 1988, *Ap. J.*, **334**, 70.
Henriksen, R. N., 1987, *Ap. J.*, **314**, 33.
Ho, C. M. & Huang, L. S., 1982, *J. Fluid Mech.*, **119**, 443.
Ho, C. M. & Huerre, P., 1984, *Ann. Rev. Fluid Mech.*, **16**, 365.
Kochhar, R. K. & Trehan, S. K., 1988, *M. N. R. A. S.*, **234**, 123.
Königl, A., 1980, *Phys. Fluids*, **23**, 1083.

Laing, R. A., 1981, *Ap. J.*, **248**, 87.

Laing, R. A., 1984, In *Physics of Energy Transport in Extragalactic Radio Sources*, eds. Bridle, A. H. & Eilek, J. A. (NRAO: Green Bank, WV), p. 90.

Laing, R. A. & Bridle, A. H., 1987, *M. N. R. A. S.*, **228**, 557.

Landau, L. D. & Lifshitz, E. M., 1959, *Fluid Mechanics*, (Pergamon Press: Oxford).

Lerche, I. & Wiita, P. J., 1980, *Ap. Sp. Sci.*, **68**, 207.

Linfield, R. & Perley, R. A., 1984, *Ap. J.*, **279**, 60.

Maslowe, S. A., 1981, In *Hydrodynamic Instabilities and the Transition to Turbulence*, eds. H. L. Swinney & J. P. Gollub (Springer-Verlag: New York), p. 181.

Mattingly, G. E. & Chang, C. C., 1974, *J. Fluid Mech.*, **65**, 541.

Mattingly, G. E. & Criminale, W. O., 1972, *J. Fluid Mech.*, **51**, 233.

Michalke, A., 1969, *J. Fluid Mech.*, **38**, 765.

Miksad, R. W., 1972, *J. Fluid Mech.*, **56**, 695.

Miksad, R. W., 1973, *J. Fluid Mech.*, **59**, 1.

Misner, C. W., Thorne, K. S. & Wheeler, J. A., 1973, *Gravitation*, (Freeman: San Fransisco).

Nepveu, M., 1982a, *Astr. Ap.*, **105**, 15.

Nepveu, M., 1982b, *Astr. Ap.*, **112**, 223.

Norman, M. L. & Hardee, P. E., 1988, *Ap. J.*, **334**, 80.

Norman, M. L., Smarr, L., Winkler, K.-H. A. & Smith, M. D., 1981, *Ap. J.*, **247**, 52.

Norman, M. L., Smarr, L., Winkler, K.-H. A. & Smith, M. D., 1982, *Astr. Ap.*, **113**, 285.

Norman, M. L., Winkler, K.-H. A. & Smarr, L., 1983, In *Astrophysical Jets*, eds. Ferrari, A. & Pacholczyk, A. G. (Reidel: Dordrecht, Netherlands), p. 227.

Norman, M. L., Winkler, K.-H. A. & Smarr, L., 1984, In *Physics of Energy Transport in Extragalactic Radio Sources*, eds. Bridle, A. H. & Eilek, J. A. (NRAO: Green Bank, WV), p. 150.

Payne, D. G. & Cohn, H., 1985, *Ap. J.*, **291**, 655.

Perley, R. A., Bridle, A. H. & Willis, A. G., 1984a, *Ap. J. Suppl.*, **54**, 291.

Perley, R. A., Dreher, J. W. & Cowan, J. J., 1984b, *Ap. J.*, **285**, L35.

Perley, R. A., Willis, A. G. & Scott, J. S., 1979, *Nature*, **281**, 437.

Plashko, P., 1979, *J. Fluid Mech.*, **92**, 209.

Ray, T. P., 1981, *M. N. R. A. S.*, **196**, 195.

Ray, T. P., 1982, *M. N. R. A. S.*, **198**, 617.

Ray, T. P. & Ershkovich, A. I., 1983, *M. N. R. A. S.*, **204**, 821.

Roberts, B., 1987, *Ap. J.*, **318**, 590.

Roberts, F. A., Dimotakis, P. E. & Roshko, A., 1982, In *Album of Fluid Motions*, ed. A. Van Dyke, (Parabolic Press: Stanford, California).

Roy Choudhury, S. & Lovelace, R. V. E., 1984, *Ap. J.*, **283**, 331.

Roy Choudhury, S. & Lovelace, R. V. E., 1986, *Ap. J.*, **302**, 188.

Sarazin, C., 1986, *Rev. Mod. Phys.*, **58**, 1.

Sato, J., 1956, *J. Phys. Soc. Japan*, **11**, 702.

Sato, J., 1959, *J. Phys. Soc. Japan*, **14**, 1797.

Saunders, R., Baldwin, J. E., Pooley, G. G. & Warner, P. J., 1981, *M. N. R. A. S.*, **197**, 287.

Spitzer, L., 1962, *Physics of Fully Ionized Gases*, (Interscience: New York).

Spulak, R. G. & Burns, J. O., 1984, In *Physics of Energy Transport in Extragalactic Radio Sources*, eds. Bridle, A. H. & Eilek, J. A. (NRAO: Green Bank, WV), p. 265.

Stewart, P., 1971, *Ap. Sp. Sci.*, **14**, 261.

Stuart, J. T., 1971, *Ann. Rev. Fluid Mech.*, **3**, 347.

Synge, J. L., 1957, *The Relativistic Gas* (North-Holland Publ. Co.: Amsterdam).

Trussoni, E., Massaglia, S., Bodo, G. & Ferrari, A., 1988, *M. N. R. A. S.*, **234**, 539.

Turland, B. D. & Scheuer, P. A. G., 1976, *M. N. R. A. S.*, **176**, 421.

Waggett, P. C., Warner, P. J. & Baldwin, J. E., 1977, *M. N. R. A. S.*, **181**, 465.

Widnall, S. E., Bliss, D. B. & Tsai, C. Y., 1974, *J. Fluid Mech.*, **66**, 35.

Williams, A. G. & Gull, S. F., 1984, *Nature*, **310**, 33.

Winant, C. D. & Browand, F. K., 1974, *J. Fluid Mech.*, **63**, 237.

Wirth, A., Smarr, L. & Gallagher, J. S., 1982, *Astr. J.*, **87**, 602.

Zaninetti, L., 1986a, *Astr. Ap.*, **156**, 194.

Zaninetti, L., 1986b, *Astr. Ap.*, **160**, 135.

Zaninetti, L. & Van Horn, H. M., 1988, *Astr. Ap.*, **189**, 45.

7

Numerical Simulations of Radio Source Structure

A. G. WILLIAMS
Mullard Radio Astronomy Observatory, Cavendish Laboratory, Madingley Road, Cambridge, CB3 0HE, UK.

7.1 Gross morphology

A crucial morphological feature of extended extragalactic radio sources is that (see Chapter 2) there are actually two fundamentally distinct classes of object, in which the weaker, Fanaroff & Riley (FR) class I, sources are characterized by quasi-continuous luminous jets which often become distorted as they interact with the inter-galactic medium (Fig. 2.3), whereas the more powerful, FR II, sources have a simple, linear, double-lobed structure, with the brightest emission occurring in compact hotspots at the edge of each lobe (Fig. 2.8). A central challenge for any theoretical model of extragalactic radio source structure is therefore to reproduce this observed dichotomy, and to identify the factor, or factors, that determine which type of extended structure develops. Furthermore, we might hope that an improved theoretical understanding of the Fanaroff & Riley classification would enable us to make improved, dynamical, estimates for the, observationally badly determined, physical parameters of jets in radio sources.

Although a complete model for radio source structure would probably have to involve variations in both the strength and direction of the central engine, and a non-uniform external medium, considerable insight into their gross morphology can be gained from axisymmetric simulations of steady jets in a constant ambient medium. This model, which we shall refer to as the basic model, has the additional virtue of being less computationally expensive than fully three-dimensional simulations, permitting a wider range of jet parameters to be investigated. Furthermore, uncertainties in the emission mechanism and the effects of projection permit the easy introduction of *ad hoc* arguments into models of detailed features. The approach followed below has therefore been to concentrate firstly on the gross structural properties of the radio source population as a whole, before trying to account for their more detailed features with modifications to the basic model. As a fundamental aim of this chapter is to forge a link between maps of radio sources and their underlying flow pattern, we use the term 'jet' for both the observed and the simulated structures.

7.1.1 Hotspots

It is generally agreed that the existence of compact hotspots in FR II sources implies high Mach number flows, and most numerical simulations have concentrated on this region of parameter space (Norman *et al.* 1982; Wilson & Scheuer 1983; Williams & Gull 1984). The results of these simulations enable the density ratios of jets in high Mach number FR II sources to be constrained, since two qualitatively different structures result according to whether the density ratio of the jet is greater than or less than unity. If, on one hand, the jet is lighter than the confining medium, then the advance speed of the jet-shock is much less than the speed of the jet, from one-dimensional arguments (see *e.g.*, Andersen 1982). A large proportion of the kinetic energy of the jet is therefore converted to thermal energy at the jet-shock, and large quantities of post-jet-shock material flow transversely out of the way of the jet, inflating a cocoon. If, on the other hand, the jet density is greater than the material on which it impinges, then relatively little energy is lost, and the ex-jet material continues to move forwards in a "lump" trailing behind the jet-shock. Heavy jets therefore do not produce either cocoons or strong hotspots, implying that the jets in FR II sources must be lighter than the surrounding medium.

The structure at the head of a supersonic jet is remarkably similar to that of the original Blandford & Rees (1974) hotspot, and is now widely known to comprise three distinct features – bow shock, jet shock and contact surface (see Fig. 7.1). The bow-shock adopts the form of the detached shock due to a blunt body, in this case the contact surface, which advances supersonically into the intergalactic medium. This is the same structure as the shock wave formed by a supersonic plane or missile. The flow pattern is, however, more interesting than most aeronautical applications because, unlike the nose cone of a supersonic plane, the contact surface is free to change

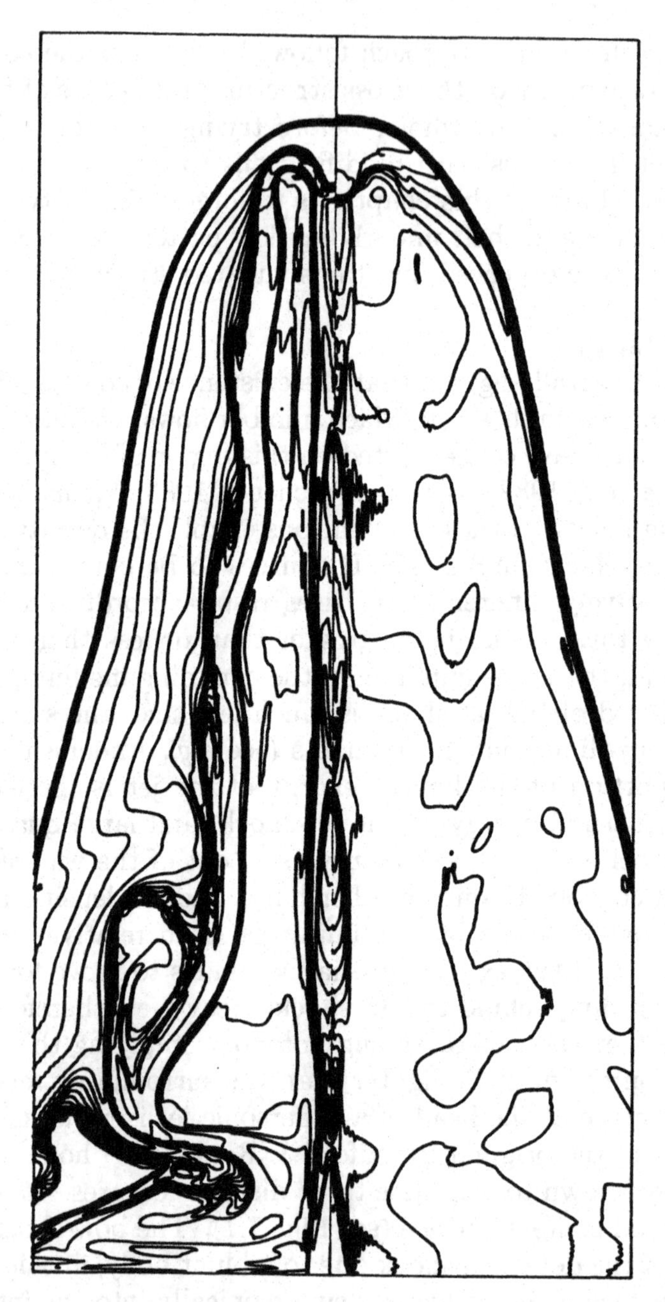

Fig. 7.1. The distribution of density (left) and pressure (right) in an axisymmetric simulation of a $\eta = 0.01$, $\mathcal{M} = 100$ jet. The contours are logarithmic with ratio 1.5 : 1.

shape according to the distribution of pressure in the hotspot, and actually shows considerable time-variation in the simulations! The structure of the jet-shock, which numerical simulations show usually contains a triple point, is a little more complicated. In one interpretation (Norman *et al.* 1982), the triple shock is caused by the Mach reflection of a strong incident shock, similar to the Mach discs formed at the exhausts of mildly supersonic nozzles, but this structure seems to be more convincingly explained by noticing that, seen from the point of view of the material in the jet, the contact surface also marks the edge of a blunt body, in this case the shocked intergalactic medium. A stand-off shock will therefore be formed in the jet, which will be normal on-axis, defocussing the flow. The segment of normal shock on-axis, sometimes described as a Mach disc, therefore arises naturally in this explanation as a consequence of the blunt "plug" of shocked intergalactic medium in the path of the jet. The ingoing conical shock which meets the stand-off shock at a triple point (actually, of course, a triple ring in cylindrical geometry), seems to be caused, independently of the stand-off shock, by the increase in thermal pressure as the jet nears the hotspot, which is sufficiently rapid, in high Mach number jets, to create an ingoing shock. This interpretation is confirmed by simulations of a jet in a decreasing atmosphere (§7.3.4), which clearly show that a normal shock can be formed at the center of the jet even in the absence of an ingoing "incident shock".

7.1.2 **Backflow**

A novel feature of the simulations, first reported by Norman *et al.* (1982), is the existence, particularly in low density, high Mach number jets, of strong backflow. The jet material, which is still travelling forwards after passing through the jet-shock, is bent back towards the source of the jet by the pressure gradient between the shocked intergalactic medium at the head of the cocoon (*i.e.*, the material responsible for the stand-off shock in the jet) and the thermal pressure of the material in the cocoon (*i.e.*, the material responsible for the ingoing shock).

The strength of the backflow depends both on the density ratio and Mach number of the jet. The dependence of backflow speed, relative to the fixed intergalactic medium frame, on the density ratio is simply a consequence of the dependence of hotspot advance speed on density ratio. For sufficiently light jets, however, the hotspot is, to a good approximation, stationary and the dependence of the backflow strength on density ratio is very weak. The dependence of backflow strength on the Mach number of the jet is due to the pressure gradient between the hotspot and lobes which scales as $\mathcal{M}_j^2$. Thus only low density, high Mach number jets produce appreciable backflow. We can quantify the amount of backflow expected by using the steady flow form of Bernoulli's equation for polytropic gases (see Chapter 3)

$$v^2 + \left(\frac{2}{\Gamma - 1}\right) c_S^2 = \text{const.} \tag{7.1}$$

In the frame of the hotspot (which is approximately at rest for light jets), we obtain, for a strong shock,

$$v_b^2 = v_j^2 \left[\left(\frac{\Gamma - 1}{\Gamma + 1}\right)^2 + \Gamma \left(\frac{2}{\Gamma + 1}\right)^2 \left(1 - \left(\frac{P_h}{P_j}\right)^{\frac{1-\Gamma}{\Gamma}}\right)\right] \tag{7.2}$$

where the first term comes from the velocity of the flow immediately downstream of the jet-shock. Using the fact that the Mach number downstream of a strong shock is given by

$$\mathcal{M}_{j,d} = \sqrt{\frac{\Gamma - 1}{2\Gamma}} \tag{7.3}$$

and Mach number varies with pressure in a steady isentropic flow as

$$\frac{P}{P_0} = \left(1 + \frac{1}{2}(\Gamma - 1)\mathcal{M}^2\right)^{-\frac{\Gamma}{\Gamma-1}} \tag{7.4}$$

we obtain an expression for the Mach number of the backflow

$$\mathcal{M}_b^2 = \left(\frac{2}{\Gamma - 1}\right) \left[\left(1 + \frac{(\Gamma - 1)^2}{4\Gamma}\right) \left(\frac{2\Gamma \mathcal{M}_j^2}{\Gamma + 1}\right)^{\frac{\Gamma-1}{\Gamma}} - 1\right]. \tag{7.5}$$

7.1.3 The location of radio emission

We now turn to the question of how to associate these flow patterns with the observed radio emission. Although some particle re-acceleration may occur due to turbulence at the contact surface, spectral index "maps" (*e.g.*, Alexander *et al.* 1984) suggest that most of the relativistic electrons are accelerated to synchrotron energies at the hotspot. The presence of knots in jets implies that the jet-shock almost certainly contributes to the radio emission. The key question is, therefore, whether particle acceleration also takes place at the bow-shock. The published literature contains both views; Wilson & Scheuer (1983) assumed that both shocks were responsible for radio emission, whereas Smith *et al.* (1985) argued that the weakness of the magnetic field in the ambient intergalactic medium makes radio emission from the bow-shock unlikely. The numerical simulations contain several additional reasons for supposing that radio emission is restricted to material inside the contact surface. First, the axial ratios of many Classical Doubles (Leahy & Williams 1984) are too large for the bow-shock to be a significant source of radio-emitting particles. Furthermore, the distortions in the "bridges" of these sources can be simply accounted for with a model involving the asymmetric deflection of light material inside the contact surface (cf. §7.3.6). Third, high resolution maps of the hotspots in Cygnus A (Scheuer, private communication) reveal localized emission resembling the endpoints of jets rather than the broad umbrella of emission expected from a bow-shock. Finally, the relatively uniform surface-brightness of the lobes contrasts sharply with distribution of pressure behind the bow-shock, which the numerical simulations show falls sharply.

7.1.4 **Cocoon dynamics**

The numerical simulations show that the width of the cocoon depends both on the density ratio and on the Mach number of the jet, with the broadest cocoons being formed by low density, high Mach number jets. An analytic expression for the cocoon width in terms of the jet parameters can be derived by assuming that the flow is axisymmetric and that the hotspot advance speed (v_h) is given by one-dimensional ram-pressure balance. The ambient medium is also taken to be constant and the jet is assumed to be both steady and in rough pressure balance with the lobes. Finally assuming that the waste products from the hotspot expand isentropically to fill a uniform cylinder, we obtain

$$\frac{R_c}{R_j} = \left(1 + \left(\frac{\rho_{\mathrm{ext}}}{\rho_j}\right)^{\frac{1}{2}} \left(\frac{P_h}{P_j}\right)^{\frac{1}{\Gamma}}\right)^{\frac{1}{2}} . \tag{7.6}$$

We can also obtain the (normal) Mach number of the jet from the jump conditions at the jet-shock

$$\frac{P_h}{P_j} = \frac{2\Gamma\mathcal{M}_j^2 - (\Gamma - 1)}{\Gamma + 1}. \tag{7.7}$$

The significance of this model is that two observational parameters (the ratios of cocoon width to jet width and hotspot pressure to lobe pressure), derivable from radio observations alone, determine the Mach number and density ratio of the jet.

7.1.5 The relation between FR I and FR II sources

One of the central problems in the theory of extragalactic radio sources is the relation of source morphology to jet parameters. In the context of axisymmetric models, Norman *et al.* (1984a) have suggested that the crucial distinction between FR I and FR II sources is that the fastest-growing pinch instability is longitudinal ("ordinary") in FR I jets, whereas it is partly transverse ("reflective") in FR II jets. In more recent work Norman & Winkler (1985) have linked this idea to the intuitively appealing criterion that jet stability requires that the jet be supersonic with respect to the *sum* of the internal and external sound speeds, leading to the relation (Payne & Cohn 1985)

$$\mathcal{M} = 1 + \sqrt{\eta} \tag{7.8}$$

where η is the ratio of jet to ambient density.

Another factor which may be related to the stability of jets is the inflation of the cocoon itself. Fig. 7.2 shows that light, low-Mach number jets are naked over most of their length, and may therefore be vulnerable to disruption by the intrusion of dense intergalactic medium into the jet channel. Furthermore, we show in §§7.3.1 and 7.3.3 how cocoons may be *stripped* from a jet by the action of cross-winds or galactic atmospheres. Whichever mechanism is more important, FR II sources (invisible jets, hotspots, lobes) can be identified with supersonic jets, and FR I sources (luminous jets, no lobes) can be identified with subsonic ones.

7.1.6 Application to Cygnus A

The model for the width of the lobes can be used to make dynamical estimates for the parameters of jets in Classical Doubles. This has previously only been possible for jets in bent FR I sources where it might in any case be expected that the jet parameters

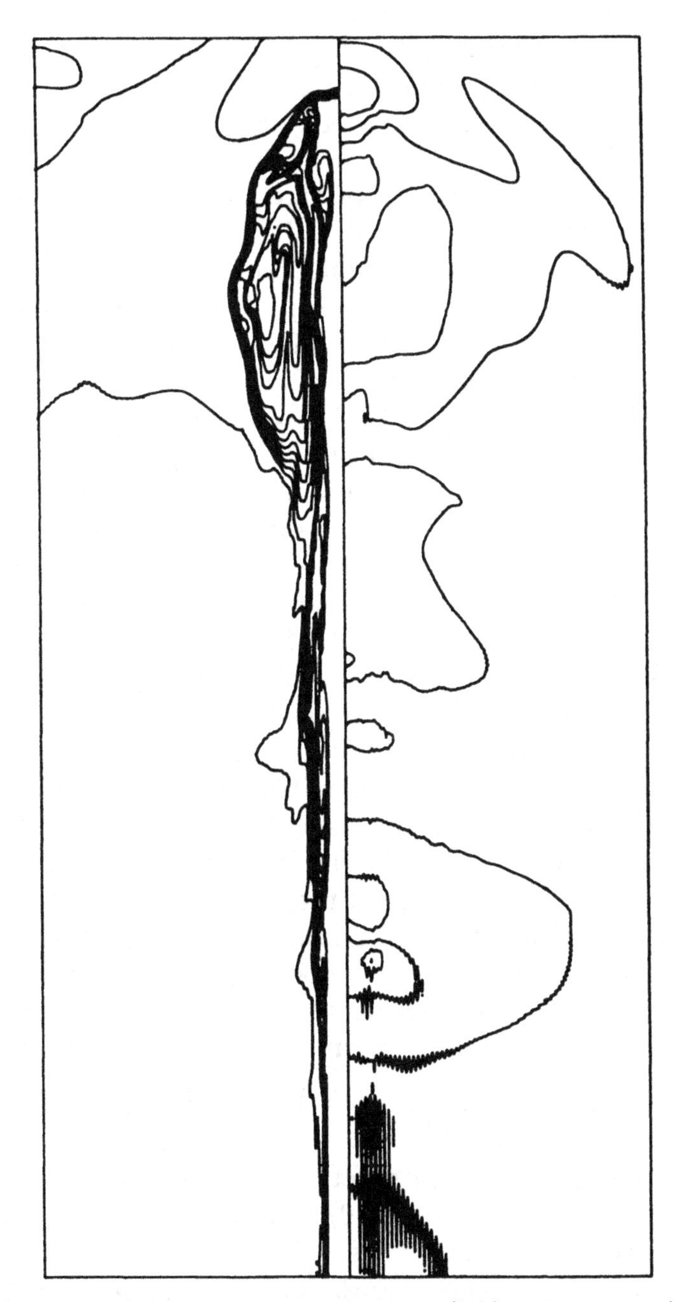

Fig. 7.2. The distribution of density (left) and pressure (right) in an axisymmetric simulation of a $\eta = 0.1$, $\mathcal{M} = 0.6$ jet. The contours are logarithmic with density ratio $1.2 : 1$ and pressure ratio $1.05 : 1$.

will be altered by their interaction with cluster gas. We will use data from the "archetypal" Classical Double, Cygnus A, despite the fact that, although having the best data, it may be in some respects atypical of the FR II population due to its unusual location at the centre of a cluster of galaxies. Recent high resolution maps of the hotspots (Scheuer, private communication) nevertheless reveal hotspot pressures 1.6×10^{-9} J m^{-3}, making the usual minimum energy assumptions (filling factor of unity, no energy in relativistic protons). Taking the lobe pressure to be 2×10^{-11} J m^{-3} (Alexander *et al.* 1984), and a ratio of lobe width to hotspot size of about 30, produces Mach numbers of around 8, and density ratios of 2.4×10^{-4} ($\Gamma = 5/3$) or 8.8×10^{-4} ($\Gamma = 4/3$). These density ratios unfortunately lie below the values that can at present be simulated numerically, because the computational cost of modelling low density jets increases both due to their reduced hotspot advance speeds and because huge grids are required to contain their cocoons.

The X-ray data, which fixes the density of the external medium at 9.4×10^{-24} kg m^{-3} (Arnaud *et al.* 1984) can be used to put some hard figures to the model. The inferred jet density is 2.3×10^{-27} kg m^{-3} ($\Gamma = 5/3$) or 8.3×10^{-27} kg m^{-3} ($\Gamma = 4/3$). If charge neutrality is provided by protons, this implies a proton number density in the lobes of 0.41 m^{-3}. Naive substitution in the non-relativistic jump conditions at the shock yields jet velocities of 9.6×10^{8} m s^{-1} ($\Gamma = 4/3$), suggesting that, with due regard to the simplifications (normal shocks, axisymmetry, 1-d ram-pressure balance, minimum energies, isentropic flow), a relativistic model may be needed.

We can also use the dynamical model to investigate the composition of the jets. The number density of relativistic electrons in a minimum energy synchrotron plasma is given approximately by

$$n \approx 2 \times 10^{18} \left(1 - \frac{1}{2\alpha}\right) \frac{4}{7} u_{\min} \sqrt{\frac{B_{\min}}{\nu_L}} \ \mathrm{m}^{-3} \tag{7.9}$$

where α is the low frequency spectral index and ν_L is the low frequency cutoff. Winter *et al.* (1980) give $\alpha \sim 0.6$ and Alexander *et al.* (1984) give minimum energies in the lobes $\sim 6 \times 10^{-11}$ J m^{-3}. Using this formula, we estimate the number density of relativistic electrons in the lobes to be $\sim 27/\gamma_L$ m^{-3}, where γ_L is the Lorentz factor at which the low-frequency spectrum is truncated. Since we would require $\gamma_L \sim 65$ in order to reconcile the density of rela-

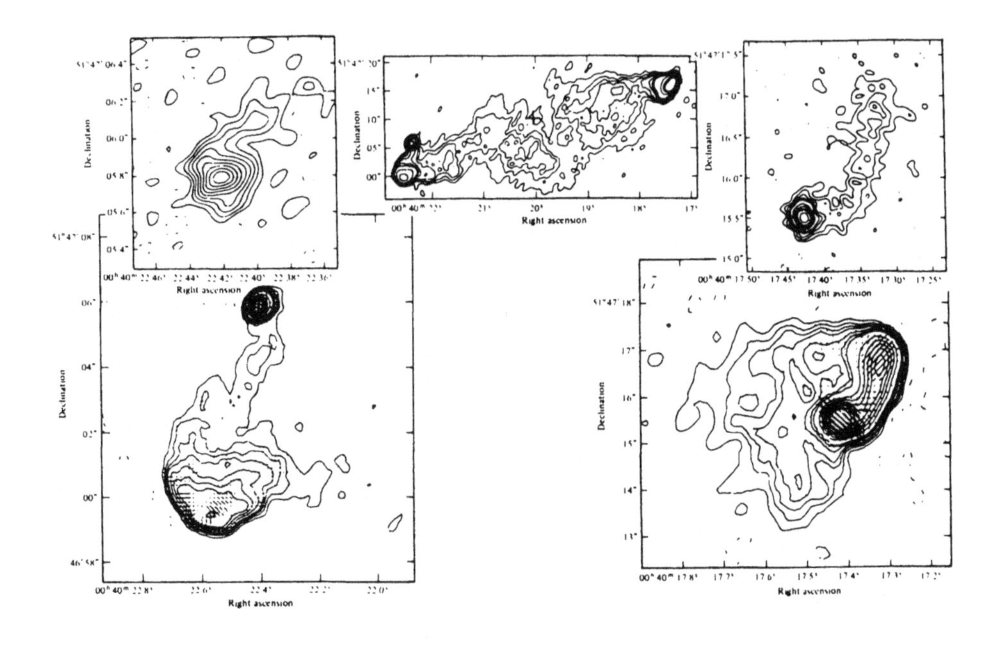

Fig. 7.3. Multifrequency VLA maps of 3C 20 from Laing (1982).

tivistic electrons with the proton number density, a proton-electron plasma is probably excluded by the model.

7.2 Variation in the direction of the jet

7.2.1 Hotspots

Although axisymmetric models can account for much of the gross morphology of the radio-source population, many detailed features display considerable asymmetry. Recent observations with sub-kiloparsec resolution show, for example, that the hotspots in powerful extragalactic radio sources often contain asymmetric internal double structure (*e.g.*, Cygnus A, Alexander *et al.* 1984; 3C 351, Kronberg *et al.* 1980; 3C 20, Laing 1982), in which one subcomponent is usually extremely compact, of size less than 1 kpc, while the other, more diffuse subcomponent, appears to point away from the first, and is usually further from the nucleus (Fig. 7.3).

In the context of Blandford & Rees' now-accepted jet model of radio sources, authors differ in their interpretation of where the jet first deposits its energy. Laing (1982) has suggested that the compact subcomponent is the current point of impact of the jet on the surrounding medium and that the diffuse subcomponent either marks a previous hotspot created at a time when the direction of the jet was different, or is produced by material escaping from the compact hotspot (this model is also discussed by Lonsdale & Barthel 1984). Conversely, Kronberg & Jones (1982) propose that the diffuse subcomponent occurs at the end of the jet and that the compact subcomponent might arise as a short-lived instability in the contact surface between the lobe and the shocked intergalactic medium (*e.g.*, Norman *et al.* 1982). Begelman *et al.* (1984) suggest that the subcomponents may either result from jitter in the orientation of the jet or be simultaneous points of impact of a jet that has become subdivided, whereas Norman *et al.* (1982) and Smith *et al.* (1985) propose that the double hotspot structure might be interpreted as the signature of the triple-shock configuration found at the working-surface of axisymmetric jets.

In this section, we describe the effect of instantaneously varying the direction of the jet inside a previously inflated cocoon (Williams & Gull 1985). Although a plausible cause of asymmetry in astrophysical jets is slow precession of the primary collimator (Ekers *et al.* 1978), we could not attempt to model this in detail because of the high computational cost of providing high resolution in all three directions. Many of the features that might result from precession can nevertheless be seen in the models – different rates of precession correspond, for example, to different epochs in the simulations. The principal difference between the flow of a jet whose direction is suddenly changed by 10° (Fig. 7.4), and an axisymmetric jet of comparable parameters, is that the asymmetric jet is decollimated in two stages producing a pair of hotspots rather than terminating at a single working-surface. A compact hotspot, recessed from the leading edge of the cavity, is formed where the jet first meets the cocoon wall. Downstream of the shock, the jet recrosses the cocoon meeting the opposite wall at a secondary hotspot or "splatter-spot" (third time step in Fig. 7.4). In its early stages the secondary jet-shock is sufficiently oblique for the jet to form a tertiary hotspot near the apex of the cocoon, although it later becomes less oblique and begins to resemble the normal shock of axisymmetric simulations with subsonic outflow (fourth time step in Fig. 7.4), while later still, the primary hotspot dominates the dynamics, produc-

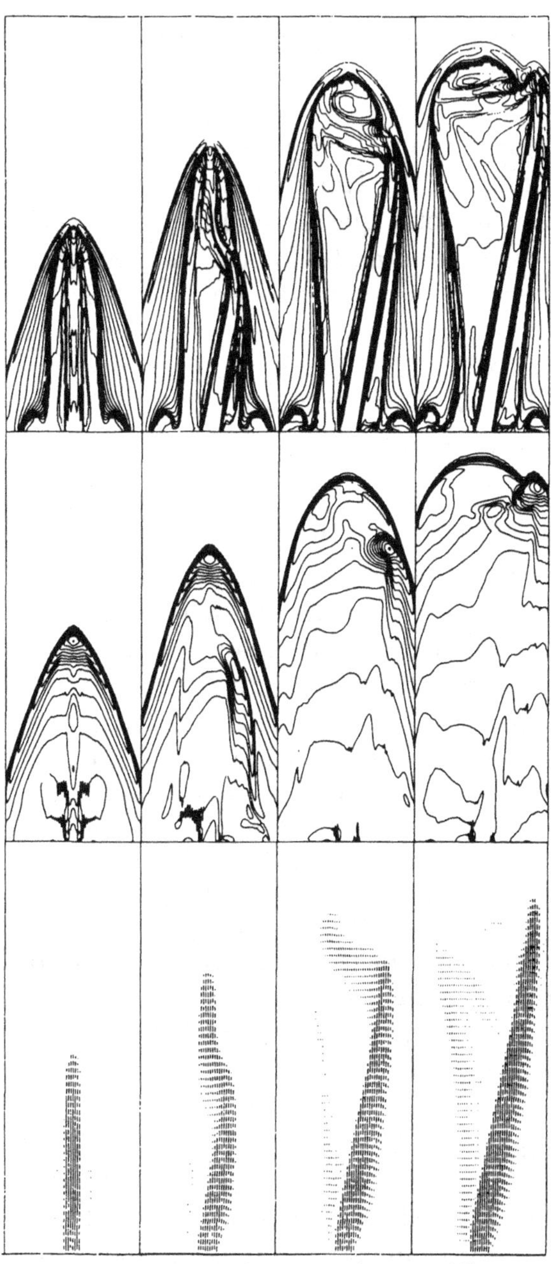

Fig. 7.4. The distribution of density (top panel), pressure (middle panel) and projected velocity vectors (lower panel) in the $z = 0$ plane for the 10° simulation, shown at $t = 14.1$, $t = 19.4$, $t = 27.7$ and $t = 33.1$, from left to right. The density contours are logarithmic with contouring ratio 1.4 : 1, and the pressure contours are logarithmic with contouring ratio 1.3 : 1.

ing its own axisymmetric backflow which cuts off the flow to the secondary hotspot.

7.2.2 Parameter evolution at oblique shocks

We can use the Rankine-Hugoniot relations in conjunction with the formula for a quasi-one-dimensional isentropic rarefaction to give an analytical model for the net parameter evolution at an oblique shock (characterized by angle β) followed by a rarefaction wave. A special case of this model was developed in §7.1.2 to describe the backflow and cocoon parameters of an axisymmetric jet. In general, however, the jet-shock will be oblique, with downstream Mach number given by

$$\mathcal{M}_{j,d}^2 = \frac{2 + (\Gamma - 1)\,\mathcal{M}_{j,u}^2}{2\Gamma \mathcal{M}_{j,u}^2 \sin^2\beta - (\Gamma - 1)} + \frac{2\mathcal{M}_{j,u}^2 \cos^2\beta}{2 + (\Gamma - 1)\,\mathcal{M}_{j,u}^2 \sin^2\beta}. \quad (7.10)$$

This expression, which is plotted in Fig. 7.5(a) for a $\Gamma = 5/3$ gas, shows that the downstream flow is subsonic for shock angles greater than about 65°, and is supersonic for angles smaller than this. It is tempting to associate this fluid-dynamical distinction with the difference between "hotspots" and "knots". In any case the pressure immediately downstream of the shock is given by

$$\frac{P_{j,d}}{P_{j,u}} = \frac{2\Gamma \mathcal{M}_{j,u}^2 \sin^2\beta - (\Gamma - 1)}{\Gamma + 1}. \quad (7.11)$$

If we assume that the post-shock flow is isentropic, then its pressure and Mach number are related by equation (7.4) and we can thus obtain an expression for the final Mach number of the flow in terms of its post-shock Mach number and the ratio of the post-shock pressure to the final pressure, which is taken to be the same as the pre-shock pressure

$$\mathcal{M}_c^2 = \left(\frac{2}{\Gamma - 1}\right) \left[\left(1 + \frac{1}{2}(\Gamma - 1)\,\mathcal{M}_{j,d}^2\right) \left(\frac{P_{j,d}}{P_c}\right)^{\frac{\Gamma-1}{\Gamma}} - 1 \right]. \quad (7.12)$$

The final Mach number, plotted in Fig. 7.5(b), is supersonic for all upstream Mach numbers and shock angles, but is, nevertheless, always reduced by the interaction, with the greatest proportional reduction occurring in high Mach number jets with normal shocks.

We can also determine the evolution of the density ratio and the

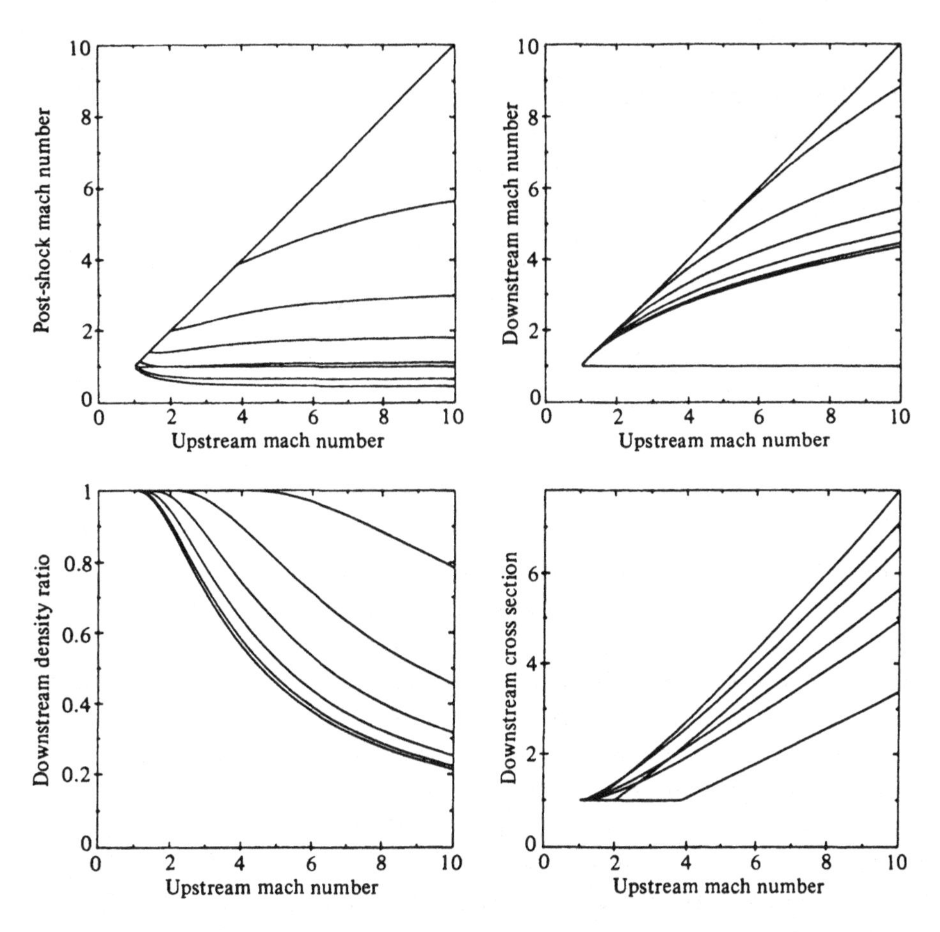

Fig. 7.5. Parameter evolution at an oblique shock followed by an isentropic expansion for shock angles of 15°, 30°, 45°, 60°, 75° and 90° in a $\Gamma = 5/3$ gas. The small-angle shocks produce large post-shock and downstream Mach numbers ($\mathcal{M}_{j,d} = \mathcal{M}_{j,u}$ and $\mathcal{M}_{j,d} = 1$ lines are also shown). The downstream density ratio is also a monotonically decreasing function of shock angle. The order of increasing final cross-sectional areas for an upstream Mach number of 10 is 15°, 90°, 75°, 30°, 60° and 45°.

cross-section of the jet. Using the post-shock density

$$\frac{\rho_{j,d}}{\rho_{j,u}} = \frac{(\Gamma + 1)\,\mathcal{M}_{j,u}^2 \sin^2\beta}{2 + (\Gamma - 1)\,\mathcal{M}_{j,u}^2 \sin^2\beta} \tag{7.13}$$

we can calculate the final density of the flow, given by

$$\rho_c = \rho_{j,d}\,(P_c/P_{j,d})^{1/\Gamma}, \tag{7.14}$$

which is plotted in Fig. 7.5(c). This shows that the nett effect of each shock and rarefaction is to decrease the density of the jet, with the greatest decrease occurring in high Mach number jets with normal shocks.

At the shock, the area of the jet is decreased by the cosine of the deflection angle θ, where

$$\tan\theta = \frac{2\left(\mathcal{M}_{j,u}^2 \sin^2\beta - 1\right)\cot\beta}{2 + \mathcal{M}_{j,u}^2\left(\Gamma + \cos^2\beta\right)}. \tag{7.15}$$

The area-Mach number relation can then be used to determine the final cross-section of the flow, giving

$$\frac{A_c}{A_{j,d}} = \frac{\mathcal{M}_{j,d}}{\mathcal{M}_c}\left(\frac{1 + \frac{1}{2}(\Gamma - 1)\mathcal{M}_c^2}{1 + \frac{1}{2}(\Gamma - 1)\mathcal{M}_{j,d}^2}\right)^{\frac{\Gamma+1}{2(\Gamma-1)}}. \tag{7.16}$$

This expression, plotted in Fig. 7.5(d), shows that the cross-section of the flow is always increased in such an interaction, with the largest increase occurring in high Mach number jets with *intermediate* angle shocks. This is simply a consequence of the fact that subsonic post-shock flow can only be re-accelerated to supersonic speeds through a convergent-divergent nozzle. The greatest increase in the cross-section of the jet therefore occurs for transonic post-shock flows. In the simulations, the secondary hotspot (formed at the second impact with the walls of the cocoon) is much larger, and has a lower pressure, than the hotspot formed where the jet first meets the cocoon wall, reproducing the asymmetric double structure found in the hotspots of powerful extragalactic radio sources. The model for parameter evolution provides a natural explanation for both aspects of this asymmetry. A Mach 10 jet, for example, bent by a 45° oblique shock, ultimately produces, for isentropic downstream flow, a Mach 5.5 jet with cross-section 8 times greater than the pre-shock value.

7.2.3 Lobe drag in asymmetric jets

An important difference between the simulations in Figs. 7.1 and 7.4 concerns the distribution of pressure at the head of the lobes. Because it advances uniformly into the intergalactic medium, the head of an axisymmetric cocoon consists of a nearly uniform region of high pressure, somewhat larger than the jet itself. If, however, the working-surface moves around the head of the lobes, as in Scheuer's (1982) "Dentist's Drill" model, then not all parts of the cocoon need advance at the same *instantaneous* rate, and localized subcomponents, separated by regions of lower pressure, can be formed. The existence of multiple subcomponents in the hotspots of Classical Doubles, and absence in axisymmetric simulations, therefore provides circumstantial evidence for variation in the direction of the jet. The time-averaged lobe advance speed in such sources should be scaled down to represent some averaging of the hotspot pressure over the head of the lobes. If the advance speed corresponding to the pressure of inter-component material may be neglected, each subcomponent's advance speed should be scaled down by the ratio of the area over which it tracks to the area of the subcomponent. This means that a higher energy density in the more compact subcomponent may therefore be consistent with both subcomponents having similar *averaged* advance speeds. Taking the pressure in the lobes as a lower limit for the averaged hotspot pressure we can obtain an upper limit for the amount of variation in the direction of the jet in Cygnus A. Since the ratio of peak hotspot pressure to lobe pressure $\sim \sqrt{80}$, the ratio of instantaneous to average hotspot advance speed is ~ 9. Its age is then correspondingly increased to 6×10^7 years, an order of magnitude in excess of the spectral age determination of 6×10^6 years (Winter *et al.* 1980).

7.2.4 **Knots**

Many extragalactic radio sources, of varying sizes and luminosities, possess jets that contain substantial variations in emissivity (*e.g.*, NGC 6251, Perley *et al.* 1984; M 87, Biretta *et al.* 1983; Cen A, Burns *et al.* 1984; Her A, Dreher & Feigelson 1984) – see Fig. 7.6. Various mechanisms, involving inhomogeneities in the jet and the surrounding medium, have been proposed to account for their knotty structure. Rees (1978) has suggested that the regularly-spaced series of eight knots in the M 87 jet (Fig. 7.7) might be caused by periodic variations in the speed of the jet, and Wilson (1983) has carried out numerical simulations of this model. Because supersonic relative velocities are needed, this model requires very large fractional variations in the speed of the jet if it is to account for knots in low power, and presumably low Mach number, jets like M 87. Unless the proportional variation in jet speed is a strongly decreasing function of Mach number, we would

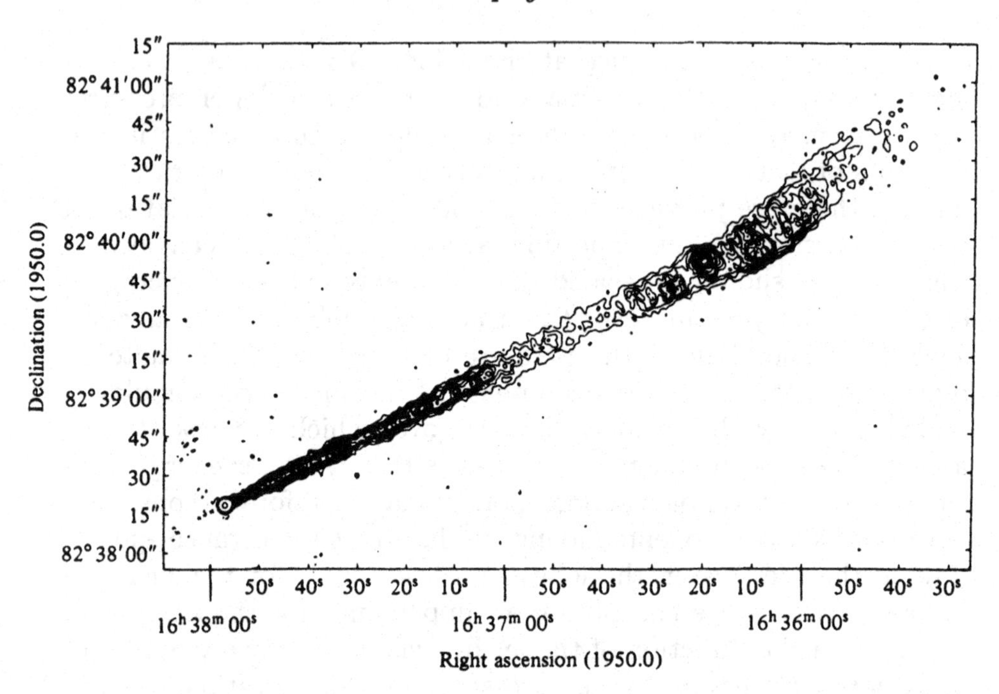

Fig. 7.6. 1.5 GHz VLA map of the jet in NGC 6251 from Perley *et al.* (1984).

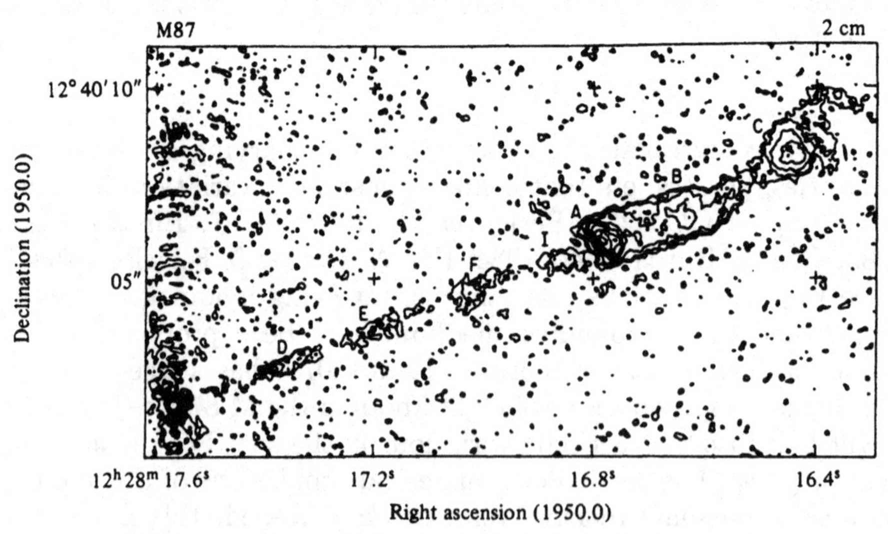

Fig. 7.7. 15 GHz VLA map of the jet in M 87 from Biretta *et al.* (1983).

expect this mechanism, if it were operating, also to produce strong regularly-spaced knots in the more powerful sources, whereas, in fact, the jets in Classical Doubles are often invisible against background emission from the lobes (Bridle & Perley 1984). An alternative hypothesis has been proposed by Blandford & Königl (1979), who pointed out that the knots in M 87 appear to be smaller than the total extent of the jet, have asymmetric brightness distributions and are not exactly collinear. They suggested that knots might be caused by dense clouds lying in the path of the jet, which could dissipate some of the energy of the jet in bow shocks while the speed of the jet remained supersonic with respect to the clouds. It is, however, difficult to obtain, with this model, an uncontrived explanation for the quasi-periodic spacing of the knots in M 87.

A third model has been proposed by Sanders (1983), who suggested that variations in the pressure of the external medium could create reconfinement shocks in the jet. Falle & Wilson (1985) have carried out high resolution numerical simulations which suggest that reconfinement shocks can be produced in atmospheres consistent with the X-ray data for M 87. Models involving the growth of surface instabilities in the jet have also been proposed (Hardee 1979), and Norman *et al.* (1984b) and Woodward (1986) have carried out numerical simulations of the growth of pinching and kinking modes. A further possibility suggested by Norman *et al.* (1982) is that internal shocks in the jet, induced by their interaction with backflowing material in the cocoon, might be a possible explanation for knots in the M 87 jet.

In this section, we present an alternative model in which the regularly-spaced strings of knots are attributed to variation in the direction of a low-density jet inside a cocoon which has been compressed onto the surface of the jet, in the manner of Scheuer's (1974) model C. Fig. 7.8 shows the results of numerical simulations of a planar jet whose direction was changed suddenly by a small angle, while surrounded by a denser, stationary ambient medium. Note that the model correctly reproduces both the quasi-periodic spacing and the asymmetric "side-to-side" structure of knots found in extragalactic radio sources (*e.g.*, Cen A, Fig. 7.9). The asymmetric jet thermalizes a fraction of its kinetic energy and is redirected by the knot of enhanced pressure formed where it meets the denser surrounding intergalactic medium. If the oblique shock makes only a small angle to the flow, parameter evolution to lower

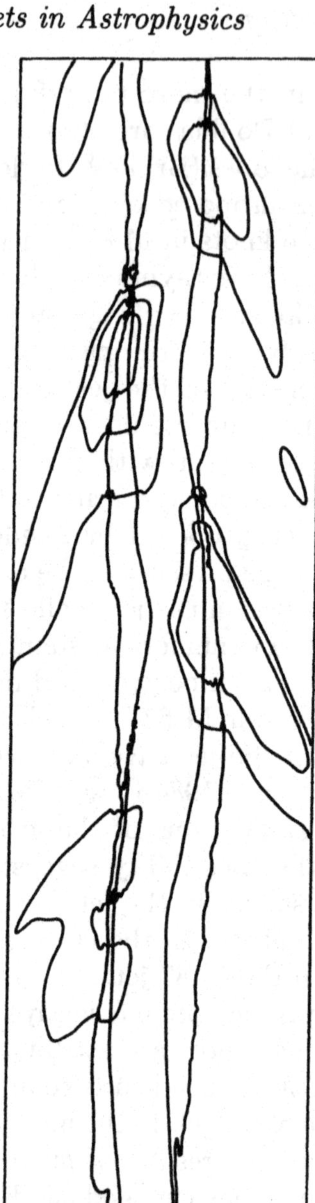

Fig. 7.8. Superposed contours of Mach number, density and pressure for a 2-d planar simulation of a $\eta = 0.01$, $\mathcal{M} = 3$ jet, initially surrounded by ambient intergalactic medium, whose direction was changed instantaneously by 5°. The inner channel marks the boundary of supersonic flow, and the edge of the jet. The outer channel shows the position of the contact discontinuity, and the presence of knots is revealed by the pressure contours.

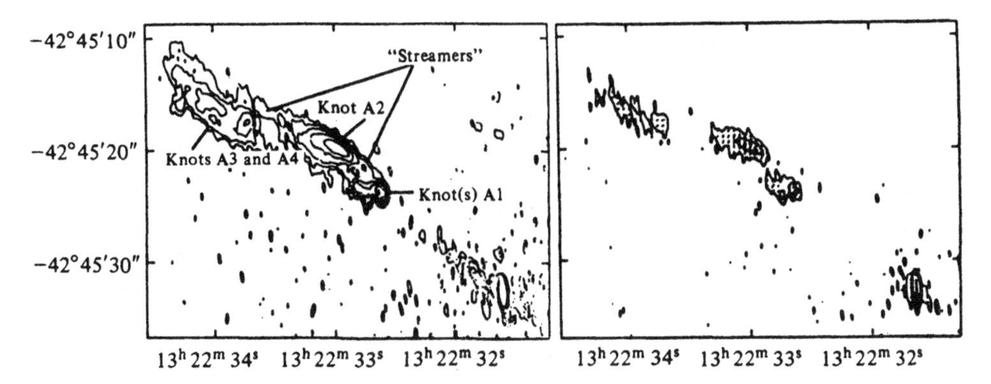

Fig. 7.9. 5 and 15 GHz VLA maps of the inner jet in Cen. A from Burns *et al.* (1984).

Mach numbers is slow, and a string of many knots may be formed before the jet is disrupted. It has been suggested by Biretta *et al.* (1983) that because the knots in M 87 have synchrotron minimum pressures that are too large for confinement to be provided by the thermal pressures derived from X-ray Bremsstrahlung observations (Schreier *et al.* 1982), magnetic confinement of the jet (*e.g.*, Benford 1978) may be required. Our simulations suggest that such modifications to the standard jet model may be unnecessary as lateral ram-pressure balance can, particularly for low-density jets, provide quasi-stationary confinement of significantly over-pressured knots. The over-pressure found in the knots will depend on the Mach number and deflection angle of the jet; over-pressures of four to eight times ambient pressure were common amongst the jets we studied.

7.3 The effect of the external medium

7.3.1 Cross-winds and radio-trail sources

Unlike the more powerful extragalactic radio sources, which possess simple collinear features, weaker sources display a wide range of complex structures with twin jets emerging from the nucleus of the parent galaxy which bend and twist as they interact with the intergalactic medium; extreme examples are the radio-trail sources (*e.g.*, NGC 1265, O'Dea & Owen 1986; 3C 129, Downes 1980) – see Fig. 2.9. Despite the fact that dynamical arguments suggest that these jets have supersonic Mach numbers, given from ram-pressure bending by $\sqrt{R_{\rm curv}/R_j}$, the presence of a cross-wind in

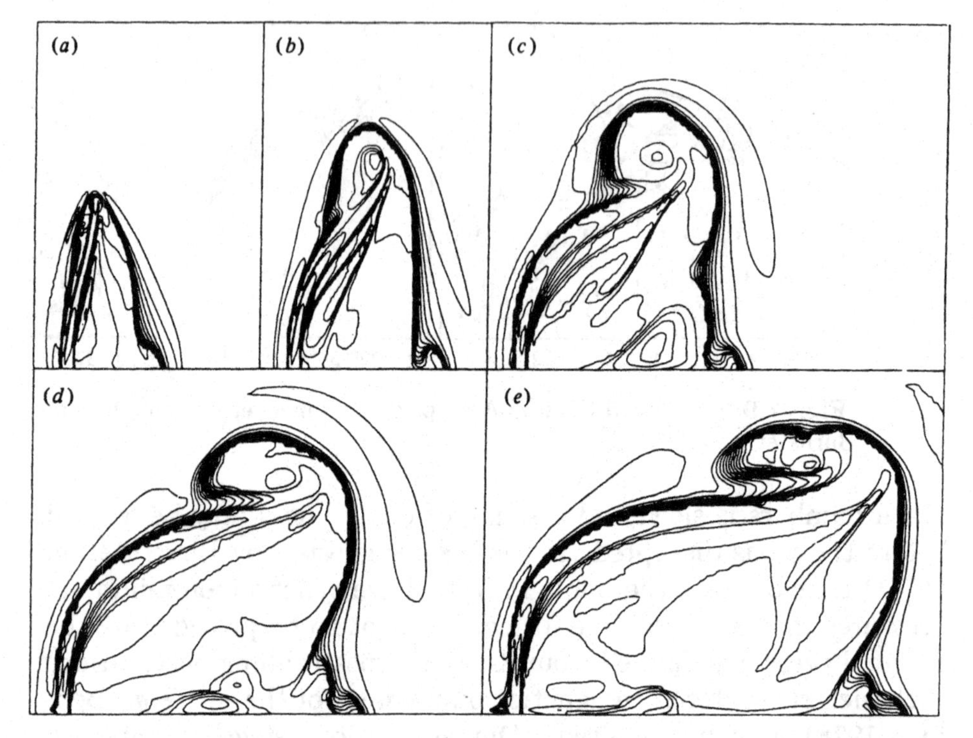

Fig. 7.10. The distribution of density in the $z = 0$ plane of a fast, light jet at various epochs. The density contours are logarithmic, contouring ratios 1.5 : 1. (a) $t = 11.2$, (b) $t = 20.9$, (c) $t = 34.3$, (d) $t = 49.4$, (e) $t = 65.3$.

these sources seems to fundamentally alter the structure produced by the jet. Unlike the invisible jets of Classical Double sources, which can propagate hundreds of kiloparsecs before terminating in compact hotspots, the jets in radio-trails often appear to display an abrupt transition from supersonic to subsonic flow. Near the galaxy, the jets are luminous, knotty and well-collimated, whereas further from the galaxy, they expand rapidly and become turbulent. In this section, we suggest how the general features of axisymmetric jets are modified by the presence of a cross-wind, and why their resulting morphology, in particular the transition from supersonic to subsonic flow, is so different from that of Classical Doubles.

Fig. 7.10 shows the distribution of density in the central plane of a $\eta = 0.1$, $\mathcal{M} = 10$ jet at various epochs after switching on the jet, while Fig. 7.11 shows the distribution of pressure and velocity vectors corresponding to the density distributions in Fig. 7.10(b)

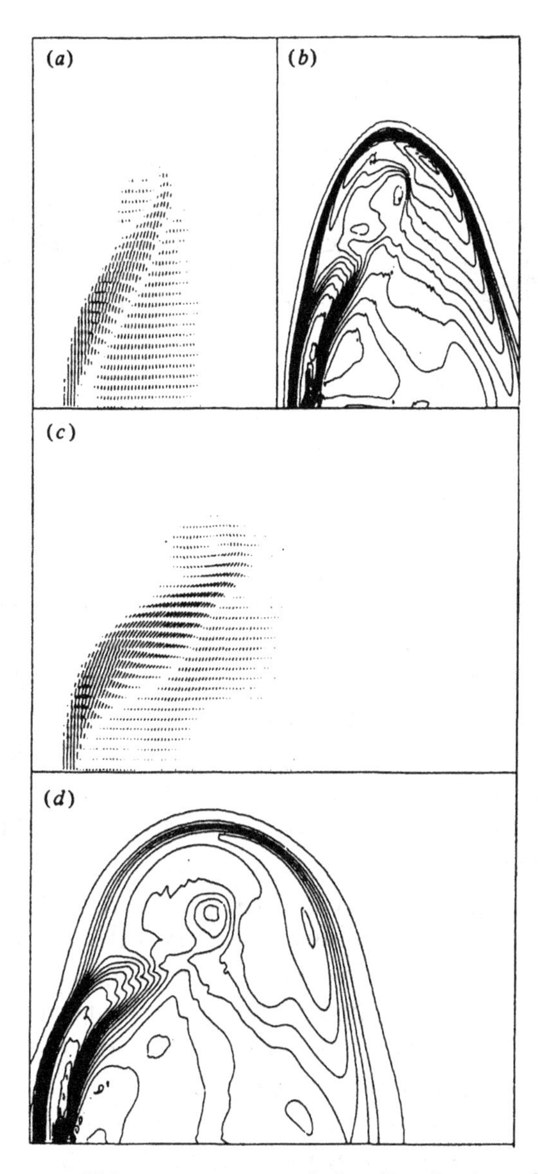

Fig. 7.11. Velocity vectors and the distribution of pressure in the $z = 0$ plane of a light jet at $t = 20.9$, (a) and (b), and at $t = 34.3$, (c) and (d). The contours are logarithmic, contouring ratio 1.15 : 1.

and (c). In the early stages, the structure of the jet is similar to that of an axisymmetric jet, and contains a hotspot, cocoon and backflow. Near the head of the jet, a cocoon is inflated upstream, while further from the hotspot, the pressure in the cocoon is smaller

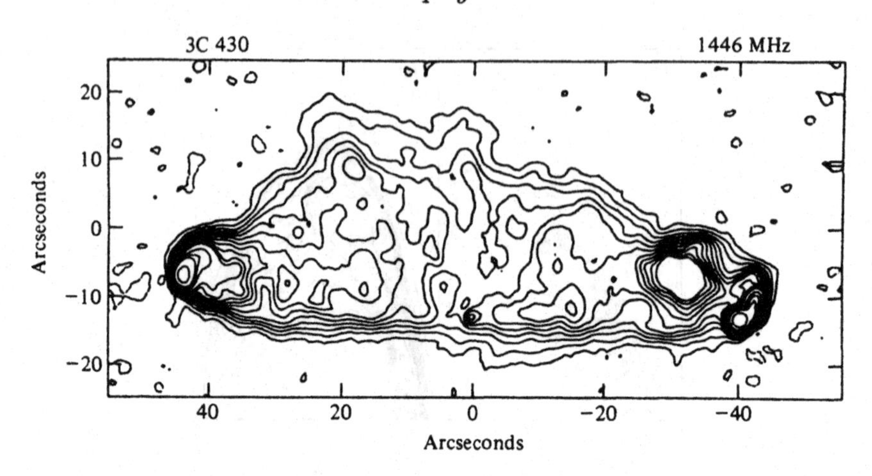

Fig. 7.12. 1446 MHz map of 3C 430 from Spangler *et al.* (1984).

and the contact surface becomes wrapped round the jet. Upstream of the jet the cocoon is ram-pressure confined by a thin spindle of shocked intergalactic medium, whereas downstream, the expansion of the contact surface lags behind that of the bow-shock, in the manner of axisymmetric numerical models, and is thermally confined by the pressure of shocked intergalactic medium. Backflow from the hotspot is swept downwind into the asymmetric cocoon, resembling the structure in 3C 430 – a small Classical Double whose asymmetric lobes may be caused by ram-pressure (Fig. 7.12).

A striking and unexpected feature of the later epochs shown in Fig. 7.10 is that the jet collides obliquely with the *downwind* wall of its cocoon, bending its tip upwind in a hook-like feature. A possible example of a source displaying this re-entrant morphology is IC 708 (Fig. 7.13), whose structure has previously been attributed to a gravitational interaction (Vallée *et al.* 1981). A further point to note about the simulations is that the jet-shock decreases in strength as the jet is bent downwind, in agreement with the observation that developed radio-trail sources do not possess hotspots (compare Fig. 7.11(b) and (d)). The principal cause of this seems to be the increase in the cross-sectional area of the jet as it is bent.

7.3.2 Parameter evolution in bent jets

We now turn to the question of how the parameters of the jet change as it is bent by the cross-wind, and how this parameter evolution is related to the gross structural properties of radio-trail

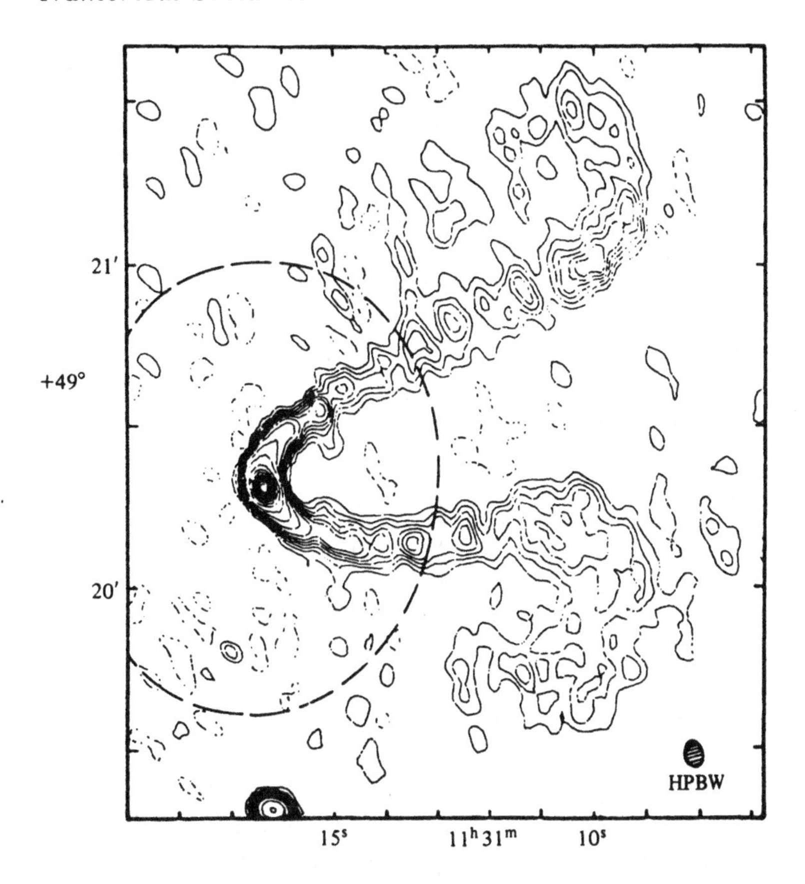

Fig. 7.13. 5 GHz VLA map of IC 708 from Vallée *et al.* (1981). Reproduced by permission of J. Vallée.

sources. In answer to the first question, the simulations suggest that the bending of a steady jet is accompanied not, as one might think, by a series of shocks, but, instead, by a rarefaction wave. Although a small initial deflection is caused by an oblique shock as the jet enters the computing grid, the density contours in Fig. 7.10 show that, thereafter, the bending of the, still supersonic, jet is accompanied by an expansion wave. Although at first surprising; this is, in fact, a straightforward consequence of the decrease in the ram-pressure of the cross-wind as the jet bends downstream.

Near the hotspot, the jet parameters are changed by the increase in thermal pressure there. Provided, however, that the tip of the jet is swept sufficiently far downwind, as it is in developed radio-trail sources, the dynamics of the bending jet will be dominated instead

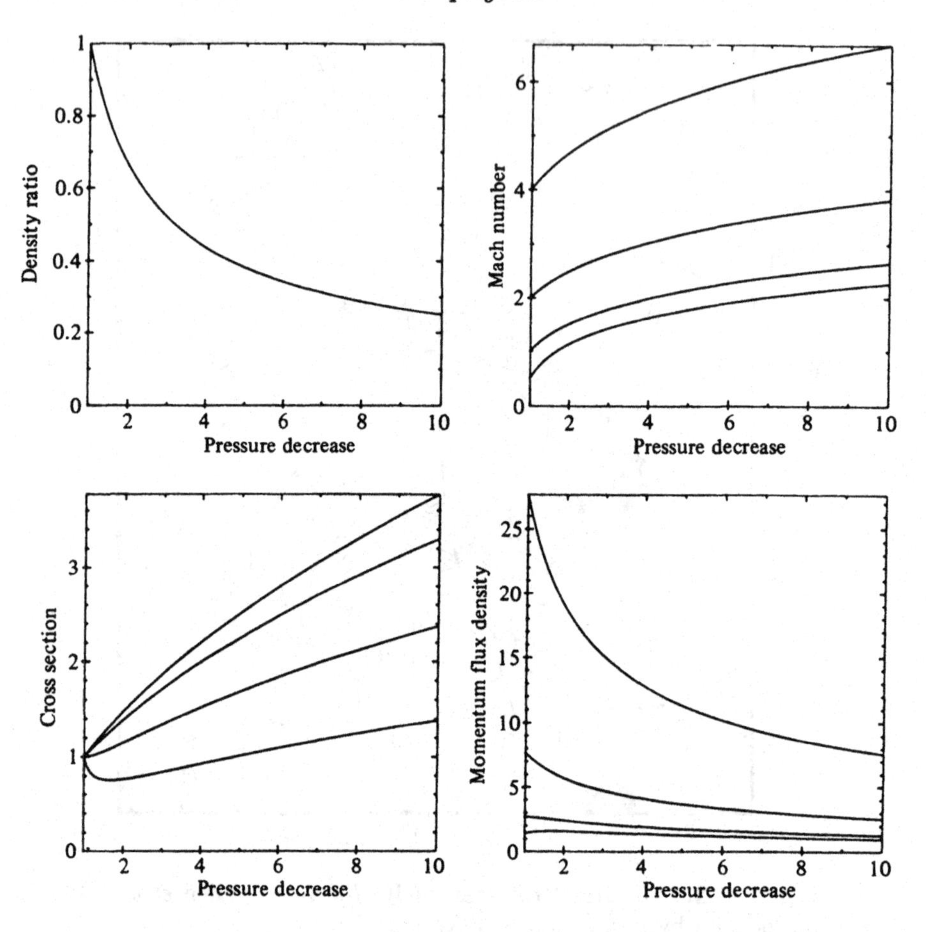

Fig. 7.14. Parameter evolution over a decade decrease in pressure through isentropic expansion for initial Mach numbers of 0.5, 1, 2 and 4. Higher Mach number flows produce larger final cross-sections.

by the decrease in the stagnation pressure of the cross-wind. We can then use the relations for isentropic quasi-one-dimensional flow to obtain an analytical model for the parameter evolution in a bent jet. The pressure in the jet is related to its stagnation pressure through equation (7.4) and its cross-section is related to the area of the throat by

$$\left(\frac{A}{A_0}\right)^2 = \frac{1}{\mathcal{M}^2}\left[\left(\frac{2}{\Gamma+1}\right)\left(1+\frac{1}{2}(\Gamma-1)\,\mathcal{M}^2\right)\right]^{\frac{\Gamma+1}{\Gamma-1}} . \qquad (7.17)$$

Taking the pressure in the jet to be the independent variable, Fig. 7.14(a), (b) and (c) show the evolution in the density, Mach number

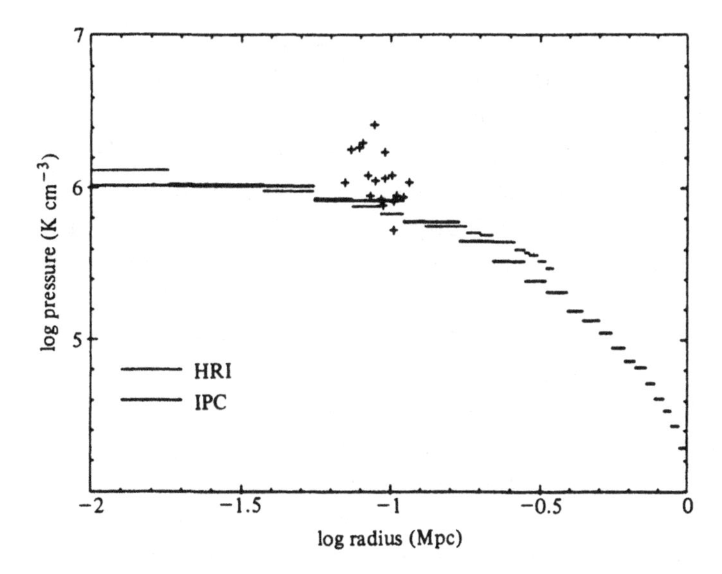

Fig. 7.15. Radial pressure variation in the Cygnus A cluster, derived from *Einstein* observations, together with radio synchrotron minimum pressures from Arnaud *et al.* (1984).

and cross-section of four jets with initial Mach numbers of 0.5, 1, 2 and 4. An order of magnitude decrease in pressure results in a small increase in the jet's Mach numbers and cross-sectional areas. Note that the cross-section of the $\mathcal{M} = 0.5$ jet initially decreases in the de Laval nozzle. The evolution of momentum flux density, $P + \rho v^2$ (Fig. 7.14(d)), which shows the greatest decrease in high Mach number jets, is consistent with the lack of strong hotspots in radio-trail sources.

7.3.3 Galactic atmospheres and bridges

Although many features of extragalactic radio sources can be successfully reproduced by simulations of the basic model, observations of X-ray bremsstrahlung (Fabbiano *et al.* 1984; Arnaud *et al.* 1984) suggest that, rather than being uniform, the environments of Classical Double radio sources consist of extended gaseous atmospheres in which pressure and density decrease with radius on the galactic and the cluster scale (Fig. 7.15). Since the flow pattern created by an axisymmetric jet depends both on its Mach number and its density ratio (Norman *et al.* 1982), we should expect the dynamics of extragalactic radio sources to be influenced by the effect of a decreasing atmosphere. This approach has already been followed by Baldwin (1982) and Saunders (1982) for source models in which lateral confinement of the lobes is provided by ram-pressure balance. By extending Scheuer's (1974) model A to include a power-law variation in external density, they showed that evolutionary tracks could be found in the P-D diagram (Baldwin 1982) which satisfy both the decrease in number density and in luminosity of sources larger than about 300 kpc.

Numerical simulations of a jet in a decreasing atmosphere (Figs. 7.16 and 7.17) show that two qualitatively different structures are formed according to whether or not the cocoon of waste products, created at the hotspot, extends back along the entire length of the jet. Note that even jets which produce cocoons over their entire length in a constant ambient medium, may have their cocoons squeezed out to larger radii by the effects of buoyancy. The waste products in Fig. 7.16 are created at sufficiently low-pressure for the cocoon to be crushed near the source of the jet, confining the backflow from the hotspot in the manner of reflective boundary conditions in the symmetry plane of the basic model. The parameters of the jet then evolve rapidly, because it is in lateral pressure balance with the confining medium until, further from the galaxy, the jet enters its cocoon. If, on the other hand, the cocoon pressure is sufficiently large, as in our simulations of high Mach number, low density jets (Fig. 7.17), the variation in the density and pressure of the ambient medium has little effect, and the jet is surrounded by a cocoon over its entire length. In this case, because the pressure gradient inside the cocoon is small, evolution of the jet parameters is relatively slow. In terms of the density *ratio*, parameter evolution will take place vertically in the (η, $\mathcal{M}$) plane (see Sumi & Smarr 1984), although the situation is complicated by the fact that, once the jet becomes significantly over-pressured, its behaviour is no longer well-described by the two-parameter (η, $\mathcal{M}$) model. Fig. 7.17, for example, shows that the jet is surrounded by a cocoon even though its density is greater than the density of the intergalactic medium at the head of the jet.

7.3.4 Cocoon dynamics in a decreasing atmosphere

The model for cocoon size (§7.1.4) can be used to determine the conditions under which sources have their cocoons crushed near

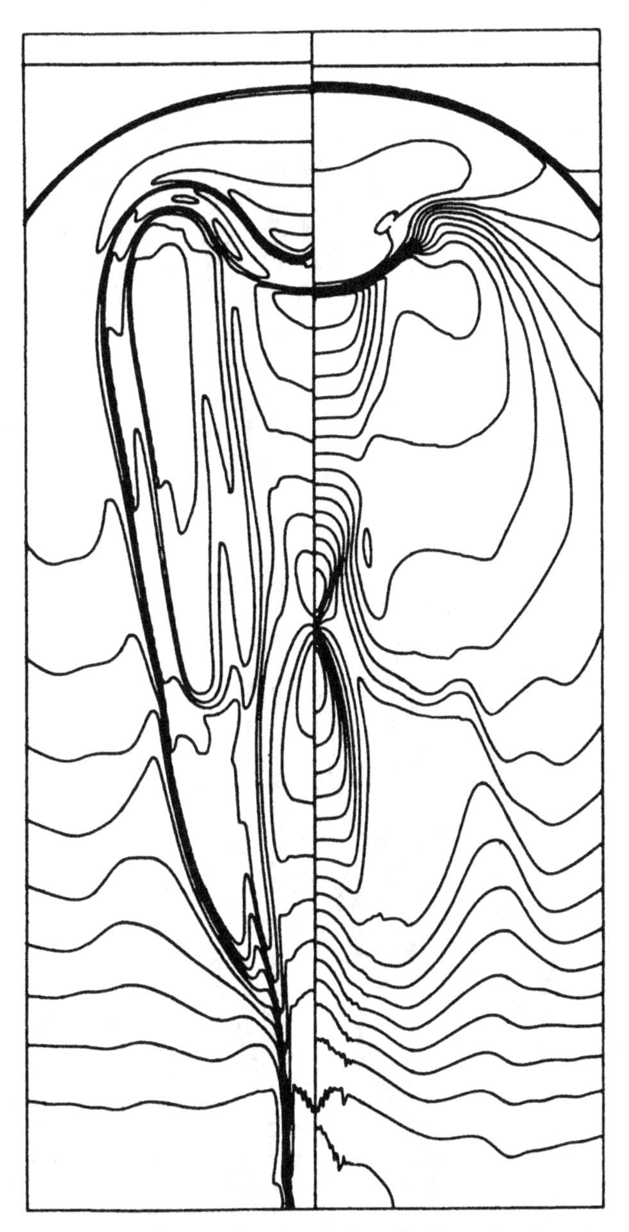

Fig. 7.16. The distribution of density (left) and pressure (right) in axisymmetric simulations of jets in an isothermal atmosphere. The gravitational field was assumed to be fixed in space, and to act wholly in the z-direction. The isothermal atmosphere was simulated using a King model, and was chosen to allow a decrease of about two decades in density and pressure across the computing grid. The contours were logarithmic with density ratio 1.5 : 1 and pressure ratio 1.3 : 1. The jet parameters were $\eta = 0.1$, $\mathcal{M} = 3$.

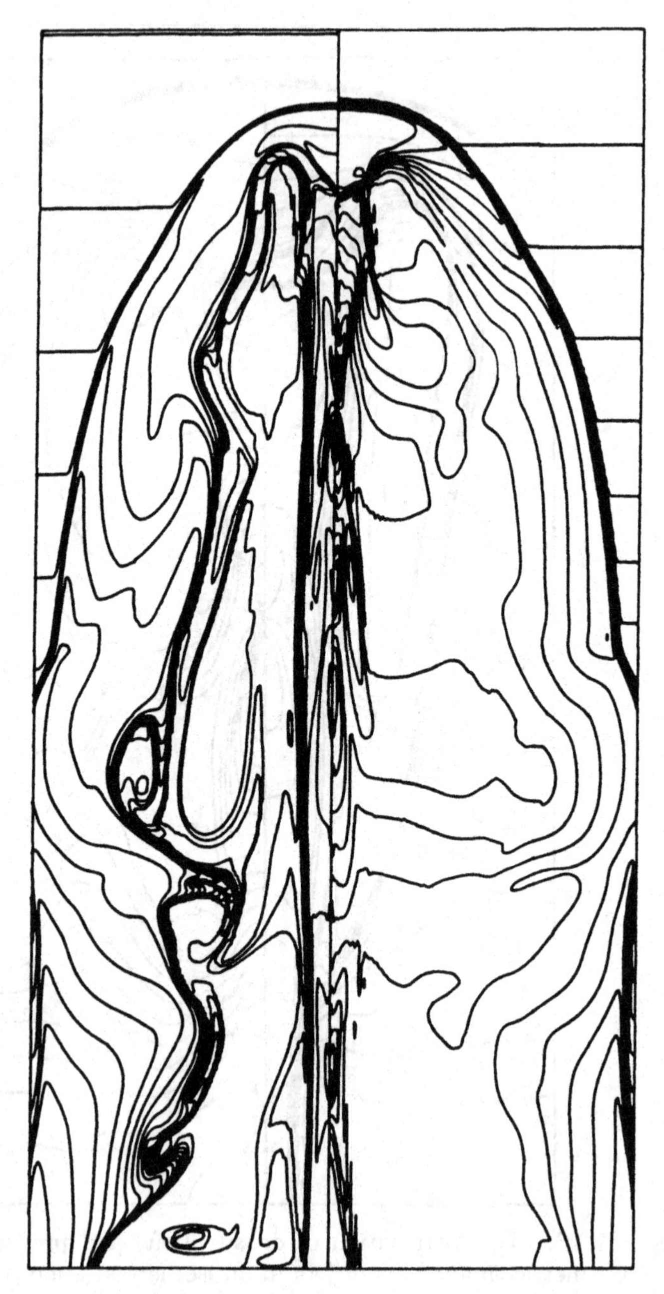

Fig. 7.17. As Fig. 7.16, with $\eta = 0.01$, $\mathcal{M} = 10$.

the galaxy. Since the radius of the lobes is given by

$$R_c = R_j \left(\frac{P_h}{P_c}\right)^{\frac{1}{2\Gamma}} \left(\frac{v_j}{v_h}\right)^{\frac{1}{2}} \tag{7.18}$$

the (half) length of sources with axial ratio λ is then given by

$$D = \frac{R_j}{\lambda} \left(\frac{P_h}{P_c}\right)^{\frac{1}{2\Gamma}} \left(\frac{v_j}{v_h}\right)^{\frac{1}{2}} . \tag{7.19}$$

We can use this relation to determine the variation in cocoon pressure with source size. If we assume that the product $R_j P_h^{\frac{1}{2\Gamma}}$, the ratio v_j/v_h, and the axial ratio of the source, all remain approximately constant, the cocoon pressure is then proportional to $D^{-2\Gamma}$, a strongly decreasing function of source size. Clearly, the cocoon pressure cannot decrease indefinitely without falling below the thermal pressure in the surrounding medium. Whereas in Classical Doubles, such as Cygnus A, the cocoon pressure is still large enough to prevent lobe squeezing (cf. synchrotron and bremsstrahlung pressures, Arnaud *et al.* 1984), the decrease in cocoon pressure with source size in giant radio galaxies, such as NGC 6251 (Fig. 7.18), may be sufficient for their inner lobes to be squeezed away from the galaxy, leaving the jet naked. In weaker sources with lower hotspot pressures, cocoon squeezing may take place long before they become giants.

7.3.5 **Axial ratios**

Simulations of the basic model reveal two significant discrepancies between the model and observations. Whereas the distribution of axial ratios in the radio source population is more-or-less independent of source size (Saunders 1982), simulations of jets in a constant ambient medium produce axial ratios that are strongly correlated with the length of the source. In fact, after an initial period of lateral expansion, the width of the cavity remains more-or-less constant. An associated discrepancy is that whereas observations suggest that radio jets expand with cone angles in the range $3° - 15°$ (Bridle & Perley 1984), simulations of the basic model produce jets whose mean width remains more-or-less constant over its length at any time. The simulations in Figs. 7.16 and 7.17 show that both difficulties may be significantly ameliorated by the inclusion of a decreasing atmosphere. Instead of remaining constant as in simulations of the basic model, the width of the (supersonic)

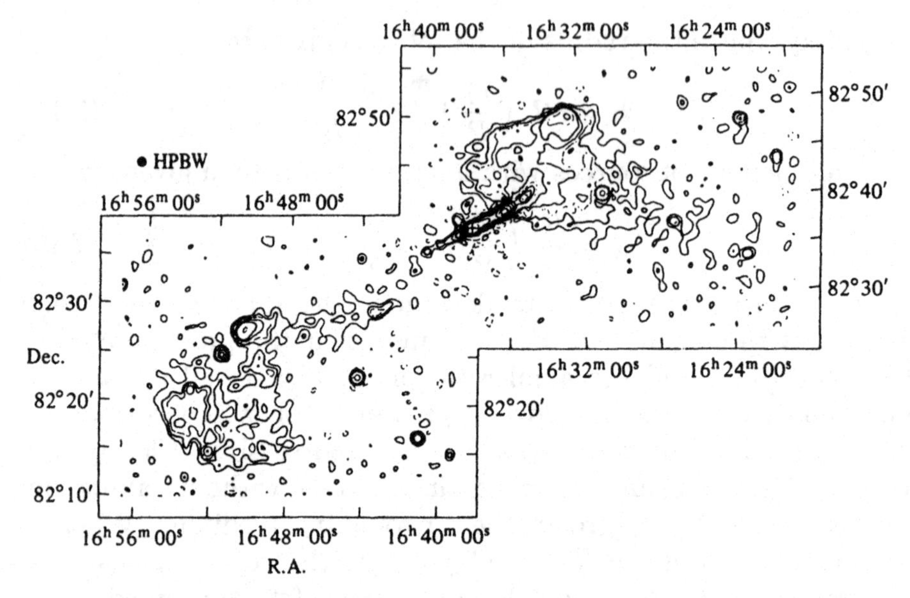

Fig. 7.18. 610 MHz map of NGC 6251 from Willis *et al.* (1982).

jet and the lobes then increase due to the decreasing pressure of the confining medium, with a consequent reduction in the rate of growth of the axial ratio.

7.3.6 Bridge distortions

Simulations of jets in a decreasing atmosphere show that, like Scheuer's model C, the cocoon can be pinched-off near the nucleus. Breaking axial symmetry changes the problem more fundamentally, and can produce bridges that bend away from the galaxy. Large-scale fully three-dimensional simulations will be needed to confirm the effect, but three possible symmetry-breaking mechanisms (Fig. 7.19) have been suggested:

1. The source axis is not parallel to the axis of a spheroidal gas distribution, giving rise to the bridge distortions in 3C 274.1 (Fig. 7.20).
2. Motion of the source galaxy through the surrounding intergalactic medium creates a ram-pressure gradient across the bridge and also distorts the gaseous halo of the galaxy into an egg-shape. This mechanism may be responsible for the bridge distortion in 3C 430 (Fig. 7.12).
3. If the source axis has varied considerably in the past, the old

(*a*)

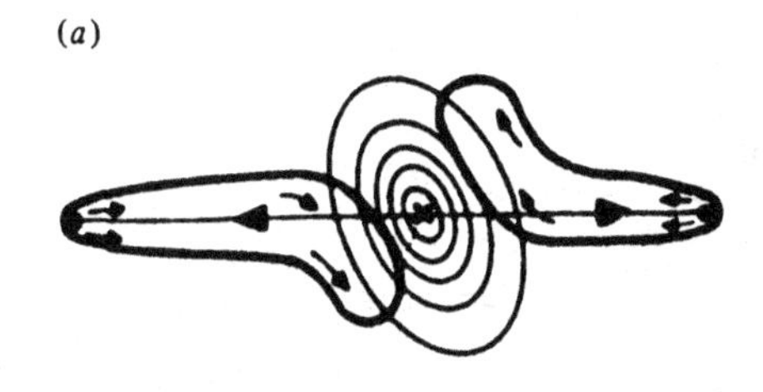

(*b*)

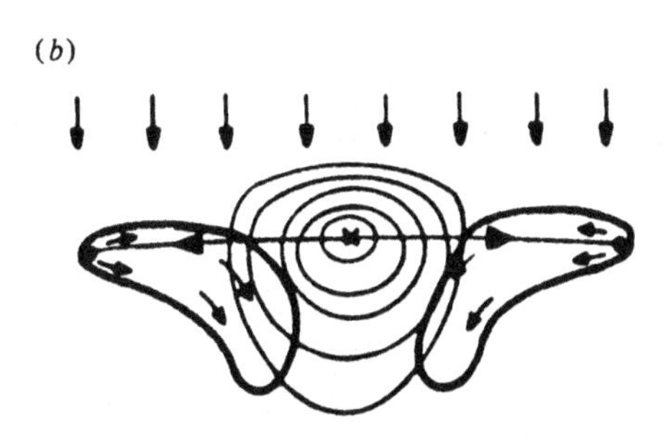

(*c*)

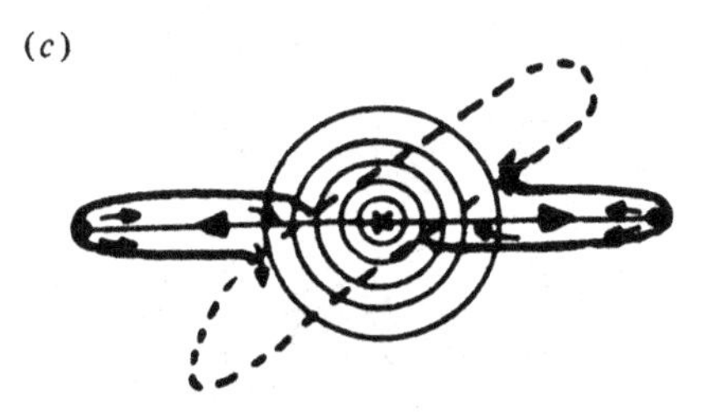

Fig. 7.19. Sketch diagrams of the three symmetry-breaking mechanisms discussed in the text. Each diagram shows the expected X-ray contours (thin lines), radio core (x), the outline of the radio source (thick line), and arrows giving the flow pattern. (a) Non-aligned spheroidal gas distribution leading to a type 2 distortion; (b) Galaxy motion leads to type 3 distortion; (c) Old cavity channels backflow (dashed outline) leading to type 2 distortion.

cocoon provides a channel into which backflow from the new bridge may be distorted. Such a mechanism may explain the extreme distortions in 3C 315 (Fig. 7.21).

This model may go some way towards accounting for the fact that at least 60% of low-redshift radio galaxies have distorted bridges

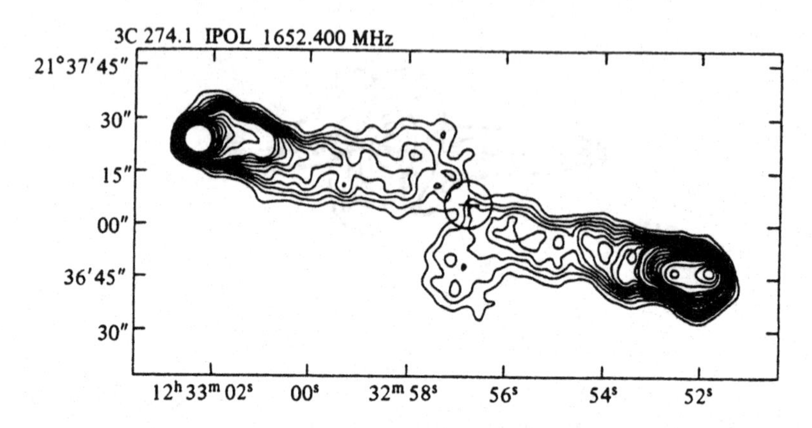

Fig. 7.20. 1652 MHz map of 3C 274.1 from Leahy & Williams (1984).

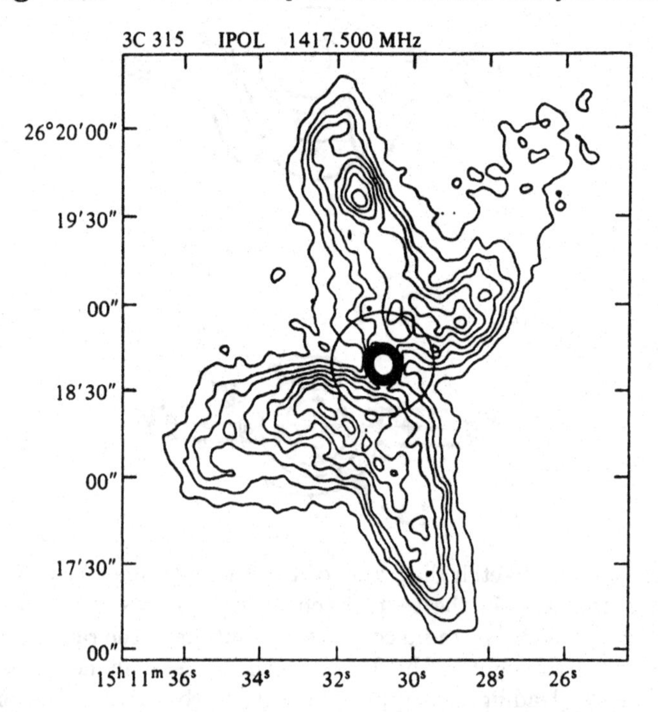

Fig. 7.21. 1417 MHz map of 3C 315 from Leahy & Williams (1984).

(Leahy & Williams 1984). As Laing (private communication) has pointed out, we cannot appeal to a model such as that applied to NGC 326 by Ekers *et al.* (1978) in which the striking rotational symmetry is attributed to motion of hotspots at the end of a pair of precessing beams, since randomly oriented precessing sources should appear X-shaped or S-shaped with equal probabilities, while only X-shaped sources are found amongst FR II sources.

7.4 Conclusions

The author's work described in this chapter, which was carried out at MRAO between 1982 and 1986, was motivated by a desire to try to understand how the diverse, and often strikingly beautiful, maps of extragalactic radio sources that were then pouring off the VLA could be produced.

Although computational constraints restricted both the number and resolution of simulations that could be performed, the basic model of a light jet interacting with a dense external medium successfully reproduced a wide variety of radio source structures with relatively few variable parameters.

Not only did the basic model reproduce the observed Fanaroff & Riley classification with supersonic jets producing simple, linear double-lobed structures containing compact hotspots, but analytical models suggested by the results of the simulations allowed the Mach number and density ratio of the jet to be deduced from the ratio of cocoon width to jet width and hotspot pressure to lobe pressure. Applying this model to Cygnus A produced density ratios of the order of $10^{-3.5}$ and Mach numbers of order 10.

While much of the gross morphology of radio sources could be reproduced by axisymmetric simulations, closer inspection of the maps suggested that genuinely three-dimensional effects would be crucially important in accounting for detailed features. And so it turned out. Even the relatively low-resolution three-dimensional simulations it was possible to run on a Vax 11-780 contain qualitatively new and quite different features to those produced by axisymmetric simulations.

Varying the direction of the jet inside a previously inflated cocoon, for example, reproduced the multiple hotspot structure first noted by Laing, and a simple model for the evolution of jet parameters at an oblique shock proved able to account for the asymmetric double hotspot structure commonly found in these sources.

Simulations of a jet in a cross-wind, on the other hand, showed that the same jet parameters that produced Classical Doubles in a stationary external medium, could produce Radio Trail Sources when acted on by a cross-wind, and a variety of other types of bridge distortions could be attributed to atmospheres, asymmetries and motions in the external medium.

Symbols used in Chapter 7

Symbol	Meaning
A	cross-sectional area
A_0	cross-sectional area at stagnation point ('throat')
A_c	cross-sectional area of cocoon
$A_{j,d}$	cross-sectional area post jet-shock
$B_{\rm min}$	'minimum energy' magnetic field
c_S	sound speed
D	half-length of source
$\mathcal{M}$	Mach number
$\mathcal{M}_b$	Mach number of backflow
$\mathcal{M}_c$	Mach number in cocoon
$\mathcal{M}_j$	Mach number of jet
$\mathcal{M}_{j,d}$	Mach number post jet-shock
$\mathcal{M}_{j,u}$	Mach number pre jet-shock
n	number density of relativistic electrons
P	pressure
P_c	pressure in cocoon
P_h	pressure in hotspot
P_j	pressure in jet
$P_{j,d}$	pressure post jet-shock
$P_{j,u}$	pressure pre jet-shock
P_0	pressure at 'static point'
R_c	radius of cylindrical cocoon
$R_{\rm curv}$	radius of jet curvature
R_j	radius of jet
t	time
$u_{\rm min}$	minimum energy density
v	velocity
v_b	velocity of backflow
v_h	advance speed of hotspot
v_j	velocity of jet material
α	frequency spectral index ($S \propto \nu^{-\alpha}$)

β	angle of oblique shock
γ_L	cutoff energy in particle spectrum
Γ	ratio of specific heats
η	ratio of jet to ambient density
θ	angle of flow deflection in shock
λ	axial ratio
ν_L	low frequency cutoff to spectrum
ρ	density
ρ_c	density in cocoon
ρ_{ext}	density in external medium
ρ_j	density in jet
$\rho_{j,d}$	density post jet-shock
$\rho_{j,u}$	density pre jet-shock

References

Alexander, P., Brown, M. T. & Scott, P. F., 1984, *M. N. R. A. S.*, **209**, 851.

Andersen, J. D., 1982, *Modern Compressible Flow*, (McGraw-Hill: New York).

Arnaud, K. A., Fabian, A. C., Eales, S. A., Jones, C. & Forman, W., 1984, *M. N. R. A. S.*, **211**, 981.

Baldwin, J. E., 1982, In *IAU Symposium 97*, eds. Heeschen, D. S. & Wade, C. M. (Reidel: Dordrecht, Netherlands), p. 21.

Begelman, M. C., Blandford, R. D. & Rees, M. J., 1984, *Rev. Mod. Phys.*, **56**, 255.

Benford, G., 1978, *M. N. R. A. S.*, **183**, 79.

Biretta, J. A., Owen, F. N. & Hardee, P. E., 1983, *Ap. J. Lett.*, **274**, L27.

Blandford, R. D. & Königl, A., 1979, *Ap. Lett.*, **20**, 15.

Blandford, R. D. & Rees, M. J., 1974, *M. N. R. A. S.*, **169**, 395.

Bridle, A. H. & Perley, R. A., 1984, *Ann. Rev. Astr. Ap.*, **22**, 319.

Burns, J. O., Clarke, D., Feigelson, E. D. & Schreier, E. J., 1984, In *Physics of Energy Transport in Extragalactic Radio Sources*, eds. Bridle, A. H. & Eilek, J. A. (NRAO: Greenbank, WV), p. 255.

Downes, A. J. B., 1980, *M. N. R. A. S.*, **190**, 261.

Dreher, J. W. & Feigelson, E. F., 1984, *Nature*, **308**, 43.

Ekers, R. D., Fanti, R., Levi, C. & Parma, P., 1978, *Nature*, **276**, 588.

Fabbiano, G., Miller, L., Trinchieri, G., Longair, M. & Elvis, M., 1984, *Ap. J.*, **277**, 115.

Falle, S. A. E. G. & Wilson, M. J., 1985, *M. N. R. A. S.*, **216**, 79.

Hardee, P. E., 1979, *Ap. J.*, **234**, 47.

Kronberg, P. P., Clarke, J. N. & van den Bergh, S., 1980, *Astr. J.*, **85**, 973.

Kronberg, P. P. & Jones, T. W., 1982, In *IAU Symposium 97*, eds. Heeschen, D. S. & Wade, C. M. (Reidel: Dordrecht, Netherlands), p. 157.

Laing, R. A., 1982, In *IAU Symposium 97*, eds. Heeschen, D. S. & Wade, C. M. (Reidel: Dordrecht, Netherlands), p. 161.

Leahy, J. P. & Williams, A. G., 1984, *M. N. R. A. S.*, **210**, 929.

Lonsdale, C. J. & Barthel, P. D., 1984, *Astr. Ap.*, **135**, 45.

Norman, M. L., Smarr, L., Winkler, K.-H. A. & Smith, M. D., 1982, *Astr. Ap.*, **113**, 285.

Norman, M. L., Winkler, K.-H. A. & Smarr, L., 1984a, In *Physics of Energy Transport in Extragalactic Radio Sources*, eds. Bridle, A. H. & Eilek, J. A. (NRAO: Greenbank, WV), p. 150.

Norman, M. L., Smarr, L. & Winkler, K.-H. A., 1984b, In *Numerical Astrophysics: A Festschrift in Honor of James R. Wilson*, eds. Centrella, J., Le Blanc, J. M. & Bowers, R. L. (Jones and Bartlett: Boston).

Norman, M. L. & Winkler, K.-H. A., 1985, *Los Alamos Science*, **12**, 38.

O'Dea, C. P. & Owen, F. N., 1986, *Ap. J.*, **301**, 841.

Payne, D. G. & Cohn, H., 1985, *Ap. J.*, **291**, 655.

Perley, R. A., Bridle, A. H. & Willis, A. G., 1984, *Ap. J. Suppl.*, **54**, 292.

Rees, M. J., 1978, *M. N. R. A. S.*, **184**, 6lp.

Sanders, R. H., 1983, *Ap. J.*, **266**, 73.

Saunders, R. D. E., 1982, Ph.D. Thesis, University of Cambridge.

Scheuer, P. A. G., 1974, *M. N. R. A. S.*, **166**, 513.

Scheuer, P. A. G., 1982, In *IAU Symposium 97*, eds. Heeschen, D. S. & Wade, C. M. (Reidel: Dordrecht, Netherlands), p. 163.

Schreier, E. E., Gorenstein, P. & Feigelson, E. D., 1982, *Ap. J.*, **261**, 42.

Smith, M. D., Norman, M. L., Winkler, K.-H. A. & Smarr, L., 1985, *M. N. R. A. S.*, **214**, 67.

Spangler, S. R., Myers, S. T. & Pogge, J. J., 1984, *Astr. J.*, **89**, 1478.

Sumi, D. M. & Smarr, L., 1984, In *Physics of Energy Transport in Extragalactic Radio Sources*, eds. Bridle, A. H. & Eilek, J. A. (NRAO: Greenbank, WV), p. 168.

Vallée, J. P., Bridle, A. H. & Wilson, A. S., 1981, *Ap. J.*, **250**, 66.

Williams, A. G. & Gull, S. F., 1984, *Nature*, **310**, 33.

Williams, A. G. & Gull, S. F., 1985, *Nature*, **313**, 34.

Willis, A. G., Strom, R. G., Perley, R. A. & Bridle, A. H., 1982, In *IAU Symposium 97*, eds. Heeschen, D. S. & Wade, C. M. (Reidel: Dordrecht, Netherlands), p. 141.

Wilson, M. J., 1983, *Ph. D. Thesis*, University of Cambridge.

Wilson, M. J. & Scheuer, P. A. G., 1983, *M. N. R. A. S.*, **205**, 449.

Winter, A. J. B., Wilson, D. M. A., Warner, P. J., Waldram, E. M., Routledge, D., Nicol, A. T., Boysen, R. C., Bly, D. W. J. & Baldwin, J. E., 1980, *M. N. R. A. S.*, **192**, 931.

Woodward, P. R., 1986, In *Astrophysical Radiation Hydrodynamics*, eds. Winkler, K.-H. A. & Norman, M. L. (Reidel: Dordrecht, Netherlands), p. 245.

8

The Production of Jets and Their Relation to Active Galactic Nuclei

PAUL J. WIITA
Department of Physics and Astronomy, Georgia State University, Atlanta, GA 30303, USA.

8.1 Introduction

The actual initiation of jets is a subject that remains extremely difficult to discuss in a detailed and convincing manner despite the obvious importance of this fundamental topic. There are two main reasons for this. The first is a lack of unequivocal observations. Although VLBI measurements have provided structural information on scales corresponding to $\lesssim 0.1$ pc in extragalactic sources, the phenomena that govern the beginnings of jets almost certainly occur on scales at least two to three orders of magnitude smaller. Other observations, especially of X-ray and optical variability, are indubitably important and provide useful constraints on models; however, they do not yield information that is clearly interpretable in a model-independent fashion.

The second generic difficulty has to do with the certainty that the physical processes involved in producing jets are extraordinarily complex. The core of the picture, that accretion onto a supermassive black hole (SMBH) of somewhere between 10^6 and $10^{10} M_\odot$ is at the heart of beam generation as well as the other properties of active galactic nuclei (AGN), has been commonly accepted for about a decade. However, in attempting to add details to this picture, astrophysicists find that general relativity, hydrodynamics, plasma physics and radiation transport all form thick blobs on their palettes, and the portraits which emerge from combining them in different proportions are, not surprisingly, rather different. On the basis of current observations and theoretical understanding, choosing between several of these candidate views is extremely hard from a scientific standpoint and such choices might better be deemed art criticism. Furthermore, the probability that several blends of "colours" are involved in different types of AGN is high.

Because this chapter is addressing such a broad and difficult aspect of this subject the approach cannot possibly be as quantitative as that found in several other chapters. The dearth of directly applicable data also forces this discussion to be somewhat more speculative than most other parts of this volume. §8.2 summarizes some characteristics of the most popular models for central engines. Attention will be directed to the ways in which beams are supposedly produced in those models. Models based upon hydrodynamic mechanisms are treated first, and in somewhat more detail, than those dominated by magnetohydrodynamics. The emphasis in treatment is determined more by our current knowledge and the level of development of the scenarios than by their viability probability. In §8.3 some of the relationships between small scale jets and the matter in the central regions of AGN are discussed, with an emphasis different from those in Chapter 4. A few conclusions are given in §8.4. In conformity with the great majority of literature in this area, cgs units are used throughout this chapter.

To close this introduction, it should be pointed out that while the possibility of eventually sorting out all of the difficulties with powerhouse models does exist, the "galactic weather forecasting" analogy of Phinney (1985) must be borne in mind. Although the gross features of our weather are determined by solar insolation and the density of the earth's atmosphere, no one can argue that the full complexities of the weather are properly understood, nor can one possibly claim that additional knowledge of these details will give us much significant information concerning the solar interior. Similarly, many of the observational details of jets and other properties of AGN may be able to shed very little light on the ultimate nature of the powerhouse. We may have to rely on self-consistency and aesthetics in choosing which picture(s) to buy.

8.2 Central engine models

The literature on models for various aspects of AGN is immense, even after removing a number of extremely implausible suggestions. Thus it is not possible to do justice to all, or even most, of the serious ideas that have been put forward. Recent reviews that cover some of the same material discussed in this section are

Begelman (1985); Begelman, Blandford & Rees (1984); Blandford (1986); Phinney (1985); Rees (1984a,b; 1986) and Wiita (1985). Kundt (1987) stresses and Wiita (1985) covers some of the more unorthodox proposals such as magnetoids, supermassive discs and white holes; while some of those ideas deserve more attention, they will not get it here.

8.2.1 Why supermassive black holes?

Many lines of evidence support the argument that SMBHs are at the cores of all AGN. Recent reviews of this evidence include Wiita (1985); Blandford (1986); and Trimble & Woltjer (1986), and here we will just briefly mention some of these points.

8.2.1.1 Rapid variability

Significant changes are often seen over days in the optical and over hours in the X-ray parts of the spectrum, strongly suggesting extremely compact active regions. Smaller, but still significant, optical variability over periods of tens of minutes in BL Lacertae objects has recently been firmly detected by Miller *et al.* (1989). The radius of the event horizon of a black hole of mass M_{BH} and angular momentum parameter a is

$$R_s = 1.48 \times 10^5[1 + (1 - a^2)^{1/2}](M_{BH}/M_\odot) \text{ cm}, \tag{8.1}$$

and in the absence of relativistic effects the minimum time for variability would be $\Delta t \approx 2cR_s$. Relativistic beaming effects could allow for the variations we observe at a distance apparently to be faster than those seen by an observer in close proximity to the SMBH (cf. Chapter 4). However, the inclusion of these (always uncertain) effects would nominally allow even higher mass objects to be at the centre.

8.2.1.2 Emission line widths

The broad emission line regions of quasars and Seyferts frequently correspond to spreads in velocities of greater than 5000 km s^{-1}. The shapes of these lines and the fact that most of the continuum emission escapes strongly imply that they are due to a large number of small clouds. The relative strengths of different lines enable densities of between 10^{10} and 10^{13} cm^{-3} to be determined and X-ray absorption measurements indicate column depths of about 10^{22} to 10^{23} cm^{-2}, implying radii of about 10^{11} to 10^{13} cm and temperatures of $\sim 10^4$ K (*e.g.*, Matthews & Capriotti 1985; Krolik 1988). Both variability measurements and photoionization models relying on the observed AGN continuum imply that this broad line region (BLR) is located within about $(L_{46})^{1/2}$ pc from the very centre of the AGN, where $L_{46} = L_{\substack{ionizing\\continuum}} / 10^{46}$ erg s^{-1}. While an entire industry has grown up around the manufacture of detailed models of the formation and motion of these BLR clouds, for our current purposes it does not matter whether they are infalling, outflowing or orbiting: the large velocities at those rather small distances indicate the presence of contained masses consistent with very massive SMBHs. Recent spectroscopic observations of broad line variability in the Seyfert galaxies Akn 120 (Peterson *et al.*. 1985) and NGC 5548 (Peterson & Ferland 1986; Peterson 1987) indicate that the BLR is more compact (a few light weeks) than normally assumed; in this case the line widths imply SMBHs of $\sim 10^8 M_\odot$. Some of the possible interactions of jets with the BLR will be discussed in §8.3.3.

Narrow emission lines (velocity spreads of 300-1500 km s^{-1}) are normally assumed to originate well outside the compact core, and it has been convincingly demonstrated that a large fraction of galaxies contain unusual low intensity emission line regions within their inner kpc or so. Such "LINERs" (*e.g.*, Keel 1985) have now been shown frequently to exhibit broad wings around their narrow lines; these wings have been used to support the claim that SMBHs are extremely common in galactic nuclei (Filippenko & Sargent 1985). Further, the marginally detected relative redshift of broad Balmer emission lines with respect to forbidden narrow lines can be attributed to the general relativistic gravitational redshift expected from the proximity of the BLR clouds to a SMBH (*e.g.*, Peterson 1987, 1988).

8.2.1.3 Radio observations

Several types of radio observations strongly support the SMBH idea. First, the frequency of good alignments between small scale and large scale jets (cf. §4.4) implies that their source has a good memory of direction for at least 10^6 yr. This long term stability is most naturally attributed to the spin axis of a SMBH. Second, regardless of which model for relativistic beaming is adopted, the probability of very rapid bulk motion fits naturally into a picture where a deep relativistic gravitational well is available. It is very difficult to conceive of a mechanism that can accelerate significant quantities of matter to velocities comparable to c without involving relativistic potential wells, which, in the case of an AGN, almost certainly has to be a SMBH. Finally, VLBI observations of NGC 1275 have shown a very small elongated central component perpendicular to the small-scale jet; this may well correspond to optically thick synchrotron emission from a disc extending $\sim 10^3 R_s$ from an $\sim 10^9 M_\odot$ BH (Readhead *et al.* 1983).

8.2.1.4 Power and efficiency

The tremendous power emitted by quasars, up to 10^{48} erg s^{-1}, provides yet another indirect argument in favour of this picture. The standard argument notes that outward radiation pressure drives out any infalling matter when the luminosity exceeds the Eddington limit

$$L_E = 4\pi GM/\kappa \approx 1.3 \times 10^{38} M_{BH}/M_\odot \text{ erg s}^{-1}; \qquad (8.2)$$

the opacity is expected to be due to electron scattering at the conditions encountered in an AGN, so that κ is taken as a constant. This corresponds to the conversion of $\dot{M}_E \equiv L_E/c^2$ completely into radiant energy.

Since nuclear fusion allows less than 1% of mass to be converted into energy, relying on that source demands incredibly high mass fluxes (hundreds of solar masses per year). However, accretion onto BHs can in principle allow the radiation of up to 42% (although more probably no more than 32%; see §8.2.2.1) of the rest mass energy of infalling matter, substantially reducing the necessary influx. Since some outbursts apparently correspond to the entire rest energy of a star, even extremely enhanced supernova rates seem incapable of providing a viable explanation, while the tidal disruption and swallowing of a massive star captured by a SMBH or a super-flare associated with a disc around one might possibly fit the bill. Still, it should never be forgotten that the huge energy requirements depend upon the now well-confirmed claim that quasars are at the large distances demanded by the cosmological interpretation of their redshifts and upon the less certain assumption that the overall emission is roughly isotropic.

8.2.1.5 Stellar dynamics

Steep rises of the velocity dispersion of stars towards the centres of galaxies and steep gradients in stellar velocities across the galactic nuclei attributed to rotation provide dynamical evidence in favour of massive, non-stellar, objects at the centres of a growing number of galaxies. Continuation of the rising surface brightness of galaxies towards their very centres also supports this hypothesis. The difficulties in obtaining high spatial resolution and removing the blurring effects of seeing on the spectra are tremendous, and mean that early conclusions that the active galaxy M 87 contains a $3 \times 10^9 M_\odot$ BH (Young *et al.* 1978; Sargent *et al.* 1978) are weak in that anisotropic velocity distributions can also explain the observations (*e.g.*, Binney & Mamon 1982). However, the case based on velocity dispersions and rotation curves for SMBHs of around $10^{6-7} M_\odot$ in nearby non-active galaxies such as M 31 and M 32 is quite impressive (Dressler 1984; Tonry 1984; Dressler & Richstone 1988). The evidence in favour of a similarly massive object at the Galactic Centre will be mentioned in §10.5.

8.2.1.6 Evolution of galactic cores

Although it would be impossible to convincingly compute every possible fate for a dense star cluster at the centre of a nucleus, the general consensus (*e.g.*, Begelman & Rees 1978; Wiita 1985) is that such a cluster is unstable and much of it will eventually collapse into a SMBH. Various massive objects supported by rotation and/or magnetic fields could certainly exist, and might well play roles in energizing AGN; however, they are probably transient phenomena and unlikely to be the dominant prime mover. Recent numerical simulations of the evolution of very dense clusters of solar mass sized compact objects explicitly illustrate the formation of a trapped surface, *i.e.*, black hole (Shapiro & Teukolsky 1985a,b; Kochanek *et al.* 1987), adding strength to this theoretical argument.

Some very plausible evolutionary scenarios for AGN in terms of the various types of accretion to be discussed in §§8.2.2 and 8.2.3 have been proposed (*e.g.*, Blandford 1986). Such pictures involve the growth of BHs in the vast majority of galaxies, with quasars and bright Seyferts corresponding to those SMBHs that are currently being fed at high rates (comparable to the Eddington limit), and

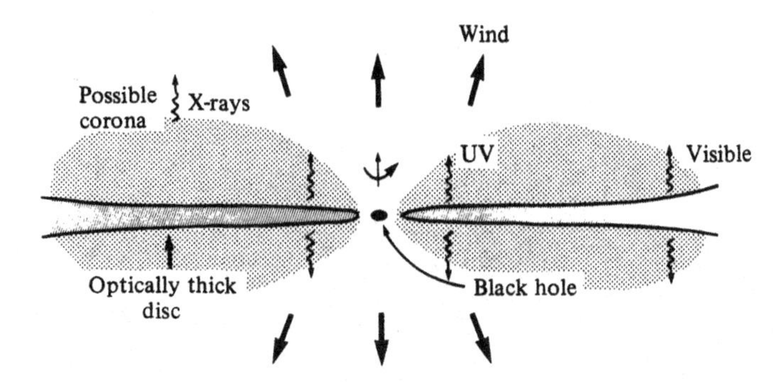

Fig. 8.1. Relatively low accretion rates and weak magnetic fields yield optically thick, but geometrically thin discs.

weak Seyferts, LINERS and radio galaxies being associated with subcritically fed holes. Although greatly simplified, such a scenario seems to fit observations of the cosmological evolution of the co-moving densities of Seyferts, quasars and radio galaxies surprisingly well (Blandford 1986).

8.2.2 Hydrodynamic beam production

One large class of models assumes that magnetic fields do not play a dominant dynamical role in the extraction of energy from matter in the vicinity of a SMBH, nor are they important in the acceleration and collimation of beams (Fig. 8.1 and 8.2).

In these pictures, radiation from an accretion disc, either as direct thermal emission from its surface due to dissipation of energy within the disc, or as non-thermal reprocessed emission, is responsible for most of the observed activity. Such emission might occur in an optically thin corona above and below a disc or it might involve inverse-Compton scattering of photons off relativistic electrons. The question of the production and exact nature of the AGN continuum spectra will not be discussed in any detail in this chapter. In this area too, a wide range of explanations still seems possible (Begelman 1985, 1988; Stein & O'Dell 1985; Krishan & Wiita 1986; Zdziarski 1986; Stein 1988).

8.2.2.1 Thin accretion discs

It is likely that the gas finding itself in the vicinity of a SMBH will have significant angular momentum, and if the mass infall rate is less than about $\dot{M}_E/2$, but is not exceedingly small,

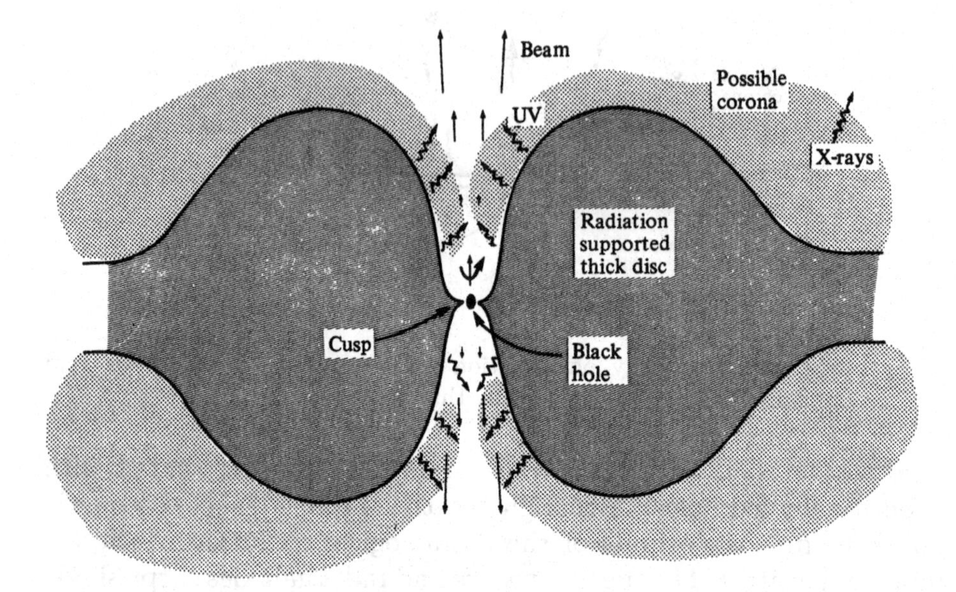

Fig. 8.2. High accretion rates can yield optically and geometrically thick radiation supported discs.

then it should evolve in such a way that most of the mass spirals inward in a thin disc while most of the angular momentum wends its way outward (*e.g.*, Shakura and Sunyaev 1973). Under these circumstances the great majority of the released binding energy can be radiated away from the disc and it can remain geometrically thin (Fig. 8.1). The assumption that the total disc mass is much less than that of the central body, along with the thinness of the discs, implies that their vertical and radial structures can be approximately decoupled. Although both molecular and radiative viscosities are far too small to drive significant accretion or release much energy, the assumption that turbulent motions and/or tangled magnetic fields can and will produce important macroscopic viscosities has been widely accepted. The standard models for accretion discs grossly simplify the picture and assume that the shear stress is related to the pressure by a constant viscosity parameter: $t_{\phi r} = \alpha^* P$. Nevertheless, our lack of a fundamental understanding of these processes has precluded the formation of fully self-consistent models.

In the case of a Schwarzschild black hole, the last stable circular orbit around a body of mass M is at (*e.g.*, Lynden-Bell 1969)

$$R_{ms} = 3R_s = 6GM/c^2. \tag{8.3}$$

The formula for the angular momentum per unit mass on a circular orbit is

$$l = [GMR^2/(R - 1.5R_s)]^{1/2}, \tag{8.4}$$

while that for the binding energy of a mass m in such an orbit is

$$\epsilon m = mc^2\{1 - (R - R_s)\,[R(R - 1.5R_s)]^{-1/2}\}. \tag{8.5}$$

The maximum efficiency with which mass can be converted to energy during the accretion process is equal to the binding energy of the last stable circular orbit; using equations (8.4) and (8.5) one has

$$l_{ms} = \sqrt{12}GM/c \tag{8.6}$$

and

$$\epsilon_{ms} = (1 - \sqrt{8/9}) = 0.05719. \tag{8.7}$$

Unstable orbits can exist down to the marginally bound radius, $R_{mb} = 2R_s$, where $\epsilon = 0$.

For a rotating (Kerr) BH, its angular momentum produces an effective repulsion of material orbiting in the same sense with the result that the radius of the last stable orbit is reduced and the binding energy is increased. It is natural to expect that even if a BH somehow formed without net angular momentum, it would be spun up via the addition of angular momentum from the accreted matter. It is believed that this process will not continue until the BH becomes an extreme Kerr BH, with angular momentum parameter $a = 1$, because of the preferential capture of counter-rotating photons emitted by the accreting matter; rather the spin- up should halt at $a = 0.998$ (Thorne 1974), where the efficiency $\epsilon \sim 0.32$.

Although a wide range of temperatures would be characteristic of the emission from the material spiralling into a SMBH, the maximum flux and highest temperatures both occur at only a few R_s. Such a disc has a maximum temperature of roughly (Begelman 1985)

$$T_{\text{max}} \approx 4.5 \times 10^7 (\alpha^* M_{BH}/M_\odot)^{-1/4}\ \text{K}, \tag{8.8a}$$

if radiation pressure dominates, while if the disc is dominated by gas pressure then

$$T_{\text{max}} \approx 8.7 \times 10^7 (\alpha^* M_{BH}/M_\odot)^{-1/5} \dot{m}^{2/5}\ \text{K}, \tag{8.8b}$$

where $\dot{m} \equiv \dot{M}/\dot{M}_E$. Gas pressure should be more important if $\dot{m} \lesssim 0.2(\alpha^* M_{BH}/M_\odot)^{-1/8}$. As long as the electron scattering opacity dominates free-free opacity and the disc is optically thick then the observed radiation will be a diluted blackbody with a temperature closer to

$$T_{\rm disc} \approx T_{\rm max}/\tau_{\rm disc}^{1/4} \approx 1.4 \times 10^7 \dot{m}^{1/4} (M_{BH}/M_\odot)^{-1/4} \text{ K}. \qquad (8.9)$$

Evidence for the existence of accretion discs around SMBHs of between $\sim 2 \times 10^8$ and $4 \times 10^9 M_\odot$ has come from detailed analyses of the IR-UV spectra of six powerful QSOs (Malkan 1983). Power laws plus single blackbody components do not fit the observed continua well, while the broader additions to the spectra that arise from theoretical models of geometrically thin and optically thick accretion discs do a good job of modelling the "big blue bumps", or near UV excesses above power-laws that characterize many quasar spectra. However, the weak dependences of $T_{\rm disc}$ on M_{BH}, $\dot{M}$ and α^* mean that any values for disc properties or black hole masses derived from QSO spectra must be treated gingerly. Despite this *caveat*, recent fits by Madau (1988) and Sun & Malkan (1988) to AGN spectra are impressive indeed.

Another approach towards determining SMBH masses involves comparing X-ray variability and emission linewidths (cf. §§8.2.1.1-.2). There is a good correlation between the shortest observed X-ray variability time, assumed to be proportional to the hole's mass, and a virial mass (derived from velocities obtained from [O III] linewidths and distances obtained from ionization parameters) if $0.003 \lesssim L/L_E \lesssim 0.03$ for QSOs and Seyfert 1's (Wandel & Mushotzky 1986). However, this analysis depends on uncertain selection effects and many reasonable, but by no means fully established, assumptions; while suggestive, it cannot yet be considered definitive.

Some of the variability might be due to unstable modes of oscillation in thin accretion discs. A long-wavelength instability that leads the inner parts of thin discs to break up into alternately dense and rarefied rings was discovered quite early (Lightman & Eardley 1974), and was later shown to be the limiting case of one of two families of instabilities to which ordinary α^*-discs were subject (Shakura & Sunyaev 1976); the other thermal instabilities have growth rates nearly independent of wavelength. Radiation pressure dominated thin discs are also unstable to axisymmetric instabilities unless the shear is proportional only to gas pressure and not the total pressure (*e.g.*, Camenzind *et al.* 1986).

Although thin accretion discs obviously define preferred directions for emission of matter that could then be collimated into beams, no well-developed models for beam production in the absence of dynamically controlling magnetic fields have been proposed for them. If thin discs are capable of generating significant outflows then they probably must do so through winds expelled from a corona above and below the disc. Such a corona could be produced through mechanical heating such as the dissipation of shocks or turbulence (*e.g.*, Liang & Price 1977). Alternatively, X-rays emitted from the very central region could Compton heat more distant portions of a thin disc which flares in thickness at larger radii (cf. Fig. 8.1); this might then produce a wind (Begelman, McKee & Shields 1983). It has been shown that a wind emerging from the outer portions of an accretion disc may be able to assist in the collimation of a more intense beam emitted from the very central portions (Smith & Raine 1985; Sol *et al.* 1989), which in turn is probably generated through one of the mechanisms to be discussed below.

8.2.2.2 Radiation supported thick accretion discs

When the accretion rate approaches and surpasses $\dot{M}_E$ not all of the energy generated by viscosity within the disc can be radiated from the surface of a thin disc. Two distinct possibilities for the structure of the accreting material then emerge: (1) the angular momentum is lost very efficiently through very strong turbulence or some other mechanism and then the accretion flow becomes quasi-spherical; (2) if the value of α^* is very small then the angular momentum loss remains gradual and the direction defined by the spin axis of the black hole and infalling gas retains its influence. The first possibility implies that essentially spherical supercritical winds are likely to be driven away from the SMBH (to be discussed in §8.2.2.3); the latter implies that a geometrically thick accretion disc, supported by radiation pressure and rotation in comparable amounts, will form around the SMBH (Lynden-Bell 1978; Paczyński & Wiita 1980); see Fig. 8.2.

8.2.2.2.1 Thick disc structure

Because the ratio of thickness to distance from the central object, h/r, is proportional to $\dot{m}$, it is clear that such supercritical discs would be geometrically thick, and are often called tori or doughnuts. For discs to be in hydrostatic equilibrium (*i.e.*, $|v_r| \ll c_s$, the local speed of sound)

$$|v_r| \approx c_s^2/v_\phi \approx \alpha^* c_s h/r. \tag{8.10}$$

For thin discs $h/r \ll 1$, so α^* of order unity is acceptable, but for thick discs, $h/r \approx 1$, so $\alpha^* \ll 1$ is necessary. Even at subcritical accretion rates, the clumping instabilities that affect thin discs could imply a time-averaged bloated structure in the inner regions (Paczyński & Wiita 1980). The recognition that radiation supported rings or tori of fluid could exist around BHs if the fluid had a non-Keplerian angular momentum distribution, and that the inner radii of these discs could penetrate R_{ms} (*e.g.*, Abramowicz *et al.* 1978), led to the suggestion that accreting material could form a pair of very steep vortices along the rotation axes. Such vortices could conceivably lead to collimated jets in QSOs and radio galaxies (Lynden-Bell 1978).

Since the vertical and radial parts of the disc structure cannot be separated, complete detailed models cannot be constructed analytically, and even numerical simulations (Hawley *et al.* 1984; Eggum *et al.* 1985; Clarke *et al.* 1985) must involve major simplifications. Nevertheless, it was shown that the boundary of such thick discs could be constructed with one free function (Paczyński & Wiita 1980; Jaroszyński *et al.* 1980). For example, the specific angular momentum (l) or equivalently the disc thickness (h), could be specified as a function of cylindrical radius (r), instead of just specifying a single free variable (usually chosen as α^*) for a thin disc. Even though $l(r)$ is a free function, basic stability requirements greatly restrict its allowed forms. Of course, in the presence of a BH, the appropriate relativistic generalizations of angular velocity and angular momentum should be used, *e.g.*,

$$\Omega = v^\phi/v^t, \qquad l = -v_\phi/v_t, \tag{8.11}$$

and, for a static distribution, the angular momentum is not constant on cylinders, but rather on curved Von Zeipel surfaces (Chakrabarti 1985a).

The large radiation pressure forces the rotational force to provide extra outward support close to the BH so that the specific angular momentum in a thick disc starts out at a value above the relativistic generalization of the Keplerian value, and rises more slowly than does $l_{\rm Kep}$; in the innermost regions of the accretion disc $l \approx$ constant must be true (Abramowicz *et al.* 1978). The two curves intersect at the point of maximum pressure and density in the equator of the torus, which typically lies at only a few R_s. Beyond that distance $l < l_{\rm Kep}$, but at large distances they should asymptotically agree, and the thick disc would eventually merge onto a "thin" disc of constant thickness $2\dot{m}R_s$. As long as the equation of state is assumed to be barotropic (*i.e.*, pressure solely a function of energy density) then Chakrabarti (1985a) shows that stability requirements and the above considerations imply that a very reasonable parameterization is

$$l = (\lambda - \lambda_0)^n + l_{\rm in} \qquad \text{with} \qquad \lambda^2 \equiv l/\Omega \tag{8.12}$$

where $0 \le n \le \frac{1}{2}$, and $l_{\rm in}$ is the specific angular momentum at the inner edge, $r_{\rm in}$ of the disc. That edge, which locates the cusp through which matter falls into the BH, must lie between the closest marginally stable orbit, R_{ms} (which is $3R_s$ for Schwarzschild BHs) and R_{mb}.

The most surprising claim from these analyses was that these radiation pressure supported discs can evince super-Eddington luminosities while remaining in mechanical equilibrium. These models assume that the radiative flux from the surface of the disc is locally near the Eddington rate, *i.e.*, $F_{\rm rad} = cg_{\rm eff}/\kappa$, and then the integrated flux from the highly non-spherical surface can exceed L_E. Thus thick, radiation supported disc models may be of relevance to quasars, Seyferts and BL Lacs, but almost certainly cannot be the primary explanation for optically faint radio galaxies.

As $\dot{M}$ rises the cusp at the inner edge of the disc proceeds inward from R_{ms} towards R_{mb}, and the radiated luminosity can increase beyond L_E. However, because ϵ drops concomitantly the accretion rate $\dot{M}$ rises faster than L (Paczyński & Wiita 1980). It turns out that radiation emitted from a torus around a rapidly rotating Kerr BH is $\sim 30\%$ higher than that from a disc around a Schwarzschild BH with the same accretion rate (Jaroszyński *et al.* 1980). Even though the existence of thick discs does depend on relativistic effects, *i.e.*, $r_{\rm in} < R_{ms}$, some useful approximations can be made using a purely Newtonian treatment, where the key parameter is the ratio of inner and outer disc radii. The maximum luminosity (for non-rotating BHs) is (Abramowicz *et al.* 1980)

$$L_{\max}/L_E \approx 2\ln(r_{\rm out}/r_{\rm in}) - 2.44, \tag{8.13}$$

as long as $r_{\rm in}/r_{\rm out} < 0.01$, with $r_{\rm out}$ the radius at which the disc is effectively geometrically thin.

8.2.2.2.2 Consistency and stability

Relatively simple thick disc models can be self-consistent only if: (1) the mass of the disc is small compared to the mass of the BH; (2) viscous processes, not nuclear fusion, generate the vast majority of the energy; and (3) the disc remains in mechanical equilibrium. Condition (1) is probably violated if $M_{BH} > 10^7 M_\odot$ (Wiita 1982b), although that does not preclude more complicated models where the disc's self-gravity is not ignored. Such models are possible, and probably not very much different in the inner regions of the disc, but values of $M_{\rm disc}/M_{BH}$ as low as ~ 0.01 can imply significant changes in the structure of the outer disc (Abramowicz *et al.* 1984). Allowing for the changes in disc structure induced by the disc's own gravity this constraint probably only rules out radiation supported thick discs around extraordinarily massive ($> 10^{11} M_\odot$) BHs (*e.g.*, Begelman 1985).

Simple thick discs around objects of less than 100 $M_\odot$ are unlikely, since fusion reactions would then probably generate more energy than viscous dissipation (Wiita 1982a,b). As far as AGN are concerned constraint (2) probably only serves to keep the temperature of the core of the disc below $\sim 10^8$ K (Begelman 1985). The possibility that significant nuclear fusion could occur in accretion discs in AGN without leading to disruption has recently been explored by Chakrabarti *et al.* (1987); they show that p-p nucleosynthesis could proceed smoothly as an important energy source but that CNO or helium fusion is not likely to be stable. Fully self-consistent discs with ongoing fusion reactions ought to be calculated before we can be confident that they play no significant role.

The stability of thick discs is a most important question that has received tremendous attention over the past few years. Too high an effective value of α^* would imply rapid collapse of the disc. But a direct application of constraint (3) is extremely model dependent in that our lack of knowledge of the applicable viscosity under these conditions means that we cannot make any convincing argument in either direction concerning this point (Wiita 1982b; Begelman 1985). A thick disc could be disrupted if material jets driven by radiation pressure from its funnels were too efficient in removing matter from the disc; however, this process is probably not tremendously important (Narayan *et al.* 1983), although here too more detailed calculations are needed.

Recently Papaloizou & Pringle (1984, 1985) have discovered a set of virulent instabilities that apparently afflict thick accretion discs. These instabilities are global and non-axisymmetric (and thus were overlooked in earlier, purely axisymmetric analyses); they grow on dynamical (orbital) time scales. The original calculations were confined to constant entropy, incompressible tori with constant specific angular momentum (Papaloizou & Pringle 1984; Blaes 1985a; Hanawa 1986) and so were not clearly important for "real" cases. Extensions of this work showed that even incompressible tori are subject to the same problem (Blaes 1985b). The physical nature of the instability was clarified when it was shown that even uniform entropy thin rings with arbitrary angular momentum distributions were subject to this type of non-axisymmetric instability. The rapid growth of the instability, however, demands a good reflecting boundary at either the inner or outer edge of the disc (Goldreich & Narayan 1985), which may well be lacking. Although all of the above work was done in a Newtonian framework, similar results were shown to hold when a "pseudo-Newtonian" (Paczyński & Wiita 1980) potential was employed (Blaes & Glatzel 1986), and the necessary fully relativistic framework has been given (Blandford *et al.* 1985) although it is not yet completely explored.

Further analysis indicated that the most general mode is a combination of the modes described above which are driven by compressibility and additional Kelvin-Helmholtz-like instabilities excited in regions of the disc where there were maxima in the ratio of vorticity to surface density (Papaloizou & Pringle 1985). The most detailed analytic treatment to date clarifies and changes the physical nature of these instabilities (Goldreich *et al.* 1986). It now appears that the modes are best understood as edge waves propagating backwards at the inner edge of the torus and forwards at the outer edge; they are coupled together at the corotation point which must occur at the pressure maximum. These modes can grow most rapidly when the torus is incompressible since the edge is sharply defined, but in a more realistic compressible case such modes can also propagate along the "natural soft edge provided by the density gradient" (Goldreich *et al.* 1986). However, the growth rates are somewhat less for thicker tori. For (Newtonian) angular velocity variations as $\Omega \propto r^{-q}$, $q = 3/2$ corresponds to Keplerian rotation, while $q = 2$ corresponds to constant specific angular momentum. The rapidly growing dynamical modes apply for $\sqrt{3} < q < 2$.

The importance of these instabilities has recently been demonstrated by numerical simulations (Zurek & Benz 1986; Hawley 1988). If an isentropic pressure supported torus is initiated with a constant specific angular momentum distribution then it very rapidly becomes unstable. But non-linear effects that cannot be treated by analytical computations come into play (*e.g.*, Kojima 1986), and such instabilities appear to provide an effective viscosity which rapidly redeploys the angular momentum into a distribution with $q \sim \sqrt{3}$ (Zurek & Benz 1986). Much more work on this question is required, as relativistic effects and wider ranges of initial assumptions must be incorporated. But at this point is appears reasonable to conclude that the dynamical instabilities do not necessarily disrupt thick accretion discs; rather, they may force them into very specific forms dictated by the viscosity provided by that self-same mechanism.

A very important recent calculation has finally included the effect of accretion from the thick tori onto the BH on the growth of these instabilities, albeit only in the two-dimensional approximation of an accretion annulus (Blaes 1987). It is found that when the inner edge of the flow crosses the critical cusp radius the flow into the hole is transonic; this removes the strong reflection necessary for the instability and stabilizes all of the modes that could be calculated. This is a preliminary result which might overstate the stabilizing effect of accretion on tori with respect to the full three-dimensional situation, but is of great interest and clearly must be amplified upon.

8.2.2.2.3 Beam formation

The basic geometry of thick discs, with their rather narrow funnels containing an extremely high radiation density, immediately suggested a natural way to produce beams (Lynden-Bell 1978; Paczyński & Wiita 1980). Early calculations using these models were promising, in that test particles within the funnel are definitely in a non-stationary zone: they either must fall into the BH or they must be expelled, and this cut-off point occurs very far down into the funnel. The net flux of radiation within the funnel is directed both upward and inward towards the axis, so both acceleration and collimation are to be expected.

The acceleration of a particle by radiation pressure within a beam is determined by (Jaroszyński *et al.* 1980; Abramowicz & Piran 1980)

$$(1-\frac{1}{z})(\frac{d\gamma_b}{dz}) = -\frac{1}{2}\frac{\beta}{z^2} + \gamma_b^2\{L_{\rm eff}\left[\left(1-\frac{1}{z}-\beta\right)^2 + (1-f)\,\beta\right]\left(1-\frac{1}{z}\right)^{-1} + u\,(1+f)\,\beta\}, \quad (8.14)$$

where u is the energy density of the uncollimated radiation, z is the height above the equator in units of R_s, $0 \leq f \leq 1$ is a factor describing the deviation of u from isotropy, and $L_{\rm eff}$ is the effective outgoing luminosity, approximately given by

$$L_{\rm eff} = L(2\pi/\phi^2), \quad (8.15)$$

with ϕ the opening angle of the funnel and L the total luminosity of the radiation emitted within the funnel. Because particles accelerated in the lower parts of the funnel feel the isotropic component of the radiation field as a drag force, efficient acceleration only starts to occur at a distance up the funnel of $\sim 100R_s$ and a terminal velocity is reached at heights of a few times that value (Abramowicz & Piran 1980). Mildly relativistic velocities ($\beta \lesssim 0.8$) are achieved for ordinary plasma in a wide variety of models constructed using different disc models (Abramowicz & Piran 1980; Sikora & Wilson 1981) but if the plasma is of lower mass, *i.e.*, predominantly electrons and positrons, then γ_bs around 3 or 4 are possible. However, this type of beam is not terribly well collimated, with half opening angles typically exceeding 10°. This is because the particle trajectories tend to diverge from the paths determined in the often narrower funnels once above the point where the funnels spreads out. Such optically thin beams, based on test particle calculations, cannot carry a large fraction of the total disc luminosity (Sikora

& Wilson 1981; Narayan *et al.* 1983). Although the fluxes and Lorentz factors involved may not be adequate for all types of jets and the collimations are not as good as often seen, this type of beam model may still have astrophysical significance is some cases (Wiita *et al.* 1982).

Computations of optically thick flows within accretion funnels are far more difficult, especially because at some point the distinction between funnel wall and accelerated material becomes fuzzy. Crude models that include the continuous loss of hot mass into the funnel from the walls and allow for moderately optically thick flows have been produced (Calvani & Nobili 1983; Calvani *et al.* 1983); they yield extremely hot beams, with $10^7\text{K} \lesssim T_b \lesssim 10^9\text{K}$ with $\gamma_b \sim 2$ for ordinary plasma. However, these models rely upon considering the flow in a one-dimensional approximation and also require arbitrary, and very unsure, assumptions concerning mass loss from the funnel walls, so any conclusions must be regarded as extremely uncertain. More detailed radiation transport calculations in mildly optically thick funnels have recently been calculated and distinctly anisotropic emission and polar outflows are produced (Madej *et al.* 1987). Still, these preliminary results do not indicate that extremely high velocities or very narrow beams are likely to emerge.

If the flows are extremely optically thick then the radiation gas mixture acts like an adiabatic fluid with polytropic index, $\Gamma = 4/3$, and such an outflowing mixture in a funnel could conceivably reach very high outflow velocities if the sonic point is close enough to the SMBH (Fukue 1982). However, Lu (1986) has shown that while extremely relativistic terminal velocities are possible under these idealized circumstances they can only be achieved if the enthalpy of the gas is extraordinarily high. Only if the sonic point for the flow is within $4R_s$ can β exceed 0.9, and it is extremely difficult to see how such a situation can possibly be attained. A general treatment by Chakrabarti (1985b, 1986) has also shown that in principle very rapid, very narrow beams can form, where the collimation is primarily engendered by the angular momentum contained in the fluid approaching the SMBH. These rotating wind solutions are very interesting but so far have employed very specialized disc models. They deserve to be tied to more detailed disc models, so the possibility that such narrow beams can really form in this way can be fairly evaluated.

Numerical hydrodynamical models for the formation of thick accretion discs and possible beams have been attempted, but in all cases significant approximations have had to be made so as to keep the computations tractable. Eggum *et al.* (1985) neglect general relativistic effects and assume an approximately constant value of $\alpha^* \sim 0.1$ for an $\dot{m} = 4$ calculation for an $M = 3M_\odot$ BH. They find that complex convection cells form inside the disc and block accretion flow along the disc midplane and the majority of the radiation is trapped and swallowed by the BH. Low density material near the photosphere is accelerated to $\beta \sim 0.3$ in a conical outflow and the outgoing mass flux is very small. Calculations incorporating full general relativistic effects, but assuming the infalling matter has fixed values of specific angular momentum and neglecting all viscosity save that generated by numerical shock smoothing have been performed by Hawley (1985). He finds that if l is in the right range to allow accretion tori to form, a complex series of shocks and convection patterns are set up, which do apparently stabilize in an axisymmetric computation. Published results to date do not yet clarify the likely strength and opening angle of ejected matter, because, among other reasons, the computational zone does not extend far enough away from the BH.

In conclusion, thick, radiation supported accretion discs do have many points in their favour, and, despite the uncertainties caused by the discoveries of the Papaloizou-Pringle instabilities, they probably do occur if circumstances are such that large amounts of matter rain upon a BH. But note that if the arguments of Wandel & Mushotzky (1986) are borne out by further data and analysis, then such supercritical conditions are, at best, infrequent. Nonetheless, assuming that supercritical mass flows are available then there is also little question that radiation supported tori can produce moderately powerful, reasonably well collimated, mildly relativistic beams. However, observations of extremely narrow jets (at least on much larger scales) and the probability that Lorentz factors of greater than five are needed for some sources, mean that they are not the most promising explanation of the origin of such jets.

8.2.2.3 **Windy models**

Assorted models for powerful, rapid outflows have been proposed that in principle are independent of the details of an accretion flow. While none of them have received the attention of other ideas discussed in this chapter, these approaches are intriguing.

8.2.2.3.1 **Funnel winds**

Models for beam formation based upon results taken from solar wind theory have also been suggested for AGN (Ferrari *et al.* 1984a,b, 1985, 1986). These wind type flows can be quasi-spherical or confined within funnels, either produced by radiation supported tori (§8.2.2.2) or ion supported tori (§8.2.3.1); the key new feature is that multiple critical points are found to exist in the flow so that the solutions are steady and transonic, but very possibly multiple. Details of this picture are too complex to develop here, but a summary of the exciting results is worthwhile. In a simplified Newtonian version of this wind scenario, a highly supersonic flow can be generated if the flow tube shape varies appropriately and there is significant non-thermal deposition of energy within the fluid (Ferrari *et al.* 1984a). In particular, if the flow expands rapidly, as it might on emerging from the outer part of an accretion disc funnel, and it is also acted upon by electromagnetic, plasma or MHD waves, then transonic solutions are easily achieved just above or within the disc region. Changes in the cross-section would naturally result in shock discontinuities and associated compression, heating, and particle acceleration (cf. Chapter 9). An important result is that degenerate solutions can exist; *i.e.*, for given boundary conditions while there is always one continuous solution there are often additional discontinuous solutions. Which branch is followed depends sensitively on the history of the flow, indicating that numerical experiments must be treated very cautiously.

This basic idea was expanded to include relativistic motion of an optically thin wind within an accretion funnel (although the temperature is assumed to be non-relativistic) and again multiple critical points are found (Ferrari *et al.* 1985). Many solutions of the quasi-two-dimensional relativistic Navier-Stokes equations can be found if a lengthy, albeit reasonable, list of assumptions is made. A key claim is that if $L < L_E$ such flows are typically accelerated to mildly relativistic velocities ($\beta \lesssim 0.28$) very close to the disc. Perhaps more interestingly, for hot winds emerging from funnels of highly supercritical thick discs, $\beta > 0.9$ is possible, in contrast to the conclusions based upon purely radiative acceleration discussed in §8.2.2.2.3. Another interesting result is that an increase in the collimation of the radiation field leads to the critical points descending deeper into the funnel even though the acceleration is not purely radiative; such a shift both increases the geometric collimation of the beam by the funnel and increases the amount of energy the radiation field deposits into the supersonic part of the flow, thereby raising the terminal velocity. However, it must not be forgotten that these calculations depend on a large number of parameters and simplifications; in particular, only essentially isothermal flows are accelerated very efficiently, and the relevance of these assumptions to real situations is anything but clear. Nonetheless, when rotational (cf. Chakrabarti 1985b) or magnetic (*e.g.*, Siah 1985; Clarke *et al.* 1986; Siah & Wiita 1987) effects are included the collimation and acceleration which are essentially produced by this wind-type acceleration should be even better.

8.2.2.3.2 **Supercritical Comptonized winds**

Another type of wind scenario assumes that no funnel can survive if the accretion rate is high enough. Rather, the BH would essentially be smothered (Shakura & Sunyaev 1973; Meier 1982) and a quasi-spherical wind could be expelled (Becker & Begelman 1986a,b). Applications of this type of wind to AGN are motivated by the evidence emerging from some Broad Absorption Line Quasars that mass outflows exceeding $\dot{M}_E$ are probable (*e.g.*, Drew & Boksenberg 1984). If the scattering optical depth below a critical sonic radius, r_c, is so high that the flow velocity of the wind exceeds the diffusion velocity of the photons then the radiation is trapped inside r_t and is advected along with the flow. Such supercritical winds require a mass loss such that

$$r_c \ll r_t \equiv \frac{1}{2} R_s \dot{m}. \tag{8.16}$$

The great majority of this radiation is then available to accelerate the wind through adiabatic cooling. Under these circumstances the radiation and matter will be coupled by multiple Compton scatterings and must essentially act as a single adiabatic fluid.

These models depend upon the assumption that nearly all of the energy is deposited in a thin layer of matter near the base of the flow, at r_i, perhaps through turbulent or shock dissipation. As long as $r_i \ll r_t$ significant acceleration and a relativistic terminal velocity can be achieved. The luminosity seen by an observer at infinity is roughly given by (Becker & Begelman 1986b)

$$L_i/L_E \approx (2r_c/R_s)^{-1/3}\dot{m}, \tag{8.17}$$

and this can greatly exceed L_E as long as the supercritical outflow traverses a sonic point which is deep in the relativistic potential. The emergent spectrum can have a colour temperature consistent with those argued for QSO blue bumps while the Compton scattered hard tail can be made to fit the overall QSO X-ray spectra.

This approach is indubitably interesting, but does require rather extreme conditions that are not likely to be frequently encountered. This is particularly true in that the mechanical heating assumed is not likely to be very efficient, thereby implying a tremendous excess supply of energy is required. These calculations have also been performed assuming spherical symmetry, so that it is not at all clear if this idea can have any productive role in beam formation.

A brief mention of a related way in which radiation pressure might drive jets is in order. If the fluid is so optically thick that radiation is trapped and somehow $P_{\rm rad} > \rho_{\rm matter}c^2$ then relativistic outflows could be produced in a "cauldron" (Begelman & Rees 1983). Any collimation achieved would depend upon the distribution of the matter at larger scales and is unlikely to be very good.

8.2.3 Magnetohydrodynamical models

The other major category of central engine and beam models consider the magnetic fields to be of dominating importance. The existence of dynamically significant magnetic fields on relevant scales is consistent with polarization measurements (cf. §§4.3.2, 4.3.3 and 8.3.1), and many explanations for the continuum spectrum, such as the synchrotron self-Compton model, also require $B \sim 10^{2-4}$ gauss in the region very close to the SMBH. A detailed understanding of magnetohydrodynamical processes under such extreme conditions is obviously even more difficult to obtain than a similar level of understanding of purely hydrodynamical events, but a great deal of progress has been made over the past few years, and a few plausible scenarios have emerged.

8.2.3.1 Extraction of the black hole's rotational energy

As long as one ignores quantum evaporation, which is certainly a reasonable thing to do when considering SMBHs, a Schwarzschild BH is purely a sink of mass and energy; however, the same is not true for Kerr-Newman BHs. Energy can be extracted from a rotating BH if particles and/or fields penetrate its ergosphere, the region between the event horizon and the static limit for rotating and/or charged BHs (Penrose 1969). In principle it is possible to remove energy equivalent to a significant fraction of the mass of an extreme ($a = 1$) Kerr BH, leaving an irreducible mass of $M_{\rm ir} = M/\sqrt{2}$; the available energy is thus an immense $[(\sqrt{2}-1)/\sqrt{2}]Mc^2$, or 29%, of the BH's rest-mass energy. This energy is most easily understood as being extracted if a particle breaks up within the ergosphere, with one piece falling into a negative energy state of the BH, while the other emerges with more energy than the initial one had to begin with; after such an event the BH is left rotating less rapidly.

8.2.3.1.1 **"Pure" Penrose processes**

Applications of this idea in astrophysically useful situations are few and they are all based upon greatly simplified pictures. It has been analyzed in the context of "inverse-Compton" (Piran & Shaham 1977) and pair production (Kafatos & Leiter 1979) mechanisms near rapidly rotating black holes. An infalling photon that collides with a proton or electron fairly deep within the ergosphere can knock the particle into a negative energy orbit and emerge with a high energy as seen by an observer at infinity, but the huge flux of γ-rays produced by this process does not agree with observed AGN spectra. It is also a problem that in the absence of large electromagnetic fields, such a process could only work if the incident particle starts with a relativistic velocity, which means the efficiency is very low. It has recently been pointed out that if a BH is immersed in an electromagnetic field then negative energy states can be opened up to even slowly moving particles (Dhurandhar & Dadhich 1984; Wiita *et al.* 1983) and the efficiency of this process is not necessarily negligible (Wagh *et al.* 1985). The efficiency of this "magnetic Penrose process" is given roughly by (Parthasarathy *et al.* 1986)

$$\epsilon \approx 10^{-16}\mu B(M_{BH}/M_{\odot}), \tag{8.18}$$

with μ the charge per unit mass of the escaping particle. Although ϵ is essentially unity for reasonable values of the parameters if the disintegration is near the static limit, no convincing self-consistent model of the particle break-up picture has yet emerged. Further, it appears as if the emitted particles tend to be concentrated towards the equator and not the poles of the BH, so direct beam formation by this mechanism is unlikely.

8.2.3.1.2 **Black hole magnetodynamics**

A more promising way to extract the energy of a rotating BH via magnetic fields exists, and one particular model will be discussed in §8.2.3.2. Although even the simplest general set of BH MHD solutions depends upon an extraordinarily complicated evaluation of the equations of axisymmetric magnetohydrodynamics in a curved background spacetime, several things are the same in all types of magnetospheres. Phinney (1983) has illustrated the basic analogy between: (a) a conductor rotating in a uniform magnetic field with and without surrounding plasma and the subsequent induced charge density and energy loss and (b) a BH surrounded by a magnetic field and plasma. In the absence of an external plasma and currents, an electric field of magnitude $aBR_s^2/(Mr^2)$ and an azimuthally circulating Poynting flux are induced; however, no net radiation escapes if the field is aligned with the spin axis. If the magnetic axis is misaligned by an angle Υ then the BH feels an aligning torque and surface currents dissipate energy on the horizon at a rate proportional to $(\Omega_{BH}BR_s^2\sin\Upsilon)^2$ (Damour 1978) and the loss can be made analogous to that of an ordinary rotating conductor if the surface impedance of the horizon is identified as $Z_H = 4\pi/c = 377$ ohms.

Allowing a circuit, presumably in the form of an accretion disc with associated currents, to interact with the BH means that its rotational energy can be tapped and delivered as work at large distances. In one simplified picture the disc acts like a unipolar inductor, generating an electric field in the inertial frame and driving a DC Poynting flux. If charged particles of either sign can leave the disc to form a force-free magnetosphere above and below it then, in classical notation,

$$\mathbf{F}_{\mathrm{tot,EM}} = \rho_c\mathbf{E} + \mathbf{j} \times \mathbf{B} \approx 0, \tag{8.19}$$

with a space charge ρ_c yielding the divergence of $\mathbf{E}$ and the current distribution $\mathbf{j}$ modifying $\mathbf{B}$ (Blandford 1976; Blandford & Znajek 1977). A related view of this situation is that the large electric potentials actually cause a vacuum breakdown which generates electron-positron pairs which in turn set up the force-free magnetosphere. If this is physically possible then the energy and any generated beams apparently can be focused along the rotation axes.

However Phinney (1983) argues that if the accretion disc produces any significant quantity of > 1 MeV γ-rays then the electron-positron pair density produced by photonic collisions in the magnetosphere would exceed that allowed for a force-free magnetosphere. Under these circumstances a perfect MHD approximation becomes applicable and drift forces transfer energy from fields to bulk kinetic energy. While the complexity of the MHD equations is such that no complete solution is likely to be found, some general results can be obtained. This is because four integrals of the motion exist (the energy at infinity/unit rest mass; angular momentum at infinity/unit rest mass; ratio of rest mass to magnetic flux in a flux tube; and the angular velocity of magnetic field lines, Ω_F), and boundary conditions at the BH horizon are known. A disc is necessary to confine the magnetic field that threads the hole's horizon and magnetic flux is advected into the hole by the infalling matter until $B^2/8\pi \approx P_{\rm disc}$. Once that condition is achieved any subsequent accretion would probably be by a slow "drip" process accompanied by magnetic reconnection, so no more flux is added to the BH. There should be a region between the BH horizon and infinity where particles are created, since some fall into the hole, while others are accelerated out to infinity. Physically, this could be accomplished by the photon-photon collisions mentioned above or by flares from the disc. The ratio Ω_F/Ω_{BH} determines the type of emergent wind. Optimal extraction of energy from the BH is conceivable if $\Omega_F \approx (1/5)\Omega_{BH}$, for then there is near "impedance matching" between the magnetosphere load and the BH dynamo, and the power radiated is 64% of the conceivable maximum (which would occur if $\Omega_F \approx \frac{1}{2}\Omega_{BH}$, but is hard to arrange) (Phinney 1983). It also turns out that the maximum power can be extracted from a BH with $a \sim 0.85$, for while raising the spin of the hole increases the effective EMF, at very high values of BH angular momentum the field is excluded from the horizon because of the strong dragging of inertial frames (Phinney 1986).

The recently developed "membrane" technique, based upon the 3+1 split of spacetime into space + time, for treating BH physics provides the mathematical and conceptual framework which allows extensions of calculations such as the ones just discussed to be performed (Macdonald & Thorne 1982; Thorne 1986; Thorne *et al.* 1986). For example, using this approach it is surprisingly simple to demonstrate that the mass of the hole decreases due to both spindown and ohmic heating exactly at the rate at which energy is deposited in a resistive load between an assumed perfectly conducting disc and the BH; *i.e.*, (Thorne 1986)

$$\frac{dM}{dt} = T_{BH}\frac{dS_{BH}}{dt} + \Omega_{BH}\frac{dJ}{dt} = -\frac{dE_L}{dt}, \tag{8.20}$$

where T_{BH} and S_{BH} are the temperature and entropy of the BH given by

$$T_{BH} \approx 6 \times 10^{-8}\mathrm{K}\,(M/M_\odot), \tag{8.21a}$$

and

$$S_{BH} \approx 1 \times 10^{77} k_B (M/M_\odot)^2 \tag{8.21b}$$

and $dE_L/dt = I^2 R_L$, with I the current generated by this battery-like BH through the plasma of total resistance R_L. Although the application of this technique to specific cases through numerical simulation has begun (Macdonald 1984), much more remains to be done if sufficiently realistic models along these lines are to be produced.

The ability of magnetospheric models to produce and accelerate beams with relatively high ratios of kinetic power to photonic emission is a strong point and highly desirable for ordinary radio galaxies. While all explicit beam shapes are based upon extremely simplified assumptions, there is no reason not to expect that quite well collimated, relativistic jets can emerge, carried outward by the powerful Poynting fluxes. Hence this general approach must be considered a favourite explanation for the origin of jets at this stage, but the high degree of uncertainty involved in any application to date should be borne in mind.

8.2.3.2 Ion supported tori

As mentioned above, thick accretion discs supported by radiation pressure will always have $L \gtrsim L_E$, while many active galactic nuclei, particularly in radio galaxies, emit $L \ll L_E$ in the optical and UV bands where the bulk of the disc radiation should emerge. However a geometrically thick disc, with the natural abil-

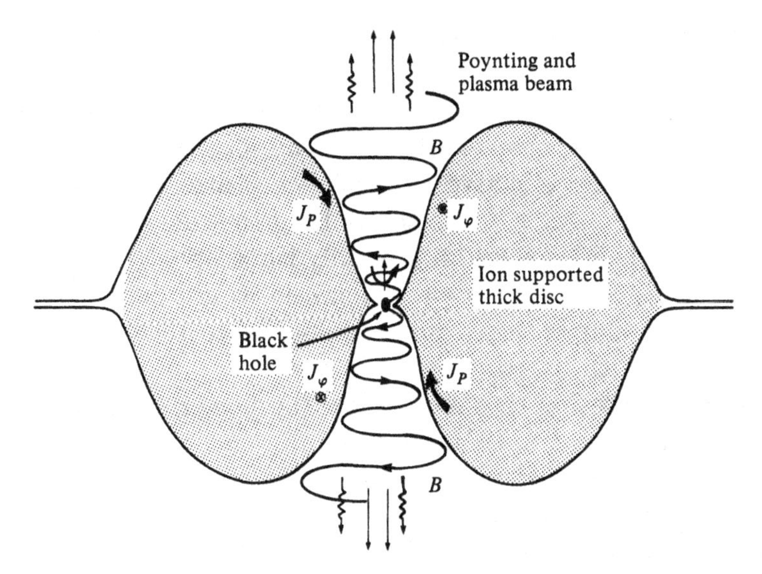

Fig. 8.3. Very low accretion rates and strong magnetic fields can produce optically thin, but geometrically thick ion-supported discs that tap the black hole's rotational energy.

ity to launch and collimate beams, *can* radiate much less than L_E if it is supported by ions whose temperature is much higher than that of the electrons (Rees *et al.* 1982); see Fig. 8.3.

Such a two-temperature disc could be established in the inner regions of an accretion flow (where $r/R_s < m_p/m_e$) if the bulk motions connected with the infall are easily randomized and if the plasma spirals inward on a time scale shorter than its cooling time so that the energy of the ions is not shared with the electrons. Under these circumstances the ion temperature can lie close to the value expected from the virial theorem while the electron temperature would stabilize at a much lower value:

$$T_p \approx m_p c^2 R_s/(k_B r); \qquad T_e \approx m_e c^2. \tag{8.22}$$

These ion-supported tori might exist if the densities are low enough and this translates into an upper limit on the accretion rate (Rees *et al.* 1982)

$$\dot{m} \lesssim 50(v_{\text{infall}}/v_{\text{free-fall}})^2. \tag{8.23}$$

If collective effects are inefficient at coupling ions to electrons so that Coulomb collisions dominate the energy exchange, then this picture is viable, and extremely subcritical accretion can still produce bloated geometries.

Approximate models of such discs can be made with some additional assumptions. For example, if the ions are hot, but non-relativistic, then the effective specific heat ratio is $\sim 5/3$ and if $\alpha^* \approx$ constant then $\rho \propto r^{-3/2}$ so the bulk of the mass of the torus is at large radii. The ions would be roughly isentropic in such a model and the pressure and temperature maxima would occur very close to the BH. The shapes would be very similar to those of radiation supported tori, with narrow funnels along the spin axis of the BH.

Because the accretion rate is very low the power emitted through viscous heating would also be very small, and the production of any powerful jets requires extracting part of the spin energy of the BH by means of magnetic fields, as discussed above. Let us now consider some specific results relevant to the context of ion supported tori. If B_{pH} is the poloidal field in the flux tube that intersects the hole's horizon, and a is the Kerr parameter, a small amount of matter flowing along that tube into the negative energy orbits inside the ergosphere of the BH can produce an external positive Poynting flux of luminosity (Rees *et al.* 1982)

$$L_{EM} \lesssim (B_{pH} R_s a)^2$$
$$\approx 10^{39} (M_{BH}/M_\odot) \dot{m} B_{pH}^2 / (P_{\rm gas} \alpha^*) \text{ erg s}^{-1}, \qquad (8.24)$$

where $P_{\rm gas}$ is the maximum gas pressure in the torus. Typical values might be $M_{BH} \sim 10^8 M_\odot$, $\dot{m} \sim 10^{-4}$, and $\alpha^* \sim 0.01$ with $P_{\rm gas} \approx B_{pH}^2$, so powers of $\sim 10^{44}$ erg s^{-1} are envisioned. The magnetic field would be predominantly toroidal and confined to the funnels, while large toroidal currents and return poloidal currents would flow in a thin surface layer of the ion supported disc. The Poynting flux would rapidly become infiltrated with a plasma and the domination of the toroidal fields would produce a self-collimated beam (*e.g.*, Chan & Henriksen 1980; Blandford & Payne 1982).

Such an extremely efficient scenario fits both evolutionary (Blandford 1986) and "unified" models of AGN where M_{BH}, $\dot{M}$ and the viewing angle to the beam, θ, determine the observed category of object (*e.g.* Blandford 1984), and this is indubitably a large part of its appeal. Early high accretion phases could produce standard thin or thick accretion discs that would be very bright optically and correspond to quasars or Seyferts. During this phase the SMBH would both grow in mass and spin up to a high value of a. But when the available supply of plasma dwindles then an ion supported torus could take over; spin energy could slowly be extracted from the SMBH. Now the energy output in relativistic, low density beams would far exceed $\dot{M}c^2$, but the optical emissivity would be very low: a typical strong radio galaxy could be produced.

However this particular picture requires that the magnetic field associated with the infalling matter must be very small, for otherwise the external fields would probably yield a very efficient outward transfer of both energy and angular momentum, thereby restricting L_{EM} to less than about $\dot{M}c^2$. While it is possible for this to be true, and it is also possible that the required minimal mechanical coupling between the beam and the torus be achieved, it is by no means clear that this is likely. Yet another uncertain requirement of this model is that it demands that no collective plasma processes are efficient enough to couple the ions and electrons in the torus on a much shorter than a Coulombic interchange time scale.

While this model is definitely ingenious and has been very widely cited, its complexities and uncertainties undoubtedly explain why it has not been elaborated upon until very recently. Because the temperatures involved are so high, electron-positron pair production ought to be considered in conjunction with such discs and it turns out that its effect can be very important (Begelman *et al.* 1987; Tritz & Tsuruta 1989). Two basic mechanisms of pair production are possible: pair creation by photon produced by collisional processes in the thermal plasma will be important since such high temperatures are involved, but a nonthermal "cascade" initiated by photo-photon collisions in the presence of magnetic fields may dominate. Regardless of the assumed mechanism of pair production, it is likely that the large additional number of leptons in the innermost part of the flow produces many more collisions with the ions so that the Coulomb scatterings are more frequent and the ions' energy is quickly shared with the electrons. Begelman *et al.* (1987) argue that this strongly cooled inner part of the flow yields a small optically thick but geometrically thin disc that could couple to the two-temperature accretion flow further out if $\dot{m} \gtrsim 0.1$. If thermal pairs predominate, such a dual-type model appears to be explicitly unstable, with the inner annulus expanding outward since the photons it emits can cool the outer part of the flow. This mod-

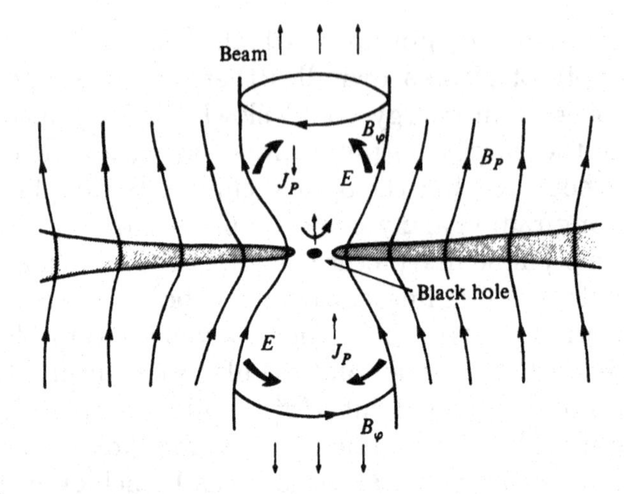

Fig. 8.4. Magnetic fields can also dominate the energy extracted from a geometrically thin, optically thick disc.

ified model implies that abrupt transitions between types of flows with and without cool inner discs can occur even though changes in the accretion rate are only small and continuous. Such a shift in the nature of the flow might provide the explanation for some types of sudden flares and spectral changes in AGN.

The combination of the above uncertainties, coupled with the possibility that instabilities similar to those discussed above for radiation supported thick accretion discs might afflict these ion supported tori, means that we cannot have real confidence in this entire class of model yet. However, its promise remains high and justifies intensive study.

8.2.3.3 Magnetized disc models

Another class of models assumes that magnetic fields are of extreme importance in the production of beams, but do not rely upon extracting energy from the BH. In these scenarios the accretion disc's fields are used to accelerate and collimate charged particles (Fig. 8.4).

An interesting dynamo mechanism which yields two steady collimated beams of ultrarelativistic protons can be produced if a disc is assumed to be both conductive and magnetized (Lovelace 1976). The key assumption is that the differential rotation in the inwardly spiralling matter tends to amplify the z-component of the field, and for simplicity it is assumed that B_ϕ is non-vanishing only within the disc. Then a large potential between the inner ($r_{\rm in}$) and outer ($r_{\rm out}$) edges of the disc of

$$V \approx 10^8 B_* M_{BH}/M_\odot \ln(r_{\rm out}/r_{\rm in}) \text{ volts} \tag{8.25}$$

is generated, where $B_* = \max(B_z)$, is expected to be $\sim 10^3$ gauss. A plasma capable of conducting a return current back to the disc is assumed to exist so that charge neutrality can be maintained in the disc. If B_z is anti-parallel to the angular momentum vector of the disc, then the potential on the disc implies that electric fields $\approx (v_\phi/c)B_z$ are produced; this greatly exceeds the gravitational attraction of protons to the disc, but is weak with respect to the poloidal magnetic field. In this case, a self-consistent space-charge limited flow of protons will yield a current, $I \approx cV$ (Lovelace 1976) off both faces so that the total power in the two beams is

$$L_b \approx 10^{22} B_*^2 \,(M_{BH}/M_\odot)^2 \,[\ln(r_{\rm out}/r_{\rm in})]^2 \text{ erg s}^{-1}. \tag{8.26}$$

On the other hand, if the magnetic field and angular momentum vectors are parallel rather than anti-parallel, electrons rather than protons are preferentially stripped off the disc, but similar voltages and powers can be achieved (Lovelace *et al.* 1979). In this latter case, electromagnetic cascades should produce an overall neutral e^+e^- beam, triggered by primary γ-rays. Although the voltages traversed are huge ($\sim 10^{21}$ V for reasonable parameters) the particles do not reach energies corresponding to those voltage drops because of Compton and synchrotron losses (especially for the electrons). Note that in this scenario, the energy of the BH is not extracted and the efficiency for mass to energy conversion is the same as for that in thin discs, *i.e.*, $\lesssim 0.32$. This type of neutral beam may be more stable than other beams (Lovelace *et al.* 1979), but there are several strong constraints on such beams.

It has been shown that an e^+e^- beam is almost certainly very optically thick to scattering (and therefore cooling or annihilation) at the scale of 10^{15} cm if it is powerful enough to drive a strong double source (Rees 1984b). In particular, it is necessary that

$$\gamma_b \gtrsim 3.5(L_{+-}/10^{45}\text{erg s}^{-1})^{1/3}\phi^{-2/3}(r/10^{15}\text{cm})^{-1/3}, \tag{8.27}$$

where γ_b is the bulk Lorentz factor of a lepton beam of power L_{+-} and opening angle ϕ. The effective random Lorentz factor in such a beam is almost certainly reduced to $\gamma \sim 1$ through either synchrotron or Compton losses. The conclusion is that the bulk of the random motion must either be efficiently converted to ordered flow, or predominantly converted to a narrow beam of photons (of opening angle $\max[\phi, \gamma_b^{-1}]$), at least by the time the small distances of 10^{-3} pc are reached. But lepton beams will also suffer severe Compton drag from the ambient radiation field, indicating that such high values of γ_b are hard to maintain over extended distances (Rees 1984b).

A different approach involves investigation of self-similar solutions to the perfect MHD equation

$$\mathbf{E} + \mathbf{v} \times \mathbf{B} = 0 \tag{8.28}$$

for flows driven from Keplerian discs (Blandford & Payne 1982). Several specialized assumptions allow analytic solutions to be obtained. For example, if the Alfvén speed ($\propto B/\sqrt{\rho}$) scales with the Keplerian velocity and the specific angular momentum and energy fluxes scale with those for Keplerian discs, then the rate of mass loss per logarithmic radial interval is independent of radius. In this situation, $\rho \propto r^{-3/2}$, while $B \propto r^{-5/4}$, and the sound speed, $c_s \propto r^{-1/2}$, so the pressure is proportional to $r^{-5/2}$. These centrifugally driven winds exist if the poloidal component of the magnetic field makes an angle $\lesssim 60°$ with the disc's surface (Blandford & Payne 1982; Shibata & Uchida 1985). An interesting feature of these winds is that at large distances the toroidal field component becomes significant, and eventually collimates the outflow. These beams confine their power mainly to a central core, while the bulk of the angular momentum and flux are removed near the beam walls. Such a configuration would have a unique radio structure signature, but whether it would be observable has not been worked out. High resolution studies of relatively close sources, such as M 87, are most likely to be useful in this regard if any such structure extends to the parts of the jet well outside the nucleus.

Somewhat related calculations have recently been performed within the framework of supermassive discs supported by rotation and magnetic fields (Camenzind 1986a,b; Kundt 1987).

Recently, a more elaborate treatment of MHD discs around BHs has been presented (Lovelace *et al.* 1986; Lovelace *et al.* 1987). Assuming a perfectly conducting disc and neglecting viscosity the MHD equations of motion and Maxwell's equations can be combined into one massive second-order non-linear partial differential equation of the Grad-Shafranov type (Lovelace *et al.* 1986). Within this framework a general family of thin magnetized discs around a Schwarzschild BH cannot have total magnetic energy greater than $\sim 36\%$ of their gravitational binding energy and sensible magnetic field configurations are expected to have major effects on the disc structure. The field tends to compress the disc vertically and since the magnetic force on the disc matter acts outward in the inner regions and inward in the outer regions of the disc, the azimuthal velocity of the disc matter in the inner part may be far less than the Keplerian value.

Generalizing the above model to viscous, resistive MHD thin discs leads to some very interesting results. A force-free plasma is again assumed to lie outside the disc, so the claim that this situation is not stable should be borne in mind (Phinney 1983, 1986); nonetheless, those claims are not overwhelmingly convincing, so that these new calculations are worth summarizing. Combining Ampere's Law and Ohm's law with Faraday's Law leads to the induction equation (Lovelace *et al.* 1987):

$$\partial \mathbf{B}/\partial t = \nabla \times (\mathbf{v} \times \mathbf{B}) + \eta \nabla^2 \mathbf{B}, \tag{8.29}$$

with $\eta = c^2/(4\pi\sigma)$, the magnetic diffusivity; η is probably fairly large, being determined by macroscopic MHD instabilities. Solutions are found by breaking the flux function inside the disc, $\Psi \equiv rA_\phi$, with A_ϕ the toroidal component of the vector potential, into odd and even functions of z. Magnetic stresses due to field-line "leakage" from the disc as well as radiative power-losses and unipolar induction effects can be incorporated into thin disc equations of vertical structure, angular momentum balance, and energy balance in a natural fashion. Then both Ψ and the toroidal magnetic field within the disc can be solved for in terms of Fourier expansions.

Outside of the disc, far field solutions of the "pulsar equation" can be found using a Green's function technique. One class of such solutions, which appears to be produced by sensible boundary conditions, are self-collimated electromagnetic beams. Because this collimation is due to the magnetic pinch effect, it does not depend on the existence of a narrow vortex. These electromagnetic beams are naturally relativistic and have radii smaller than the speed-of-light cylinder for the disc, in contrast to the non-relativistic MHD outflows mentioned above (Blandford & Payne 1982). Such beams carry axial and toroidal current densities as well as initially non-zero charge densities, but as they propagate outwards they will become charge neutralized before the current is neutralized.

It turns out that the emerging power is subject to a global consistency requirement which limits the fraction of the accretion power carried off by these beams. If Ψ is an odd function of z this fraction is very small, but if Ψ is an even function of z it appears that this fraction can approach unity. Lovelace *et al.* (1987) argue that quasars and other optically active AGN might correspond to cases where Ψ is odd and the vast bulk of the emission would be quasi-thermal radiation, while radio jet sources could be related to situations where Ψ is even and the bulk of the accretion energy is carried off by the electromagnetic beams. This approach is new and complex, yet, as always, somewhat oversimplified. The detailed properties of the emerging beams must be calculated and some attempts to compare them with observations must be made. Until that is done and the basic nature of the magnetosphere is clarified it will be too soon to pass judgement on this most intriguing possibility.

8.3 Observations of jets on extremely small scales

Unfortunately, even the highest resolution radio maps do not probe deep enough into an AGN to give a crisp picture of what is going on at the $\lesssim 10^{15}$ cm scale over which jets are probably being formed, nor is structural information on that scale likely to emerge from higher frequency observations. Neither the possible advent of longer radio baselines provided by an orbiting antenna (*e.g.*, RADIOASTRON) nor the currently contemplated optical interferometers would change this picture in a significant fashion, except perhaps for the very nearest active galaxies (see §4.3.2). Several questions concerning the very core AGN relevant to jet formation can, however, be partially addressed with current or planned observational capabilities, and the rest of this chapter will consider some of these.

8.3.1 Better evidence for black holes

Despite all of the evidence for SMBHs discussed in §8.2.1, the proof that they are responsible for nuclear activity remains merely circumstantial. Several of the arguments may be greatly improved in the near future. For example, the Hubble Space Telescope should be able to clearly examine the stellar spectra of the innermost few parsecs of nearby galaxies thanks to its diffraction-limited resolution. This would give convincing dynamical values for the masses of their cores. Improved spectroscopy of the BLR will put more constraints on the total size of that region. More sophisticated analyses of the differential origins in radii of specific broad lines could also give a handle on the mass distributions within galactic nuclei. Also, many models that rely on dense electron-positron plasmas in the near vicinity of the SMBH would produce γ-rays of significant intensities. In particular, broadened annihilation lines ought to be produced, and the Gamma Ray Observatory should be capable of detecting them, at least in the brightest sources.

Perhaps the clearest evidence for SMBHs would arise from the detection of binary periods due to black holes in a mutual orbit. Many active galaxies are almost certainly the product of galactic mergers, and if two galaxies housing nuclear SMBHs merge, then the smaller galaxy's hole will sink to the core of the larger galaxy over just a few orbital periods (Begelman, Blandford & Rees 1980). An orbit decaying over time scales $\gtrsim 10^9$ yr can then be established until gravitational radiation leads to a rapid coalescence. The lifetime of typical SMBH binaries with orbital periods of $T_{\rm orb}$ years and masses of about $10^6 M_\odot$ is roughly $4 \times 10^7 T_{\rm orb}^{8/3}$ yr, which is a significant fraction of the total expected active lifetime of $\sim 3 \times 10^8$ yr (Blandford 1986), although it decreases with increasing BH masses. If both SMBHs retain, or reform, accretion discs, then there is a reasonable possibility that eclipsing binaries (of easily detectable periods, $T_{\rm orb} \lesssim 1$ yr) could be found. This is because it is likely that between 10^{-4} and 10^{-2} of Seyfert and LINER nuclei host such binary accretion discs and a careful monitoring of their optical continua could provide a new, and extremely strong, argument in favour of SMBHs (Blandford 1986). An apparent occultation of the nucleus of the Seyfert galaxy NGC 4151 in February 1983 has been reported (Meaburn *et al.* 1985) and the observers argue that it is best explained as the blockage of a central source of $\lesssim$ 3 A.U. by a dark cloud of diameter $\sim$ 6 A.U. It is not clear if this observation can be made consistent with the binary black hole hypothesis, but close monitoring of NGC 4151 is clearly warranted. A binary pair of SMBHs has also been proposed for OJ 287 (Sillanpää *et al.* 1988).

On the other hand, because of the inexorable pull of BHs, fluid flows around BHs do not appear to be capable of supporting long-term variations with short periods comparable to the horizon crossing time. Therefore, convincing observations of such stable, short-term periods would be powerful evidence in favour of spinar or magnetoid models. The only source where there is any real evidence for minute-scale variability is the BL Lac object OJ 287 (Valtaoja *et al.* 1985). Although a 15.7 minute period was apparently fairly stable over more than a year, that group also found possible periodicities at 13.0 and 33.0 minutes, but another group found that the strongest periodic variation was at 22.8 minutes with other peaks in the Fourier spectrum at 42.2, 12.0 and 7.7 minutes (Carrasco *et al.* 1985). These multiple periodicities are probably best explained as a series of "hot spots" on an accretion disc spiralling into a SMBH, which could provide short-lived, quasi-periodic variability (Carrasco *et al.* 1985).

8.3.2 Direct evidence concerning extremely small jets

The VLBI data concerning pc-scale jets has been discussed and interpreted in Chapters 1 and 4 but a few particular points concerning what that data might say with respect to jet formation models should be made here. While current VLBI maps frequently do show clearly elongated features that are best classified as jets, rarely if ever is the resolution across the jet good enough to enable a good measurement of the jet opening angle to be made. The size of the opening angle and its variation with distance from the nucleus might be determined for some sources by RADIOASTRON, and could have a bearing on the choice of formation model. Models relying on winds from thin accretion discs are likely to produce rather wide beams; beams emerging from radiation supported thick discs are likely to have intermediate opening angles ($3^\circ \lesssim \theta \lesssim 15^\circ$) while truly narrow beams can probably only be produced by models where magnetic forces dominate the focusing even on extremely small scales. In light of the very narrow jets often observed on multi-kpc scales, a great deal of intermediate scale focusing would be necessary for either of the first two general classes of models. While purely hydrodynamical processes in an appropriately flattened and dimpled interstellar medium make this possible to some degree (*e.g.*, Wiita & Siah 1986), it is not likely to be good enough in most cases, and good focusing probably requires magnetic dominance at some distance (*e.g.*, Mitteldorf 1987).

Explicit predictions of the cross-sectional shapes of beams emerging from different models of the powerhouse are usually not made, and when they are, they tend to be rather vague. In general, it is expected that beams from radiation supported discs or winds off thin discs would be somewhat denser and perhaps a bit faster towards the axis, probably implying a limb darkened structure. Several of the models involving magnetically dominated flows yield beams which are much denser towards the edges, so that one would naively predict an edge-brightened structure. However, there are many caveats and uncertainties involved here, and it would probably be worthwhile to direct some effort toward pinning down predictions in this regard. Although it does not appear that currently possible observations will be able to discriminate between these options, RADIOASTRON or a lunar-based VLBI system might have the necessary resolution for nearby sources.

Polarization measurements of VLBI sources are beginning to appear (Roberts *et al.* 1984; Wardle *et al.* 1986) and have been mentioned in Chapters 1 and 4. As they improve, they will give information on magnetic field orientation for nuclear jets similar to currently available data on extended jets, and will provide an extremely interesting tool. Strong polarizations and indications of predominantly toroidal magnetic fields would most easily fit models where the beams are formed and collimated by magnetic effects but it is not clear if they could distinguish between fields anchored or formed in large accretion discs or those threading a BH's horizon. If the magnetic field is apparently aligned with the flow at early stages it might argue in favour of models that advect and stretch the field and do not rely upon magnetic collimation, but some types of magnetically dominated flows could also exhibit this topology. Moreover, the effects of depolarization and Faraday rotation, as well as the real possibility of the physical separation between the strongest fields and the densest radio emitting plasma (*e.g.*, Clarke *et al.* 1986) all make it very difficult to apply any such observations in a model-independent fashion. Still, the variability to be expected from sub-pc scale motions means that polarization measurements might allow viewing angles, and thus three-dimensional velocities, for nuclear flows to be determined (Phinney 1985).

Very firm evidence for highly relativistic bulk motion ($\gamma_j \gtrsim 5$) would cause severe difficulties for any method (wind or radiation supported disc funnel) that relied upon radiation pressure for acceleration, regardless of whether the beam were to be predominantly ordinary plasma or mostly leptonic. Conversely, strong evidence in favour of significant mass flows down *nuclear* jets, which might eventually emerge from multi-frequency VLBI polarization measurements, would be negative for the ion-supported thick disc or other models that depend upon the dominant energy flow being in the form of electromagnetic Poynting flux. Large mass fluxes at greater distances do not cause such a problem because of the probability of entrainment of ambient matter (cf. §§5.4 and 8.3.3). The possible future detection of extended collimated γ-rays would also probably argue in favour of dynamo type low density jets, but the implications of such observations have not yet really been explored.

8.3.3 Interactions between jets and the ambient nuclear medium

Because of the huge powers carried by jets they must have significant impacts on the material through which they move at all length scales, many of which were discussed in previous chapters. Here we shall concentrate on several ways in which jets may interact with the surrounding nuclear plasma, other than those already alluded to in our discussion of beam formation.

It has been proposed that essentially all aspects of the BLR are produced by jets (Norman & Miley 1984; Barthel *et al.* 1984). The pressure exerted by a jet can be estimated as

$$P_j = 10^{-4}\left(\frac{L}{10^{44}\text{erg s}^{-1}}\right)\left(\frac{4\pi}{\Omega}\right)\left(\frac{1\text{ pc}}{d}\right)^2 \times \left(\frac{10^5\text{ km s}^{-1}}{v}\right)^2 \text{dyne cm}^{-2}, \qquad (8.30)$$

which is comparable to that necessary to confine putative broad line clouds for a jet of solid angle $\Omega \sim 10^{-2}$ steradians. In this picture initially uniform material would be swept up by the jet into a cocoon which could be unstable to Rayleigh-Taylor type fragmentation. Those fragments could produce many (or all) of the BL clouds if the cloud densities are $\sim 10^{10}$ cm^{-3}. Further, much of the jet's mechanical energy would be dissipated if interactions with the surrounding medium produce shocks, and this could yield the majority of the local UV and X-ray emission necessary to excite the clouds. In this scenario the cocoon would further function as a natural screen, thereby providing a possible explanation of superluminal motions. One-sided jets could be caused by one-sided blocking of a jet by massive molecular clouds (Norman & Miley 1984). Recent indications of the necessity for higher density BL clouds seem to make this rather extreme picture less attractive, as has the growing list of superluminal sources.

Still, this interaction may be responsible both for "stopping" superluminal components, as in BL Lac, and the rather small overall radio extent of most high redshift quasars (Barthel 1986). Further, the small core polarizations revealed by VLBI (Wardle *et al.* 1986) may imply the need for an extended depolarizing screen, which could be provided by this material.

Looking at this question from an essentially opposite point of view it can be shown to be very unlikely that the intercloud medium is capable of thermally confining jets at scales much below one parsec, although at larger distances this is certainly possible (Rees 1984b); considering the recent conclusions that the BLR is smaller and the cloud densities higher than typically assumed, this argument is only strengthened. Once out in the line emitting region the jets are unlikely to be able to destroy most clouds passing through them; rather, if the cloud density were high enough the jet could be completely dissipated in the course of its interaction with them (Rees 1984b). Of course, one possible explanation for the knotty structure in VLBI scale jets is the interaction of jets with ambient clouds (*e.g.*, Blandford & Königl 1979), while another would involve shocks propagating down the jet, presumably due to fluctuations in power or velocity (cf. Chapter 4). This is another area where the improved resolution of a RADIOASTRON could be very useful, for the impact of jets upon broad line clouds would probably lead to sharp edges facing the nucleus, while internal shocks would more likely lead to sharper leading edges.

Rees (1984b) has also stressed that flows on sub-pc scales are likely to be much more dissipative than those on larger scales, primarily because the ratio of the radiative cooling time ($\propto d^2$) to the dynamical time ($\propto d$) scales with d, rather than being independent of distance as is true for larger jets. These inner-most jets are likely to lose their internal pressure more easily and should waste more energy on being bent in comparison to their larger scale incarnations. Unfortunately, the micro-arcsecond optical resolution needed to see the visible synchrotron radiation emitted through these interactions, similar to that frequently observed on kpc scales (*e.g.*, van Breugel *et al.* 1984), is unlikely to be achieved.

The possibility that jets passing through the ambient medium will drive large circulation currents that in turn act back upon the jets has been examined from several viewpoints. Fluid flowing along a beam will interact with the surrounding medium through surface instabilities which can provide an effective viscous coupling between the two (cf. Henriksen 1985). The external fluid might be drawn up along the jets and, if so, is likely eventually to separate from them and then descend at greater distances; the circulation would be completed by flowing towards the jets near the equatorial plane. Small asymmetries in these external fluid flow patterns can then significantly alter the density and pressure of the surrounding gas near the equatorial region in such a way as to affect the jets. In general, any asymmetry in jet strength will be amplified, and this process can even cause one of the pair of jets to choke off: this is the so-called "clam-shell' mechanism (Icke 1983).

This idea has been modified to stress the role of material entrained by the jets (Allen 1986). If a blob of material is swallowed and dispersed by one jet the net flow speed in that channel is reduced, since the momentum is shared by a larger mass. Allen argues that Kelvin-Helmholtz perturbations will grow more easily in the denser and slower jet. If so, that implies that yet more material is likely to be engulfed on that side, amplifying the asymmetry; further, this increase in entrained matter would increase the radio emission from that choked channel. Such instabilities might act on a wide range of scales and could plausibly explain one-sided sources without the need for Doppler favouritism. If one side of the jet remains relativistic, while the other engulfs enough matter to become subrelativistic, a broader and more diffuse radio structure should emerge from the subrelativistic, choked side.

There are a number of radio sources which have roughly equidistant lobes but exhibit detailed anti-symmetry on the two sides of the central galaxy. It has been argued that such sources are activated in alternate directions, with a single jet "flipping" from one side to another (Rudnick & Edgar 1984; Rudnick 1985). The "clam-shell" type of mechanisms provide a conceivable way of explaining these "flip-flop" sources, although exactly how the switch from one direction to the other might occur with some quasi-fixed periodicity is unclear. Models involving the orbit of the powerhouse within the surrounding medium might also suffice to explain some "flip-flop" sources (Wiita & Siah 1981; Saikia & Wiita 1982), in that the jet that has to propagate through a shorter path-length in the confining medium could escape while the one moving through more material might be temporarily blocked, or much reduced in strength. Half an orbit later, the jet would be expected to escape in the opposite direction and would be stifled on the originally powerful side. However, before spending a great deal of time worrying about this problem, it should be noted that statistical evidence indicates that such candidates for "flip-flop" sources are probably not very common (Ensman & Ulvestad 1984).

8.4 Conclusions

Evidence in favour of supermassive black holes lurking at the centre of all active galactic nuclei continues to mount, as does evidence that jets are launched very close to those central engines. While it is not yet possible to make a clear choice between competing models for beam formation, those requiring that a dominant role be played by magnetic fields are growing in sophistication and popularity. Still, none of the models have been developed to the point where clear predictions concerning the structure and emission properties of jets are made, so that confrontations with observations are not as restrictive as we would desire. Further, even if those predictions are firmly made, the observations needed to make clear choices between models will be extremely difficult to obtain and to interpret without bias. Even though some models may not stand the test of time, the possibility that more than one basic mechanism may be involved should not be overlooked.

Related to the possibility of multiple jet formation mechanisms being operative, the hypothesis that most galaxies house SMBHs and go through active phases of different types, depending upon the available fuel supply, is also gaining favour. The interactions of sub-parsec jets with the certainly non-uniform surrounding medium is bound to be extremely complex, and investigations of this subject are really just beginning. Significant advances in this area will probably only emerge from increasingly detailed numerical simulations advancing in tandem with improved theoretical understanding of the relevant physics. Uncertainties in our knowledge of the processes occurring in the region where severely warped space-time is invaded by extraordinarily hot plasmas are likely to remain for quite a long time, and a definitive answer to the question "Where *exactly* do jets come from?" is likely to remain elusive.

I am grateful to many colleagues for conversations on this subject and for relevant preprints. This work was supported in part by NSF grant AST87-17912 and by a Smithsonian Institution Foreign Currency Research Grant. I thank the Indian Institutes of Science and Astrophysics and the Tata Institute of Fundamental Research for their hospitality while I was beginning work on this chapter.

Symbols used in Chapter 8

Symbol	Meaning
a	angular momentum parameter for BH
$\mathbf{A}$	magnetic vector potential
Å	Ångstrom unit
$\mathbf{B}$, B	magnetic field
B_{pH}	poloidal component of the field intersecting the BH horizon
c_s	local speed of sound
d	distance along a beam or jet
$\mathbf{E}$	electric field
f	radiation anisotropy factor
$\mathbf{F}$	force
F_{eff}	radiative flux
g_{eff}	effective gravitational acceleration
h	disc half-thickness
I	current
$\mathbf{j}$	current density
$\mathbf{J}$, J	angular momentum of BH
k_B	Boltzmann's constant

l specific angular momentum
$l_{\rm Kep}$ specific angular momentum of a particle on a Keplerian orbit
L total luminosity
$L_{\rm eff}$ effective luminosity
L_E Eddington luminosity
L_{EM} Poynting luminosity
L_i luminosity at infinity
L_{+-} lepton beam luminosity
$\dot{m}$ specific accretion rate, $\dot{M}/\dot{M}_E$
$\dot{M}$ accretion rate
$\dot{M}_E$ Eddington (critical) accretion rate
$M_\odot$ solar mass
M_{BH} mass of black hole (BH)
$M_{\rm disc}$ mass of accretion disc
$M_{\rm ir}$ irreducible mass of BH
n power law for specific angular momentum
$P_{\rm disc}$ pressure in the disc
$P_{\rm gas}$ gas pressure
$P_{\rm jet}$ pressure exerted by the jet
$P_{\rm rad}$ radiation pressure
q power law for angular velocity
r cylindrical radial coordinate
r_c critical radius (sonic point in wind flow)
r_i energy injection radius
r_t radiation trapping radius
$r_{\rm in}$ inner edge of accretion disc
$r_{\rm out}$ outer edge of accretion disc
R_L electrical resistivity of plasma
R_{mb} radius of marginally bound orbit
R_{ms} radius of marginally stable orbit
R_s radius of event horizon
S_{BH} entropy of BH
$t_{\phi r}$ component of shear tensor
$T_{\rm orb}$ orbital period
$T_{\rm disc}$ average black body temperature radiated by a disc
T_e electron temperature

T_{BH}	temperature of BH
$T_{\max}$	maximum temperature in an accretion disc
T_p	proton temperature
u	energy density in radiation
$\mathbf{v}$	three-velocity
v^μ	four-velocity
v_ϕ	azimuthal component of velocity
v_r	radial component of velocity
V	voltage drop
z	cylindrical axial coordinate
Z_H	impedance of the BH horizon
α^*	viscosity parameter
β	bulk speed/c
Δt	time scale of variability
ϵ	efficiency of conversion of mass to energy
Υ	angle between spin and magnetic axes for BH
γ	Lorentz factor for random velocities of particles
γ_b	Lorentz factor for bulk flow of beam
γ_j	Lorentz factor for bulk flow of jet
Γ	polytropic (adiabatic) index
η	magnetic diffusivity
κ	opacity
μ	charge per unit mass
ρ	mass density
ρ_c	space charge density
σ	electrical conductivity
ϕ	opening angle of beam
τ	optical depth
Ψ	magnetic flux function
Ω	angular velocity, or solid angle
Ω_F	angular velocity of magnetic field lines
Ω_{BH}	angular velocity of BH

References

Abramowicz, M. A., Calvani, M. & Nobili, L., 1980, *Ap. J.*, **242**, 772.

Abramowicz, M. A., Curir, A., Schwarzenberg-Czerny, A. & Wilson, R. E., 1984, *M. N. R. A. S.*, **209**, 279.

Abramowicz, M. A., Jaroszyński, M. & Sikora, M., 1978, *Astr. Ap.*, **63**, 221.

Abramowicz, M. A. & Piran, T., 1980, *Ap. J. Lett.*, **241**, L7.

Allen, A. J., 1986, *Ap. J.*, **301**, 44.

Barthel, P. D., 1986, In *Quasars: IAU Symposium 119*, eds. G. Swarup & V. K. Kapahi (Reidel: Dordrecht, Netherlands), p. 181.

Barthel, P., Norman, C. & Miley, G., 1984, In *VLBI and Compact Radio Sources: IAU Symposium 110*, eds. R. Fanti, K. Kellermann & G. Setti (Reidel: Dordrecht, Netherlands), p. 237.

Becker, P. A. & Begelman, M. C., 1986a, *Ap. J.*, **310**, 534.

Becker, P. A. & Begelman, M. C., 1986b, *Ap. J.*, **310**, 552.

Begelman, M. C., 1985, In *Astrophysics of Active Galaxies and Quasi-Stellar Objects*, ed. J. S. Miller (University Science Books: Mill Valley), p. 411.

Begelman, M. C., 1988, In *Active Galactic Nuclei*, eds. H. R. Miller & P. J. Wiita (Springer Verlag: Berlin), p. 411.

Begelman, M. C., Blandford, R. D. & Rees, M. J., 1980, *Nature*, **287**, 307.

Begelman, M. C., Blandford, R. D. & Rees, M. J., 1984, *Rev. Mod. Phys.*, **56**, 255.

Begelman, M. C., McKee, C. F. & Shields, G. A., 1983, *Ap. J.*, **271**, 70.

Begelman, M. C. & Rees, M. J., 1978, *M. N. R. A. S.*, **185**, 847.

Begelman, M. C. & Rees, M. J., 1983, In *Astrophysical Jets*, eds. A. Ferrari & A. G. Pacholczyk (Reidel: Dordrecht, Netherlands), p. 215.

Begelman, M. C., Sikora, M. & Rees, M. J., 1987, *Ap. J.*, **313**, 689.

Binney, J. & Mamon, G. A., 1982, *M. N. R. A. S.*, **200**, 361.

Blaes, O. M., 1985a, *M. N. R. A. S.*, **212**, 37p.

Blaes, O. M., 1985b, *M. N. R. A. S.*, **216**, 553.

Blaes, O. M., 1987, preprint.

Blaes, O. M. & Glatzel, W., 1986, *M. N. R. A. S.*, **220**, 253.

Blandford, R. D., 1976, *M. N. R. A. S.*, **176**, 465.

Blandford, R. D., 1984, *Eleventh Texas Symposium on Relativistic Astrophysics*, ed. D. S. Evans, *Ann. New York Acad. Sci.*, **422**, 303.

Blandford, R. D., 1986, In *Quasars: IAU Symposium 119*, eds. G. Swarup & V. K. Kapahi (Reidel: Dordrecht, Netherlands), p. 359.

Blandford, R. D., Jaroszyński, M. & Kumar, S., 1985, *M. N. R. A. S.*, **215**, 667.

Blandford, R. D. & Königl, A., 1979, *Ap. Lett.*, **20**, 15.

Blandford, R. D. & Payne, D. G., 1982, *M. N. R. A. S.*, **199**, 883.

Blandford, R. D. & Znajek, R., 1977, *M. N. R. A. S.*, **179**, 433.

Calvani, M. & Nobili, L., 1983, In *Astrophysical Jets*, eds. A. Ferrari & A. G. Pacholczyk (Reidel: Dordrecht, Netherlands), p. 189.

Calvani, M., Nobili, L. & Turolla, R., 1983, *Mem. Soc. Astron. Ital.*, **54**, 703.

Camenzind, M., 1986a, *Astr. Ap.*, **156**, 137.

Camenzind, M., 1986b, *Astr. Ap.*, **162**, 32.

Camenzind, M., Demole, F. & Straumann, N., 1986, *Astr. Ap.*, **158**, 212.

Carrasco, L., Dultzin-Hacyan, D. & Cruz-Gonzalez, I., 1985, *Nature*, **314**, 146.

Chakrabarti, S. K., 1985a, *Ap. J.*, **288**, 1.

Chakrabarti, S. K., 1985b, *Ap. J.*, **288**, 7.

Chakrabarti, S. K., 1986, *Ap. J.*, **303**, 582.

Chakrabarti, S. K., Jin, L. & Arnett, W. D., 1987, *Ap. J.*, **313**, 674.

Chan, K. L. & Henriksen, R. N., 1980, *Ap. J.*, **241,** 534.
Clarke, D. A., Karpik, S. & Henriksen, R. N., 1985, *Ap. J. Suppl.*, **58,** 81.
Clarke, D. A., Norman, M. L. & Burns, J. O., 1986, *Ap. J. Lett.*, **311,** L63.
Damour, T., 1978, *Phys. Rev. D.*, **18,** 3598.
Dhurandhar, S. V. & Dadhich, N., 1984, *Phys. Rev. D.*, **29,** 2712.
Dressler, A., 1984, *Ap. J.*, **286,** 97.
Dressler, A. & Richstone, D. O., 1988, *Ap. J.*, **324,** 701.
Drew, J. E. & Boksenberg, A., 1984, *M. N. R. A. S.*, **211,** 813.
Eggum, G. E., Coroniti, F. V. & Katz, J. I., 1985, *Ap. J. Lett.*, **298,** L41.
Ensman, L. M. & Ulvestad, J. S., 1984, *Astron. J.*, **89,** 1275.
Ferrari, A., Habbal, S. R., Rosner, R. & Tsinganos, K., 1984a, *Ap. J. Lett.*, **277,** L35.
Ferrari, A., Rosner, R., Trussoni, D. & Tsinganos, K., 1984b, In *VLBI and Compact Radio Sources: IAU Symposium 110*, eds. R. Fanti, K. Kellermann & G. Setti (Reidel: Dordrecht, Netherlands), p. 233.
Ferrari, A., Trussoni, E., Rosner, R. & Tsinganos, K., 1985, *Ap. J.*, **294,** 397.
Ferrari, A., Trussoni, E., Rosner, R. & Tsinganos, K., 1986, *Ap. J.*, **300,** 577.
Filippenko, A. V. & Sargent, W. L. W., 1985, *Ap. J. Suppl.*, **57,** 3.
Fukue, J., 1982, *Publ. Astron. Soc. Japan*, **34,** 163.
Goldreich, P., Goodman, J. & Narayan, R., 1986, *M. N. R. A. S.*, **221,** 339.
Goldreich, P. & Narayan, R., 1985, *M. N. R. A. S.*, **213,** 7p.
Hanawa, T., 1986, *M. N. R. A. S.*, **223,** 859.
Hawley, J. F., 1985, *Bull. A. A. S.*, **17,** 857.
Hawley, J. F., 1988, *Ap. J.*, **326,** 277.
Hawley, J. F., Smarr, L. L. & Wilson, J. R., 1984, *Ap. J.*, **277,** 296.
Henriksen, R. N., 1985, In *Physics of Energy Transport in Extragalactic Radio Sources*, eds. A. Bridle & J. Eilek, (NRAO: Green Bank), p. 211.
Icke, V., 1983, *Ap. J.*, **265,** 648.
Jaroszyński, M., Abramowicz, M. A. & Paczyński, B., 1980, *Acta Astronomica*, **30,** 1.
Kafatos, M. & Leiter, D., 1979, *Ap. J.*, **229,** 46.
Keel, W. C., 1985, In *Astrophysics of Active Galaxies and Quasi-Stellar Objects*, ed. J. S. Miller (University Science Books: Mill Valley), p. 1.
Kochaneck, C. S., Shapiro, S. L. & Teukolsky, S. A., 1987, *Ap. J.*, **320,** 73.
Kojima, Y., 1986, *Prog. Theoret. Phys.*, **75,** 1464.
Krishan, V. & Wiita, P. J., 1986, In *Quasars: IAU Symposium 119*, eds. G. Swarup & V. K. Kapahi (Reidel: Dordrecht, Netherlands), p. 419.
Krolik, J. H., 1988, In *Active Galactic Nuclei*, eds. H. R. Miller & P. J. Wiita (Springer Verlag: Berlin), p. 19.
Kundt, W., 1987, In *Astrophysical Jets & their Engines: Erice Lectures*, ed. W. Kundt (Reidel: Dordrecht, Netherlands), p. 1.
Liang, E. T. P. & Price, R. H., 1977, *Ap. J.*, **218,** 247.
Lightman, A. P. & Eardley, D., 1974, *Ap. J. Lett.*, **187,** L1.
Lovelace, R. V. E., 1976, *Nature*, **262,** 649.
Lovelace, R. V. E., MacAuslan, J. & Burns, M., 1979, In *Particle Acceleration in Astrophysics*, eds. J. Arons, C. Max & C. McKee (AIP Conf. Proc. No. 56), p. 399.

Lovelace, R. V. E., Mehanian, C., Mobarry, C. M. & Sulkanen, M. E., 1986, *Ap. J. Suppl.*, **62,** 1.

Lovelace, R. V. E., Wang, J. C. L. & Sulkanen, M. E., 1987, *Ap. J.*, **315,** 504.

Lu, J. F., 1986, *Astr. Ap.*, **168,** 346.

Lynden-Bell, D., 1969, *Nature*, **223**, 690.

Lynden-Bell, D., 1978, *Phys. Scripta*, **17,** 185.

Macdonald, D. A. & Thorne, K. S., 1982, *M. N. R. A. S.*, **198,** 345.

Macdonald, D. A., 1984, *M. N. R. A. S.*, **211,** 313.

Madau, P., 1988, *Ap. J.*, **327**, 116.

Madej, J., Loken, C. & Henriksen, R. J., 1987, *Ap. J.*, **312,** 652.

Malkan, M. A., 1983, *Ap. J.*, **268,** 582.

Matthews, W. G. & Capriotti, E. R., 1985, In *Astrophysics of Active Galaxies and Quasi-Stellar Objects*, ed. J. S. Miller (University Science Books: Mill Valley), p. 185.

Meaburn, J., Ohtani, H. & Goudis, C. D., 1985, In *Active Galactic Nuclei*, ed. J. E. Dyson (Manchester University Press: Manchester), p. 184.

Meier, D. L., 1982, *Ap. J.*, **256,** 681.

Miller, R. H., Carini, M. T. & Goodrich, B. D., 1989, *Nature*, **327,** 627.

Mitteldorf, J. J., 1987, unpublished thesis, University of Pennsylvania.

Narayan, R., Nityananda, R. & Wiita, P. J., 1983, *M. N. R. A. S.*, **205,** 1103.

Nobili, L., Calvani, M. & Turolla, R., 1985, *M. N. R. A. S.*, **214,** 161.

Norman, C. & Miley, G., 1984, *Astr. Ap.*, **141,** 85.

Paczyński, B. & Wiita, P. J., 1980, *Astr. Ap.*, **88,** 23.

Papaloizou, J. C. B. & Pringle, J. E., 1984, *M. N. R. A. S.*, **208,** 721.

Papaloizou, J. C. B. & Pringle, J. E., 1985, *M. N. R. A. S.*, **213,** 799.

Parthasarathy, S., Wagh, S. M., Dhurandhar, S. V. & Dadhich, N., 1986, *Ap. J.*, **307,** 38.

Penrose, R., 1969, *Riv. Nuovo Cimento*, **1,** 252.

Peterson, B. M., 1987, *Ap. J.*, **312,** 79.

Peterson, B. M., 1988, In *Active Galactic Nuclei*, eds. H. R. Miller & P. J. Wiita (Springer Verlag: Berlin), p. 38.

Peterson, B. M. & Ferland, G. J., 1986, *Nature*, **324,** 345.

Peterson, B. M., Meyers, K. A., Capriott, E. R., Foltz, C. B., Wilkes, B. J. & Miller, H. R., 1985, *Ap. J.*, **292,** 164.

Phinney, E. S., 1983, In *Astrophysical Jets*, eds. A. Ferrari & A. G. Pacholczyk (Reidel: Dordrecht, Netherlands), p. 201.

Phinney, E. S., 1985, In *Astrophysics of Active Galaxies and Quasi-Stellar Objects*, ed. J. S. Miller (University Science Books: Mill Valley), p. 453.

Phinney, E. S., 1986, Lecture at George Mason University Workshop on *Supermassive Black Holes.*

Piran, T. & Shaham, J., 1977, *Ap. J.*, **214,** 268.

Readhead, A. C. S., Hough, D. H., Ewing, M. S., Walker, R. C. & Romney, J. D., 1983, *Ap. J.*, **265,** 107.

Rees, M. J., 1984a, *Ann. Rev. Astr. Ap.*, **22**, 471.

Rees, M. J., 1984b, In *VLBI and Compact Radio Sources: IAU Symposium 110*, eds. R. Fanti, K. Kellermann & G. Setti (Reidel: Dordrecht, Netherlands), p. 207.

Rees, M. J., 1986, In *Quasars: IAU Symposium 119*, eds. G. Swarup & V. K. Kapahi (Reidel: Dordrecht, Netherlands), p. 1.

Rees, M. J., Begelman, M. C., Blandford, R. D. & Phinney, E. S., 1982, *Nature*, **295,** 17.

Roberts, D. H., Potash, R. I., Wardle, J. F. C., Rogers, A. E. E. & Burke, B. F., 1984, In *VLBI and Compact Radio Sources: IAU Symposium 110*, eds. R. Fanti, K. Kellermann & G. Setti (Reidel: Dordrecht, Netherlands), p. 35.

Rudnick, L., 1985, In *Physics of Energy Transport in Extragalactic Radio Sources*, eds. A. Bridle & J. Eilek (NRAO: Green Bank), p. 35.

Rudnick, L. & Edgar, B. K., 1984, *Ap. J.*, **279,** 74.

Saikia, D. J. & Wiita, P. J., 1982, *M. N. R. A. S.*, **200,** 83.

Sargent, W. L. W., Young, P. J., Boksenberg, A., Shortridge, K., Lynds, C. R. & Hartwick, F. D. A., 1978, *Ap. J.*, **221,** 731.

Shakura, N. I. & Sunyaev, R. A., 1973, *Astr. Ap.*, **24,** 337.

Shakura, N. I. & Sunyaev, R. A., 1976, *M. N. R. A. S.*, **175,** 613.

Shapiro, S. L. & Teukolsky, S. A., 1985a, *Ap. J. Lett.*, **292,** L41.

Shapiro, S. L. & Teukolsky, S. A., 1985b, *Ap. J.*, **298,** 58.

Shibata, K. & Uchida, Y., 1985, *Publ. Astron. Soc. Japan*, **37,** 31.

Siah, M. J., 1985, *Ap. J.*, **298,** 107.

Siah, M. J. & Wiita, P. J., 1987, *Ap. J.*, **313,** 623.

Sikora, M. & Wilson, D. B., 1981, *M. N. R. A. S.*, **197,** 529.

Sillanpää, A., Haarala, S., Valtonen, M. J., Sundelius, B. & Byrd, G. G., 1988, *Ap. J.*, **325,** 628.

Smith, M. D. & Raine, D. J., 1985, *M. N. R. A. S.*, **212,** 425.

Stein, W. A., 1988, In *Active Galactic Nuclei*, eds. H. R. Miller & P. J. Wiita (Springer Verlag: Berlin), p. 188.

Sol, H., Pelletier, G. & Asséo, E., 1989, *M. N. R. A. S.*, **237,** 411.

Stein, W. A. & O'Dell, S. L., 1985, In *Astrophysics of Active Galaxies and Quasi-Stellar Objects*, ed. J. S. Miller (University Science Books: Mill Valley), p. 381.

Sun, W.-H. & Malkan, M. A., 1988, In *Active Galactic Nuclei*, eds. H. R. Miller & P. J. Wiita (Springer Verlag: Berlin), p. 220.

Thorne, K. S., 1974, *Ap. J.*, **191,** 507.

Thorne, K. S., 1986, In *Highlights of Modern Astrophysics*, eds. S. L. Shapiro & S. A. Teukolsky (Wiley: New York), p. 103.

Thorne, K. S., Price, R. H. & Macdonald, D., eds., 1986, *Black Holes, The Membrane Paradigm* (Yale U. Press: New Haven).

Tonry, J., 1984, *Ap. J. Lett.*, **283,** L27.

Trimble, V. & Woltjer, L., 1986, *Science*, **234,** 155.

Tritz, B. G. & Tsuruta, S., 1989, *Ap. J.*, **340,** 203.

Valtaoja, E., *et al.* 1985, *Nature*, **314,** 148.

van Breugel, W., Heckman, T. & Miley, G., 1984, *Ap. J.*, **276,** 79.

Wagh, S. M., Dhurandhar, S. V. & Dadhich, N., 1985, *Ap. J.*, **290,** 12.

Wandel, A. & Mushotzky, R. F., 1986, *Ap. J. Lett.*, **306,** L61.

Wardle, J. F. C., Roberts, D. H., Potash, R. I. & Rogers, A. E. E., 1986, *Ap. J. Lett.*, **304,** L1.

Wiita, P. J., 1982a, *Comm. Ap.*, **9,** 251.

Wiita, P. J., 1982b, *Ap. J.*, **256,** 666.

Wiita, P. J., 1985, *Phys. Rep.*, **123,** 117.

Wiita, P. J., Vishveshwara, C. V., Siah, M. J. & Iyer, B. R., 1983, *J. Phys. A.*, **16,** 2077.

Wiita, P. J., Kapahi, V. K. & Saikia, D. J., 1982, *Bull. Astron. Soc. India*, **10,** 304.

Wiita, P. J. & Siah, M. J., 1981, *Ap. J.*, **243,** 710.

Wiita, P. J. & Siah, M. J., 1986, *Ap. J.*, **300,** 605.

Young, P. J., Westphal, J. A., Kristian, J., Wilson, C. J. & Landauer, F. P., 1978, *Ap. J.*, **221,** 721.

Zdziarski, A. A., 1986, *Ap. J.*, **305,** 45.

Zurek, W. H. & Benz, W., 1986, *Ap. J.*, **308,** 123.

9

Particle Acceleration and Magnetic Field Evolution

JEAN A. EILEK
Physics Department, New Mexico Tech, Socorro, NM 87801, USA.

PHILIP A. HUGHES
Astronomy Department, University of Michigan, Ann Arbor, MI 48109, USA.

9.1 Introduction

Our interferometric images of radio sources reflect the synchrotron emissivity arising from their relativistic electrons and magnetic fields. These trace the underlying plasma flow, albeit imperfectly. The local dynamical evolution of the particles and fields is determined by their transport from the nuclear source, and by their *in situ* dynamics. This chapter presents the physics necessary for an understanding of current theories of particle acceleration and magnetic field evolution. It describes these theories and attempts to assess whether or not they provide an adequate account of the inferred particle spectra, energetics and magnetic field geometry of extragalactic jets.

It was shown in Chapter 3 that some sources have severe lifetime problems, in that the time for the electrons to be carried out to the lobes (even with a jet speed $\sim c$) is longer than their radiation lifetime (the upper limit of which is the lifetime to Compton losses on the 3 K background) and that the surface brightness and spectral index distributions do not decay as fast as would be expected in a constant velocity, expanding flow. These problems may be overcome by the local reacceleration of the radiating particles. Further, simple estimates of convection of flux-frozen magnetic field out from the core predict that the convected field decays significantly; however, this is probably offset by *in situ* amplification of the magnetic field by some dynamo process. The basic question is how to account for the observed luminosity and radiation spectrum, either by using energy convected from the nucleus – as bulk flow energy, some fraction of which is converted locally to electrons and magnetic fields – or using energy stored in currents and their associated magnetic fields.

Charged particles undergo a net energy gain in the presence of electric fields. These fields can be direct, as in reconnection events, current sheets or the unipolar dynamos believed to be associated with rotating compact objects (pulsars, black holes). Quasi-perpendicular shocks also can support DC fields. Or the electric fields can be stochastic; such fluctuating electric fields are likely to occur near shocks, or in the turbulent, diffuse plasmas of radio jets.

The distribution function of cosmic rays in jets (and in the ISM, or a supernova remnant) probably consists of a cool component with a significant, power-law, high-energy tail. This is definitely not a thermalized distribution (which would be described by a Maxwell-Boltzmann distribution, with an exponential high-energy tail). Most processes which energize a plasma, in fact, tend to distribute any additional energy equally among all of the particles. This will *heat* the plasma, but will not produce the strong high-energy tail characteristic of these relativistic particles. For proper *acceleration* we need some process which will select some small fraction of the entire particle population (such as a slightly suprathermal tail) and transfer all of the injected energy to this population. In §9.2 we describe the acceleration of charged particles by interaction with turbulent plasma waves, and in §9.3 we discuss acceleration by shocks – processes that seem capable of producing a power-law, high-energy tail from a slightly relativistic seed population.

These two common acceleration mechanisms in diffuse astrophysical plasmas (turbulence and shocks) appear to be much more effective at accelerating electrons if the particles are already somewhat relativistic. Thus, these processes might better be called *reacceleration.* If these processes are believed to be involved in true acceleration of electrons from a cool, thermal distribution into a high-energy, power-law tail, then some other mechanism must be acting to *inject* the electrons at mildly relativistic energies. These injection mechanisms usually involve direct electric field acceleration, and often are also mechanisms that heat the entire plasma. In §9.4 we present the argument that one or more injection mechanisms must be operating in jets, and in §9.5 we discuss possible injection processes.

Magnetic fields arise from currents or from changing electric fields. An important astrophysical dynamo (we use this term to refer to any process which generates the currents) comes from the inductive effects of fluid flow. A flow across existing magnetic field lines gives rise to an opposite Lorentz force on the electrons and the ions, driving a current which can amplify the initial magnetic field (or maintain it at a constant level, in the presence of dissipative or turbulent losses). Both ordered and turbulent flows can maintain large-scale field structures. This is believed to be the origin of the solar and planetary magnetic fields, as well as the field in the galactic disc. In these situations – and probably in radio jets as well – the dynamo process converts fluid kinetic energy to magnetic energy. Thus, it is another *energization* process. Turbulent dynamos are also believed to occur in laboratory plasmas. There, the plasma can spontaneously generate turbulence, which dissipates some field energy and allows relaxation to the minimum-energy state. This relaxation process may determine the large-scale field structure in some jets, or the structure of filaments within jets. In §9.6 we describe both turbulent dynamos, and the process known as Taylor relaxation.

To conform with the literature in this area, we use cgs units throughout the chapter.

9.2 Stochastic acceleration

9.2.1 History and fundamentals

The classic work of Fermi (1949) described the scattering of test particles by randomly moving scattering centres (such as interstellar clouds). Consider a particle of velocity $v = \beta c$ and Lorentz factor $\gamma = (1 - \beta^2)^{-1/2}$, moving amidst scatterers which have a random velocity V and average separation L (see for instance Achterberg 1986). A particle gains energy in this process, because the rate of head-on collisions, $\nu_+ = (v + V)/L$, in which the particle gains energy, is a bit greater than the rate of overtaking collisions, $\nu_- = (v - V)/L$, in which the particle loses energy. Since the mean momentum change per collision is $\gamma m V$, the net rate of change (in the relativistic limit $v \sim c >> V$) is

$$\frac{dE}{dt} = (\nu_+ - \nu_-)\,\gamma m c V \approx \frac{2V^2}{Lc}E, \tag{9.1}$$

which is second-order in V/c. If the system involves only head-on collisions – as with scattering in the converging flow across a shock – the rate is

$$\frac{dE}{dt} = \nu_+ \gamma mcV \approx \frac{V}{L} E, \tag{9.2}$$

which is first-order in V/c. This process (either first- or second-order) leads to an acceleration rate which has the functional form $dE/dt \approx E/\tau$, where τ is some characteristic acceleration rate; any process which yields this form is referred to as a 'Fermi process'.

We must, however, ask a bit about the microphysics: how do the particles interact with the scattering centres? Early discussions assumed that magnetic fields and density enhancements acted as magnetic mirrors. Most recent work, however, assumes that the particles interact with MHD turbulence, usually represented as a random field of MHD or plasma waves. The particles interact with the waves through electromagnetic resonances, thus allowing energy and momentum transfer (the details of this are in §9.2.2, below).

Turbulent acceleration requires waves whose energy comes from some external source. If the plasma dynamics is dominated by the fluid motions (say, $\beta_K = 4\pi\rho v^2/B^2 > 1$), the most likely sources of waves are large-scale MHD instabilities. Surface instabilities of a jet propagating through a background medium (see Chapter 6) will drive long-wavelength MHD waves directly, as well as driving fluid turbulence. In a subsonic, high-Reynolds number jet (see Chapter 5) these are the instabilities which initially create a turbulent boundary layer, and eventually lead to the growth of fully developed fluid or MHD turbulence. In a supersonic flow, shocks will be set up, and MHD turbulence is often seen near shocks.

A different situation arises if $\beta_K < 1$. In this case, much of the energy transport is in the form of fields and currents. Current driven and resistive MHD instabilities can lead to MHD and plasma turbulence (*e.g.*, Spicer 1982; Furth 1985). An especially interesting example of this is known in the laboratory: a low-β_K plasma will spontaneously relax to a state of low magnetic energy (Taylor 1986). This relaxation appears to proceed *via* self-generated MHD turbulence, which destroys magnetic energy by the local reconnection of field lines.

9.2.2 MHD waves and their interaction with particles

Long wavelength MHD waves are likely to be the ones responsible for the acceleration of relativistic particles, for two reasons. They are the ones that are likely to be externally generated by fluid-like instabilities; and they are the ones that can resonate with the particles. Most applications picture a mixed plasma consisting of some relativistic electrons and ions (whose high energies are in isotropic rather than directed, streaming motion) and some cool, inertial plasma. In this picture the waves are carried by the mixed plasma, and can interact selectively with the relativistic electrons, thus accelerating or maintaining a significant high energy tail on the electron distribution. (If the plasma contains only relativistic particles, the wave modes still exist; if the waves are generated externally, their dissipation by the plasma will be more of a heating process, but is still effective as an accelerator.) Note that the interaction between particles is mediated by electromagnetic fields – it is not necessary for the plasma to be collisional in order to support waves. In this section we will assume that some relativistic particles exist, and will look at their interaction with waves, their acceleration and the evolution of their distributions. In §9.4 we will return to the question of the 'injection' of non-relativistic particles.

A magnetized plasma can support two types of long wavelength waves: Alfvén and magnetosonic. Alfvén waves are circularly polarized, transverse waves which propagate along or at an angle to the magnetic field. In the low-amplitude limit they are not compressive. These waves look like transverse oscillations of the field lines, much like an electromagnetic wave and the restoring force can be thought of as the tension in the field lines. We find the wave speed by deriving a dispersion relation – that is, by assuming that $\mathbf{B}$, $\mathbf{E}$ and the plasma velocity vary as $e^{i(\mathbf{k}.\mathbf{x}-\omega t)}$. Using this with Maxwell's equations, Ohm's law (in the infinite conductivity limit) and the hydrodynamic equation of motion, gives the relation (*e.g.*, Krall & Trivelpiece 1973):

$$\omega^2 = \frac{k^2 c_A^2 \cos^2\theta}{1 + c_A^2/c^2}, \tag{9.3}$$

where $c_A = (B^2/4\pi\rho)^{1/2}$ is the Alfvén speed, and θ is the angle between $\mathbf{k}$ and $\mathbf{B}$. Alfvén waves exist for frequencies $\omega < \Omega_p = eB/m_p c$, the ion gyrofrequency. For frequencies $\Omega_p < \omega < \Omega_e$

(where Ω_e is the electron gyrofrequency), these waves turn into whistlers and approach electron-cyclotron waves as $\omega \to \Omega_e$.

Magnetosonic waves are compressive waves which propagate across or at an angle to the magnetic field. In the limit of $\theta \to \pi/2$ (propagation across the field), they become mostly longitudinal waves; the restoring forces thus are plasma pressure and magnetic pressure. In a warm plasma (with $c_S^2 = k_B T/m_p$), the dispersion relation has two branches, the fast and slow magnetosonic modes (Krall & Trivelpiece 1973):

$$\omega_\pm^2 = \frac{1}{2}\frac{k^2}{1 + c_A^2/c^2}\left[c_A^2 + c_S^2 \pm \left(\left(c_A^2 + c_S^2\right)^2 - 4c_A^2 c_S^2 \cos^2\theta\right)^{1/2}\right]. \quad (9.4)$$

The fast magnetosonic wave is the one of higher frequency, corresponding to the (+) solution in equation (9.4). Equations (9.3) and (9.4) apply to internally subrelativistic plasmas; the more general, relativistic forms are given by Barnes & Scargle (1973).

These waves affect the relativistic particles through their fluctuating electric fields, and this interaction is strongest when the wave and particle are in resonance – when the wave and particle speeds are well matched, so that energy and momentum transfer are the most effective.

The fundamental resonance in a magnetized plasma – and the relevant one for Alfvén waves – is the cyclotron resonance (also called the Doppler resonance), given by

$$\omega - k_\parallel v_\parallel - n\Omega^* = 0 \quad (9.5)$$

where $k_\parallel$ and $v_\parallel$ are the components of wavenumber and velocity parallel to the magnetic field, and $\Omega^* = \Omega/\gamma$ is the relativistic gyrofrequency. In this resonance, the resonant particles see the wave Doppler-shifted to a multiple of their gyrofrequency. The most important resonance for electron interaction with Alfvén waves is the $n = -1$ resonance, while for protons $n = +1$ is the most important (Melrose 1968). Energy and momentum transfer occur because a particle gyrates in phase with the rotating ΔB of the circularly polarized Alfvén wave, and a coherent Lorentz force persists for many gyrations. For relativistic particles, this resonance condition can be written $pk = \Omega m/\alpha\,(\cos\psi - c_A/c)$ if ψ is the particle pitch-angle and $\alpha = \cos\theta = k_\parallel/k$. Thus, there is a correspondence between the particle energy and the wavelength with which it can resonate. The particle does not 'see' all turbulent wavenumbers, but only those above $k_{\text{res}}(p) = \Omega m/p$.

This resonance has important consequences for particle trapping and pitch-angle isotropization. An anisotropic distribution of relativistic particles will generate resonant Alfvén waves at a rate proportional to the anisotropy and the relative streaming speed (*e.g.*, Melrose & Wentzel 1970). Unless the wave damping is strong – which is unlikely in an ionized plasma – the self-generated waves will scatter the particles, maintaining isotropy and reducing the streaming speed to $\sim c_A$ (or c_S: see Holman *et al.* 1979).

If the Alfvén waves are externally generated, they can transfer energy to the particles as well as scatter them, although the acceleration rate is slower than the scattering rate by a factor of (c_A/c) (compare also equations 9.1 and 9.2).

The $n = 0$ limit of the cyclotron resonance is

$$\omega = k_{\parallel} v_{\parallel} \tag{9.6}$$

which is the Landau (or Čerenkov) resonance. This describes a wave which the resonant particles see Doppler shifted to zero frequency. The particle energy gain can be understood as arising from the force of interaction between the particle's magnetic moment, and the gradient of the wave field, as the particle 'rides' the wave. This is the important resonance for acceleration by long-wavelength magnetosonic waves. The resonance depends only on pitch-angle, so that all wavenumbers can interact with a given particle energy. An interesting detail of this acceleration mechanism is that it is effective only when cyclotron-resonant scattering maintains a pitch-angle isotropy. This is because the rate of energy gain for a given particle depends on the component of momentum perpendicular to the magnetic field, but the energy gain all goes to the component parallel to the field (cf. Achterberg 1981). Thus, without pitch-angle scattering to redistribute the energy gained, the acceleration rate will be much slower and will not necessarily be Fermi-like. Results quoted below and in the literature generally assume isotropy. Acceleration by long wavelength magnetosonic waves is also related to what is called magnetic pumping – in which rapid pitch-angle scattering allows adiabatic compression (betatron pumping) to work (cf. Melrose 1980), and in which the acceleration rate is again Fermi-like.

9.2.3 Wave spectra and energetics

The rate of particle acceleration, by either resonance, depends on the energy spectrum of the waves. The total wave energy density must reflect the balance between driving and dissipation mechanisms. This is a complicated subject, and must be treated either *via* analytic approximations or numerical investigations. For all but the lowest wave amplitudes, the turbulence is a nonlinear phenomenon – and for the largest amplitudes, the concept of linear wave modes is probably not valid. Further, there are few measurements of MHD or plasma turbulence – such measurements are quite difficult in the laboratory. Some data exist for space plasmas; we are not aware of any direct measurements of waves and/or turbulent spectra for radio jets (which is not surprising, since the resonant wavenumbers are well below interferometer resolution). We thus must rely on theory (usually linear) and on numerical simulations.

The wave driving will come from the fluid and plasma instabilities described above. The energy spectrum of the waves is defined by $\Delta B^2 = 8\pi \int W(k)dk$ so that $W(k)$ is the wave energy density per unit wavenumber. The wave dissipation, $dW(k)/dt = -\gamma(k)W(k)$, can be calculated explicitly for Alfvén and magnetosonic waves in the linear limit (cf. Lacombe 1977, or Eilek 1979). The main dissipation of Alfvén waves in an ionized plasma is relativistic particle acceleration, and this energy loss rate depends on the particle number density: $\gamma(k) \propto n_{\rm rel}$. Magnetosonic waves lose energy to relativistic particle acceleration as well, but also to heating (both through the Landau resonance, equation 9.6, and through collisional effects at low wavenumbers) of any thermal plasma which is present. The Landau heating/acceleration depends on the energy densities: $\gamma(k) \propto u_{\rm rel} + u_{th}$. Conductive and viscous damping on the thermal gas can also be important for magnetosonic waves (Eilek 1979). Thus, in sources with a significant amount of entrained thermal gas, much of the magnetosonic wave energy will go to heating the thermal plasma rather than to accelerating the relativistic component.

If the waves are driven externally, at a rate $I(k)$ per wavenumber, the wave spectrum can be estimated: $W(k) \approx I(k)/\gamma(k)$ (cf. Eilek

1979; Eilek & Henriksen 1984; Scott *et al.* 1980). The resultant wave spectrum depends on the nature of the driving and damping processes, so that this approach does not predict a unique, 'universal' spectrum. This approach can be used only (a) when the energy input covers all wavenumbers of interest; and (b) when $W(k)$ is low enough that terms describing two-wave interactions, $\propto W(k)^2$, are less important than direct-driving and damping terms.

Turbulent spectra in the nonlinear limit, or with driving restricted to a narrow wavenumber range, must be found from dimensional arguments (Kolmogorov 1941; Kraichnan 1965) or numerical simulations (*e.g.*, Pouquet *et al.* 1976; De Young 1980; Grappin *et al.* 1983; Shebalin *et al.* 1983). A great range of turbulent spectra is possible, depending on conditions in the system. As very little work has been done on particle acceleration by nonlinear turbulence, we merely refer the reader to the references cited.

9.2.4 Evolution of the particle spectra

Once the wave spectrum is known, we must ask what effect the turbulent acceleration by these waves has on the particle spectrum. If particle energy losses are important, such as those due to synchrotron radiation, we are interested in the relative acceleration and loss rates for a given particle energy. More generally, we are interested in the particle spectrum that results from some combination of acceleration and loss processes. In particular, as power-law spectra are generally inferred and/or assumed in the sources, we would like to determine the conditions necessary to create or maintain a power-law spectrum.

The Fermi acceleration model predicts a power-law particle spectrum, if the particles also leak out of the system at an energy-independent rate. Let the acceleration rate be given by $dE/dt = E/\tau$, and let the escape probability be dt/T in time dt (we again follow Achterberg 1986). If particles are injected at a rate R, at energy E_0, the time required for one particle to reach energy E is $t(E) = \tau \ln(E/E_0)$. But the fraction of particles remaining at time t after injection is $e^{-t/T}$. Thus, the particle spectrum is given by $n(E)\,dE = n(t(E))\,dt$, if $n(t(E))$ is the number of particles with age $t(E)$. This gives

$$n\left(E\right)=\frac{R\tau}{E_0}\left(\frac{E}{E_0}\right)^{-(1+\tau/T)}. \tag{9.7}$$

This is a power-law spectrum; but the spectral index depends sensitively on two parameters (T and τ) which can vary greatly from source to source (although Burn (1975) argues for a self-regulating mechanism). We note in particular the need for steady injection and steady leakage, in order to get the form (9.7).

In order to consider the evolution of the particle spectrum more generally, we use a kinetic theory approach (we follow Livshitz & Pitaevskii 1981, and Blandford 1986). Let $f\left(\mathbf{p}\right)$ be the distribution function (with time dependence implicit). $f\left(\mathbf{p}\right)d\mathbf{p}$ is the number of particles in the element $d\mathbf{p}$ of momentum space. The fundamental description of the behaviour of this distribution function is the Boltzmann equation:

$$\frac{\partial f}{\partial t}+\mathbf{v}\cdot\frac{\partial f}{\partial \mathbf{x}}+m\mathbf{a}\cdot\frac{\partial f}{\partial \mathbf{p}}=\left(\frac{\partial f}{\partial t}\right)_{\text{coll}}+\left(\frac{\partial f}{\partial t}\right)_{\text{diff}}. \tag{9.8}$$

$\mathbf{v}$ and $\mathbf{a}$ refer to the velocity and acceleration of a particular particle. The 'diffusion' term describes spatial diffusion, while the 'collision' term contains all the physics of collisions and scattering. A stochastic acceleration process may be thought of as a diffusion in momentum space, characterized by a diffusion coefficient $D_{\mathbf{pp}}$, so that

$$\left(\frac{\partial f}{\partial t}\right)_{\text{coll}}=\frac{\partial}{\partial \mathbf{p}}\cdot D_{\mathbf{pp}}\cdot\frac{\partial f(\mathbf{p})}{\partial \mathbf{p}}. \tag{9.9}$$

This diffusion coefficient must contain the details of the wave spectrum and the wave-particle interactions. If we assume isotropy we can write $f(\mathbf{p})d\mathbf{p}=4\pi p^2 f(p)dp$. Adding a radiative loss term so that the losses for one particle are $dp/dt=-b\left(p\right)$, we have the Boltzmann equation for the case that only collisions contribute to $\partial f/\partial t$:

$$\frac{\partial f\left(p\right)}{\partial t}=\frac{1}{p^2}\frac{\partial}{\partial p}\left(p^2 D\left(p\right)\frac{\partial f}{\partial p}+p^2 b\left(p\right)f\left(p\right)\right). \tag{9.10}$$

This is one form of the so-called Fokker-Planck equation. (Many authors add arbitrary leakage and source terms as well.) This equation is also commonly written in terms of the energy distribution, $n\left(E\right)=4\pi p^2 f\left(p\right)dp/dE$ (still assuming isotropy). When relativistic, $E = cp$ and the basic equation can be written

$$\frac{\partial n\left(p\left(E\right)\right)}{\partial t} = -\frac{\partial}{\partial E}\left(\tilde{A}\left(E\right)n\right) + \frac{\partial^2}{\partial E^2}\left(\tilde{D}\left(E\right)n\right) + \frac{\partial}{\partial E}\left(\tilde{b}\left(E\right)n\right) \quad (9.11)$$

where $\tilde{D}\left(E\right) = c^2 D\left(p\left(E\right)\right)$, $\tilde{A}\left(E\right) = \partial\tilde{D}/\partial E + 2\tilde{D}/E$, and $\tilde{b}\left(E\right) = cb\left(p\left(E\right)\right)$. The first and second terms are often described as 'first-order' and 'second-order' Fermi acceleration.

This basic equation (9.10) or (9.11) describes the evolution of the particle spectrum, subject to particular acceleration processes (described in $D\left(p\right)$) and loss processes (described in $b\left(p\right)$). The nature of the solutions is sensitive to initial and boundary conditions, as well as any assumed 'injection' or leakage terms and the forms of $D\left(p\right)$ and $b\left(p\right)$. We present a few basic types of solutions.

9.2.4.1 Particle spectra with no acceleration

The most basic solutions of equations (9.10) or (9.11) are those without acceleration, but with radiation losses. Synchrotron losses are the most interesting: $b(p) = b_0 p^2$, with $b_0(\psi) \propto \sin^2\psi$ (where ψ is the particle pitch-angle). The energization necessary to offset synchrotron losses is usually treated by assuming an arbitrary source of particles. The particle source can "inject" particles initially or continually; the injection spectrum is usually taken to be a power-law, $f(p) \propto p^{-(2+\delta)}$ (*e.g.*, Kardashev 1962). The subsequent evolution of $f(p)$ depends on whether or not pitch-angle scattering is important.

Consider initial injection of a power-law spectrum. If there is no pitch-angle scattering, the distribution function at each pitch-angle evolves separately. If the power-law is injected at $t = 0$, then after a time t, $f(p, \psi)$ at a given ψ maintains the injected power-law for $p < 1/b_0(\psi)t$ and drops exponentially to zero at $p > 1/b_0(\psi)t$. This cutoff is just the energy to which a particle which started with infinite energy evolves in time t. If the particle distribution is isotropic in ψ, the net distribution function (integrated over ψ) is a broken power-law: $f(p) \propto p^{-(2+\delta)}$ for $p < 1/b_0(\pi/2)t$, and $f(p) \propto p^{-(3+\delta)}$ for $p > 1/b_0(\pi/2)t$. Alternatively, if pitch-angle scattering is rapid, a given particle will move through all possible pitch-angles in a time short compared to its synchrotron lifetime. In this limit, the net distribution function evolves as if it had a mean pitch-angle, $\overline{\psi}$; it then steepens exponentially for $p > 1/b_0(\overline{\psi})t$.

The case opposite to that of initial injection is continual injection, at an assumed steady rate, of the same power-law spectrum. For this case, models with and without pitch-angle scattering both find that the particle distribution function steepens in the same way (from $2+\delta$ to $3+\delta$ in the power-law exponent) at $p = 1/(b_0(\pi/2)t$.

We pointed out above that pitch-angle scattering is likely to be important in radio sources, and that the relativistic particle spectrum is likely to be close to isotropic. In particular, the anisotropy created by synchrotron radiation (which depletes particles at high ψ) will generate resonant Alfvén waves which tend to maintain $f(p,\psi)$ isotropic. Melrose (1970) has shown that for times $t < 1/b_0(\pi/2)p$, this process maintains the anisotropy at a low level, $O(c_A/c)$. We are not aware of any work that has established the level of anisotropy for longer times (or higher energies). At present, most authors concerned with particle spectra at high energies or long times work in one or the other limit (*e.g.*, Myers & Spangler 1985).

9.2.4.2 Particle spectra with stochastic acceleration

Acceleration can be included directly in (9.10) or (9.11) if the $D\,(p)$ coefficient is specified. The second-order Fermi test-particle model can be interpreted as (Blandford 1986)

$$D\,(p) = \frac{p^2 < V^2 >}{3Lc}. \tag{9.12}$$

More physically, the details of the wave-particle interaction should be included in evaluating $D\,(p)$. Acceleration due to long-wavelength magnetosonic waves in the quasi-linear limit results in a diffusion coefficient (Kulsrud & Ferrari 1971; Achterberg 1981; Eilek 1979), if pitch-angle isotropy is assumed,

$$D\,(p) = a_1 \frac{c_M^2}{c} \frac{p^2}{B^2} \int_0^\infty k W_{MS}\,(k)\,dk, \tag{9.13}$$

where a_1 is a numerical constant of order unity and $c_M = \omega/k$ is the magnetosonic wave speed. We notice that all wavenumbers contribute to the acceleration, with the lowest ks (longest wavelengths) being the most important, and that the momentum dependence of Fermi acceleration is retained ($D\,(p) \propto p^2$, equation (9.12)).

Acceleration by Alfvén waves, through the cyclotron resonance, leads to a diffusion coefficient (Lacombe 1977; Eilek 1979)

$$D(p) = \frac{2\pi e^2 c_A^2}{c^3} \int_{k_{\rm res}(p)}^{\infty} W_A(k) \frac{1}{k} \left[1 - \left(\frac{c_A}{c} + \frac{\Omega m}{pk} \right)^2 \right] dk, \quad (9.14)$$

and this picks out the lowest *resonant* wavevector, $k_{\rm res}(p)$. This leads to a form for Alfvén-turbulent acceleration that is quite different to that for Fermi acceleration.

With $D(p)$ known, (9.10) can be solved for the evolution of $f(p)$. As this is a diffusion equation, the particle spectrum can be thought of as diffusing and convecting in momentum space and a wide range of solutions are possible (cf. Borovsky & Eilek 1986, for a discussion of the properties of this equation and some numerical solutions).

Consider, for instance, Fermi acceleration in the presence of synchrotron losses; let $D(p) = D_0 p^2$ and $b(p) = b_0 p^2$. We would expect that synchrotron losses dominate the spectral shape for $p > p_c = D_0/b_0$, while Fermi acceleration determines the spectrum below p_c. This is in fact the case. Borovsky & Eilek (1986) found that the spectrum of a closed system approaches a steady solution peaked at p_c, namely $f(p) \propto p^2 e^{-p/p_c}$ (see also Schlickheiser 1984). This is not a power-law solution. On the other hand, at low energies where synchrotron losses are not important, and with injection at some low p_0, a power-law solution is recovered (Borovsky & Eilek 1986; see also Achterberg 1979, who adds leakage). This power-law will be truncated at $p \sim p_c$; classical Fermi acceleration cannot maintain a power law at all energies if the system suffers from synchrotron losses. (We note that the same conclusion will apply to shock acceleration, as long as it can be described as a first-order Fermi process.) The classical Fermi power-law solution (equation (9.7) and equation (9.18) in the case of shock acceleration) includes steady leakage of particles from the system, and does not include synchrotron losses.

An interesting variant, which differs from the usual Fermi acceleration model, is the case of $D(p) = D_0 p^3$. This can arise from Alfvén wave acceleration with a wave spectrum, $W_A(k) \propto k^{-3}$; Eilek & Henriksen (1984) found that this will be the state to which a coupled system of fluid turbulence, Alfvén waves and relativistic particles evolves in times $t > t_{sy}$. Such a steep wave spectrum is also characteristic of Alfvénic turbulence in which the velocity and magnetic fluctuations are strongly correlated, as is found in the solar wind (Grappin *et al.* 1982). This particular form of $D(p)$ allows the turbulent acceleration to balance the synchrotron losses at all energies. The kinetic equation (9.10) admits solutions in this case which are linear combinations of $f_1(p) = p^{-b_0/D_0}$ and $f_2(p) = p^{-4}$ (Lacombe 1977; Eilek & Henriksen 1984).

9.2.4.3 Overview of the section

In this section we have seen that MHD turbulence can accelerate relativistic electrons, if the necessary cyclotron-resonant Alfvén waves exist in the plasma. A major question, then, is whether this mechanism can produce the observed power-law spectra. We saw that simple, macroscopic arguments suggest that it can (as in the derivation of equation 9.7). However, more detailed investigations of the microphysics shows that power-laws are only one possible result of either Fermi or Alfvén wave acceleration, and may require rather special conditions: such as steady injection and leakage, and the absence of synchrotron losses, for Fermi-type acceleration; or the presence of synchrotron losses and a particular wave driving mechanism for Alfvénic acceleration. We will see in the next section that shock acceleration can also give power-law spectra, but again, *only if the microphysical conditions are right.* Thus it appears that power-laws are more common in nature than in our current theories.

9.3 Acceleration by shocks

9.3.1 The physics of collisionless shock waves

As with the waves discussed above, shocks may exist in a collisionless plasma – the necessary interaction between the particles being mediated by electromagnetic fields. Indeed, although shocks are commonly discussed in terms of the jump conditions applying to the magnetohydrodynamic variables (see §4.3.3), they may also be understood as 'steepened' plasma wave-packets (see below) and they possess structure that can be understood only in terms of the electromagnetic interaction between the different species.

9.3.1.1 Fundamentals

Suppose a wavepacket comprising waves of various wavenumber $\mathbf{k}$, satisfying a dispersion relation $\omega = \omega(\mathbf{k})$, forms a large amplitude structure with some profile propagating in a plasma. Now three plasma waves may interact resonantly when $\omega_3 = \omega_1 + \omega_2$ and $\mathbf{k}_3 = \mathbf{k}_1 + \mathbf{k}_2$, allowing a transfer of energy from some wavenumber to wavenumbers both above and below this. In the case of the component waves of a wavepacket, such resonant interactions between waves (lying on the same dispersion curve) redistribute the energy in wavenumber space: this corresponds to a change in the Fourier components of the wavepacket, and hence to a change in its profile. In particular, it may correspond to a steepening of one edge of the wavepacket to form a shock.

For a typical dispersion relation, $d\omega/dk$ is approximately constant for $k < k_i$, where $k_i \approx \omega_e/c$ in terms of the electron plasma frequency, but it diverges from this behaviour at high wavenumber, allowing waves of different wavenumber to propagate at different speeds. The three-wave interaction discussed above may allow the transfer of energy from the Fourier components corresponding to the main part of the shock transition, to higher wavenumber. These waves of shorter wavelength will lag behind the transition or run ahead of it, according to whether $d\omega/dk$ decreases or increases compared to the constant value of small wavenumber. In the former case, a train of waves downstream of the transition occurs (see Fig. 9.1(A)); in the latter case, an upstream wavetrain results.

These wavetrains may 'decay' through three wave interactions with waves from the ever-present spectrum of noise in the plasma, or may be 'damped' by plasma processes that limit the magnetic field gradient of the shock, and hence limit the dispersion that will occur. Such processes (*e.g.*, current driven instabilities) are the mechanism of flow energy randomization in laboratory and space shocks and tend to allow shocks with a thickness (often many ion-gyroradii) that is longer than the 'inertial length' – *i.e.*, the wavetrain scale of a few oscillations, each of characteristic width $\approx 1/k_i$.

As with low-amplitude magnetosonic and Alfvén waves, which are compressive and non-compressive respectively, the magnetic field of the plasma is or is not compressed by the shock, according to whether it lies in the plane of the shock, or parallel to the shock normal. Shocks are classified by the angle, Θ, between the magnetic field carried into the shock by the upstream flow, and the normal to the shock: 'perpendicular' ($\Theta = \pi/2$, magnetic field in the plane of the shock), 'quasi-perpendicular', 'quasi-parallel', and 'parallel'

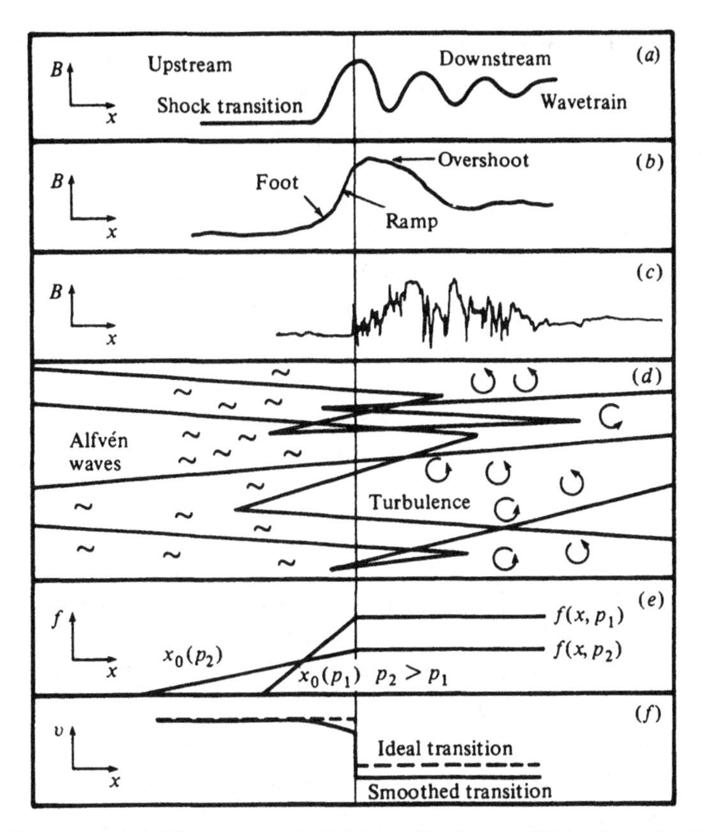

Fig. 9.1. (a) The magnetic field profile for a collisionless shock with infinitesimal dissipation. (b) The magnetic field profile for an almost perpendicular, supercritical shock, showing a smooth transition between upstream and downstream states. (c) The magnetic field profile for an almost parallel shock, showing a highly turbulent downstream flow. (Shocks with $\beta_P \geq 1$ have a similar character.) (d) A schematic of the 'test-particle' Fermi-process, in which charged particles are scattered by the turbulence of the downstream flow, and by self-generated waves upstream. (e) A schematic of the distribution function for the process shown in panel (D); there are fewer high-energy particles, but they penetrate further into the upstream flow. (f) A schematic showing how the flow velocity is modified by the accelerated particles; these slow the upstream flow, reducing the strength of the 'subshock', but increase the overall compression.

($\Theta = 0$, magnetic field parallel to the shock normal). Perpendicular shocks tend to be laminar (*i.e.*, there is a well defined transition from upstream to downstream states); parallel shocks tend to be turbulent. These should not be confused with *oblique* shocks, in which the upstream flow velocity is at an angle to the shock normal, and only that component of velocity along the normal need be subsonic downstream. The other two important parameters are the *Mach number* of the flow, $\mathcal{M}$, defined in terms of the flow velocity and c_A, c_S or c_M – the Alfvén, sound and fast magnetosonic speeds of the plasma (*e.g.*, $\mathcal{M}_A = v/c_A$); and β_P, the ratio of gas pressure to magnetic pressure ($\beta_P = P_{\rm gas}/P_B$). At low Mach number shocks tend to be laminar, while at high Mach number significant fluctuations in magnetic field and particle density occur near the transition, and downstream. If $\beta_P \leq 1$, shocks tend to be laminar. If $\beta_P >> 1$, shocks tend to be turbulent – there are large amplitude fluctuations in magnetic field. The character of a particular shock depends on exactly where in (Θ, $\mathcal{M}$, β_P) space it lies.

9.3.1.2 The internal structure of collisionless shocks: evidence from interplanetary shocks

No study has yet been made of the internal structure of relativistic shocks (although Barnes & Suffolk 1971, and Barnes 1983 have formally demonstrated that Alfvén and fast-magnetosonic waves in collisionless, relativistic plasmas steepen to form shocks), and we must rely on studies of nonrelativistic structures – principally the Earth's bow-shock – applying the results to high-energy situations with caution.

Shocks with $\beta_P << 1$ or $>> 1$ are rarely observed, and their theory has not been extensively explored. The most notable feature of $\beta_P >> 1$ shocks (*e.g.*, Formisano *et al.* 1975) is the absence of a smooth transition in the magnetic field (even when the shock is almost perpendicular). The downstream magnetic field can exhibit rapid fluctuation, with peak amplitude energy density comparable to the upstream dynamic pressure. It has been suggested that in this domain no steady shock can form, and that dissipation is associated with many lower-β_P 'shocklets'. $\beta_P << 1$ shocks may be similar to their $\beta_P \sim 1$ relatives (*e.g.*, Greenstadt *et al.* 1975, for quasi-perpendicular shocks, and Greenstadt 1974, for quasi-parallel shocks). Numerical simulations (*e.g.*, Quest 1986) suggest that, at least at low Mach number, they are more unsteady than $\beta_P \sim 1$ shocks – but that a similarity persists both in the 'snapshot' and

time-averaged senses. The best understood shocks are those with $\beta_P \sim 1$, and we shall now discuss these in some detail.
$\beta_P \sim 1$, *Quasi-perpendicular Shocks*: At low Mach number the magnetic field gradient of the shock supports a current that drives plasma waves. These waves scatter electrons, and so provide an anomalous resistivity (see §9.5.1). The dissipation associated with this resistivity is the mechanism whereby a fraction of the upstream flow energy is randomized. The level of waves excited is such as to limit the current, and hence the magnetic field gradient, so that a laminar transition somewhat broader than the wave train scale (see above) results (Morse & Greenstadt 1976). The increase in the electron density at the transition produces a rise in the electrostatic potential Φ there, and this slows the protons, heating them adiabatically (Morse 1973, 1976); the ions are not slowed by a scattering process – they are simply 'held back' by the electrons. (See Winske (1984) for a comprehensive review of this field.)

This process may be described using 'fluid' equations for both the electrons and the protons. However, above a *lower critical Mach number*, $\mathcal{M}_A^*$, it is found that the fluid equations do not admit a shock solution. This is because for $\mathcal{M}_A > \mathcal{M}_A^*$ the electrostatic potential of the shock is inadequate to slow the ions sufficiently: they cannot form a subsonic downstream flow. An additional mechanism for ion randomization (heating) is required, and this turns out to involve reflection of a subset of the ions at the transition. It is therefore necessary to study the ions by kinetic theory (*i.e.*, follow their orbits), although a fluid description of the electrons still suffices. $\mathcal{M}_A^* \sim 2-3$, the value depending on the resistivity of the plasma.

The role of ion reflection has been discussed by Forslund & Freidberg (1971) and Morse (1976); numerical simulations and their interpretation have been presented by Leroy *et al.* (1981), (1982) and Leroy (1983). A self-maintaining structure arises as follows: because a subset of the ions are reflected at the shock, the mean flow velocity V_x into the shock is less than that of the far upstream flow, $V_{x\infty}$. The incoming ions therefore have velocity $v_x > V_x$. The magnetic field is carried with the local mean velocity V_x, and in crossing this, the incoming ions experience a Lorentz force

$$\frac{d}{dt}v_y = -\frac{e}{m_p c}\left(v_x - V_x\right) B \tag{9.15}$$

which gives them a negative value of v_y – a velocity in the plane of the shock. The Lorentz force associated with this negative v_y acts to decelerate the ions in the x-direction according to

$$\frac{d}{dt}v_x = -\frac{e}{m_p}\frac{\partial}{\partial x}\Phi + \frac{e}{m_p c}v_y B, \tag{9.16}$$

a process capable of reversing the sign of v_x for the slowest ions – *i.e.*, reflecting them. (By itself, the electrostatic potential Φ is inadequate to do this.) The reflected ions reduce the mean flow speed to V_x in the foot region (Fig. 9.1(B)) – causing a build-up of particles responsible for the initial rise of field and particle density there – and so complete the cycle whereby the shock is maintained. On a subsequent approach to the shock, the previously reflected ions will have been energized by the electric field of the shock (cf. §9.5.3), and so have velocity high enough that they are 'transmitted' to the downstream flow.

From the point of view of tapping upstream flow energy to heat particles, the important conclusion to draw from the above is that *supercritical* shocks involve the formation of two distinct ion streams. A bimodal ion velocity distribution arises as a consequence of the transmission of part of the incoming ion distribution, and the reflection and subsequent transmission of the rest. A shock is possible because the mean velocity of the two distinct subsets of ions is subsonic, and satisfies the shock jump conditions (see §4.3.3). Ultimately this bimodal distribution will relax (most likely through scattering by plasma waves arising from an ion-ion counterstreaming instability); however, it may be many hundreds of ion-gyroradii downstream before this occurs, and the upstream flow energy can truly be regarded as thermalized. Even then, the distribution may not be Maxwellian (see Fig. 3 of Greenstadt & Fredricks 1979; Greenstadt *et al.* 1980). In such supercritical shocks the heating of the electrons depends on the value of β_P. If $\beta_{P,e} < 1$, Ohmic heating through anomalous resistivity in the ramp dominates, as for the subcritical shocks; if $\beta_{P,e} > 1$, they are heated adiabatically – *i.e.*, by compression (Leroy *et al.* 1982). In either case it is the gyrating ions that determine the scale of the transition.

The simulations by Leroy *et al.* found no steady structure above an *upper critical Mach Number* ($\mathcal{M}_A^{**} \sim 10$), but recent studies by Quest (1985, 1986) show that the behaviour at high Mach number depends on the resistivity of the electron fluid. As $\mathcal{M}_A$ increases above $\mathcal{M}_A^{**}$ the shock structure becomes time-dependent, and can eventually reach a periodic state in which alternately 100% of the incoming ions are transmitted and reflected. This happens only if the resistive diffusion length, defined as $l_r = \eta c^2/4\pi v_u$, is $l_r = 0$, or $l_r \geq 0.2c/\omega_p$. For other values of l_r a steady structure results; the number of reflected ions asymptotes to $\sim 40\%$, and this fraction is always sufficient to allow the necessary randomization of the upstream flow energy. Quest argues that the highly unsteady structure might arise because of the neglect of the electron mass (and hence inertia); if l_r is zero (or too large) downstream waves may be able to catch up with the shock and destabilize it, unless a small resistivity allows a fast enough downstream flow speed to prevent this. Further study is needed to clarify the behaviour of high Mach number quasi-perpendicular shocks.

$\beta_P \sim 1$, *Quasi-parallel Shocks*: An understanding of these structures has been elusive, because they exhibit considerable magnetic turbulence extending over a scale of many ion-gyroradii. Nevertheless here, as with the perpendicular shocks, numerical simulations have given insight into the processes at work.

A simulation by Kan & Swift (1983) shows that for $\mathcal{M}_A \leq \mathcal{M}_A^* \sim 3$ there is laminar transition in the magnetic field, together with a distribution of waves *upstream* which are identified as 'whistlers'. These whistler waves are generated by the dispersion of the transition from upstream to downstream states (see §9.3.1.1). Randomization of the ion flow energy is accomplished simply by the nonadiabatic scattering of the ions by these whistlers that stand almost stationary in the frame of shock.

Above the critical Mach number ($\mathcal{M}_A^*$) the character of the downstream flow changes dramatically, due to the onset of a 'firehose' instability. This arises because if magnetic field lines are bent, the parallel component of velocity of the particles 'tied' to these field lines exerts a centrifugal force that enhances the bending. This effect is countered by the resistance to bending of both the perpendicular component of gas-pressure, and the field line tension; thus the instability occurs only if

$$p_{\parallel} > p_{\perp} + \frac{B^2}{4\pi} \tag{9.17}$$

(see, *e.g.*, Ichimaru 1973). In supercritical, quasi-parallel shocks the scattering of ions by whistler waves enhances $p_{\parallel}$ to the point that inequality (9.17) is satisfied; perturbations to the magnetic field then grow to form fast magnetosonic waves (and some whistlers), in the vicinity of, and upstream of, the transition. Kan & Swift's simulations show the development of the firehose instability at the transition, its extension to the upstream flow, and a settling to marginal stability; there is then a steady production of waves that are advected into the downstream flow where they give rise to the observed magnetic turbulence (see Fig. 9.1(C)).

The nature of the electron flow energy randomization has been addressed by Quest *et al.* (1983). These authors model a supercritical, quasi-parallel shock, but include the electron dynamics. They find that the electrons are heated adiabatically; *i.e.*, in a way that is consistent with a simple compression (due to the electrostatic potential jump) at the transition.

Kan & Swift find a population of very energetic ions for supercritical Mach number. However, it does not appear that these are a good candidate seed population for the Fermi process (see below), because they are found only in the upstream flow. They are probably energized by reflection from propagating, large-amplitude waves (cf. §9.5.3).

A review of the distribution of waves and particles upstream of a collisionless shock – as typified by the Earth's bow shock – may be found in Tsurutani & Rodriguez (1981), and the papers that are introduced by that review.

9.3.2 The shock-Fermi process

The processes discussed above apply to the thermal particles, and the way in which their energy is modified by a shock. We now address a scheme whereby a small number of suprathermal particles are accelerated to a power-law energy distribution that may contain a significant fraction (≥ 10 %) of the flow energy randomized in a shock. This mechanism, a variant of the first-order Fermi process discussed in §9.1, and hence referred to as the *shock-Fermi process*, has been amply tested by observation within the solar system (*e.g.*, Ellison 1985). It has been applied, with a greater or lesser degree of success, to areas as diverse as solar flares (*e.g.*, Achterberg & Norman 1980), interplanetary shock waves (*e.g.*, Eichler 1981), stellar wind terminal shocks (*e.g.*, Webb, Forman & Axford 1985), supernova shocks (*e.g.*, Blandford & Ostriker 1980; Cavallo 1982; Moraal & Axford 1983; Bogdan & Völk 1983; Heavens 1984c) and the galactic wind terminal shock (*e.g.*, Jokipii & Morfill 1985).

9.3.2.1 The microphysics of the Fermi process

In its simplest form – that is, without considering the back action of the accelerated particles on the shock structure – the shock-Fermi process was first described by Axford, Leer & Skadron (1977), Krymsky (1977), Blandford & Ostriker (1978) and Bell (1978a,b). A formulation of the problem using classical macroscopic thermodynamics has been presented by Michel (1981). The reader wanting to see a more detailed discussion of the shock-Fermi process is refered to reviews by Drury (1983) and Blandford & Eichler (1987).

From the discussion of §9.3.1 it should be apparent that the transition thickness of a collisionless shock wave is a few thermal-ion gyroradii. Particles of sufficient energy (*e.g.*, those in the suprathermal tail of the energy distribution) will have gyroradii that are so much larger than this scale that they can propagate across the transition without 'seeing' it. Thus if some such particles can be scattered back and forth between upstream and downstream flows, the only energy change that they will suffer will be that associated with the change of reference frame (defined by the upstream and downstream scattering centres). Alternatively, we can say that these particles are scattered back and forth by approaching (converging) magnetic mirrors – but mirrors that converge as long as the shock persists! We have seen that the downstream flow is likely to be highly turbulent (particularly for parallel shocks) and this turbulence may scatter the suprathermal particles, giving the plasma an effective diffusion coefficient, κ. (We use κ to avoid confusion with the momentum space diffusion coefficient, D.) A fraction of the particles will diffuse away downstream, whilst others will diffuse (will be scattered back) into the upstream flow. Similarly, scattering centres in the upstream flow will deflect the 'reflected' particles, limiting the speed with which they can stream against the upstream flow, and ensuring that they are ultimately carried back through the transition (or overtaken by the shock – an alternative point of view). In fact it is not necessary for the scattering centres (at least upstream) to be preexisting plasma turbulence associated with the thermal shock. Those particles scattered into the upstream flow may themselves generate waves by virtue of counterstreaming against the unshocked material. Particles that have crossed the transition three times may diffuse away downstream or again be carried upstream; clearly there will be fewer and fewer particles that have crossed five, seven and so on times, gaining energy each crossing; there will thus be many low energy particles (having made only a few crossings before diffusing away downstream) and a small number of high energy particles (having made many crossings before being lost). This is illustrated in Fig. 9.1(D).

A simple argument shows that a power-law energy (or momentum) distribution arises from this mechanism: particle acceleration is associated with a change in momentum $dp/dt = \Delta p/t_{\rm cross} = (4/3) \cdot (v_u - v_d)/v \cdot (p/t_{\rm cross})$. This means that the acceleration time scale is $\tau = (3/4)vt_{\rm cross}/(v_u - v_d)$. The number of particles lost per crossing of the shock is $(4v_d/v)n$, so the loss rate due to diffusion is $dn/dt = -(4v_d/v)(n/t_{\rm cross})$, and the time scale characterizing particle loss is $T = t_{\rm cross}/(4v_d/v)$. From equation (9.7) we can immediately derive the energy spectral index

$$\delta = 1 + \frac{\tau}{T} = \frac{2+r}{r-1}, \tag{9.18}$$

where r is the compression ratio of the shock ($r = v_u/v_d$). In momentum space the corresponding index is $3r/(r-1)$. Note the important result that index is independent of $t_{\rm cross}$; indeed, it depends only on the 'strength' of the shock. (A much fuller discussion of this type of argument may be found in Bell (1978a,b) and Peacock (1981); in particular, Bell formally demonstrates that Fermi acceleration at non-parallel shocks – in the frame of which there must be an electric field – produces this power-law spectrum, as long as it is possible to transform into a frame in which the electric field vanishes – *i.e.*, as long as the angle between the magnetic field and the shock normal is not too great.)

9.3.2.2 The test-particle theory of the Fermi process

As in the study of stochastic acceleration (§9.2.4), a general analysis of this process requires the use of the Boltzmann equation (9.8), which can be written for a one-dimensional, steady state flow:

$$\frac{\partial f}{\partial t} \equiv 0 = -v\frac{\partial f}{\partial x} + \frac{1}{3}\frac{\partial v}{\partial x}p\frac{\partial f}{\partial p} + \frac{\partial}{\partial x}\left(\kappa\frac{\partial f}{\partial x}\right). \tag{9.19}$$

v is the flow speed, p is the particle momentum and f is the distribution function, defined in §9.2.4. On the right-hand side of (9.19) the first term describes convection of the particles at speed v, the second term describes their momentum change at the transition (where the speed changes) and the third term describes diffusion. The factor of 1/3 comes from averaging over the angles at which the particles are scattered across $x = 0$.

It is important to realize that equation (9.19) is a 'macroscopic' or 'thermodynamic' equation. The collision term of equation (9.8) has been taken to be zero, and **v** and **a** have been identified with the *fluid* velocity and acceleration. Thus the significant assumption has been made that the microphysics works to ensure that the particles are strongly coupled to the flow.

In the idealization that the velocity changes discontinuously at $x = 0$, the second term can be set to zero in both upstream and downstream flows, and the resulting equation integrated over x to yield $vf - \kappa\partial f/\partial x = constant$. This describes the flux of particles on either side of the transition; a second integration shows that $f(x)$ takes the form

$$f(x) = A + B exp\left(\int v/\kappa dx\right), \tag{9.20}$$

where A and B are constants. Obviously the diffusion coefficient is positive, and $f \to \infty$ as $x \to \infty$ unless $B = 0$. Thus on the downstream ($x > 0$) side of the shock, the distribution function (density) must be a constant, f_d (n_d). To determine f on the upstream side of the shock ($x < 0$) we need to know $\kappa(x, p)$. Bell (1978a) assumed that the scattering in the upstream flow arises from Alfvén waves, generated by the counterstreaming of high energy particles that return from the downstream flow, and found

$$f(x,p) - f(-\infty,p) = \frac{f_d(p) - f(-\infty,p)}{1 - x/x_0(p)}. \tag{9.21}$$

The distribution of particles is shown schematically in Fig. 9.1(E) and has several important consequences:

The higher energy particles penetrate far upstream before they are scattered back into the downstream flow; although it was ignored in the above analysis, wave damping will become important upstream, and reduce the wave intensity to a point where no significant scattering of particles occurs. There is thus a critical energy above which particles will escape freely upstream; this sets a limit to the energy attainable by this acceleration mechanism. Another factor that determines the maximum attainable energy is the shock lifetime. Doubling a particle's energy requires $p/\Delta p$ ($\approx c/v_u$) cycles, each of time L/c, where $L = \kappa_u/v_u$ is the distance a particle diffuses before being overtaken by the shock; thus the acceleration time scale is $\approx \kappa_u/v_u^2$, or more precisely

$$\tau_{\rm acc} = \frac{3}{v_u - v_d}\left(\frac{\kappa_u}{v_u} + \frac{\kappa_d}{v_d}\right) \tag{9.22}$$

(*e.g.*, Achterberg 1986). The shock must persist for a time $\geq t_{\rm acc}(p)$ if a steady spectrum at momentum p is to be attained. Shock geometry also is important; the mechanism breaks down if the shock exhibits significant curvature on a scale near to or less than the diffusion length scale or gyroradius at a given momentum. The maximum attainable energy has been investigated for supernova shocks by Lagage & Cesarsky (1983a,b) and Heavens (1984a). Biermann & Strittmatter (1987) have investigated the maximum attainable energy in the context of shocks in AGN and jets. They assume that both electron and proton energies are limited by synchrotron radiation energy loss, and that the turbulence spectrum can be characterized by the gyroradius of the most energetic protons. It appears from this model that 3×10^{14} Hz is a strong upper limit to the radiation spectrum from shock-Fermi accelerated electrons.

High energy particles penetrating the upstream flow modify the shock structure. In particular, these particles tend to decelerate the thermal plasma causing a weaker shock than would otherwise be the case. We shall come back to this point in §9.3.3.

9.3.2.3 The particle momentum spectrum downstream

Because there is no natural length scale in the simple case that the velocity profile of the shock is taken to be a step-function, we can determine the spectrum of accelerated particles in the downstream flow without detailed knowledge of κ there. Integrating (9.19) from $x = -\infty$ to $x = \infty$, with $\partial f/\partial x = 0$ at each endpoint, and $\partial v/\partial x = (v_d - v_u)\,\delta(x)$, we get

$$\frac{df_d(p)}{dp} = \frac{3v_u}{v_d - v_u}\frac{f_d(p) - f(-\infty, p)}{p}, \tag{9.23}$$

which has the solution

$$f_d(p) = Cp^{-q} + qp^{-q}\int^{p} f(-\infty, p')\,p'^{q-1}dp' \tag{9.24}$$

where $q = 3v_u/(v_u - v_d)$, and C is a constant of integration. C must be zero so that $f_d(p) \to 0$ as $f(-\infty, p) \to 0$.

(1) *Monoenergetic Injection*: $f(-\infty, p) = f_0\delta(p_0)$: Substituting this into equation (9.24) we see that $f_d(p) \propto p^{-q},\ \ p > p_0$. A monoenergetic distribution of particles is Fermi accelerated by a shock to a power-law distribution. The index q depends on the ratio of upstream to downstream velocities, *i.e.*, on the shock compression. For a very weak shock ($\mathcal{M} \sim 1$) $v_d \sim v_u$ and $q >> 1$. The compression for a stronger shock depends on the equation of state; for a gas with ratio of specific heats $\Gamma = 5/3$, $v_u/v_d \to 4$ for a strong shock, in which case $q = 4$; in terms of the energy spectral index $\delta = q - 2$, this corresponds to $\delta = 2$. The shock-Fermi mechanism in the thermodynamic limit readily produces the 'universal' spectral index value discussed in Chapters 1 through 4. We warn the reader against applying the above with a compression ratio appropriate to a relativistic equation of state, because for all but the weakest shocks, that implies relativistic flow speeds; as we shall show in §9.3.4, such speeds require the above analysis to be modified.

(2) *Power Law Injection*: $f(-\infty, p) = f_0 p^{-s}$: If $s < q$ it follows from (9.24) that $f_d(p) \propto p^{-s}$. A flat spectrum power-law distribution passes through the shock without change of form. In other words, the Fermi mechanism cannot steepen a power-law that is flatter than the distribution resulting from monoenergetic injection. If $s > q$ most particles reside at low energy, and this is therefore similar to the monoenergetic injection case; low energy particles will be accelerated to a power-law, to which the high energy particles of the original distribution will be a perturbation. Thus $f(p) \propto p^{-q}$ is a good approximation in this case.

This test-particle analysis assumes that the magnetic field is ordered. A preliminary study of shock-Fermi acceleration in a turbulent magnetic field has been made by Achterberg (1988). The energy spectrum of accelerated particles always exhibits a cutoff, because particles are 'trapped' by the largest scale structure of the magnetic field and they are energized only while the shock crosses each domain. With an 'open' field topology (many long field lines) the spectral slope is similar to that of the uniform field case over much of momentum space, but the particle number density is reduced (by up to an order of magnitude); this is because acceleration is predominantly within large cells, but these occupy only some fraction of the flow volume. For a 'closed' field topology (many small cells of field) the particle spectrum is a 'hump' near the injection momentum; here, acceleration is associated with small cells which inhibit free scattering across the shock and severely limit the energy attained.

9.3.2.4 The particle energy spectrum downstream

Observations usually provide information on energy spectra, which we can derive from $n(E)dE = n(p)dp = f(p)4\pi p^2 dp$, where $E^2 = p^2c^2 + m_r^2c^4$, m_r being the rest mass of either proton or electron. Applying this to the case of monoenergetic injection yields (with $E_0 \equiv p_0$ and $\delta = q - 2$)

$$n\,(E)\,dE = (\delta - 1)\,(E_0^2 - m_r^2c^4)^{\frac{\delta-1}{2}}\;E\,(E^2 - m_r^2c^4)^{-\frac{\delta+1}{2}}\;dE. \quad (9.25)$$

When $E >> m_r c^2$ the spectral form is $n \propto E^{-\delta}$ for both species. Figure 1 of Bell (1978b) shows $n(E)$ for the injection of equal numbers of protons and electrons; in this case there are more protons than electrons at relativistic energy and the total energy in protons is about an order of magnitude more than that in electrons. However, it is not clear that electrons and protons are injected with equal efficiency in the shocks of jets.

9.3.3 The non-linear theory of the Fermi process

If we consider the influence of the accelerated particles on the (thermal) subshock, considerable complications arise: the mixture of thermal and accelerated particles has an adiabatic index that, in general, is less than the 5/3 value of the asymptotic upstream flow (assuming that this is not relativistic; see §4.3.3); this makes the plasma more 'compressive', and thus increases the value of v_u/v_d.

However, it is this ratio that determines the energy spectral index, δ, hence the number of high energy particles, and hence the effective adiabatic index. Thus the structure of the subshock needs to be determined self-consistently with δ. But furthermore, if the plasma becomes more compressive, the ratio v_u/v_d can exceed 4, causing δ to become less than 2. The accelerated particle spectrum is then dominated by those at high energy, and the total energy is infinite in the absence of a cutoff. In any real system, we expect particles above some energy $E_{\rm max}$ to readily diffuse away; but if the spectrum is hard, this 'leakage' will carry away a significant flux of energy and influence the jump conditions (cf. optically thin radiative shocks, where energy loss by radiation allows the shock to be more compressive). This influences the compression, hence the spectral slope and hence the amount of energy residing at the top of the spectrum that may be lost. Thus leakage also needs to be built into a self-consistent model.

These points have been discussed by Eichler (1984) and by Ellison & Eichler (1984). The latter authors use a Monte-Carlo technique to scatter particles in the frame of the thermal subshock, determining the adiabatic index self-consistently, allowing for the back reaction of the accelerated particles and leakage. In this treatment, *all* the ions are subject to the Monte-Carlo scattering, including those that form the thermal subshock; it is seen that a subset of these ions get injected quite naturally into a shock-Fermi process (see §9.3.5). Observations of accelerated protons at the Earth's bow shock support this model (Ellison & Möbius 1987) that puts a significant fraction of the incoming energy into a power-law particle distribution.

The *structure* of shocks with significant cosmic ray pressure has been investigated by many authors (*e.g.*, Drury & Völk (1981); Axford, Leer & McKenzie (1982); McKenzie & Völk (1982); Webb (1983a); Völk, Drury & McKenzie (1984); Achterberg, Blandford & Periwal (1984); Heavens (1984b); Eichler (1985); Bogdan & Lerche (1985) and Webb *et al.* (1985)). The theme of these papers is to treat the thermal and cosmic ray gases (and sometimes waves) as fluids; although these models provide little or no information on the spectrum of accelerated particles, they support the idea that the shock-Fermi mechanism is both efficient and self-regulating (in that the more high energy particles stream ahead of the shock, the more the incoming gas is slowed, inhibiting further acceleration). The models show that, as noted in §9.3.2.2, the back action of the accelerated particles slows the upstream thermal matter, producing a smooth decline in velocity leading into a weakened subshock (Fig. 9.1(F)); they also suggest that in certain Mach number regimes, the subshock may disappear. Falle & Giddings (1987) present a model in which the subshock disappears; when this happens a transition in the cosmic ray pressure may be maintained in a particle distribution whose number decreases due to leakage, but whose energy content is maintained by the continued acceleration of the remaining particles. However, the disappearance of the subshock may depend on the unphysically weak dependence of diffusion coefficient on momentum ($\kappa \propto p^{1/4}$) adopted by these authors.

Investigations of the *spectrum* of shock-Fermi accelerated particles have been made by Eichler (1979); Blandford (1980); Ellison, Jones & Eichler (1981); Drury, Axford & Summers (1982); Webb (1983b); Heavens (1983); Webb, Drury & Biermann (1984) and Ellison & Eichler (1985). The general conclusions are that the spectrum is steepened at low momentum, because these particles do not scatter far from the subshock, which they see as weakened by the back action of the accelerated particles ($\delta \to \infty$ as $v_d \to v_u$), and flattened at high momentum, because these particles see the full compression of the flow – which is enhanced because of the reduced adiabatic index. A study by Bell (1987) clearly shows a concave spectrum, the spectral index at a given momentum being determined by the interplay of these effects for a given injection rate. In particular, if the particle distribution extends to high enough momentum, the low momentum spectrum can be of the steepness typically observed in radio sources ($\delta \sim 2.5$) because of the smoothing of the subshock over the scale of these lower momentum particles.

Some of the above cited authors consider acceleration in the presence of synchrotron radiation losses, and Heavens & Meisenheimer (1987) compute the synchrotron emission from a finite region containing particles accelerated by a strong shock with momentum-independent diffusion coefficient. The general characteristics of the frequency spectrum (the energy spectrum varies with distance from the shock) are: an $\alpha = 0.5$ power-law below a frequency corresponding to some low momentum – here particles cross the finite region without significant synchrotron loss; an $\alpha = 1$ power-law at intermediate frequencies – here, the higher the particle momentum, the less volume they occupy, because they lose energy before reaching the region's boundary; an almost exponential cutoff at high frequency – here synchrotron radiation loss readily wins over acceleration. This modelling has clear application to the type of model described by Marscher & Gear (see §4.3.3). It has been used to explain the spectrum of a hotspot in the jet of 3C 273 (Meisenheimer & Heavens 1986), and to derive estimates of the magnetic field and electron mean-free-path in that source. Other attempts to apply the theory of Fermi acceleration at shocks include those by Bregman (1985) (the knots of M 87's jet and quasar accretion shocks), Pelletier & Roland (1986) (the hotspots of Cygnus A) and Biermann & Strittmatter (1987) (the sharp infra-red cutoff in AGN; see §9.3.2.2).

9.3.4 **Fermi acceleration at relativistic shocks**

In the limit of relativistic flow velocity two complications arise: because the streaming speed is comparable to the speed of random motion, the pitch-angle distribution of particles scattered back and forth in the vicinity of the thermal shock is highly anisotropic; and the randomization of the upstream flow energy leads to a downstream state that can be internally relativistic, *i.e.*, the equation of state may change across the shock.

Bell's test particle arguments have been extended to the relativistic case by Peacock (1981). This work adopted an approximation for the pitch-angle anisotropy and a model for the flow energy randomization. Although later work suggests that Peacock underestimated the spectral index of the accelerated electrons (particularly for flow speed $\sim c$), the study provides a valuable exploration of a range of both upstream temperatures ($0 < T_u < \infty$) and flow speeds ($v_u < c$).

Kirk & Schneider (1987a) have studied Fermi acceleration at relativistic shocks, retaining pitch-angle diffusion in the Boltzmann equation (equation (9.19) refers to the isotropic part of the distribution function) and hence determining rigorously the pitch-angle distribution of the accelerated particles. They apply this theory in the limit that the upstream flow is fully internally relativistic (*i.e.*, $\Gamma = 4/3$; this corresponds to the $T_u = \infty$ of Peacock). For $0.8c \leq v_u \leq 0.98c$ they find that $1.3 \geq \alpha \geq 0.63$.

Heavens & Drury (1988) have complemented the work of Kirk & Schneider, by considering the case of a cold upstream flow (which corresponds to the $T_u = 0$ of Peacock). (With $\mathcal{M}_u >> 1$, the shock is strong, and in the nonrelativistic case we would expect $\alpha = 0.5$.) They modify Kirk & Schneider's technique, in order to consider general (and more realistic!) power spectra for the scattering waves, and allow anisotropy in the pitch-angle *diffusion coefficient.* As the upstream flow speed is increased, they find that α falls from 0.5 to 0.35 (because of the increased compression ratio associated with the changing equation of state of the downstream electrons), but then increases (for $v_u > 0.9c$) to ~ 0.63 (as relativistic effects reduce the compression ratio). A similar calculation for a pair-plasma shows that $0.5 < \alpha < 0.6$, for $0 < v_u/c < 0.98$. The authors note the agreement between their result for $v_u \sim c$ and that of Kirk & Schneider; they conjecture that for ultrarelativistic shocks $\alpha \sim 0.63$, independently of the particle species and scattering spectrum.

It should be borne in mind that the above discussion is based entirely on a test particle analysis, and that much work remains to be done to produce the non-linear theory that has advanced our understanding of the non-relativistic case. A first step has been made by Kirk & Schneider (1987b) who use a Monte-Carlo method to confirm their previous analytic results, and consider the effects of radiation loss.

9.3.5 Threshold energy for the Fermi process

The origin of the ions that are Fermi-accelerated is now understood. The Monte-Carlo simulations of Ellison & Eichler (1984) show that the Fermi mechanism is capable of extracting ions from the 'thermal pool' and imparting them with a significant fraction of the incoming energy. Recent simulations of parallel shocks by Quest (1988) clearly show (for the first time in simulations) thermal ions energized by repeated passages of the thermal shock. It is certainly not necessary to inject an independent population of ions for the Fermi process to work.

However, the situation of the electrons is less clear: the simulations treat the electrons as a charge-neutralizing fluid, and do not address the problem of their acceleration from the 'thermal pool'. For an electron to be Fermi-accelerated, it must be able to scatter between the upstream and downstream flows, suffering negligible deviation at the thermal subshock (which it should see as a discontinuity). This implies a minimum energy for the electrons that can be accelerated, and an estimate of this has been made by Peacock (1981): a particle of mass m and charge q receives an impulse per unit charge of $\approx \gamma_u m v_u / q$ on reaching the shock; for a plasma of 'cosmic abundance', much of the energy and momentum is carried by Helium nuclei, and the shock must be capable of randomizing their streaming motion – thus to estimate the impulse per unit charge delivered by the shock, we can set $m = 4m_p$ and $q = 2e$, getting a value of $2\gamma_u m_p v_u / e$; this will be the impulse per unit charge delivered to an electron (assumed relativistic) of momentum per unit charge $\gamma_e m_e c/e$, and if this impulse is to be a small perturbation to the momentum, we see that

$$\gamma_e >> \gamma_{e,\mathrm{crit}} = 2\gamma_u \frac{v_u}{c} \frac{m_p}{m_e} \tag{9.26}$$

is needed. This is similar, $0\,(m_p/m_e)$, to the minimum energy necessary for the electrons to resonate with Alfvén waves, and hence scatter between upstream and downstream flows (see §9.4). This argument does not preclude the Fermi acceleration of electrons from the suprathermal tail of a distribution, but does imply that such electrons must already possess highly relativistic energy.

9.4 The injection problem

We have seen in §§9.2 and 9.3 that all stochastic acceleration processes share one requirement. There must be waves which can interact with the particles *via* the cyclotron resonance, so that the particles are scattered in pitch-angle. This scattering is necessary for both Alfvén and magnetosonic turbulent acceleration, and for shock acceleration.

9.4.1 Resonances with nonrelativistic particles

Alfvén waves satisfy the cyclotron resonance (9.5) for relativistic particles. We saw (in §9.2) that higher particle energies resonate with lower frequency (longer wavelength) waves. But Alfvén waves exist only for frequencies $\omega < \Omega_p = eB/m_p c$ – so there is a minimum particle energy which 'sees' the Alfvén waves. For protons, this minimum energy (setting $\omega = \Omega_p$ in equation (9.5)) is

$$\gamma_{\mathrm{min,p}} = 1 + 2\frac{c_A^2}{c^2 \mu^2} \tag{9.27}$$

(if μ is the cosine of the pitch-angle, $v_{\parallel} = v\mu$). Thus, all protons with $v_{\parallel} > 2c_A$ can see the waves. For electrons, however, the limiting energies are higher. In high density plasma (for which $c^2\mu^2/c_A^2 >> m_p^2/m_e^2$) the minimum energy is

$$\gamma_{\mathrm{min,e}} = 1 + \frac{m_p}{m_e}\frac{c_A^2}{c^2\mu^2}, \tag{9.28}$$

so that the corresponding minimum kinetic energy for an electron is $\frac{1}{2}m_p c_A^2/\mu^2$; thus resonance occurs only if $v_{\parallel} > (m_p/m_e\mu)^{1/2}\, c_A$. For a low density plasma ($c^2\mu^2/c_A^2 << m_p^2/m_e^2$) the limit is higher:

$$\gamma_{\mathrm{min,e}} = \frac{m_p}{m_e}\frac{c_A}{c\mu}. \tag{9.29}$$

Thus, the acceleration of low energy (say, thermal) particles by any mechanism that depends on resonant Alfvén waves is quite different for electrons and protons. (9.27) shows that a significant number of thermal protons can resonate with Alfvén waves (those with $v/v_{Ti} > c_A/v_{Ti}$), and can (in principle) be accelerated. Thus, a significant high energy tail can be formed from an initially thermal distribution. But thermal electrons have less chance of acceleration. In a low density plasma, $\gamma_{\mathrm{min,e}}$ is already mildly relativistic, and thermal electrons are unaffected by the waves. The high density limit will allow a thermal electron tail to be accelerated (noting that $v_{Te}^2 = (m_p/m_e)\, v_{Tp}^2$) but this limit requires quite high densities, $n > 10^{-3} B_{\mu G}^2$ cm^{-3}.

Now and then, whistler waves (which exist for $\Omega_p < \omega < \Omega_e$) have been suggested as a way to extend the range of resonant electron energies to lower, possibly thermal energies. The resonance condition, (9.5), does allow lower energy electrons to interact with the whistlers as long as $\omega < \Omega_e$. Whistlers are most commonly generated internally to the plasma *via* pitch-angle anisotropies, and thus they seem less attractive for true particle acceleration. However, the appearance of whistlers upstream of the shock in the simulations of Kan & Swift (1983; §9.3.1) is intriguing.

9.4.2 The need for injection

This is the so-called injection problem: in the low density conditions likely to exist in radio sources, no acceleration mechanism that depends on Alfvén wave resonance can accelerate electrons significantly from a subrelativistic, thermal population. The usual shock and turbulent acceleration models are truly reacceleration models, in that they start with already relativistic electrons, and either accelerate these further or maintain them against losses. Therefore, one of two things must be operating in the sources. One is that the plasma may be created fully internally relativistic in the core (with the average electron energy above $m_e c^2$). The other is that, if the plasma starts subrelativistic, the initial acceleration must be a two-stage process, with a first ('injection') stage moving the electrons up in energy above the minimum (equations (9.28) or (9.29)), and a second stage, either turbulent or shock-Fermi acceleration, moving the electrons up to the observed highly relativistic energies.

We note that some models of the central engine and initial creation of the beam would produce a plasma that is initially internally relativistic (cf. Chapter 8). Some dynamo and hot accretion disc models predict that the beam plasmas are dominantly relativistic electron-positron pair plasmas (*e.g.*, Lovelace 1976; Fabian *et al.* 1986). It has also been suggested (Williams 1986; Chapter 7) that even the large-scale jets, at least in the more energetic, Type II sources, are pair plasmas. It has further been suggested that acceleration of protons to relativistic energies in the central engine (for instance, by shocks and/or turbulence in the accretion flow) could produce > 100 MeV electrons and positrons *via* pion production and decay (Kazanas & Ellison 1986). These models would not require injection, as we define it here; reacceleration by shocks and turbulence would be enough.

Most work, however, has considered how to accelerate particles out of a thermal background (at a subrelativistic temperature). Thus it seems that acceleration is a two-stage process. The most likely first stage mechanism seems to involve a DC electric field, which is probably capable of accelerating electrons out of a thermal background into a significant suprathermal tail. We discuss these mechanisms in the next section.

9.5 Electrodynamic acceleration

The discussion up to now has focussed on stochastic electric fields and their effect on accelerating relativistic particles. Different possibilities arise when DC electric fields can be maintained in the plasma.

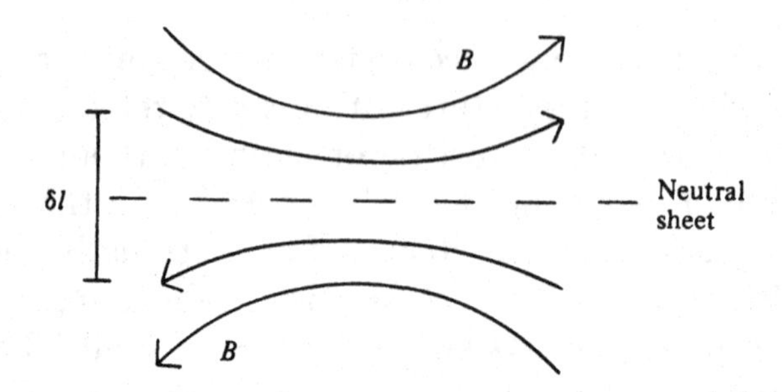

Fig. 9.2. The configuration for magnetic reconnection. An electric field $\mathbf{E} = (c\eta/4\pi)\ \nabla \times \mathbf{B}$ and a current $\mathbf{j} = \mathbf{E}/\eta$ exist in the neutral sheet.

Currents must exist wherever in the plasma the magnetic field has a nonzero curl: from Maxwell's relation, $\mathbf{j} = (c/4\pi)\nabla \times \mathbf{B}$. Jets may carry global currents if a large-scale circuit is established. This might consist of, for instance, a current flowing out along the jet and returning either on the jet surface, or in the surrounding plasma (*e.g.*, Benford 1978). Even in the absence of a global circuit, fluid motions may lead to a disordered magnetic field, and hence support local currents. The energy source for these currents and the associated stored field energy may be a central dynamo (cf. Chapter 8), or it may be the local fluid motions within the jet, which maintain the fields and currents through inductive effects (cf. §9.6).

Currents provide both a steady energy source and an energy storage mechanism. The steady energization occurs through Ohmic losses, which dissipate the electrodynamic power of the dynamo locally in the plasma. As we see below, this dissipated energy usually goes to heating but can go to particle acceleration. The magnetic field/current system also stores energy, which can be described either as the magnetic energy density, $B^2/8\pi$, integrated over the volume, or as the inductive energy of the circuit, $\frac{1}{2}I^2L$ if $L \approx l/c$ is the inductance of the circuit (of length l).

The stored energy can be released if the circuit is broken, or if resistive effects destroy the magnetic field energy. This is a reconnection event. As a simple picture, consider a region of plasma in which the magnetic field reverses direction over a small spatial distance (Fig. 9.2).

This could be the two feet of a looped solar flux tube, the equatorial sheet of the Earth's magnetotail, or adjacent 'eddies' in a turbulent magnetoplasma. In this situation, ideal (non-resistive) MHD breaks down, and resistive effects become as important as flux-freezing. The magnetic field lines can 'diffuse' relative to the fluid and can 'reconnect' with their neighbours. In the process the magnetic energy is converted to heat, to bulk plasma flow, and to some particle acceleration (*e.g.*, Spicer 1982). This process can proceed explosively, in a 'tearing mode' instability, or can proceed in a quasi-steady fashion, if driven by converging plasma flows across the neutral sheet.

This type of disruptive release of stored magnetic energy is believed to be important in solar and stellar flares. Also, the spontaneous relaxation to a minimum energy state which is observed in laboratory plasmas (cf. §9.6) probably proceeds *via* a similar, quasi-steady destruction of magnetic energy. The rate at which this energy is released (and thus the maximum possible luminosity from such a region) will be the rate at which the field energy can be destroyed: $dE_B/dt \approx E_B c_A/l$, if E_B is the total magnetic energy, c_A is the Alfvén speed, and l is the size of the region.

9.5.1 Acceleration and heating by a dc electric field

The Ohmic heating associated with steady currents, and the energy release of reconnection, are transmitted to charged particles through local DC electric fields. If a particle existed in a vacuum, it would simply be accelerated up to the net potential drop of the electric field, namely $E_{\max} = e\Delta\Phi = \int \mathbf{E} \cdot d\mathbf{l}$. However, both direct collisions with other particles, and interaction with plasma waves generated by the accelerating charges limit the acceleration, and hence $E_{\max}$.

The physical basis of resistivity in a collisional plasma is the balance on the current-carrying electrons of the forces due to the electric field and due to the momentum transfer in electronic collisions. The momentum equation for a single electron is

$$m_e \frac{d\mathbf{v}}{dt} = e\mathbf{E} - m_e \mathbf{v} \nu_e(v) . \tag{9.30}$$

Here, the classical collision rate (for electrons at or below the thermal velocity, $v_{Te} = (k_B T/m_e)^{1/2}$) is

$$\nu_e(v_{Te} + v_D) \approx \nu_e(v_{Te}) = 4\pi n e^4 \ln\Lambda / m_e^{1/2} (k_B T)^{3/2} , \tag{9.31}$$

where $\ln \Lambda$ is the Coulomb logarithm (Spitzer 1962). The balance of these two forces gives rise to the drift velocity, $\mathbf{v}_D = e\mathbf{E}/m\nu_e$. The bulk of the electrons in the thermal distribution move at v_D, and their collisional energy losses result in a local Ohmic heating rate

$$nm_e v_D^2 \nu_e (v_{Te}) = \eta j^2, \tag{9.32}$$

where we have used the current density, $j = nev_D$, and have defined the resistivity

$$\eta = m_e \nu_e (v_{Te}) / ne^2. \tag{9.33}$$

Direct collisions are very infrequent in the low-density plasmas of radio jets; and thus the Ohmic heating rate from direct collisions is very small – in fact, negligible compared to other source energetics, such as radiation losses. However, high current densities can excite several microinstabilities (Buneman, ion acoustic and ion cyclotron are often mentioned). These generate collective space-charge or electromagnetic oscillations. We do not attempt to review this complex field here; the energetic reader is referred to Papadopoulos (1977) or Spicer (1982) for more discussion. When these instabilities occur, they can lead to a high energy density of plasma waves, and the scattering of the drifting electrons by the stochastic electric fields of these waves can lead to an effective collision rate, ν_{eff}, much higher than the classical, electron-ion collision rate. The resistivity is thus increased over (9.33) by $\nu_{\text{eff}}/\nu_e (v_{Te})$, and is called an 'anomalous resistivity'. The various instabilities which can lead to anomalous resistivity share one characteristic feature: a current will drive these instabilities only if the drift velocity exceeds some characteristic plasma velocity, usually v_{Te}. Now we can estimate the current density in a jet plasma from a knowledge of the field strength and the scale on which the field varies. If the current is assumed to be distributed uniformly throughout the region, these estimates generally give $v_D << v_{Te}$, so that the instabilities would not be expected to arise. However, if the current is localized along neutral sheets or in filaments, the current density may be high enough to excite anomalous effects. The situation in the solar environment is similar (cf. Spicer 1982).

In any case, whether the resistivity is classical or anomalous, Ohmic heating is just that – heating. Since the bulk of the electron population drifts at $\sim v_D$, the energy dissipation will go to heat the entire population, rather than preferentially accelerating a high-energy tail. Thus, Ohmic heating is probably not of interest as an acceleration process, except possibly in a plasma which is already fully relativistic (such as a pair plasma). However, the highest energy particles in a thermal distribution may escape this collisional limitation to some extent.

The classical electron collision rate drops as v^{-3}. Thus, above some critical velocity, v_c, the electrons do not suffer collisions, but continue to be accelerated, becoming 'runaway' electrons. This critical velocity for escaping collisions is given by $v_c^2 = 4\pi ne^3 \ln\Lambda / m_e E$. Strong electric fields therefore allow a significant fraction of the thermal population to be accelerated. (The field at which $v_c = v_{Te}$, and at which all of the thermal electrons run away, is defined as the Dreicer field, and is given by $E_D = 4\pi ne^3 \ln\Lambda / k_B T$.) However, it appears that streaming instabilities may eventually limit the acceleration of these runaways, and thus limit the maximum particle energy reached. While some calculations of anomalous resistivity predict $\nu_{\rm eff} \propto v^{-3}$ (*e.g.*, Papadopoulos 1977), so that runaways would not be stopped, other work suggests the contrary, particularly in a magnetized plasma (*e.g.*, Spicer 1982). For instance, Moghaddam-Taaheri *et al.* (1985, 1987) use numerical modelling and find that streaming instabilities do isotropize the beam and turn its energy into heat. They find these instabilities can be best delayed if some particles are allowed to 'leak out' of the runaway tail, or if the electric field increases during the runaway process.

Another interesting possibility for acceleration in a steady current system is provided by double layers. Good reviews can be found in Spicer (1982), Carlqvist (1982) and Borovsky (1986). A strong potential drop $\Delta\Phi > k_B T/e$, applied to a plasma, may localize itself over one or more small regions; charge separation on a scale comparable to the Debye length can maintain a high potential drop on this scale, and this is termed a 'double layer'. It is believed to form under the same conditions that support anomalous resistivity, namely a current density large enough that $v_D > v_{Te}$, although Borovsky (1986) has suggested that double layers may form at even lower drift velocities. It may be that the Buneman or ion acoustic instabilities can lead to local regions of charge separation and low plasma densities, rather than, or in addition to, regions of anomalous plasma turbulence (Smith & Goertz 1978). It may be that regions of turbulence and double layers coexist in a high-current density plasma. Once a double layer forms, it is an efficient way of accelerating some particles to high energy. The low density inside the double layer, and its narrow width, mean that the acceleration is essentially collisionless, and that streaming instabilities cannot develop. Charges which are not trapped in the charge-separated edges of the double layer will thus be accelerated to the energy corresponding to the full potential drop, $e\Delta\Phi$.

9.5.2 Shock drift acceleration

Another important site for the acceleration of electrons, as well as ions, by a DC electric field is found at interplanetary shocks whose normal is nearly perpendicular to the magnetic field (quasi-perpendicular shocks; cf. §9.3). An observer in the frame of a shock sees a magnetic field, $\mathbf{B}_u$, advected through the transition with velocity $\mathbf{v}_u$, and thus experiences an electric field $\mathbf{E}_s = \mathbf{v}_u \times \mathbf{B}_u = v_u B_u \sin\Theta \mathbf{i}$, where Θ is the angle between field and shock normal, and $\mathbf{i}$ lies in the plane of the transition. Furthermore, the gradient in magnetic field, ∇B, at the transition, causes the drift of charged particles in a sense perpendicular to both $\mathbf{B}$ and ∇B (because the gyroradius is larger on one side of the particle's orbit than on the other; see *e.g.*, Chen (1974)). This drift acts on both electrons and ions in the same sense as $\mathbf{E}_s$; thus both species are accelerated.

Let us consider this acceleration in terms of individual particles. A charged particle will be either reflected or transmitted by the shock. The greatest energy gain that a particle can experience occurs for reflection – corresponding to bouncing off an approaching surface. We can transform away the electric field $\mathbf{E}_s$ by working in a frame that moves perpendicular to this field (in which frame there is no plasma flow *across* the magnetic field), and in this frame we can follow the reflection that occurs because of the conservation of adiabatic invariant $v_\perp^2/B$ (*e.g.*, Chen 1974). A retransformation then tells us the energization that is in reality a consequence of the electric field. If this adiabatic invariant is indeed conserved, and since a reflected particle sees no net change in magnetic field, we would expect the acceleration to be confined to $v_\parallel$ – which is the case. (It turns out not to matter whether the particle gyroradius is much less than or much greater than the scale of the shock. The adiabatic invariant is conserved in both cases; correspondingly, transmitted particles tend to gain their energy in $v_\perp$, allowing $v_\perp^2/B$ to be conserved as B increases (Sarris & Van Allen 1974; Holman & Pesses 1983).

This energization corresponds to reflection from a mirror moving at speed $v_u \sec\Theta$, which is, in effect, the speed with which a point on a field line moves along the shock, and is thus the speed of the 'mirror' seen by an upstream particle (see Wu 1984). The particle's velocity change is then $\Delta v_\parallel = 2v_u \sec\Theta$. As $\Theta \to \pi/2$ (perpendicular shock) so $\sec\Theta \to \infty$, and one might suppose that an infinite energy would result. This is of course not so, because particles are transmitted by the shock, rather than being reflected, if $v_u \sec\Theta > (B_d/B_u)^{1/2} v_i$ (Holman & Pesses 1983), where v_i is the upstream speed of the particle. There is thus a maximum energy gain for a nonrelativistic particle given by

$$\frac{\Delta E_{\max}}{E} = \left(\frac{\Delta v_{\parallel\max}}{v_i}\right)^2 = 4\frac{B_d}{B_u} \tag{9.34}$$

which implies that *in a single encounter* with a strong, nearly perpendicular shock a particle may increase its energy by more than an order of magnitude. In particular, because of the $\sec\Theta$ factor, such a shock can have a profound effect on particles with $v_i >> v_u$.

Indeed, it is near quasi-perpendicular, interplanetary shocks that significant suprathermal electron distributions are found (*e.g.*, Sarris & Krimigis 1985). Pyle *et al.* (1984) have observed relativistic electrons apparently produced by this shock drift mechanism (sometimes referred to as the 'fast-Fermi' process). Because of the $\sec\Theta$ term, the shock drift process is most effective for shocks nearly perpendicular to the magnetic field; this process thus complements the first-order, shock acceleration mechanism discussed in §9.3, which is most effective for quasi-parallel shocks (cf. the discussion in Holman & Pesses 1983).

9.5.3 Relation of electrodynamic acceleration to the injection problem

We saw in §9.4 that stochastic processes will act only on electrons which are already slightly relativistic, and that some first stage process is necessary to accelerate electrons from a subrelativistic, thermal distribution up to energies that satisfy equation (9.28) or (9.29). The most likely candidate for this first stage injection mechanism seems to be acceleration by a DC electric field, through one of the mechanisms described in this section. Indeed, such a two stage process is commonly believed to be acting in solar flares. The picture here is that a first stage mechanism (perhaps reconnection) creates suprathermal tails up to several tens of keV, and that second stage shock or turbulent acceleration further accelerates the electrons up to the observed ~ 100 MeV energies (*e.g.*, Ellison & Ramaty 1985; Droge & Schlickheiser 1986; de Jager 1986).

This two stage picture is also attractive because electrodynamic acceleration does not seem able to produce a power-law distribution in momentum. For example, the simulations of Moghaddam-Taaheri *et al.* (1985, 1987) tend to find extended, suprathermal tails with cutoffs; interplanetary data from quasi-perpendicular shocks seem consistent with this. If stochastic acceleration acts as a second stage, we saw in §§9.2 and 9.3 that power-law distributions can be produced, at least for some range of energies, if the conditions are right.

The remaining questions, then, are how efficient electrodynamic acceleration can be, and what energies it can reach. That is, under what conditions can the first stage electrons satisfy equation (9.28) or (9.29)? These questions appear not to be fully answered as yet. For shock drift acceleration, Pioneer 10 observed slightly relativistic electrons (of a few MeV) associated with a quasi-perpendicular shock at ~ 25 AU (Pyle *et al.* 1984), and high time resolution observations by Voyager 2 (Sarris & Krimigis 1985) found electrons of energy ≥ 2 MeV (apparently accelerated from a thermal pool of energy ~ 0.4 MeV) associated with a short lived, almost perpendicular shock structure at ~ 2 AU. The simulations of Moghaddam-Taaheri *et al.* find runaway tails reaching velocities of up to $\sim 50 v_{Te}$, after which the process is shut off by streaming instabilities. Double layers may be an important injection mechanism, since the streaming instabilities are thought to be less important there. However, when, and at what strength double layers occur is not yet well understood, although Carlqvist (1982) has shown that relativistic double layers, with $e\Delta\Phi >> m_e c^2$, can exist. This suggests that double layers also may be an important injection mechanism.

9.6 Structure and maintenance of magnetic fields

9.6.1 Fundamentals

Magnetic fields arise from electric currents and from changing electric fields. The fields in radio sources might be due to a net current flowing in the source (Benford 1978); they may also be due to inductive effects in turbulent or flowing plasmas. In this section we focus on the latter process – in which a magnetic field is supported by currents which are maintained by fluid motions. We note that on the microphysical scale, the Lorentz force $(q/c)\,\mathbf{v} \times \mathbf{B}$ in a moving plasma acts oppositely on electrons and ions, and so will drive a current. We will look at the connection between magnetic and kinetic energy in two limits: the case when kinetic energy dominates ($\beta_K > 1$ – dynamos) and the case when magnetic energy dominates ($\beta_K < 1$ – relaxation). Both limiting behaviours involve the same physics. Maxwell's equations, ($c\nabla \times \mathbf{E} = \partial \mathbf{B}/\partial t$ and $\nabla \times \mathbf{B} = (4\pi/c)\,\mathbf{j}$), and Ohm's law in a moving fluid, ($\mathbf{j} = \sigma\,(\mathbf{E} + \mathbf{V} \times \mathbf{B})$), where $\mathbf{V}$ is the fluid velocity and σ is the scalar electrical conductivity,* can be combined to give the induction equation

$$\frac{\partial \mathbf{B}}{\partial t} = \nabla \times (\mathbf{V} \times \mathbf{B} - \tilde{\eta} \nabla \times \mathbf{B}), \tag{9.35}$$

where $\tilde{\eta} = c^2/4\pi\sigma$ is the resistive diffusion coefficient. We can integrate this over some arbitrary surface in the fluid, S, bounded by a path L, to find the behaviour of the magnetic flux, $\Phi_B = \int_S \mathbf{B} \cdot d\mathbf{S}$:

$$\frac{d\Phi_B}{dt} = \int_S \frac{\partial \mathbf{B}}{\partial t} \cdot d\mathbf{S} + \int_L d\mathbf{l} \cdot (\mathbf{V} \times \mathbf{B}) = -\tilde{\eta} \int_L (\nabla \times \mathbf{B}) \cdot d\mathbf{S} \tag{9.36}$$

so that in the limit $\tilde{\eta} \to 0$ (which corresponds to high plasma conductivity and no turbulence), we find $\Phi_B \to$ constant. This is the case of 'flux freezing', and in a non-rotating jet has the consequences $B_z \propto 1/r^2$ and $B_\phi \propto 1/rv_j$ (where $\hat{\mathbf{z}}$ is the direction along the jet axis, ϕ is the azimuthal angle, r is the jet radius, and v_j is the jet

* The assumption of a scalar conductivity is strictly true only in a weakly-ionized plasma, although flux-freezing can be shown to hold in the more general case of full ionization (cf. §3.1.2). The effects of a tensor conductivity have been explored for the case of solar plasmas, (*e.g.*, Spicer 1982), but have not generally been applied to radio sources.

speed). Thus, expansion of the jet must decrease the field-strength in this limit.

But equation (9.35) allows much richer behaviour. Consider the effects of turbulence. Turbulence will destroy magnetic energy, through local reconnection of disordered field lines. On scales large compared to the turbulent scale, λ_t, this effect acts like an additional resistivity and when (9.35) refers explicitly to large-scale fields, turbulent dissipation is often included as an extra diffusion coefficient, $\tilde{\eta}_{\rm turb} \approx \lambda_t v_t$. Dissipation, whether resistive or turbulent, will destroy field energy on a time scale, $t_{\rm diss} \approx l^2/\tilde{\eta}$ if l is the characteristic length scale of the field. It is often the case that the classical, Spitzer conductivity results in very long dissipation time scales, while either anomalous resistivity or turbulent dissipation will result in much shorter, astrophysically interesting, dissipation time scales.

In addition to dissipating field energy, turbulence will also stretch and amplify the field, through induction effects, and can thereby increase the magnetic field energy. Driven MHD turbulence will generally amplify the small-scale field (*e.g.*, Pouquet *et al.* 1976; De Young 1980). If the topological structure of the turbulence is favourable, the turbulence will amplify and support magnetic fields on scales larger than the turbulent scale. These effects are called turbulent dynamos. Dynamos play an important role in terrestrial and solar physics, and also in the spontaneous relaxation observed in low-β_K laboratory plasmas. The literature in these areas affords us some insight into the role of dynamos and relaxation in extragalactic plasmas. The review of extragalactic magnetic fields by Asseo & Sol (1987) references much of the literature that has appeared in this field.

9.6.2 Turbulent dynamos

The following discussion of turbulent dynamos follows Moffatt (1978). We separate the magnetic and velocity fields into large and small scales: $\mathbf{V} = \mathbf{V}_0 + \mathbf{v}$ and $\mathbf{B} = \mathbf{B}_0 + \mathbf{b}$. This is an analytical convenience, and is not necessary for dynamo action – as can be seen from our later discussion of numerical simulations. The small-scale field is assumed to fluctuate randomly in both space and time. The induction equation (9.35) can then be separated into large-scale and small-scale parts,

$$\frac{\partial \mathbf{B}_0}{\partial t} = \nabla \times [\mathbf{V}_0 \times \mathbf{B}_0 + < \mathbf{v} \times \mathbf{b} > - \tilde{\eta} \nabla \times \mathbf{B}_0], \tag{9.37a}$$

$$\frac{\partial \mathbf{b}}{\partial t} = \nabla \times [\mathbf{V}_0 \times \mathbf{b} + \mathbf{v} \times \mathbf{B}_0 + (\mathbf{v} \times \mathbf{b} - < \mathbf{v} \times \mathbf{b} >) - \tilde{\eta} \nabla \times \mathbf{b}]. \tag{9.37b}$$

The term $< \mathbf{v} \times \mathbf{b} >$ is the contribution of the fluctuating electric field to the mean magnetic field. Large-scale dynamos work when this has a net component along $\mathbf{B}_0$: *i.e.*, $< \mathbf{v} \times \mathbf{b} >= \alpha \mathbf{B}_0$. When the mean velocity field is known, and when $\alpha \neq 0$, the solutions of (9.37a) describe the growth of the large-scale field, or the steady-state field if $\partial \mathbf{B}_0 / \partial t = 0$.

The dynamo coefficient, α, can be evaluated in terms of the velocity-helicity density ($k_v = \mathbf{v} \cdot \nabla \times \mathbf{v}$), as in Moffatt; or in terms of the magnetic-helicity density ($k_B = \mathbf{A} \cdot \mathbf{B}$ if $\mathbf{A}$ is the vector potential), as in Mattheus *et al.* (1986). The dynamo coefficient is nonzero only if two requirements are satisfied. The first is that the system must have a nonzero helicity. Moffatt shows that $< \mathbf{v} \times \mathbf{b} >$ has a net component along $\mathbf{B}_0$ if, for instance, the velocity field is not invariant under coordinate inversion ($\mathbf{r} \to -\mathbf{r}$) – and this condition corresponds to $k_v \neq 0$. Physically, helicity reflects a 'knottedness' of the field or flow lines, and could be injected into a system along with angular momentum (*e.g.*, De Young 1980). The second condition for a nonzero α is that the resistivity must be finite. This turns out to allow the $\mathbf{b}$ and $\mathbf{v}$ Fourier components to be partially out of phase with each other (again, cf. Moffatt), and so allows a nonzero $< \mathbf{v} \times \mathbf{b} >$.

This formulation is supported by numerical experiments. Models of two-dimensional and three-dimensional MHD turbulence have been done by several authors (*e.g.*, Pouquet *et al.* 1976; De Young 1980; Meneguzzi et al. 1981). These calculations are not restricted to the formal scale separation of (9.37), and they account for the feedback of the magnetic field on the velocity field. These simulations verify that energy is transferred from the velocity field to the magnetic field, up to approximate equipartition levels. Cascades of energy from the driving scale to smaller scales are always seen, but if helicity is maintained, the magnetic energy also cascades to larger scales (corresponding to $\alpha \neq 0$ in (9.37)). Such numerical turbulence experiments involve injecting energy (and perhaps helicity) at a constant rate to offset dissipative losses, and searching for steady spectra. De Young (1980) finds that a few percent of the injected energy is converted to magnetic field energy in his simulations.

9.6.3 Taylor relaxation

Laboratory plasmas – which are generally low-β_K – display a particularly interesting behaviour (cf. Taylor 1986, and references therein). A plasma, set up in some initial state, passes quickly through an initially turbulent phase, and then settles into a relatively long-lived quiescent state. The structure of the final, 'relaxed' state is independent of the history, and of the boundary conditions of the container. The relaxation process proceeds through self-generated turbulence, allied with a small but finite resistivity. The turbulence allows reconnection of magnetic field lines and dissipation of magnetic energy as the plasma relaxes to a minimum energy state.

This rapid generation and evolution of turbulence has not been measured in detail, but the mechanism is believed to be understood in terms of a model proposed by Woltjer (1958) in the astrophysical context and by Taylor (1974) for laboratory plasmas. The basic logic is as follows (*e.g.*, Taylor 1986; Ting *et al.* 1986). Both the total energy, $E_B = \int (B^2/8\pi)\, d\mathbf{x}$, and the total magnetic helicity, $K = \int \mathbf{A}\cdot\mathbf{B} d\mathbf{x}$ are invariants of an ideal ($\tilde{\eta} = 0$), closed ($\mathbf{B}\cdot d\mathbf{S} = 0$ on the boundary) MHD system. If finite dissipation is allowed, it is often the case that E_B decays faster than K. When this occurs, the plasma will relax (on a turbulent dynamic time scale) to an end-state which is described by the minimum energy compatible with the initial value of the helicity.

This end-state is described using variational principles. Ignoring plasma pressure and flow (the low-β_P and low-β_K regime), we can set

$$\delta \int \left(\frac{B^2}{8\pi} - \frac{\mu}{8\pi} \mathbf{A}\cdot\mathbf{B} \right) d\mathbf{x} = 0, \tag{9.38}$$

with the Lagrange multiplier, μ, assumed to be constant over the volume. This condition is equivalent to

$$\nabla \times \mathbf{B} = \mu \mathbf{B}, \tag{9.39}$$

and is just the condition for a force-free field – one in which the current, $\mathbf{j}$, is everywhere parallel to the magnetic field, so that the associated Lorentz force ($\mathbf{j}\times\mathbf{B}/c$) is everywhere zero. The constant μ in the relaxed state must be determined by the helicity (that is, by the initial conditions).

The condition (9.39) can be used to predict the field structure of the relaxed state. Before we look at this, however, we must consider a couple of caveats for astrophysical applications. The first is that K is strictly conserved only in a closed system; more generally, $dK/dt = \int_S (\mathbf{A}\cdot\mathbf{v})(\mathbf{B}\cdot d\mathbf{S})$ (Heyvaerts & Priest 1984). Considering applications to solar flux tubes, these authors suggest that if the relaxation time is shorter than the helicity-loss time (which is likely, if the turbulent scales are much smaller than the source size), the system will evolve through a series of force-free states, with $\mu(t)$ regulated by $K(t)$. (Berger & Field 1984, have also defined a relative helicity which is conserved in an open system; they recover the force-free state from this assumption.)

A second caveat is that (9.38) does not involve the plasma pressure or velocity. This means the field structure deduced from (9.39) is not coupled to the plasma structure. We are not aware of any general extension of this work to compressive or flowing plasmas, although some efforts on the subject have been made by Hameiri & Hammer (1982), Finn & Antonsen (1983), Field (1986) and Turner (1986).

9.6.4 Consequences and applications

Thus, in both the high-β_K case (fluid dominated) and the low-β_K case (field dominated), small-scale turbulence is involved in the maintenance of large-scale magnetic fields. These fields are described by (9.37a) and (9.39), respectively. One simple solution of these equations is worth mentioning. If $\partial\mathbf{B}/\partial t = 0$, $<\mathbf{v}\times\mathbf{b}>=\alpha\mathbf{B}$ and $\nabla\times(\mathbf{V}_0\times\mathbf{B}) = 0$ (the latter is the case, for instance, in a cylindrically symmetric, nonexpanding jet), (9.37a) reduces to (9.39). The solution of this in cylindrical geometry is well known (*e.g.*, Lundquist 1950): $B_r(r) = B_0 J_0(\mu r)$ and $B_\phi(r) = B_0 J_1(\mu r)$ if J_0 and J_1 are Bessel functions of the first kind. This solution describes what is called a 'flux rope', as in Fig. 9.3.

The solution has $\mathbf{B}\approx B_0\hat{\mathbf{z}}$ on the axis, and $\mathbf{B}\rightarrow B_\phi\hat{\phi}$ away from the axis. Such a structure is reminiscent of that seen in the Venus flux ropes (Russell & Elphic 1979). Königl & Choudhuri (1985)

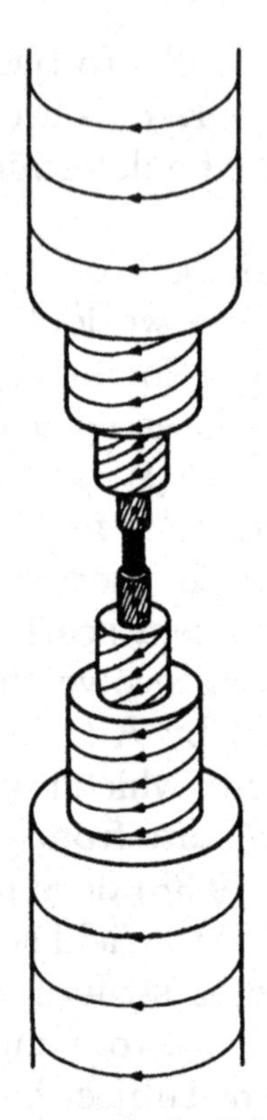

Fig. 9.3. A magnetic 'flux rope'; the field strength diminishes, and the field becomes more azimuthal with increasing radius (from Russell & Elphic 1979). Reprinted by permission from *Nature*, **279**, 616. Copyright © 1979 Macmillan Magazines Limited.

suggested the application of this to radio jets and noted that the most general solution of (9.39) in a cylindrical geometry with the radial field vanishing at the boundary consists of two modes, the cylindrically symmetric one described here and one other, helical mode.

Although these solutions are decoupled from the plasma (due to being force-free), we are of course interested in non-force-free solutions as well. In particular, suggestions of overpressure and the need for self-confinement by magnetic fields (*e.g.*, Biretta *et al.* 1983; Owen 1986) make solutions with $\mathbf{j} \times \mathbf{B} \neq 0$ of interest. We note that the relaxation models including plasma pressure studied by Finn & Antonsen (1983) and by Turner (1986) couple the plasma pressure to $\mathbf{j} \times \mathbf{B}$, and allow solutions which resemble the flux rope solutions of Fig. 9.3. Thus, overpressure, relaxed states may be consistent with plasma confinement.

9.7 Conclusions

We have set out the fundamentals of the physics that relates to particle and field dynamics in extragalactic jets, discussed current developments in this area and indicated how current theories are being applied to the interpretation of observations.

We have shown that stochastic acceleration of electrons by both Alfvén and magnetosonic waves is capable of producing the observed power-law energy spectrum, at least when account is taken of the coupling between waves, particles and the fluid whose motions are probably the ultimate energy source. Shock-Fermi acceleration also can produce this observed spectrum, and developments in non-linear theory here confirm cosmic-ray modified shocks as a viable explanation for the spectra of flow structures such as hot-spots. We do not see stochastic and shock acceleration models as competing, but rather each as relevant in different domains.

The resonant scattering of electrons by Alfvén waves, necessary for stochastic and shock acceleration, suggests that an injection mechanism is needed; although largely unexplored in this context, we believe that magnetic field reconnection, double layers and reflection from almost-perpendicular shocks are promising mechanisms.

Turbulent dynamos are capable of amplifying magnetic fields, and in the presence of helicity may produce a non-negligible large-scale field. Taylor relaxation is another mechanism that may play a role in the production of an ordered field. Both mechanisms can lead to field structures that have been suggested by observations, or considered for the role they might play in confining or stabilizing the flow.

PAH would like to thank the Astronomy Centre, University of Sussex, for their hospitality during August 1986 and Mallory Roberts for assistance in searching the geophysical literature. We thank Mark Schmitz for critically reading a draft of this chapter, and Donald Ellison and Alan Heavens for comments on §9.3. JAE was partially supported by NSF Grant AST-8316235 and PAH was partially supported by NSF Grant AST-8501093.

Symbols used in Chapter 9

Symbol	Meaning
$\mathbf{a}$	acceleration
$\mathbf{A}$	vector potential
$\tilde{A}(E)$	first-order Fermi acceleration coefficient
$b(p)$	radiative momentum loss rate per particle
$\tilde{b}(E)$	radiative energy loss rate per particle
b_o	synchrotron radiation coefficient
$\mathbf{B}$, B	magnetic field (and magnitude)
$\mathbf{B_o}$	ordered (large-scale) magnetic field
$\mathbf{b}$	turbulent (small-scale) magnetic field
B_u	upstream magnetic field
B_d	downstream magnetic field
B_ϕ	azimuthal component of magnetic field
B_z	component of magnetic field along jet axis
ΔB	magnetic field perturbation
c	speed of light
c_A	Alfvén speed
c_M	magnetosonic wave speed
c_S	sound speed
$D_{\mathbf{pp}}$	momentum space diffusion coefficient (tensor)
$D(p)$	momentum space diffusion coefficient (isotropic)
D_o	coefficient of momentum space diffusion coefficient
$\tilde{D}(E)$	second-order Fermi acceleration coefficient
e	electronic charge
$\mathbf{E}$	electric field
$\mathbf{E_s}$	electric field in a shock
E	single particle energy
E_B	total magnetic energy in a volume
E_D	Dreicer electric field
$E_{\max}$	maximum energy of a single particle
$f(\mathbf{p})$	particle distribution function
I	net current

$I(k)$	hydromagnetic wave driving rate
$\mathbf{j}$	current density
J_n	Bessel function of the first kind and order n
$\mathbf{k}$, k	wavenumber (and magnitude)
k_B	Boltzmann constant
k_v	velocity helicity density
k_B	magnetic helicity density
$k_{\parallel}$	component of wavenumber along magnetic field
$k_{\mathrm{res}}(p)$	wavenumber resonant with particle momentum p
$\mathbf{K}$	total magnetic helicity in a system
l	size of an astrophysical system
l_r	resistive diffusion length
L	average distance between scattering clouds
L	inductance of electric circuit
m	particle mass
m_p	proton mass
m_r	particle rest mass
$\mathcal{M}$	Mach number
$\mathcal{M}_u$	upstream Mach number
$\mathcal{M}^*$	lower critical Mach number
$\mathcal{M}^{**}$	upper critical Mach number
n	number density
n_{rel}	number density of relativistic particles
$\mathbf{p}$, p	particle momentum (and magnitude)
$p_{\parallel}$	gas pressure parallel to magnetic field
$p_{\perp}$	gas pressure perpendicular to magnetic field
q	momentum space particle spectral index
q	particle charge
r	jet radius
R	particle injection rate
t	time
t_{cross}	time for particle to cross shock
T	temperature
T	average particle escape time
T_u	upstream temperature
u_{rel}	energy density in relativistic particles
u_{th}	energy density in thermal gas
v	particle velocity
v_c	critical (runaway) velocity

v_d downstream fluid velocity
v_D drift velocity
v_i velocity of particle approaching shock
v_j jet velocity
v_t turbulent velocity
v_{Te} electron thermal velocity
v_{Ti} ion thermal velocity
v_u upstream fluid velocity
$v_\parallel$ component of particle velocity along magnetic field
$v_\perp$ component of particle velocity perpendicular to magnetic field
V average cloud velocity
V_x upstream pre-shock flow
$V_{x\,\mathrm{inf}}$ far upstream pre-shock flow
$\mathbf{V}$ fluid velocity
$\mathbf{V}_o$ ordered (large-scale) fluid velocity
$\mathbf{v}$ turbulent (small-scale) fluid velocity
$W(k)$ hydromagnetic wave energy density per unit wavenumber
α cosine of angle between wavevector and magnetic field
α photon spectral index
α dynamo coefficient
β particle velocity/light speed
β_K ratio of fluid kinetic energy to magnetic energy
β_P ratio of gas pressure to magnetic pressure
γ particle Lorentz factor
γ_u upstream Lorentz factor
$\gamma_{\mathrm{e,crit}}$ minimum electron Lorentz factor for shock acceleration
$\gamma_{\mathrm{min,p}}$ minimum Lorentz factor for proton resonance with Alfvén waves
$\gamma_{\mathrm{min,e}}$ minimum Lorentz factor for electron resonance with Alfvén waves
$\gamma(k)$ hydromagnetic wave damping rate
δ energy-space particle spectral index
η electrical resistivity
$\tilde{\eta}$ resistive diffusion coefficient
$\tilde{\eta}_{\mathrm{turb}}$ turbulent resistive diffusion coefficient
θ angle between wavevector and magnetic field
Θ angle between shock normal and magnetic field

κ	spatial diffusion coefficient
κ_d	downstream diffusion coefficient
κ_u	upstream diffusion coefficient
λ_t	turbulent length scale
$\ln \Lambda$	Coulomb logarithm
μ	cosine of particle pitch-angle
μ	ratio of $\nabla \times \mathbf{B}$ to $\mathbf{B}$ in force-free field
ν_e	electron collision rate
ν_{eff}	effective (anomalous) electron collision rate
$\nu_\pm$	rate of head-on (overtaking) cloud-particle collisions
ρ	mass density
σ	electrical conductivity
τ	acceleration time scale
ϕ	azimuthal angle
Φ	electrostatic potential
Φ_B	magnetic flux
ψ	particle pitch-angle
$\overline{\psi}$	mean particle pitch-angle
ω	hydromagnetic wave frequency
ω_e	electron plasma frequency
ω_p	proton plasma frequency
Ω^*	relativistic particle gyrofrequency
Ω_e	electron gyrofrequency
Ω_p	proton gyrofrequency

References

Achterberg, A., 1979, *Astr. Ap.*, **76**, 276.

Achterberg, A., 1981, *Astr. Ap.*, **97**, 259.

Achterberg, A., 1986, In *International School on Neutron Stars, Active Galactic Nuclei and Jets* (Erice).

Achterberg, A., 1988, *M. N. R. A. S.*, **232**, 323.

Achterberg, A. & Norman, C. A., 1980, *Astr. Ap.*, **89**, 353.

Achterberg, A., Blandford, R. D. & Periwal, V., 1984, *Astr. Ap.*, **132**, 97.

Asseo, E. & Sol, H., 1987, *Phys. Rep.*, **148**, 307.

Axford, W. I., Leer, E. & Skadron, G., 1977, In *Proceedings of the* 15th *International Cosmic Ray Conference*, Plovdiv, Bulgaria.

Axford, W. I., Leer, E. & McKenzie, J. F, 1982, *Astr. Ap.*, **111**, 317.

Barnes, A., 1983, *Ap. J.*, **265**, 457.

Barnes, A. & Suffolk, G. C. J., 1971, *J. Plasma Phys.*, **5**, 315.

Barnes, A. & Scargle, J. D., 1973, *Ap. J.*, **184**, 251.

Bell, A. R., 1978a, *M. N. R. A. S.*, **182**, 147.
Bell, A. R., 1978b, *M. N. R. A. S.*, **182**, 443.
Bell, A. R., 1987, *M. N. R. A. S.*, **225**, 615.
Benford, G., 1978, *M. N. R. A. S.*, **183**, 29.
Berger, M. A. & Field, G. B., 1984, *J. Fluid Mech.*, **147**, 133.
Biermann, P. L. & Strittmatter, P. A., 1987, *Ap. J.*, **322**, 643.
Biretta, J. A., Owen, F. N. & Hardee, P. E., 1983, *Ap. J. Lett.*, **274**, L27.
Blandford, R. D., 1980, *Ap. J.*, **238**, 410.
Blandford, R. D., 1986, In *Magnetospheric Phenomena in Astrophysics*, eds. R. I. Epstein & W. C. Feldman (New York: AIP), p.1.
Blandford, R. D. & Eichler, D., 1987, *Phys. Rep.*, **154**, 1.
Blandford, R. D. & Ostriker, J. P., 1978, *Ap. J.*, **221**, L29.
Blandford, R. D. & Ostriker, J. P., 1980, *Ap. J.*, **237**, 793.
Bogdan, T. J. & Völk, H.-J., 1983, *Astr. Ap.*, **122**, 129.
Bogdan, T. J. & Lerche, I., 1985, *M. N. R. A. S.*, **212**, 413.
Borovsky, J. E., 1986, *Ap. J.*, **306**, 451.
Borovsky, J. E. & Eilek, J. A., 1986, *Ap. J.*, **308**, 929.
Bregman, J. N., 1985, *Ap. J.*, **288**, 32.
Burn, B. I., 1975, *Astr. Ap.*, **229**, 409.
Carlqvist, P., 1982, *Ap. Sp. Sci.*, **87**, 21.
Cavallo, G., 1982, *Astr. Ap.*, **111**, 368.
Chen, F. F., 1974, *Introduction to Plasma Physics* (Plenum Press: New York).
de Jager, C., 1986, *Sp. Sci. Rev.*, **44**, 43.
De Young, D. S., 1980, *Ap. J.*, **241**, 81.
Droge, W. & Schlickheiser, R., 1986, *Ap. J.*, **305**, 909.
Drury, L. O'C., 1983, *Rep. Prog. Physics*, **46**, 973.
Drury, L. O'C. & Völk, H.-J., 1981, *Ap. J.*, **248**, 344.
Drury, L. O'C., Axford, W. I. & Summers, D., 1982, *M. N. R. A. S.*, **198**, 833.
Eichler, D., 1979, *Ap. J.*, **229**, 419.
Eichler, D., 1981, *Ap. J.*, **247**, 1089.
Eichler, D., 1984, *Ap. J.*, **277**, 429.
Eichler, D., 1985, *Ap. J.*, **294**, 40.
Eilek, J. A., 1979, *Ap. J.*, **230**, 373.
Eilek, J. A. & Henriksen, R. N., 1984, *Ap. J.*, **277**, 820.
Ellison, D. C., 1985, *J. Geophys. Res.*, **90**, 29.
Ellison, D. C. & Eichler, D., 1984, *Ap. J.*, **286**, 691.
Ellison, D. C. & Eichler, D., 1985, *Phys. Rev. Lett.*, **55**, 2735.
Ellison, D. C., Jones, F. C. & Eichler, D., 1981, *J. Geophys.*, **50**, 110.
Ellison, D. C. & Möbius, E., 1987, *Ap. J.*, **318**, 474.
Ellison, D. C. & Ramaty, R., 1985, *Ap. J.*, **298**, 400.
Fabian, A. C., Blandford, R. D., Guilbert, P. N., Phinney, E. S. & Guellar, L., 1986, *M. N. R. A. S.*, **221**, 931.
Falle, S. A. E. G. & Giddings, J. R., 1987, *M. N. R. A. S.*, **225**, 399.
Fermi, E., 1949, *Phys. Rev.*, **75**, 1169.

Field, G. B., 1986, In *Magnetospheric Phenomena in Astrophysics*, eds. R. I. Epstein & W. C. Feldman (AIP: New York), p. 324.

Finn, J. M. & Antonsen, T. M. Jr., 1983, *Phys. Fluids*, **16**, 3540.

Formisano, V., Russell, C. T., Means, J. D., Greenstadt, E. W., Scarf, F. L. & Neugebauer, M., 1975, *J. Geophys. Res.*, **80**, 2013.

Forslund, D. W. & Freidberg, J. P., 1971, *Phys. Rev. Lett.*, **27**, 1189.

Furth, H. P., 1985, *Phys. Fluids*, **28**, 1595.

Grappin, R., Frisch, U., Leorat, J. & Pouquet, A., 1982, *Astr. Ap.*, **105**, 6.

Grappin, R., Pouquet, A. & Leorat, J., 1983, *Astr. Ap.*, **126**, 51.

Greenstadt, E. W., 1974, In *Solar Wind Three*, ed. Russell, C. T., (Institute of Geophysics and Planetary Physics: UCLA).

Greenstadt, E. W. & Fredricks, R. W., 1979, In *Solar System Plasma Physics vol III*, eds. Lanzerotti, L., Kennel, C. F. & Parker, E. N. (Reidel: Dordrecht, Netherlands).

Greenstadt, E. W., Russell, C. T., Scarf, F. L., Formisano, V. & Neugebauer, M., 1975, *J. Geophys. Res.*, **80**, 502.

Greenstadt, E. W., Russell, C. T., Gosling, J. T., Bame, S. J., Paschmann, G., Parks, G. K., Anderson, K. A., Scarf, F. L., Anderson, R. R., Gurnett, D. A., Lin, R. P., Lin, C. S. & Reme, H., 1980, *J. Geophys. Res.*, **85**, 2124.

Hameiri, E. & Hammer, J. H., 1982, *Phys. Fluids*, **25**, 1855.

Heavens, A. F., 1983, *M. N. R. A. S.*, **204**, 699.

Heavens, A. F., 1984a, *M. N. R. A. S.*, **207**, 1p.

Heavens, A. F., 1984b, *M. N. R. A. S.*, **210**, 813.

Heavens, A. F., 1984c, *M. N. R. A. S.*, **211**, 195.

Heavens, A. F. & Drury, L. O'C., 1988, *M. N. R. A. S.*, **235**, 997.

Heavens, A. F. & Meisenheimer, K., 1987, *M. N. R. A. S.*, **225**, 335.

Heyvaerts, J. & Priest, E. R., 1984, *Astr. Ap.*, **137**, 63.

Holman, G. D., Ionson, J. A. & Scott, J. S., 1979, *Ap. J.*, **228**, 576.

Holman, G. D. & Pesses, M. E., 1983, *Ap. J.*, **267**, 837.

Ichimaru, S., 1973, *Basic Principles of Plasma Physics* (The Benjamin/Cummings Publishing Co. Inc.: Reading, MA).

Jokipii, J. R. & Morfill, G. E., 1985, *Ap. J.*, **290**, L1.

Kan, J. R. & Swift, D. W., 1983, *J. Geophys. Res.*, **88**, 6919.

Kardashev, N. S., 1962, *Soviet Astr. - AJ*, **6**, 317.

Kazanas, D. & Ellison, D. C., 1986, *Ap. J.*, **304**, 178.

Kirk, J. G. & Schneider, P., 1987a, *Ap. J.*, **315**, 425.

Kirk, J. G. & Schneider, P., 1987b, *Ap. J.*, **322**, 256.

Kolmogorov, A. N., 1941, *Compt. Rend. Acad. Sci.*, **34**, 540.

Königl, A. & Choudhuri, A. R., 1985, *Ap. J.*, **289**, 173.

Kraichnan, R. H., 1965, *Phys. Fluids*, **8**, 1385.

Krall, N. A. & Trivelpiece, A. N., 1973, *Principles of Plasma Physics* (McGraw-Hill: New York).

Krymsky, G. F., 1977, *Dokl. Akad. Nauk. SSSR*, **234**, 1306.

Kulsrud, R. M. & Ferrari, A., 1971, *Ap. Sp. Sci.*, **12**, 302.

Lacombe, C., 1977, *Astr. Ap.*, **54**, 1.

Lagage, P. O. & Cesarsky, C. J., 1983a, *Astr. Ap.*, **118**, 223.

Lagage, P. O. & Cesarsky, C. J., 1983b, *Astr. Ap.*, **125**, 249.

Leroy, M. M., 1983, *Phys. Fluids*, **26**, 2742.

Leroy, M. M., Goodrich, C. C., Winske, D., Wu, C. S. & Papadopoulos, K., 1981, *Geophys. Res. Lett.*, **8**, 1269.

Leroy, M. M., Winske, D., Goodrich, C. C., Wu, C. S. & Papadopoulos, K., 1982, *J. Geophys. Res.*, **87**, 5081.

Livshitz, E. M. & Pitaevskii, L. P., 1981, *Physical Kinetics* (Pergamon: Oxford).

Lovelace, R. V. E., 1976, *Nature*, **262**, 649.

Lundquist, S., 1950, *Ark. f. fys.*, **2**, 361.

Matthaeus, W. H., Goldstein, M. L. & Lautz, S. R., 1986, *Phys. Fluids*, **29**, 1504.

McKenzie, J. F. & Völk, H.-J., 1982, *Astr. Ap.*, **116**, 191. (Erratum in **136**, 378.)

Meisenheimer, K. & Heavens, A. F., 1986, *Nature*, **323**, 419.

Melrose, D. B., 1968, *Ap. Sp. Sci.*, **2**, 171.

Melrose, D. B., 1970, *Ap. Sp. Sci.*, **6**, 321.

Melrose, D. B., 1980, *Plasma Astrophysics* (Gordon Breach: New York).

Melrose, D. B. & Wentzel, D. G., 1970, *Ap. J.*, **161**, 457.

Meneguzzi, M., Frisch, U. & Pouquet, A., 1981, *Phys. Rev. Lett.*, **47**, 1060.

Michel, F. C., 1981, *Ap. J.*, **247**, 664.

Moffatt, H. K., 1978, *Magnetic Field Generation in Electrically Conducting Fluids* (Cambridge University Press: Cambridge).

Moghaddam-Taaheri, E. & Vlahos, L., 1987, *Phys. Fluids*, **30**, 3155.

Moghaddam-Taaheri, E., Vlahos, L., Rowland, H. L. & Papadopoulos, K., 1985, *Phys. Fluids*, **28**, 3356.

Moraal, H. & Axford, W. I., 1983, *Astr. Ap.*, **125**, 204.

Morse, D. L., 1973, *Plasma Physics*, **15**, 1262.

Morse, D. L., 1976, *J. Geophys. Res.*, **81**, 6126.

Morse, D. L. & Greenstadt, E. W., 1976, *J. Geophys. Res*, **81**, 1791.

Myers, S. T. & Spangler, S. R., 1985, *Ap. J.*, **291**, 52.

Owen, F. N., 1986, In *Quasars: IAU Symposium 119*, eds. G. Swarup & V. K. Kapahi (Reidel: Dordrecht, Netherlands).

Papadopoulos, K., 1977, *Rev. Geophys. and Sp. Phys.*, **15**, 113.

Peacock, J. A., 1981, *M. N. R. A. S.*, **196**, 135.

Pelletier, G. & Roland, J., 1986, *Astr. Ap.*, **163**, 9.

Pouquet, A., Frisch, U. & Leorat, J., 1976, *J. Fluid Mech.*, **77**, 321.

Pyle, K. R., Simpson, J. A., Barnes, A. & Mihalov, J. D., 1984, *Ap. J.*, **282**, L107.

Quest, K. B., 1985, *Phys. Rev. Lett.*, **54**, 1872.

Quest, K. B., 1986, *J. Geophys. Res.*, **91**, 8805.

Quest, K. B., 1988, *J. Geophys. Res.*, **93**, 9649.

Quest, K. B., Forslund, D. W., Brackbill, J. U. & Lee, K., 1983, *Geophys. Res. Lett.*, **10**, 471.

Russell, C. T. & Elphic, R. C., 1979, *Nature*, **279**, 616.

Sarris, E. T. & Van Allen, J. A., 1974, *J. Geophys. Res.*, **79**, 4157.

Sarris, E. T. & Krimigis, S. M., 1985, *Ap. J.*, **298**, 676.

Schlickheiser, R., 1984, *Astr. Ap.*, **136**, 227.

Scott, J. S., Holman, G. D., Ionson, J. A. & Papadopoulos, K., 1980, *Ap. J.*, **239**, 769.

Shebalin, J. S., Mattaeus, W. H. & Montgomery, D., 1983, *J. Plasma Phys.*, **29**, 525.

Smith, R. A. & Goertz, C. K., 1978, *J. Geophys. Res.*, **83**, 2617.

Spicer, D. S., 1982, *Sp. Sci. Rev.*, **31**, 351.

Spitzer, L., 1962, *Physics of Fully Ionized Gases* (Wiley Interscience: New York).

Taylor, J. B., 1974, *Phys. Rev. Lett.*, **33**, 1139.

Taylor, J. B., 1986, *Rev. Mod. Phys.*, **58**, 741.

Ting, A. C., Matthaeus, W. H. & Montgomery, D., 1986, *Phys. Fluids*, **29**, 3261.

Tsurutani, B. T. & Rodriguez, P., 1981, *J. Geophys. Res.*, **86**, 4319.

Turner, L., 1986, *I. E. E. E. Transactions on Plasma Science*, PS-14, 849.

Völk, H.-J., Drury, L. O'C. & McKenzie, J. F., 1984, *Astr. Ap.*, **130**, 19.

Webb, G. M., 1983a, *Astr. Ap.*, **127**, 97. (Erratum in **131**, 176.)

Webb, G. M., 1983b, *Ap. J.*, **270**, 319.

Webb, G. M., Drury, L. O'C. & Biermann, P., 1984, *Astr. Ap.*, **137**, 185.

Webb, G. M., Bogdan, T. J., Lee, M. A. & Lerche, I., 1985, *M. N. R. A. S.*, **215**, 341.

Webb, G. M., Forman, M. A. & Axford, W. I., 1985, *Ap. J.*, **298**, 684.

Williams, A., 1986, Ph. D. Thesis, University of Cambridge.

Winske, D., 1984, In *Chapman Conference on Collisionless Shock Waves in the Heliosphere*, Los Alamos.

Woltjer, L., 1958, *Proc. Natl. Acad. Sci. U.S.A.*, **44**, 489.

Wu, C. S., 1984, *J. Geophys. Res.*, **89**, 8857.

10

Jets in the Galaxy

R. PADMAN, A. N. LASENBY
Mullard Radio Astronomy Observatory, Cavendish Laboratory, Madingley Road, Cambridge, CB3 0HE, UK.

D. A. GREEN
National Research Council of Canada, Herzberg Institute of Astrophysics, Dominion Radio Astrophysical Observatory, P. O. Box 248, Penticton, British Columbia, V2A 6K3, Canada.

10.1 Introduction

Whilst the luminous jets of radiogalaxies and quasars are the most powerful examples of collimated outflow in the cosmos, there are many examples of jets and outflows to be found much closer to home, within our own Galaxy. These span a great range of luminosities and collimation factors, from the optically visible jets and "lobes" associated with low-mass young stellar objects, which are morphologically very similar to the classical radiogalaxies, to the poorly collimated and much less clearly defined "jets" associated with the Galactic Centre and with various supernova remnants. Galactic jet sources also include the singular object SS 433, which is known to be emitting a two-sided jet at a quarter of the speed of light. This jet is known to be associated with a binary star system, and there is some evidence that other mass-transfer binaries may also have jets.

In many cases the jet material itself is insufficiently excited to dissociate it completely, giving us a variety of spectral lines at optical, infrared and radio frequencies with which to probe the underlying kinematics, while the mere fact that these objects are close gives us greatly enhanced linear resolution. If there is a lesson to be learnt from the wide variety of systems which exhibit collimated mass-loss it is that jets are very easily formed once symmetry is broken through rotation. This has been understood for discs for some time, and has been used to explain a huge variety of astrophysical phenomena – it is now apparent that in turn jets are nearly always associated with discs, although the exact mechanisms are still generally unclear. In this chapter we summarize our current knowledge of various Galactic jets and outflows: not only are they interesting in their own right, but they can potentially tell us a great deal about their driving mechanisms, and this in turn may be applicable also to *extra*galactic jets.

10.2 Jets and outflows from protostars and YSOs

Many stars approaching the main sequence undergo a phase of rapid mass loss. This is manifested in a variety of observed phenomena, in the optical, infrared, millimetre and radio regions of the spectrum. The outflows appear to be relatively short lived: given the large (and increasing!) number of known outflows and the observed star formation rate in the Galaxy, it appears that essentially all stars undergo such a mass loss phase.

We can establish the true kinematics of the various components of the outflows from proper motions and spectroscopic radial velocities; spectroscopy additionally yields information on the temperature, density and mass of each of the various flow components. The deduced values of mass, energy and momentum fluxes constrain and test the models. These Galactic stellar jets have structures which are remarkably similar to those of many extragalactic jets, but in this case essentially all of the material connected with the jet is detectable by one or more standard techniques.

There is apparently a close association between these stellar jets and the discs on a large range of size scales thought to exist, for a variety of observational and theoretical reasons, around young stellar objects (YSOs). This is hardly surprising – almost all theoretical models of such jets invoke the disc in the collimation process in one way or another. The jets themselves may be important in carrying off angular momentum, thus allowing an accretion flow through the disc, and the formation of proto-planetary discs. Thus we may need to learn more about the jets if we are to understand the processes which control the formation of planetary systems. There is evidence also that the jets may be sufficiently energetic to disperse the dense cores in which low-mass stars form, and thereby to halt the accretion process – a theory of jets may thus turn out to be critical to our understanding of the factors determining the present day initial mass function. Because jets from protostellar sources all have velocities very much less than c, the observations are free from relativistic beaming effects and we may thus also be able to gain new insights into the physics of the powerful extragalactic jets from a study of their smaller, nearby cousins.

10.2.1 Observational evidence for collimated mass-loss

10.2.1.1 Herbig-Haro objects

These appear on the sky survey plates as "bright semi-stellar knots or irregular complexes", and are found to have low-excitation forbidden lines such as $[O^0]$, $[S^+]$ and $[N^0]$, and anomalously bright emission in Hα when compared with typical reflection nebulae (*e.g.*, Strom, Grasdalen & Strom 1974; Schwartz 1983). On the basis of both proper motion and radial velocity measurements they are found to have velocities of the order of 100 to 350 km s^{-1}, with a preponderance of negative velocity (blue-shifted) emission. Several alignments are suggestive of highly collimated outflow. These include HH 1 and 2 (Herbig & Jones 1981), where a strong infrared source lies on the line joining the two HH objects; HH 7-11 (Snell & Edwards 1981), which lie on a line passing through another infrared source; the sources in HH 46-47 (Dopita, Schwartz & Evans 1982); and the separate knots of HH 39 which lie on a line joining them all to R Mon (Jones & Herbig 1982).

Recent mm-wave observations of the source HH 7-11 (Rudolph & Welch 1988) show strong emission in the $J = 1 \rightarrow 0$ line of HCO^+ from compact sources just downwind of the optically visible HH objects. The velocity and velocity dispersion of the emission are similar to those of the ambient cloud (as determined by observations of CO) suggesting that these HH objects* are the shocked surfaces of dense stationary quiescent clumps buffeted by a strong stellar wind.

10.2.1.2 Optical jets

Deep CCD exposures in Hα and various low-excitation forbidden lines have revealed the existence of an ionized optical jet very close to the central source of the HH 34 system, as well as a number of others including those associated with DG Tau B, HH

* As will be seen later, it is clear that the catch-all term 'HH objects' actually refers to several different shock phenomena, including knots in jets, bow shocks, and shocked edges of the cavity around optical jets.

30, Haro 6-5B and the HH 46-47 system (Reipurth *et al.* 1986; Mundt, Brugel & Bührke 1987, and references therein). Individual knots in the jets closely resemble HH objects, and in a number of cases have been classified as such. Radial and tangential velocities range up to 400 km s^{-1}. The deduced mass fluxes and luminosities in the jets are relatively low – of order $10^{-9} M_{\odot}$ yr^{-1} and $0.01 L_{\odot}$ respectively.

As an example, $[S^{+}]$ emission from the jet in HH 34 is shown in Fig. 10.1. HH 34S clearly has the appearance of a bow shock associated with the working surface, and although no counter-jet can be seen, HH 34N also seems to be a bow shock from an oppositely directed jet. Fig. 10.1b shows the jet in close up. There are six bright knots with approximately equal spacing, with evidence for weaker knots both toward and away from the exciting source.

10.2.1.3 **Radio continuum jets**

A thermal jet has been observed in L 1551 (Cohen, Bieging & Schwartz 1982; Snell *et al.* 1985), oriented in the direction of the molecular outflow. This is apparently coincident with the optical jet visible to the SW of the source (a NE optical counter-jet is assumed to be obscured by a putative circumstellar disc).

10.2.1.4 **H_2O masers**

Very compact sources of 22 GHz H_2O maser emission are found to cluster in regions of massive star formation. Whilst some of these masers have radial velocities characteristic of the denser regions of the surrounding molecular cloud, spectroscopic radial velocity and VLBI proper motion studies show that often the individual masers in the cluster are expanding rapidly away from a common origin close to a luminous star or infrared source (*e.g.*, in OMC 1 and W 51, Genzel *et al.* 1981). Unlike the HH objects and optical jets, the maser sources do not seem to show any evidence for strong collimation, and the velocities involved are usually of the order of a few tens, rather than hundreds, of km s^{-1}. Because of the extreme non-thermal nature of the maser emission, and the reasonable suspicion that at any one time very little of the total gas is actually involved in production of the observed maser emission, it is difficult to estimate the mass loss rate.

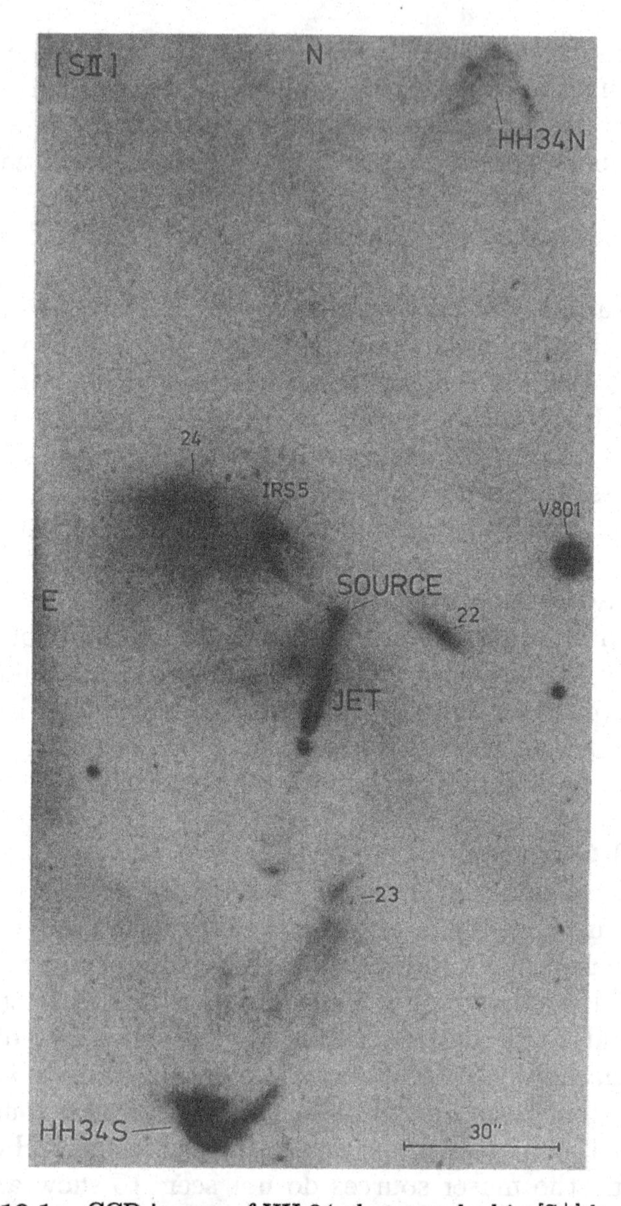

(*a*)

Fig. 10.1a. CCD images of HH 34 photographed in [S^+] by R. Mundt at the prime focus of the 3.5 m telescope on Calar Alto in Spain. A general view of the region showing apparent northern (HH 34N) and southern (HH 34S) bow shocks, as well as the jet. Photograph courtesy of R. Mundt.

10.2.1.5 Infrared absorption spectroscopy

Recent observations of the supposed pre-main sequence star

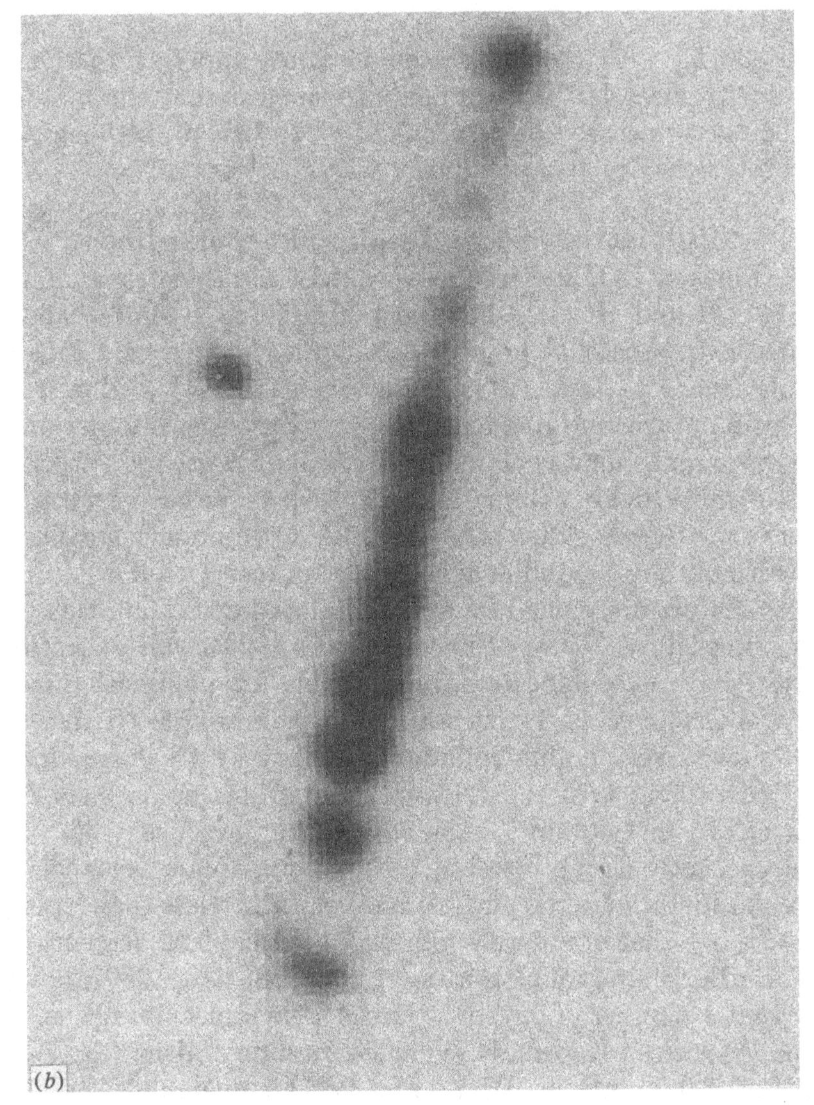

Fig. 10.1b. CCD images of HH 34 photographed in $[S^+]$ by R. Mundt at the prime focus of the 3.5 m telescope on Calar Alto in Spain. Close up of the jet. Photograph courtesy of R. Mundt.

M8E-IR show strong blue-shifted absorption in the $v = 1 \rightarrow 0$ vibro-rotational lines of CO (Mitchell *et al.* 1988). This is unambiguous evidence for *outflow* of molecular material with velocities up to 150 km s^{-1}. Using plausible assumptions about the scale of the outflow, dynamical time scales in the range 10-80 years are derived for the four components. Kinetic temperatures range from

95 to 330 K, with the lowest temperature applying to the highest velocity material. This appears to suggest that the material is still being accelerated at a radius of $\sim 4\times10^{14}$ m, although other interpretations are also possible.

10.2.1.6 Millimetre-wave and radio molecular lines

Lines of CO, and to a lesser extent of other trace molecules such as CS and HCO^+, show very clearly that almost all "protostars" or protostellar objects have an associated cold molecular outflow (see *e.g.*, Lada 1985 for a recent review). This is often bipolar in nature, and frequently shows signs of an interaction with the surrounding quiescent molecular cloud. Perhaps surprisingly even the most highly collimated outflows seldom have aspect ratios greater than 5 (even when observed with a beam much smaller than the width of the jet), and in general the sources with low luminosity central objects are much less well collimated even than this.

The blue-shifted CO outflow in NGC 2024 is an exception to the rule that molecular jets have relatively low collimation factors. Recent CO $J = 2 \rightarrow 1$ observations of this source (Richer *et al.* 1990) show a very highly collimated outflow, with the collimation factor increasing with the velocity offset from the ambient cloud. At a velocity of -30 km s^{-1} the collimation factor is ~ 30.

Until recently it was believed that the maximum velocity of the molecular material in the outflow was very much less than – perhaps only 10% of – that of the well-collimated optical jets. Recent results from the Berkeley group (Lizano *et al.* 1988; Koo 1989), however, show that a number of outflow sources have molecular gas at much higher velocities: Koo finds evidence for line widths ($\Delta v_{\rm FWZP}$) in excess of 300 km s^{-1} in HH 7-11 in CO for example. The earlier non-detection of such high velocity molecular gas in sources other than the Kleinmann-Low nebula in Orion (Kwan & Scoville 1976; Zuckerman, Kuiper & Rodriguez-Kuiper 1976) was almost certainly a sensitivity effect, and it seems reasonable to expect that more sources will show this phenomenon as the new generation of high-sensitivity high-resolution mm-wave telescopes starts to produce data. It is not yet clear, however, what the relationship of the "EHV" (extremely high velocity) gas is to the lower velocity flow or to the optical jets, although it is known that it is at least as extended.

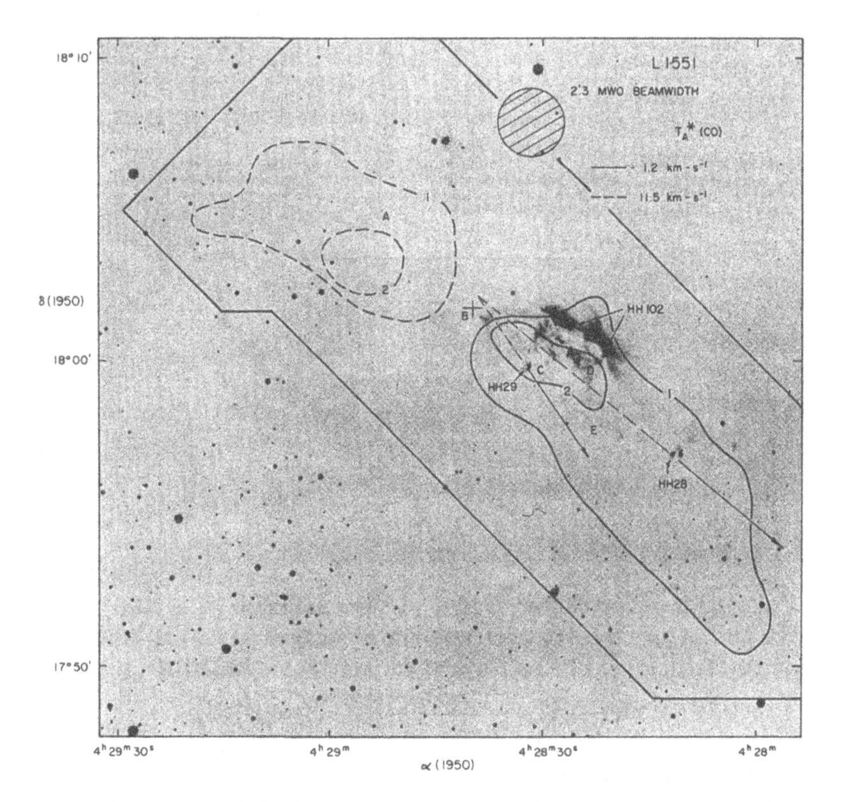

Fig. 10.2. *CO* $J = 1 \rightarrow 0$ emission in L 1551 observed with the 14 m FCRAO telescope; Figure 2 from Snell, Loren & Plambeck (1980).

Although most of the momentum and energy in the molecular flows is carried by the very highest velocity material, the bulk of the molecular gas has, however, much lower energy – velocities of a few tens of km s^{-1} are usual, while the lower luminosity sources may show velocity differences between the two lobes of only 2 – 3 km s^{-1}.

The archetypal molecular line outflow is L 1551, and the original CO map from Snell, Loren & Plambeck (1980) is reproduced in Fig. 10.2. The flow is well resolved by the beam, but is clearly not well collimated by comparison with the optical jet of Fig. 10.1. The luminosity of the central source is about $30L_{\odot}$. More detailed mapping (Uchida *et al.* 1987; Moriarty-Schieven *et al.* 1987; Moriarty-Schieven & Snell 1988) shows that the gas at any given velocity lies on an arc, with the highest velocities confined to small areas of

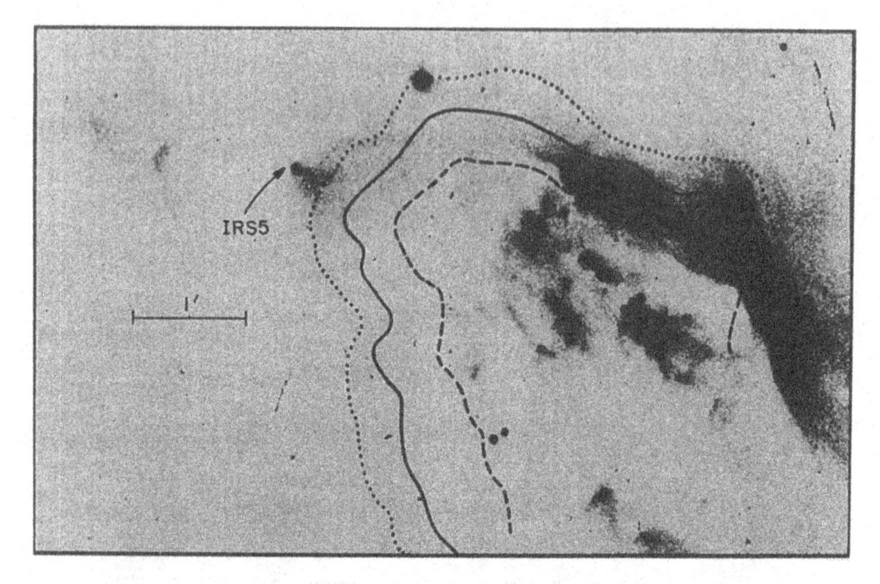

Fig. 10.3. Blue-shifted CO $J = 1 \rightarrow 0$ emission in L 1551 observed with the 14 m FCRAO telescope and reconstructed using a maximum entropy technique; Figure 7 from Moriarty-Schieven *et al.* (1987).

each lobe as shown in Fig. 10.3. The wind is apparently accelerated away from the exciting star, IRs5.

10.2.1.7 **Shocked molecular hydrogen**

The infrared vibration-rotation lines of H_2 are observed in a number of sources, with $\frac{v=1\rightarrow 0\ S(1)}{v=2\rightarrow 1\ S(1)}$ ratios indicative of shock excitation rather than fluorescence (*e.g.*, OMC 1, Beckwith *et al.* 1978; DR 21, Garden *et al.* 1986). Typical excitation temperatures are of order 2000 K, indicating a shock velocity of a few tens of km s^{-1}. Like the bipolar emission from the cold gas, the H_2 exhibits only a low degree of collimation, and apparently arises from a "working surface" at the end of the cold outflow.

10.2.1.8 **Atomic hydrogen**

Jets of neutral hydrogen have been observed in a few sources only, partly because of the extreme difficulty of the observations. High velocity H^0 was detected in NGC 2071 by Bally & Stark (1983) and more recently Bally (1986) has produced a map of the H^0 in this source. He finds that it is distributed similarly to the H_2 and along the same axis as the CO lobes. VLA observations of DR 21 (Russell

1987) show a red-shifted jet 3 pc long by 0.4 pc in width, with a very high deduced mass of $25 M_\odot$ (Russell attributes his failure to see the blue counter-jet to confusion by Galactic H^0). Lizano *et al.* (1988) have observed a number of sources with the Arecibo telescope and find high velocity H^0 in HH 7-11, at velocities up to ~ 170 km s^{-1}, but in this case the beam is too big for accurate mapping.

We see that the observations require the presence of two apparently different types of mass loss activity – highly supersonic, well collimated, partly ionized jets with relatively low mass-loss rate are most obvious in the later stages of star formation, while massive but poorly collimated cool molecular outflows are associated with the earlier stages, when the protostar is still deeply embedded in its parent molecular cloud. At least one source – L 1551 – exhibits both phenomena, and serves as a reminder that any theory which purports to explain outflows from young stellar objects must allow both types of activity to exist simultaneously.

Many of the phenomena described in this section are often collectively referred to as "signposts of star formation". In fact they are very clearly signposts of *outflow activity*, and therefore all star formation regions selected on the basis of any one of these signposts (which is most of them) are *ipso facto* undergoing a mass-loss phase. This emphasizes the universal nature of this phenomenon, and independently tells us that this phase lasts an appreciable fraction of the total time for which we are aware of the existence of the protostar.

10.2.2 **Outflow properties**

10.2.2.1 **Jet structure**

All of the main features of jet structure are shown in Fig. 10.1. The actual HH objects in this source (HH 34N, HH 34S) look like classical bow shocks, with essentially parabolic outer contours, and concave surfaces pointing back directly at the presumed exciting source. The $[S^+]$ line emission indicates that the shock is strongest on the axis of the source and at the outside edge of the bow shock (Mundt 1988), and rapidly weakens off axis. There is also a suggestion of weak shock emission arising from the walls of a cavity south of the central source and NW of HH 34S.

Closer examination of the jet itself (Fig. 10.1b) shows that it consists of at least six knots, of approximately equal intensity and spacing, with the hint of several more knots both between the jet and the source, and (barely visible on Fig. 10.1a) additional knots between the jet and the bow shock. Mundt (1986) suggests that the jet is visible because of internal shocks, which only occur when the jet is confined by external pressure. The weaker knots upstream (in the jet frame) of the bright jet may be analogous to the so-called Mach discs which arise when a pressure confined jet becomes free (Königl 1982; see Chapter 7). Mundt, Brugel & Bührke (1987) suggest that the jet from the source is initially free, and only becomes visible when pressure reconfined at some distance from the star. On the other hand Reipurth *et al.* (1986) take the view that the jet is pressure confined close to the star and only becomes visible when it exits the confining material and for one cooling length thereafter. These are quite different conclusions, which it may be possible to resolve with the aid of numerical modelling. The answer has important consequences, in that the Mundt *et al.* interpretation implies a cavity around the source of the outflow, whereas the Reipurth *et al.* view permits a collimating disc.

Assuming that the bow shock subtends a greater angle than the (largely unseen) jet, then the jet must have a projected opening angle of less than 10°, and probably somewhat less than this, indicating that for a free jet the Mach number $\mathcal{M} \gtrsim 6$, and more probably $\mathcal{M} \sim 15$. No proper motions are available yet for these HH objects, so we cannot simply obtain an estimate of the inclination, i, of the jet to the plane of the sky – a comparison of the apparent length of the optical jet and the cooling length gives $i \sim 15°$ (Reipurth *et al.* 1986), but the relatively high radial velocities measured by Mundt *et al.* suggest a somewhat higher value.

An independent estimate of the Mach number can be obtained in principle from the observed total velocity and the jet temperature, but the temperature is uncertain. Model shock calculations for the knots (see Schwartz 1983 and references therein) strongly suggest that the preshock material is mostly neutral, and that shock velocities of 70 – 100 km s^{-1} with preshock densities of order 3×10^8 m^{-3} fit the data best, while Mundt *et al.* derive typical values of (preshock) jet density in the range 2×10^7 – 1×10^8 hydrogen atoms m^{-3}. Since radiative cooling essentially turns off when the temperature falls below $(8 - 10)\times10^3$ K we might suspect that this would be a characteristic temperature for the preshock material, which would in turn yield a sound speed $c_S \sim 10$ km s^{-1}, and $\mathcal{M} \sim 20$ if the jet velocity is 200 km s^{-1}. This agrees reasonably well with the earlier estimate.

A slightly different argument gives a similar result: Observations of multiple knots of similar intensity, and the failure to detect any significant slowing of the jet through a series of shocks in most sources, suggests that the total energy of the jet is not decreased significantly by radiation in the shocks but only by adiabatic expansion. Its temperature will then be related to the shock velocity and the ratio of preshock to postshock density by $T_j = (n_j/n_s)m_p\langle v_s^2\rangle/3k_B$. Under these conditions $T_j \sim 10^4$ K and $c_S \sim 10$ km s^{-1}, as before.

For a jet which is freely expanding for most of its length, the external pressure must be $\ll n_j k_B T_j$. For a typical jet temperature of 10^4 K and mean number density of 3×10^7 m^{-3} appropriate to HH 34 (Mundt *et al.* 1987) pressure equilibrium with molecular hydrogen at a temperature of 10 – 30 K typical of molecular clouds would require a number density of only $5\times10^9 - 1.5\times10^{10}$ m^{-3}, which is close to the density of typical dark clouds, but somewhat less than that of the dense cores in which we believe stars form. The external density must certainly be no greater than required for equilibrium, which suggests that the jet has partially evacuated a cavity in the cloud (see also earlier discussion concerning extended shocks to the south of the HH 34 source).

10.2.2.2 **Wind structure**

High resolution observations of CO in L 1551 (Uchida *et al.* 1987; Moriarty-Schieven *et al.* 1987; Moriarty-Schieven & Snell 1988) indicate that in this source the CO flows in a thin shell around the edge of an evacuated cavity. (There is little evidence yet to relate this "CO cavity" to that presumed to exist around the optical jets – see previous section.) The outflow is also apparently accelerated away from the source (we will see later that this is a specific prediction of the MHD disc wind models), although Moriarty-Schieven & Snell describe a model involving geometric effects and a latitude-dependent wind velocity which can mimic acceleration on the position-velocity diagram. In addition to the shape of the emission as a function of velocity, this "accelerating shell" model predicts that for particular geometries red emission will be observed in the blue lobe, and *vice versa*, as indeed is seen.

Cabrit & Bertout (1986) catalogue the various morphologies that can be produced from pure biconical flow, as a function of the semi-opening angle of the flow θ_{max} and the angle between the jet axis and the plane of the sky, i. Patches of blue emission in the "red" lobe, and *vice versa*, can be produced in a number of ways. Unfortunately the effect of convolution of the true source distribution with the telescope beam is also to blur the distinction between their various cases, so that it is not always possible to assign a given source immediately to a particular case.

Co-rotation of the wind out to the Alfvén radius is an essential feature of most of the magnetically driven wind models. In L 1551 Uchida *et al.* (1987) suggest that the asymmetric red feature seen in the blue lobe, and corresponding blue feature in the red lobe, are evidence for a flow co-rotating with a magnetic field rooted in a disc around the source, but this is disputed by Moriarty-Schieven *et al.* (1987), who believe that it is in fact just due to density inhomogeneities in the ambient cloud, pointing out that where emission from, say, red-shifted material is seen in the blue lobe, the blue emission is also enhanced.

Many flows associated with very luminous young objects, and which have been mapped at high resolution, show an overall geometry which is apparently cylindrical rather than conical (*e.g.*, L 1551, NGC 2071, Snell *et al.* 1984; NGC 2024, Sanders & Willner 1985) suggesting that the molecular outflow is confined in some way. NGC 2024 is also remarkable for the way in which the maximum observable velocity increases linearly along the jet away from the (as yet unidentified) central source (Richer *et al.* 1990). Whether this is due to entrainment or acceleration is not yet clear however.

10.2.2.3 Energetics

Bally & Lada (1983) plot the mechanical luminosity of the molecular flow, L_w, against the luminosity L_* of the assumed driving source, and the thrust, or momentum flux Π_v, of the flow against that available in radiation from the central source (*i.e.*, L_*/c). Updated versions of these from Lada (1985) are reproduced in Fig. 10.4.

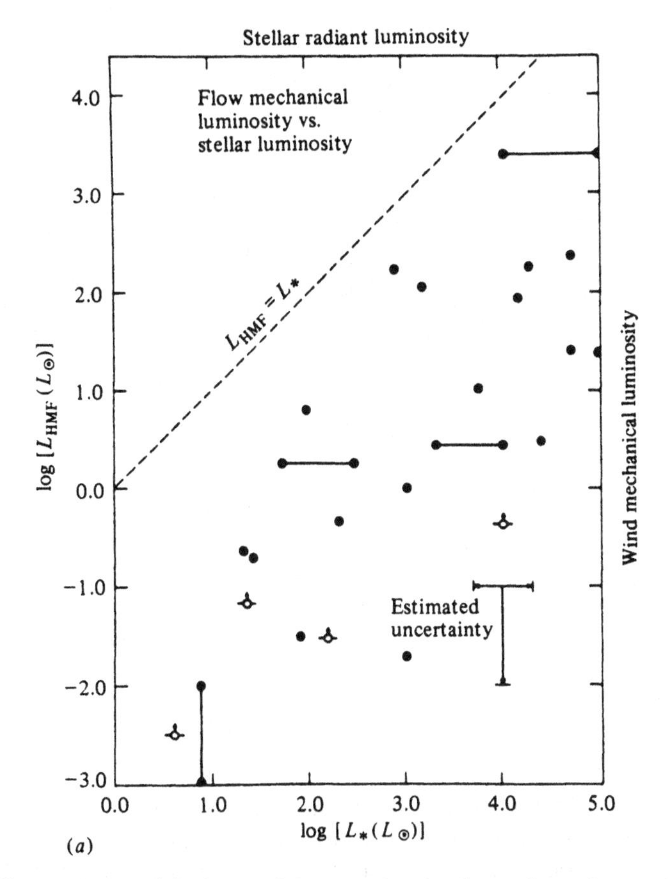

Fig. 10.4a. Mechanical luminosity in flow plotted as a function of bolometric luminosity of central source. Reproduced from Lada (1985).

On the L_w versus L_* diagram, there is a weak correlation between the luminosity of the central source and the mechanical luminosity of the flow. The mechanical luminosity is *always* less than the stellar luminosity, typically by a factor of 100 or more, but occasionally by as little as 5. A similar correlation exists on the Π_v versus L_*/c diagram. The force required to drive the flow is *always* greater than is available in radiation pressure, typically by a factor of 100 to 1000. This latter observation in particular is a very strong constraint on models for the driving sources of the outflows. We refer to the factor Π_v divided by L_*/c as the *momentum factor* of the flow, and the factor L_w/L_* as the *energy factor*.

The values of the total momentum p, energy E and dynamical time scale t_{dyn} $(= R/v_{max})$ are subject to considerable observational

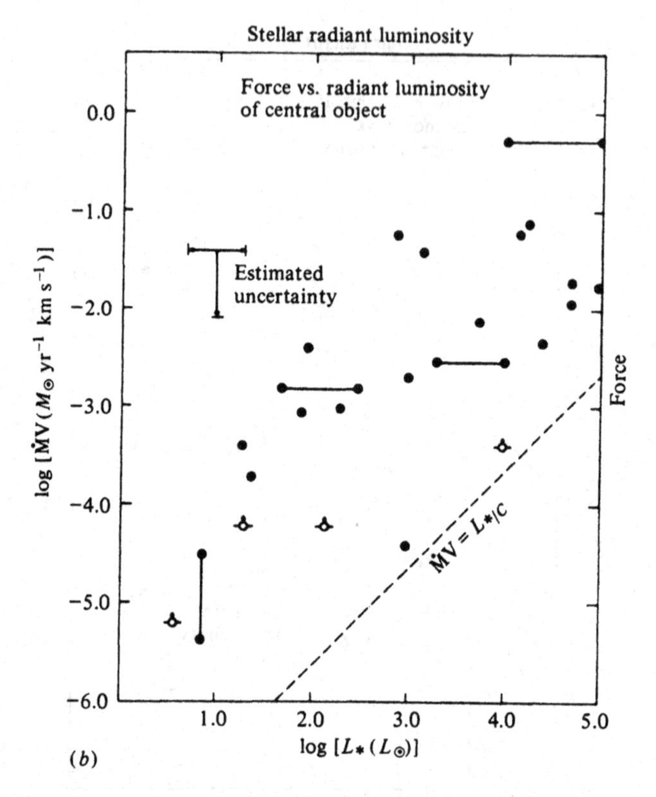

Fig. 10.4b. Estimated momentum flux (driving force) of flow, plotted as a function of momentum available in radiation from central source (L_*/c, assuming single scattering). Reproduced from Lada (1985).

uncertainties and depend strongly on the inclination of the source to the plane of the sky and on the assumed abundance of CO relative to that of H_2. More importantly, as pointed out by Dyson *et al.* (1988), the momentum and energy fluxes depend on the assumption that the apparent dynamical time scale represents the true age of the flow: If the flow is stationary then the "typical" velocity is the sound speed, and not the maximum observed flow velocity, and the calculated fluxes can be reduced by a factor which can be as large at 100.

High luminosity low surface brightness far-infrared emission has been detected in L 1551 using *IRAS* coadded survey data (Edwards *et al.* 1986). This emission is loosely associated with the molecular outflow, and appears to derive from the wind/cloud interface. The far-infrared luminosity is a factor of 35 higher than the mechanical

Table 10.1 *Dimensions, time scales and flow parameters for some representative outflows.*

Source†	R (pc)	v_{max} (km s^{-1})	t_{dyn} (yr)	L_* ($L_\odot$)	$\dot{M}$ ($M_\odot$ yr^{-1})	L_w ($L_\odot$)	Π_v ($L_\odot/c$)
Orion-KL	0.04	127	300	10000	1.6×10^{-2}	2600	2.5×10^{7}
NGC 2024	0.06	50	1150		3.5×10^{-3}	45	25000
L 1551(jet)	0.015	350	150	32	3×10^{-8}	0.02	
L 1551(CO)	0.42	30	14000	32	1.3×10^{-5}	0.2	10000
HH 34(jet)	0.004	135	2800	0.5	$<1\times10^{-9}$	<0.008	

†CO, unless noted.

References: Orion-KL: Bally & Lada 1983; NGC 2024: Sanders & Willner 1985; L 1551(jet) and HH 34(jet): Mundt, Brugel & Bührke 1987; L 1551(CO): Snell, Loren & Plambeck 1980.

luminosity of the flow – and some 18% of the bolometric luminosity of the central star – suggesting very strongly that this flow at least is *momentum-driven* (see §10.2.4).

Assuming for the present that the flows are *not* stationary, it appears that by any reasonable assessment the energy factor of the molecular flow itself is at least 0.01-0.1, while the momentum factor is of order 100-1000. Similar sums can be done for the collimated jets. In Table 10.1 we present flow parameters for a number of outflow sources.

10.2.3 The role of discs

Any theory of jets and outflows from YSOs must explain a bewildering variety of observational data:

The bipolarity evident in most sources.

The range of outflow velocities, from several hundred km s^{-1} in the ionized jets to only a few km s^{-1} in the cold outflows.

The independence of the jet velocity from the stellar luminosity.

The coexistence in some sources of a fast ionized jet with a slower cold outflow.

Regularly spaced knots observed in the optical jets.
The excess momentum flux in the flow over that available in radiation.
The observed velocity structure in the molecular material.

Additionally we might hope that such a theory would also help resolve the angular momentum problem – How does a diffuse (rotating) molecular cloud with a typical specific angular momentum $J \sim 1\times10^{16}$ m^2 s^{-1} reduce this by more than 4 orders of magnitude to form a star with $J \sim 5\times10^{11}$ m^2 s^{-1}? Magnetic braking (see Mouschovias 1981 and references therein) has been the preferred mechanism, although angular momentum transport by tidal torques has also been proposed (*e.g.*, Larson 1984). Recent observations of dense cores in the Taurus region by Heyer (1988) apparently indicate that the angular momentum problem is not resolved at the time the cores decouple from the general interstellar magnetic field. We note that there *must* be "excess" angular momentum in the system at the onset of the dynamical collapse phase if rotating accretion discs are to form as indicated by observations (see next section); therefore the resolution of the angular momentum problem is necessarily intimately bound up with the details of the collapse.

We could divide the existing theories into three types, depending on the basic acceleration mechanism. In the radiative theories the material is accelerated by radiation pressure acting directly on the gas, or on the grains embedded in it. In the hydrodynamic models a stellar wind is accelerated and collimated by thermal pressure gradients. Magnetic models take two forms – either the wind is centrifugally accelerated along magnetic field lines rooted in a rotating disc or star, or it is driven by a gradient in magnetic pressure whereby the expanding field lines carry the gas with them. In all cases, except possibly the last, the bipolar structure requires the presence of an anisotropic flattened mass distribution around the central object, which we hereinafter refer to as a disc.

10.2.3.1 **Evidence for the existence of dense discs**

Discs around YSOs seem to exist on all scales, from $< 10^{13}$ m to $> 10^{15}$ m. Flattened molecular cores centred on the molecular outflows, with their minor axes aligned with the axis of the flow, have been observed by many workers – see Torrelles *et al.* (1983) and Rodriguez (1987) for reviews. At smaller scales dense ($> 3\times10^{11}$ m^{-3}) molecular discs apparently straddle the central sources and fit well into the outer contours of the flows – for example in L 1551 Kaifu *et al.* (1984) see a disc of diameter 0.1pc rotating at 0.35 km s^{-1}, although the rotation is disputed by Batrla & Menten (1985). Little *et al.* (1985) have mapped G35.2−0.74 in the (1,1) and (2,2) lines of NH_3 and find a massive disc of diameter 0.4 pc, with a strong temperature gradient from > 100 K in the inner parts to ~ 25 K near the periphery, and a rotation gradient of 6.6 km s^{-1} pc^{-1}. In Mon R2, Bally (1986) has mapped a dense disc lying orthogonal to the major axis of the bipolar CO outflow. The velocity centroid of the CS emission changes by 1.5 km s^{-1} across the position of the central infrared cluster, and Bally interprets this as evidence for rotation about a $200M_\odot$ core. In this source, as in most purported to show rotation, the rotation period is many times the dynamical lifetime of the CO outflow. A further example of a source of this type is NGC 2071 (Bally 1982).

The evidence for true discs improves somewhat as we go to smaller size scales, and there are now a number of instances where the case for Keplerian rotation (*i.e.*, where the disc mass is insignificant compared with that of the central object, so that the rotational velocity *increases* with decreasing radius) seems secure. Sargent & Beckwith (1987) have observed a Keplerian disc of diameter 4000 AU around the relatively isolated protostar HL Tau, while Vogel *et al.* (1985) found evidence for differential rotation about IRc2 in OMC 1. In the more evolved bipolar H^+ region S 106, the infrared and radio point source is apparently embedded in dense neutral material separating the two ionized lobes (Bally, Snell & Predmore 1983). Interferometer measurements of the HCN emission from the S 106 "disc" show that the rotation gradient in this case is > 32 km s^{-1} pc^{-1} (Bieging 1984) (although more recent high resolution observations of the $HCO^+ J = 1 \rightarrow 0$ emission failed to confirm the rotation of the molecular material – Loushin *et al.* 1988). Padman & Richer (1990) have found new evidence for differential rotation in this source at radii < 2500 AU from IRs4, using the higher excitation $J = 3 \rightarrow 2$ lines of both HCN and HCO^+. Jackson, Ho & Haschick (1988) claim to have found definite rotation around the infrared source NGC 63334-I, and conclude that they are seeing a disc of about $30M_\odot$ in rotation about a central star of similar mass. Modelling work carried out at M. R. A. O. however, shows that it is impossible to reproduce the *lack* of emission at zero velocity with any purely rotational model, and it appears that this is actually an outflow source (Richer & Padman, in preparation).

Discs with dimensions of order 500 AU have been seen around pre-main sequence stars in the near-infrared using infrared speckle interferometry (HL Tau: Beckwith *et al.* 1984), and directly in the near-infrared using maximum entropy image reconstruction techniques (HL Tau: Grasdalen *et al.* 1984; L 1551: Strom *et al.* 1985), as well as in the mid-infrared using diffraction limited imaging (OMC 1-IRc2: Lester *et al.* 1985).

At scales $\lesssim 100$ AU the evidence is somewhat more indirect, although still compelling. Compact H^+ sources with a separation of 25 AU seen in L 1551 by Snell *et al.* (1985) and Bieging & Cohen (1985) have been interpreted by Rodriguez *et al.* (1986) as the ionized inner edges of an accretion disc (although a warped disc is implied since the H^+ sources do not lie in the midplane of the system). Snell *et al.* point out that the lack of a red-shifted counterpart to the jet seen in Hα in L 1551 may be due to extinction in a disc. This argument could also be made for a number of other sources. Scoville (1981) sees infrared recombination lines and CO bandhead emission from the Becklin-Neugebauer object in Orion, and interprets this as arising from the inner edge of a circumstellar disc. Recent optical observations of high excitation H^0 and Ca^+ lines in S 106, by Persson, McGregor & Campbell (1988), are indicative of dense material ($> 10^{19}$ m^{-3}) close to IRs4, and this is interpreted as clumps of neutral gas being ablated from the inner edge of an accretion disc, at a radius of order 1 AU from the star.

Further (indirect) reinforcement comes from modelling work on T Tauri stars (Adams & Shu 1986; Adams, Lada & Shu 1987). Observations show a characteristic variation in the spectra of protostellar objects, from the essentially stellar spectrum of T Tauri stars through to the cool black-body spectrum of deeply embedded objects. Objects at apparently intermediate evolutionary stages show double-peaked spectra rather than black-body spectra of intermediate temperature, and this is most easily explained by invoking a circumstellar disc which allows some light from the stellar surface to escape at the poles. Support for this view comes from the recent work of Cohen, Emerson & Beichman (1989) who find that a significant fraction of T Tauri stars have bolometric luminosities greater than that deduced from the optical spectra – a factor of 2 can arise when a disc reprocessing intercepted stellar radiation is seen pole on, while factors greater than two require some additional accretion luminosity also (Kenyon & Hartmann 1987; Adams, Lada & Shu 1987).

Finally, near-infrared polarization studies of suspected protostars (Sato *et al.* 1985; Lenzen 1987; see also Warren-Smith, Draper & Scarrott 1987) show near-infrared polarization vectors in the vicinity of the source (*i.e.*, of reflected light) to be parallel to the flattened molecular clouds, and perpendicular to the axis of the associated outflow. The visual extinction to the central source is deduced to be $A_v \sim 20^{\rm mag}$ greater than the value affecting the surrounding regions.

It is inescapable that young stars have discs. It remains to be seen, however, whether the observed discs are of fundamental importance in powering the outflows, or are just a by-product of the symmetry of the system.

10.2.4 What drives the outflows?

Several theories have been proposed. Although we cannot hope to examine all of them critically, some common strands emerge. Historically, the phenomenon of significant mass loss (in excess of that already known to occur in "ordinary" stellar winds) was first recognized in the mm-wave observations, and it was assumed that the mass loss was driven ultimately by the radiation pressure of the central object. Following the seminal paper by Bally & Lada (1983), however, it was recognized that there were real problems in obtaining the observed momentum flux, although it seemed quite possible on energetic grounds alone that radiation could provide the driving force. Thus, the development of theories to explain the jets and outflows has been dominated by the need to get apparently absurdly large amounts of momentum into the flow, when compared with that available in the (proto)starlight. Those theories that have survived are the most "efficient" or "economical", and therefore tend to appeal on aesthetic grounds also.

Later theories have mostly sought to exploit other energy reservoirs, such as the gravitational binding energy of material accreting onto the protostar. Even these theories require very large (although probably not unreasonable) accretion rates to drive the observed outflows. More recently there has been another shift in the paradigm, such that some theories now assume that the protostar is "maximally rotating" in some sense – whereas until recently the problem was how to get the angular momentum out of the system well *before* the formation of a true protostar. In this section we review the salient features of representative theories of each type. One recent theory *not* dealt with here is that of Blome & Kundt (1988), in which the jets consist of e^+e^- pair plasma formed in the magnetosphere of the rapidly rotating protostar by collisions of relativistic charged particles with photons. Although this neatly explains *why* the morphologies of stellar jets and extragalactic jets are so similar, it would seem to be overkill (in the sense of the energetics) until some other observational evidence for such high energies is found.

10.2.4.1 **Radiative models**

In at least one well-known source (L 1551, Edwards *et al.* 1986) the high value of far-infrared luminosity from the extended outflow renders a radiative drive improbable from consideration of the energetics alone, but is not sufficient to rule it out for certain. More powerful arguments come from consideration of the observable momentum flux.

The excess of momentum flux in the cold molecular outflows over that available in direct radiation from the central source was first noticed for the case of OMC 1, by Kwan & Scoville (1976). Phillips & Beckman (1980) pointed out that the difficulty can be resolved by permitting multiple scattering within the acceleration region. If the opacity is high, dust at different points within the cloud is coupled together by radiation exchange between grains even when the flow is supersonic. That is, the situation is analogous to that pertaining in stellar atmospheres: There is an outward radiation pressure gradient acting on the dust, and, if the dust-grain collision time scale is short enough, on the gas also.

Since the total kinetic energy observed in the outflow is much less than the luminosity of the star, $P \cdot dV$ work done on the dust photosphere by the photons results in negligible losses, and the momentum gain (compared with L_*/c) is limited only by "cooling" as photons escape from the source. If we assume that the escape probability on any one crossing of the photosphere by a single photon is essentially $1/\tau$, then each photon can transmit a total outward momentum of $\tau(h\nu/c)$ to the dust before being lost from the system. Thus the *minimum* optical depth required to solve the momentum problem is equal to the momentum gain. We can assume that the dust is all at a temperature of about 1000 K, since it will sublime at higher temperatures and the temperature gradient in a spherical photosphere is proportional only to $r^{-0.4}$ (Scoville & Kwan 1976). Then assuming that the minimum optical depth applies for a wavelength of 10 μm typical of this temperature, and using the standard λ^{-1} scaling law for the extinction, we obtain a minimum A_v to the star of order $20\dot{M}vc/L_*$, which is typically many thousands. This is much greater than the few tens of magnitudes usually observed, and it therefore appears that radiation pressure is not a viable mechanism for driving either the cold molecular outflows or the collimated optical jets. A slightly different presentation of radiative drives in the context of extragalactic sources is to be found in Königl (1986).

10.2.4.2 **Hydrodynamic models**

10.2.4.2.1 **Wind driven flows**

These models in general presuppose that the protostellar core develops a wind, and that it is the energy of this wind that ultimately drives the molecular outflow (and possibly also the fast jets). How is it possible for a low-mass high-velocity wind with a typical scale size of 10^{13}m to drive a high-mass low-velocity outflow with a scale size 100 to 1000 times larger? Since the wind is highly supersonic a two shock structure will develop (*e.g.*, Weaver *et al.* 1977). The wind is shocked and thermalized in a strong shock close to the star. The thermal pressure of the shocked wind drives a second highly radiative shock into the ambient cloud, which forms a cold dense shell around a hot "bubble".

If the wind cools sufficiently quickly the inner shock is also radiative and the shell is driven directly by the ram pressure of the wind. Solutions of this kind are described as momentum driven (*i.e.*, momentum conserving). On the other hand, if the wind is hot enough and fast enough it cannot cool significantly before it hits the first shock. In this case the first shock is essentially adiabatic, and the bubble fills with gas at a temperature characteristic of the shock velocity. Work is done on the shell by the pressure of the gas in the interior of the bubble, and there is no longer any requirement that momentum flux be conserved – such flows are said to be energy driven. There is clearly a close analogy between flows driven by radiation pressure acting on the dust photosphere and those driven by thermal pressure acting on the cold shell around a hot bubble.

The equations governing spherically symmetric flows are given, in slightly different forms and with slightly different scaling constants, by Königl (1982) and Dyson (1984) (see also Kahn 1987). For momentum driven flows the equation relating the shell velocity v_{sh} to the wind velocity v_w is simply

$$\dot{M}_w v_w = \dot{M}_{sh} v_{sh} \tag{10.1}$$

where $\dot{M}_w$ and $\dot{M}_{sh}$ are the mass loss rates observed in the wind and the shell respectively. Given the observational evidence that the observed cold molecular flows, which we identify with the shell in this model, have momentum fluxes greater than those seen in stellar winds, then true momentum driven flows would appear to be precluded.

For wind velocities v_w greater than some v_{crit}, cooling of the wind becomes negligible over the dynamical time scale of the bubble, and the flow becomes energy conserving:

$$v_{crit} = 230 \left\{ \left(\frac{n_{ext}}{10^9 \text{ m}^{-3}} \right) \left(\frac{\dot{M}_w}{10^{-6} M_\odot \text{ yr}^{-1}} \right) \right\}^{1/9} \text{ km s}^{-1} \tag{10.2}$$

where n_{ext} is the number density of the external medium (Dyson 1984) (a somewhat larger value is required if the swept up material mixes in with and cools the wind).

Dyson (1984) points out that it is wrong to convert the observed mechanical luminosity and momentum flux in a flow directly into those of the presumed driving wind. If the flow is momentum driven most of the energy has been radiated away, and the *original* thermal energy of the flow (hence the mechanical luminosity of the wind) must also be included in any energy balance; if the flow is energy driven, and has velocity v_f, then the momentum flux required of the wind is decreased by a factor of approximately $0.1 v_w/v_f$. That is, for momentum driven flow

$$E = 2E_{CO} v_w / v_f; \qquad p = 2p_{CO}. \tag{10.3}$$

For energy driven flow:

$$E = (77/9)E_{\rm CO}; \qquad p = (77/9)p_{\rm CO}v_{\rm f}/v_{\rm w}, \qquad (10.4)$$

where E and p are the energy and momentum of the presumed wind, and $E_{\rm CO}$ and $p_{\rm CO}$ are those of the observed molecular flow. The numerical factors in equations (10.3) and (10.4) arise from luminosity-independent efficiency factors and time scales – when these are taken into account we see that the mechanical luminosity of the flow cannot exceed about 10% of that in the wind, even in the limit $v_{\rm w}/v_{\rm f} \gg 1$.

There is a tendency for energy-driven flows to be seen as a panacea for the momentum problem. As shown by equation (10.4), the momentum gain of an energy driven flow is limited by cooling (either by radiation or due to the work done on the shell), in the same way as for radiative drives. Essentially for a momentum gain of g we require both that the fractional cooling of the bubble is less than $1/g$ in one dynamical time scale (which we here set equal to the sound crossing time of the bubble), and that the ratio of wind velocity to flow velocity be greater than g. Very large momentum gains will clearly be difficult to achieve in such a system. For example, using Kahn's (1976) empirical cooling law for the gas in the bubble, a bubble size of 0.01 pc and a shock speed of 300 km s^{-1}, we find that the hot gas radiates approximately one quarter of its energy per crossing time, which of itself will limit the momentum gain to about 4.

10.2.4.2.2 **Can stellar winds power the flows?**

Is it possible to drive the observed outflows with stellar winds of the type observed from main sequence stars? There is evidence that many O and B stars have stellar winds, with velocities of order 1000 – 2000 km s^{-1} and mass loss rates of up to $10^{-6}M_\odot$ yr^{-1}. These values are deduced from optical observations, and therefore apply to stars which are *not* deeply embedded. Further evidence for winds from embedded young objects comes from the radio observations of H^0 regions surrounding the stars, which show $\nu^{+0.6}$ spectra typical of uniform velocity outflow (Wright & Barlow 1975; Panagia & Felli 1975), and from observations which show that the velocity width of infrared recombination lines from the same sources is in excess of 100 km s^{-1} (Simon *et al.* 1979).

The mass loss rate in the stellar wind is found to be correlated with the luminosity L (*e.g.*, Van Buren 1985), such that the best fit is given by:

$$\dot{M} = 2\times10^{-13} \left(\frac{L}{L_\odot}\right)^{1.25} M_\odot \text{ yr}^{-1}. \tag{10.5}$$

Thus for a B0 star with $M = 17M_\odot$ and $L = 1.3\times10^4 L_\odot$, we find a typical mass loss rate of $3\times10^{-8}M_\odot$ yr^{-1}. Assuming a velocity of 1000 km s^{-1} for this wind, the energy and momentum fluxes would be $2.5L_\odot = 2\times10^{-4}L_*$ and $1500L_\odot/c = 0.1L_*/c$ respectively. Thus such a wind is *energetically* incapable of driving the observed CO flows. For lower mass stars the discrepancy is greater – *e.g.*, for a $30L_\odot$ star such as L 1551 IRs5, Van Buren's correlation yields $4\times10^{-5}L_*$ and $0.02L_*/c$ for the mechanical luminosity and momentum flux respectively. Thus it is clear that the CO outflows are not driven by "ordinary" stellar winds.

T Tauri star winds are apparently relatively more energetic. From a study of Hα, Hβ, Hδ and Ca$^+$ K line profiles, Kuhi (1964) found that a "typical" T Tauri star has $L = 5L_\odot$, $\dot{M} = 3\times10^{-8}M_\odot$ yr^{-1} and v_w = 225 – 325 km s^{-1}. The mechanical luminosity and momentum flux would then be $0.02L_*$ and $60L_*/c$. DeCampli (1981) points out, however, that these values may be overestimated by a factor of up to 1000, and that it is extremely difficult to convert stellar luminosity into the mechanical luminosity of the stellar wind with efficiencies > 0.1.

In fact observations of the radio spectra suggest that the mass loss rates from T Tauri stars are often overestimated. At higher resolution the wind is often seen to be highly anisotropic, and the deduced mass loss rate then decreases – recent work (Bertout 1987) suggests that the *ionized* mass loss rates in extended non-spherical sources may be as low as a few $10^{-10}M_\odot$ yr^{-1}, while Mundt *et al.* (1987) find a typical value of $2\times10^{-9}M_\odot$ yr^{-1} for a large sample of optical jets. This suggests either that the jets contain only a small fraction of the mass expelled from the star in the wind, or that they are largely neutral (there is some independent evidence for this latter hypothesis; see Snell *et al.* 1984; Snell *et al.* 1985; Russell 1987; and references therein). Unless the jets *are* mostly neutral, and therefore have mass fluxes much greater than those estimated above, they *cannot* be the driving sources for the CO outflows (cf. Mundt *et al.* 1987) because of the large momentum gain required.

Paradoxically, as the wind mass loss rate *falls* it becomes more likely that the flow will be energy driven, offering the prospect of significant momentum gain. Inverting equation (10.2) for a wind with $\dot{M}_w = 3\times10^{-8} M_\odot$ yr^{-1}, we find that the critical density of the external cloud is $\sim 3.5\times10^{12}$ m^{-3} for $v_w = 300$ km s^{-1}. For densities less than this critical value the wind is supercritical and the flow will be energy driven. This value is probably comparable to the density in the collimation region for the outflows, so we expect that most flows will initially be energy driven because of the low mass loss rate (that is, the density in the wind is low, so the cooling time is correspondingly long). Even with this factor taken into account, however, it is clear that there are severe difficulties with wind-driven flows.

Infrared recombination lines from the central source have typical widths of only 100 km s^{-1} (*e.g.*, Simon *et al.* 1979), which are much less than those arising in winds from unobscured stars. This appears to pose a further problem for wind driven flows. Dyson (1984) suggests that the Brα lines may be formed in a region where the wind is still being accelerated, thus removing this particular objection.

10.2.4.2.3 Collimation mechanisms

The uncertainties in the mass-loss rates and energy fluxes are such that we certainly cannot rule out hydrodynamic drives on these grounds alone. How could an initially isotropic stellar wind be collimated into a pair of oppositely directed "beams"? Cantó *et al.* (1981) suggested that winds from young stars can be focussed and accelerated in "interstellar nozzles" formed in interstellar toroids around the stars, while Blandford & Rees (1974) proposed an essentially similar model to explain the jets from double radio sources.

If a wind-blown bubble forms in a region of anisotropic external density gradient, it will elongate along the direction of $-\nabla n_{H_2}$. Königl (1982) describes a number of mechanisms which might then lead to the formation of deLaval nozzles and hence bipolar flows, under the condition that the density gradient is at least as steep as r^{-2}. These include the possibilities that the accelerating shell is Rayleigh-Taylor unstable (Weaver *et al.* 1977), that a rarefaction wave is generated when the shell speed approaches the sound speed in the bubble interior (Bodenheimer, Tenorio-Tagle & Yorke 1979), and that even a decelerating shell can be Rayleigh-Taylor unstable in the presence of buoyancy forces on the bubble in the gravitational field of the star (Gull & Northover 1973). However such a nozzle is finally established, it appears that there are serious problems with the stability of the subsonic section (Blandford & Rees 1974; Dyson 1987), so apparently we do not yet have the full story on these models.

Whatever the mechanism, if the observed discs are responsible for collimating the outflows, they should have a present day total outwardly directed momentum equal to the force exerted on them by the flows integrated over their lifetimes. Measurements of the interstellar molecular discs around three outflows find that the momentum in the disc is too small by an order of magnitude for the disc to have been responsible for collimating the flow (*e.g.*, L 1551: Davidson & Jaffe 1984; GL 490: Kawabe *et al.* 1984; NGC 2071: Takano *et al.* 1984). Kawabe *et al.* and Takano *et al.* argue that collimation must therefore take place on much smaller (circumstellar) scales. Bally (1986) makes a similar argument for the disc in Mon R2.

An upper limit can be found for the size of the collimating disc, on the assumption that gravitational binding energy supplies the force required to balance the outward momentum deposited by the flow (Davidson & Jaffe 1984). For the observed velocity of the disc material $\lesssim 3$ km s^{-1} in L 1551, and a mass equal to the stellar mass plus the deduced disc mass, they find that collimation must take place at a radius of $< 7.5\times10^{13}$ m (500 AU), which is consistent with the nozzle sizes required to collimate a 200 km s^{-1} wind with mechanical luminosity of $0.02L_\odot$ (Königl 1982). Rodriguez (1987) points out that projection effects and potential errors in the measurements of the disc mass may reduce the discrepancy between the measured momentum and that which would be expected if the large (0.1 pc) discs were responsible for the collimation, but it appears that in all but very few cases the collimation region is constrained to be much smaller than the interstellar toroids. If the jets and outflows are presumed to be closely related, then the existence of visible jets in some HH systems provides further evidence against the interstellar toroids as the focussing mechanism.

10.2.4.3 Magnetically mediated models

Very strong evidence that indeed bipolar outflows *are* connected with magnetic fields comes from the observations made by Reipurth of six or more collimated optical flows in the HH 1-2 region, all sharing a common position angle and lying *perpendicular* to linear emission sheets visible on deep Schmidt plates of the region (Reipurth 1989). The flows are aligned parallel to, and the sheets perpendicular to, the local magnetic field direction, as determined by the polarization vectors of nearby stars. It is, however, still possible that the orientation of the bipolar flows is determined by their location within sheets of material (which in turn are formed as a result of magnetic support in the transverse direction), rather than directly by the magnetic field in the clouds.

In this section we will ignore those models where magnetic fields are responsible for stellar winds, which then power the outflows. This leaves the larger class of models wherein magnetic forces are responsible for accelerating and channelling the flow directly. We could divide the extant models up into groups, according to whether the winds originate on the stellar surface, in the boundary layer or on an accretion disc. Here we take an alternative viewpoint in which the principal distinction is between winds driven by magnetic pressure and those which are centrifugally accelerated.

10.2.4.3.1 Centrifugally driven wind models

The suggestion that centrifugally driven winds were responsible for the observed bipolar flows from young stars can be attributed to Hartmann & MacGregor (1982). Their model envisaged that material would stream off a contracting protostar of radius $\sim 10^{12}$ m along radial field lines rooted in the star. Most of the mass loss was supposed to occur in the equatorial plane, but to be diverted towards the poles by a meridional magnetic pressure gradient in B_ϕ.

A variation on this model (Pudritz & Norman 1983, 1985) is derived from that of Blandford & Payne (1982) for winds from accretion discs around black holes. It assumes that instead of coming directly from the surface of the protostar, the wind instead originates on a centrifugally supported equilibrium disc, with a primarily poloidal field (here as elsewhere in MHD theory the field is decomposed into two parts – a pure azimuthal or toroidal field and a poloidal field consisting of axial and radial components). Gas driven off the disc surface by heating (*e.g.*, by ion-neutral slip, or by direct heating from the star) flows along the field lines away from the disc. The field lines in the disc are carried toward the star in the mid-plane as material accretes through the disc, but gently spiral out to their original distance from the rotation axis further from the star in the axial direction.

At large distances from the disc and from the flow axis, outside the Alfvén surface, the toroidal field (due to the wrapping up of the field lines) becomes very large, and the resulting magnetic pressure directed *towards* the axis exceeds any thermal pressure of the material in the flow (Blandford & Payne 1982). This magnetic stress provides both confinement and collimation for the flow.

More recently, Shu and his collaborators (Shu *et al.* 1988) have proposed a centrifugal wind model to explain the apparently different behaviours of the cold molecular outflow and of the fast optical jets. In their model, the cold atomic and molecular flow originates near the Lagrangian points of a protostar (which has been spun up close to breakup velocity by the transfer of angular momentum from an adjoining accretion disc; *i.e.*, one that fills its Roche lobe) while the ionized optical jet originates as a normal stellar wind near the poles of the protostar. Thus accretion occurs in the mid-plane and mass leaves the protostar mainly at slightly higher latitudes, along the lines of a locally quadrupolar magnetic field – that is, the field lines connect the protostellar equator to the poles. At higher latitudes still, matter is prevented from leaving the stellar surface by the predominantly meridional field, while at the poles material can again escape along the field lines as a normal stellar wind.

The cold matter outflowing at the equator makes a sonic transition near to the so-called X-point of zero effective gravity, or Lagrangian point L1, and is centrifugally accelerated out to the Alfvén surface, where it assumes an essentially ballistic trajectory. Thus in this model, as in that of Hartmann & MacGregor, the surrounding "toroid" may assume the role of deflector, loosely collimating the molecular flow, although there may also be some focussing toward the poles produced by a pressure gradient ∇B_ϕ^2. This model explains both the predominantly atomic wind seen in HH 7-11 by Lizano *et al.* (1988), and the apparent coexistence of a highly collimated ionized jet with a much less well collimated molecular and atomic flow in sources like L 1551. There are still many assumptions in this model however, not the least of which is the exact form of the magnetic field in the vicinity of the protostellar surface, while the source of viscosity – which allows *some* accretion to occur through the disc – is not addressed at all.

A similar two-component model is deduced by Kwan & Tademaru (1988) for T Tauri stars, based on [O^0] profiles, although in this case the magnetic field (presumed to be produced by ring currents in a circumstellar disc) is a collimating agent only – the jet is supposed otherwise to be a normal T Tau stellar wind.

10.2.4.3.2 **Winds driven by magnetic pressure**

The other major group of theories are derived from the work of Draine (1983) on "magnetic bubbles", which supposes that a rotating protostar magnetically coupled to its parent cloud winds up a strong toroidal field. Magnetic stresses then cause the region of enhanced field to expand away from the star essentially at Alfvénic speeds. The original model supposes that mass loss from the central object is negligible and that the bubble produced in this way is intrinsically spherical, but that collimation into a bipolar flow can occur if the bubble is formed in a region of anisotropic density gradient.

Uchida & Shibata (1985) envisage that a frozen-in field parallel to a disc axis is "wrapped up" by rotation. The disc is supposed to be in radial free fall, and the field lines are then wrapped up at the Kepler velocity, establishing a large toroidal field. The magnetic stress ∇B^2 is then essentially axially-directed, and drives a wind off the disc surface.

Pringle (1989) proposes that the strong toroidal field arises instead in the boundary layer between accretion disc and protostar. The disc is supposed to be magnetically dominated, with viscosity arising from "turbulence" set up by dynamo action. Thus magnetic loops connecting parts of the disc at different radii continually transfer angular momentum outwards through the disc, rather than into the wind as in the Uchida & Shibata models. The effect of the dynamo cycle is to maintain the field to be essentially poloidal in the disc. In the boundary layer, however, shearing takes place at a much greater rate, and the toroidal component of the field is amplified until it can diffuse out of the boundary layer region.

Regions of strong field are expected then to decouple from the gas through operation of the Parker instability, so that the Alfvén velocity rapidly becomes greater than the escape speed, and a high velocity wind can start to blow in the direction of the symmetry axis, taking with it some of the material heated in the boundary layer. Pringle points out that operation of this mechanism depends on the boundary layer being able to cool effectively, so that magnetic stresses *can* dominate.

Both for winds driven centrifugally and for those driven by magnetic pressure, the energy to power the flow comes directly from the gravitational binding energy of the disc or star. Flows are easy to start because they originate high up in the potential well (discs have low surface gravity), but the terminal velocity reflects the gravitational potential at the point where they originate – that is, the highest velocities will be seen in the centre of the flow. The star being formed in the centre of the disc is assumed to derive a significant fraction of its luminosity from accretion, which then results in a direct relationship between the luminosity of the star and the total energy and momentum flux in the outflow.

The centrifugal wind models, and in particular the disc-wind model of Pudritz & Norman, have the great appeal that they provide a very simple and direct method for transferring angular momentum out of the system, without having to invoke poorly understood viscosity mechanisms. In the case of MHD disc winds, the magnetic field lines which mediate this process are rooted in the disc and co-rotate with it out to the Alfvén radius r_A. The transfer of angular momentum outwards in the wind allows accretion through the disc onto the central object, and therefore directly resolves the "angular momentum problem": Perhaps fortuitously the energies and momenta calculated for centrifugally driven winds fall within the regions defined observationally for cold molecular outflows by Bally & Lada (1983) and Lada (1985). Thus, without being able to rule out winds driven by magnetic pressure at this stage, we now discuss centrifugally driven winds in more detail.

Lest we appear to endorse one particular theory too strongly however, we should point out that at this stage there have been no self-consistent calculations of the structure of any of these models.

The most complete of the theories to date – the centrifugally driven wind model of Pudritz & Norman – is derived from the similarity solution of Blandford & Payne, which has no real basis other than one of aesthetics and computational convenience. Although Königl (1989) has recently presented an improved self-similar model for MHD winds, we await with interest a full time-dependent numerical solution of the 3-dimensional MHD flow! Evidence is mounting (Kwan & Tademaru 1988; Shu *et al.* 1988; Lizano *et al.* 1988) that winds from YSOs may indeed have more than one component, and it is conceivable that for instance they involve a fast stellar jet *and* a disc wind of some sort.

10.2.4.3.3 **Theory of centrifugally driven winds**

The theoretical underpinnings of centrifugally driven winds are somewhat better established at this time than those for winds driven directly by magnetic pressure. Although the main variants on this theme differ considerably in detail, the basic mechanism is the same in all cases. The following summary draws heavily on the material in Pudritz (1988).

Although the accretion flow $\dot{M}(r)$ at any radius r is completely dependent on the wind, the disc rotation law $\Omega(r)$ and flux distribution $\Phi(r)$ are not well constrained by the models, and we therefore input these as independent parameters. Then the spin-down torque exerted on the disc depends only on the mass loss rate and rotation speed at the base of the field lines (*i.e.*, at $r = r_0$), and the Alfvén radius $r_{\rm A}$, where $r_{\rm A}$ represents the lever arm for acceleration, *i.e.*,

$$T_{\rm w} \approx \dot{M}_{\rm w}\Omega r_{\rm A}^2. \tag{10.6}$$

For a particle in a Keplerian orbit the rate of change of total energy with angular momentum is just $dE/d|\mathbf{L}| = \Omega$ (*i.e.*, the work done on the disc is $T_{\rm w}\Omega$). If vertical shear is dominant (so that radial shear – *i.e.*, viscous forces – can be neglected) then the wind must simultaneously provide removal of both angular momentum and gravitational binding energy to allow accretion to occur. Then the accretion flow is related to the mass loss rate by

$$\dot{M}_{\rm a}r^2 \approx \dot{M}_{\rm w}r_{\rm A}^2(r). \tag{10.7}$$

We can relate the energy and momentum to the luminosity of the central object if we assume that it generates most of its luminosity through accretion. The energy liberated in the accretion shock is just $L_* = \epsilon G\dot{M}_a M_{core}/R_{core}$, where ϵ is an efficiency term which we can take to be of the order of 0.1. Pudritz (1988) assumes that to a first approximation *all* the infrared luminosity is due to accretion onto a fully convective Hayashi core, of radius $R_{core}/R_\odot = 51.2 M_{core}/M_\odot$. Thus for this case we have

$$L_* = 6.3\epsilon \left(\frac{\dot{M}_a}{10^{-5} M_\odot \ \mathrm{yr}^{-1}} \right) L_\odot. \tag{10.8}$$

The actual luminosity from accretion depends on the evolutionary track of the central object. Larson (1969) argues that we can expect that the central object will, in the absence of rotation, undergo an extremely non-homologous "inside-out" collapse onto a core which starts to form almost immediately. The practical result is that the core radius is always less than the Hayashi radius for an $n = 3/2$ polytrope of the same mass, and the star is born much lower down the Hayashi track than would otherwise be expected. The fractional luminosity derived from accretion is therefore increased, both by reducing the core luminosity and by allowing accretion deeper in the potential well of the star.

The wind thrust and power are given in terms of the Hayashi core luminosity by

$$\frac{\dot{M}_w v_\infty}{(L_*/c)} = \frac{16}{\epsilon} \left(\frac{R_d/r_A}{0.1} \right) \left(\frac{v_\phi}{2 \ \mathrm{km \ s^{-1}}} \right) \tag{10.9}$$

and

$$\frac{L_w}{L_*} = \frac{0.077}{\epsilon} \left(\frac{v_\phi}{2 \ \mathrm{km \ s^{-1}}} \right)^2, \tag{10.10}$$

where R_d is the overall diameter of the disc. If we adopt the (very plausible) value for the mechanical advantage of $r_A/r_0 = 10$ and assume reasonable mass distributions within the disc, then these have typical values of 200 and 0.1 respectively, in line with the values found by Bally & Lada (1983).

The terminal velocity, v_∞, can be derived simply for the case where ions and neutrals are strongly coupled by collisions (Pudritz & Norman 1985). For a purely poloidal field the final velocity of gas which starts at rest and is accelerated centrifugally is just $\sqrt{2}\, r_A(r_0)\Omega(r_0)$, where we assume that $r_A \gg r_0$. For a fixed Alfvén radius r_A, and a massive disc with a uniform rotation curve, a change by a factor of 1000 in r gives a change of 100 in v_∞, while for a light disc (where most of the mass is in the central object) and Keplerian rotation, the same change gives a factor of 10 in v_∞.

The specific kinetic energy of the wind is just $\Omega(r_0)r_A^2$, which is $(r_A/r_0)^2$ times the gravitational potential at r_0. The increase in wind energy over the gravitational potential at its footpoint is possible because only a fraction $(r_0/r_A)^2$ of the disc gas is actually being expelled from the system in the wind.

For accretion flows driven by MHD disc winds the angular momentum transport can be very rapid, and accretion rates may then approach the free-fall rates of order $1\times10^{-5}M_\odot$ yr^{-1} predicted by Larson (1969) (see also Shu, Adams & Lizano (1987) for a recent review). The mass loss rates predicted for the wind scale as $(r_0/r_A)^2$ times the accretion rate, giving $\dot{M}_w \sim 10^{-9}M_\odot$ yr^{-1} for typical values of $r_A/r_0 = 10$. It is therefore necessary to assume that the bulk of the observed material in the cold molecular wind is swept up material rather than material which has been blown off the disc. This assumption is borne out by CO observations of L 1551 made by Snell & Schloerb (1985), who find that the high velocity gas observed in the outflow has a mass of $\sim 1M_\odot$, essentially equal to the apparent volume of the lobes multiplied by the external density. This is indeed what we would expect for a flow starting from rest in a uniform medium, where the material leaving the disc would catch up with, and compress and entrain, the gas starting further up the streamlines.

The disc wind models described here appear to provide theoretical answers to questions raised by many of the observations discussed earlier in this section – viz, the correlation of wind powers and thrusts with the luminosity of the central source (including the approximate magnitudes), and the solution of the angular momentum problem. The range of v_∞ for a realistic range of disc radii agrees rather well with the range from the lowest observed velocities in the cold molecular flows to those in the hot collimated jets, which as we have noted may coexist in a single source at a single epoch. The kinematic structures shown in L 1551, and in particular the apparent acceleration of the wind at large distances from the source (up to ~ 0.1 pc) can also be understood if the acceleration is mediated by magnetic fields. Finally, this model predicts that

v_∞ will not depend on L_*. This seems to be the case for the collimated jets, but is less clear for the cold molecular flows, although molecular observations may not yet be sensitive enough to detect very small amounts of material at high velocities, whilst entrainment of ambient medium may significantly alter the appearance of the outflows until the cavity has been cleared.

10.2.5 **Numerical models**

The kinematics and structure of the jets and flows have been discussed at length in §10.2.2. The HH 34 system, shown in Fig. 10.1, and the L 1551 molecular outflow shown in Figs. 10.2 and 10.3, are classical examples of their type, and in view of recent advances in 2-D and 3-D numerical hydrodynamics would appear to be ideal candidates for modelling, as discussed in Chapter 7 of this monograph.

We can characterize the idealised astrophysical jet by just two parameters – its Mach number, $\mathcal{M}$, and the ratio of the density of jet material to that of ambient material. We can additionally define some pressure gradient in the external medium if we so desire. In §10.2.2 we derived a Mach number of between 6 and 20 for the HH 34 optical jet (on the assumption that it was freely expanding), with an *initial* density ratio of about 300 if we assume that the original ambient cloud had a density $\sim 10^{10}$ m^{-3}. We have used the Williams & Gull (1984, 1985) model to calculate, for three times during its evolution, the pressure and density of a $\mathcal{M} = 5$, $\rho_j/\rho_{\rm ext} = 0.01$ jet, which offers a subjective best fit to the HH 34 system (Fig. 10.5). Strong forbidden line emission will arise in weak shocks, which show up in the figure as regions of high pressure and moderate to high density. Thus for the model shown here we expect to see the strongest line emission from the knots in the jet and the bow shock.

The simulations were devised for extragalactic sources, and do not as yet include any cooling, so the density ratio in the shocks is limited to the adiabatic strong shock value of 4. Without cooling, any prediction of the emission is bound to be only very approximate. Nonetheless the superficial similarity of the models to the "real world" seems very encouraging for our future understanding of these systems. We note that with this one minor caveat the simulations are applicable to both stellar jets and extragalactic jets,

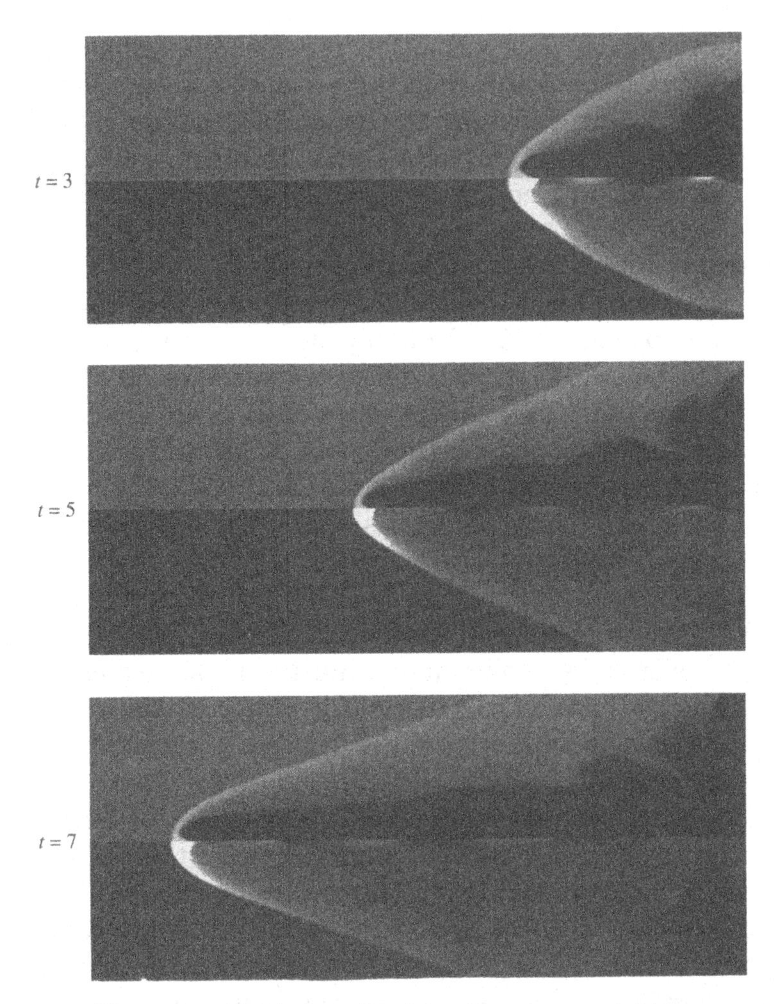

Fig. 10.5. Evolution of a $\mathcal{M} = 5$, $\rho_j/\rho_{\rm ext} = 0.01$ jet propagating into a uniform density external medium, shown at three times. The right half of each panel shows density, and the left half pressure. Simulations performed by C. I. Cox at M. R. A. O.

reinforcing the view derived from their morphologies that they may have other features in common.

Two dimensional simulations *have* now been carried out on radiative shocks in the context of Galactic jets (Blondin, Königl & Fryxell (1989), which produced some evidence that "clumpy" HH objects like HH 1 may arise as Rayleigh Taylor instabilities at the heads of overdense ("heavy") jets. This is perhaps not surprising given the huge range of densities in the molecular clouds surrounding the outflow sources: given the difficulty of measuring the jet density directly it is obviously now important that an effort be made to determine the densities of the *ambient* material in the vicinity of HH objects of both clumpy and "knotty" types, to see if the mass loss rates implied are physically reasonable.

10.2.6 Mass outflow from YSOs: Conclusions

Is it possible to find a coherent description which embraces all the observed phenomena? Probably not, but an attempt at one is presented below. Flattened dense regions form in rotating molecular clouds as material collapses preferentially along magnetic field lines. As the cloud contracts and cools, it passes through the last available isothermal equilibrium state, and starts to collapse inside out. The infalling material carries with it angular momentum and magnetic field, and both the accretion core and the forming disc spin up and start to blow off an MHD wind. At first the low-density high-Mach number wind from near the surface of the star is choked. The slower less collimated wind from the disc entrains surrounding material and starts to clear a cavity in the ambient cloud (it may also help support the disc against collapse into the midplane at larger radii).

The outflows in the lowest mass clouds (and hence lowest luminosity protostars) may never become supersonic, and these outflows will thus tend to be short and fat (analogous to the Fanaroff-Riley class I radio sources), with the flow being decelerated immediately by back pressure from the working surface. The outflow in L 1262 may be an example of this. On the other hand, the higher luminosity sources *are* associated with highly supersonic outflows, which eventually clear relatively long cylindrical cavities confined by both the external pressure and by the toroidal magnetic field in the wind. Examples of this latter class are NGC 2024 and possibly NGC 2071.

As the cavity is cleared the ram pressure required of the high velocity "jet" decreases and the hotspot starts to advance. The jet itself may be collimated either by hydrodynamic or magnetic forces or by a combination of the two. Eventually the action of the disc wind and of the backflow from the hotspot succeeds in clearing out a large cavity. The molecular material originally entrained in the wind is now confined in a thin layer at the edge of the cavity by centrifugal force and the pressure of the jet waste products, but is still being accelerated up the field lines (*e.g.*, L 1551). The visual extinction to the central source is now quite low, and both the blue-shifted optical jet and bowshock become visible. At the same time the stellar mass starts to dominate the dynamics of the system and the disc rotation becomes Keplerian at relatively high velocities (HH 34; HL Tau; S 106?). Finally the disc mass and density become so small that the remaining material is dispelled through ablation by a stellar wind and radiation flux, and the magnetic field can relax back toward the original interstellar configuration. We are then left with an isolated T Tauri star with perhaps a thin remnant accretion disc. Most of the original angular momentum of the system has been carried off by the disc wind, which has also dispersed the placental cloud.

10.3 **SS 433**

SS 433 is a complicated and fascinating object with a huge literature. Here we cannot cover all the available observations and models in detail, but concentrate on summarizing the consensus of the observational and theoretical studies (see Margon (1984) and Katz (1986) for more detailed expositions of the conventional view).

Observations at different frequencies (principally in the optical and radio) tell us about SS 433 on different physical scales in a complementary manner, and they all support the "kinematic" model (Milgrom 1979a; Abell & Margon 1979), namely that SS 433 ejects material which moves ballistically at a speed of $\sim$ 0.26 c, along two oppositely directed, well-collimated jets with opening angles of a few degrees at most. These jets sweep out a cone of half angle $\sim 20°$ with a period of $\sim$ 163 days, with the axis inclined at $\sim 80°$ to the line of sight. More detailed studies reveal $\sim$ 6.5-day periodic variations in the motions of the jets, which is easily interpreted if SS 433 is a binary system with a period of $\sim$ 13 days. Almost universally (but see Sharp, Calvani & Turolla (1984) and Kundt (1985, 1987a) for dissenting views) the system is interpreted either with the simple "kinematic" or the enhanced "dynamical" model (where the motions of the "kinematic" model are explained in terms of the precession and nutation of a component of a binary system), with accretion being generally acknowledged as the power source for the jets. Various observations, in particular the fitting of high-resolution radio-maps with the "kinematic" model, imply a distance of ~ 5 kpc for SS 433.

SS 433 is surrounded by W 50, a large non-thermal source that is usually classified as a supernova remnant (SNR), and there has been considerable discussion about whether or not there are similar objects producing jets in the Galaxy. At present, however, there is little evidence (see §10.4) that there are any other objects that closely resemble SS 433, so it remains a unique, relatively easy to study object that may shed light on many of the wider questions of accretion and jet production, collimation and propagation.

10.3.1 SS 433: Optical observations

SS 433 refers to the unusual optical stellar object at 19^h09^m 21.28^s, $+04°53'54.0''$ (1950.0). Its spectrum has an underlying strongly reddened continuum, with both a "stationary" set of strong, broad emission lines (principally Balmer and He^0, red-shifted by ~ 70 km s^{-1}, the systemic velocity of the object), and a "moving" set of emission lines (Balmer and He^0 lines in the optical, or Paschen and Brackett lines in the infrared) which are strongly red- and blue-shifted by amounts that are periodic on a time scale of ~ 163 days (Fig. 10.6). The "moving" system of lines have velocities ranging from $\sim 50,000$ km s^{-1} in the red to $\sim 30,000$ km s^{-1} in the blue, with an average of $\sim 12,000$ km s^{-1}, often with mirror-image profiles. The observed line velocities can be readily explained with the "kinematic" model outlined above, with the best-fit giving (*e.g.*, Margon 1984) a precessional period of ~ 163 days, a jet velocity of ~ 0.26 c, with one or other of the cone half-angle and the inclination of the rotation axis to the line of sight being $\sim 20°$ and the other $\sim 80°$. The ambiguity between the two angles that characterize the "kinematic" model cannot be resolved by the optical observations alone, but mapping at radio wavelengths (see below) directly resolves this problem. The apparently discrepant average velocity of the "moving" lines compared with the systemic velocity of the object inferred from the "stationary" lines is due to the fact that the large ejection velocity gives rise to an appreciable relativistic transverse, ("time-dilation") Doppler red-shift to the lines. The underlying continuum emission from SS 433 is highly reddened, and correcting for this strong absorption implies that the non-compact

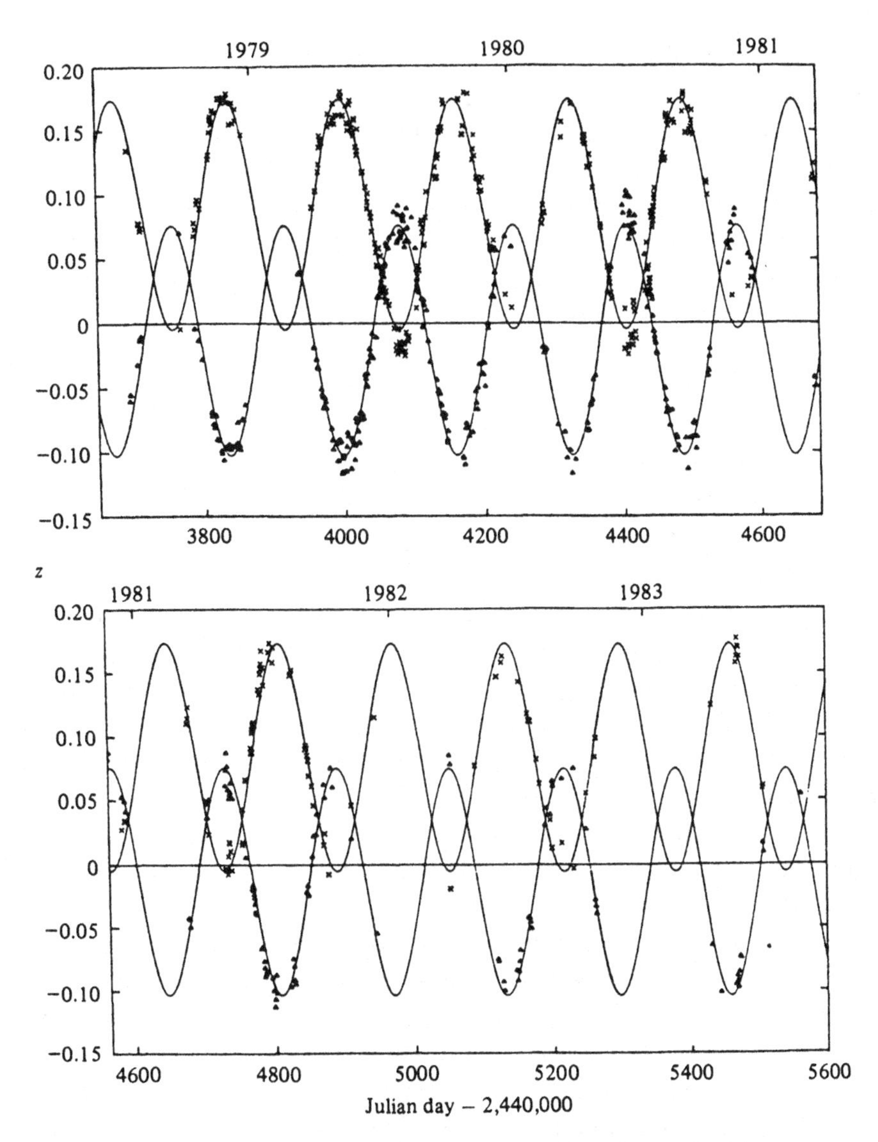

Fig. 10.6. Variation of red- and blue-shifted line velocities over several 163-day cycles showing the good fit obtained with the "kinematic" model, Figure 1 from Margon (1984).

star in the SS 433 system is intrinsically a very luminous, massive early-type star.

As well as the 163-day periodicity in the "moving" line velocities there is ample evidence for shorter period, $\sim$ 6.5 day, variations. These are readily explained by the jet-emitting object being one member of a binary system with an orbital period of ~ 13 days: there is a time variable torque on the precessing object from its companion at half the orbital period, which produces the nutational or "nodding" motion that is seen as the ~ 6.5-day periodic variations in the velocities of the "moving" lines. Both the 163- and 6.5-day periods are also apparent in photometric data of SS 433 (*e.g.*, Mazeh *et al.* 1987) as is the 13-day orbital period in the "stationary" line system. This forced precessional or "dynamical" model provides a better fit (Collins & Newsom 1987) to the observed line velocities than the simpler "kinematic" model, and requires that the object to which the jets are effectively tied is nearly spherical, most probably the blue supergiant thought to be a member of the system (see below). There is also evidence in the data for a large scale (several year) drift in the precessional period which is not at present understood.

Finally, there are residual differences in the observed and model velocities of several thousand km s^{-1} which vary on the time scale of days. These variations – "jitter" – do not vary with either the 163- or the 6.5-day periodicities (Margon 1981) and are basically anti-correlated between the red- and blue-shifted systems (Katz & Piran 1982), which implies that variations of the pointing direction of the jets, rather than variations in jet speed, are responsible. The inferred pointing variations are $\sim 2°$, which is comparable to the opening angle of the jets.

Margon *et al.* (1984) show that both the blue- and red-shifted moving line systems can disappear or reappear together from one day to the next. This implies an upper limit of $\sim 10^{13}$ m for the separation between the red- and blue-shifted emitting regions, which is comparable with the size of the emitting region implied from brightness temperature constraints (*e.g.*, Davidson & McCray 1980), and a somewhat model-dependent determination from Collins & Newsom's fitting of the "dynamical" model to "moving" line velocities.

The mass function of the binary system is $\sim 11M_\odot$ (Crampton & Hutchings 1981), and Leibowitz (1985) deduces that the system consists of an early type O or B supergiant of $20 - 50M_\odot$, and a compact object with a mass of $> 10M_\odot$ (*i.e.*, a good candidate black hole).

10.3.2 SS 433: Radio observations

SS 433 and its surroundings have been observed over a range of radio wavelengths using various instruments and techniques to look at, on various scales, the central-engine, the jets, and their possible effect on the surrounding interstellar medium. The central radio source has a non-thermal spectrum, and is highly variable on time scales from hours upwards (*e.g.*, Johnston *et al.* 1984), although the variations do not show the 163- and 13-day periodicities that are apparent at optical wavelengths. VLBI observations (Geldzahler, Downes & Shaffer 1981) imply that the bulk of the emission from the central source is from a region greater than ~ 5 milli-arcseconds, which corresponds to $\sim 4 \times 10^{12}$ m.

A significant fraction of the emission from the central source is produced on angular scales of up to a few arcseconds. This emission has a non-thermal spectrum and is highly polarized (implying synchrotron emission) and has been mapped at several wavelengths over at least one 163-day cycle with MERLIN, the European VLBI network and the VLA (see Spencer 1984; Romney *et al.* 1987; Vermeulen *et al.* 1987: and Gilmore & Seaquist 1980; Hjellming & Johnston 1981a respectively, and references therein). These studies show, on scales up to 3 arcsec (2×10^{15} m), the central source and a clumpy double "corkscrew" structure (Fig. 10.7) which varies on the time scale of days. The structures on these maps are well fitted by the ballistic flow of the "kinematic" model, and give an independent solution for the model that is in excellent agreement with that deduced from the optical data, but with the advantage that the ambiguity between the two angles in the model is resolved – the half-cone angle is $\sim 20°$ and the inclination to the line of sight is $\sim 80°$. Moreover, to get a consistent fit to the "kinematic" model it is necessary to take into account the light travel time across the source, which in turn effectively measures the distance of the source. The best-fit to the radio data provides a distance of 5 ± 0.5 kpc to SS 433 (Romney *et al.* 1987; see also Hjellming & Johnston 1981b), which is in agreement with various less accurate distance estimates available for SS 433 and W 50. The best-fit model of the radio data even tells us that rotation of the jets is right-handed.

On a much larger scale, radio observations (*e.g.*, Downes, Pauls & Salter 1986; Elston & Baum 1987) show that SS 433 is in the middle of W 50, an extended (120×60 arcmin^2) radio source (see

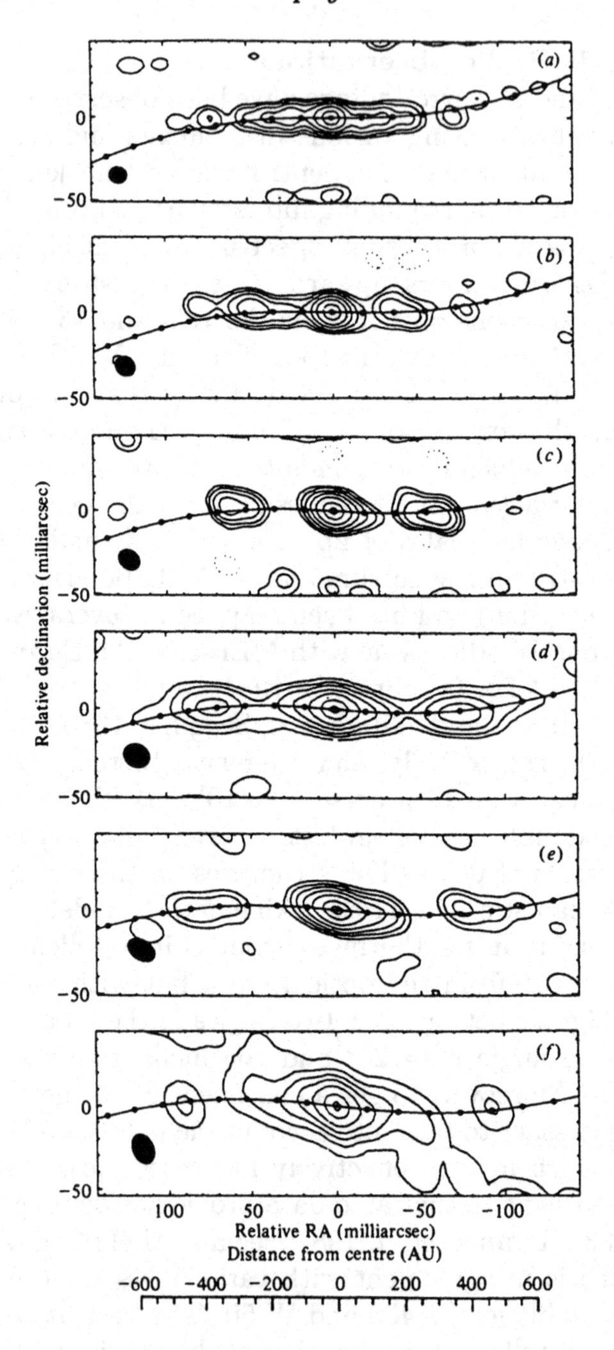

Fig. 10.7. VLBI radio images of the central ~ 0.2 arcsec of the "jets" made over a 10 day period; Figure 1 from Vermeulen *et al.* (1987). Reprinted by permission from *Nature*, **328**, 309. Copyright © 1987 Macmillan Magazines Limited.

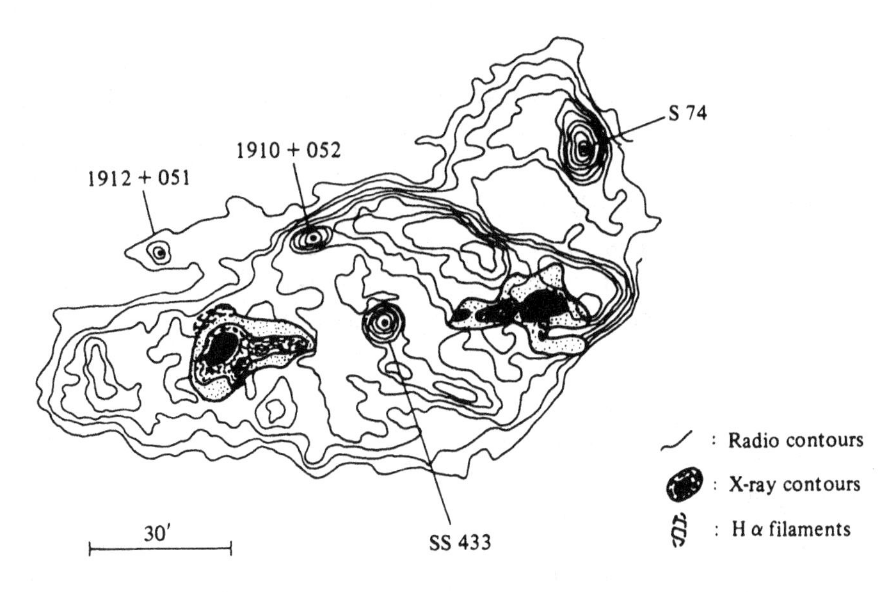

Fig. 10.8. Extent of X-ray lobes superposed on radio map of W 50 and SS 433; Figure 5 from Watson *et al.* (1983).

Fig. 10.8). The W 50 field is complicated and contains not only SS 433, but also several unrelated compact radio sources, and the foreground H^+ region S 74 which overlaps W 50 in the northwest. W 50 has generally been catalogued as a supernova remnant because of its non-thermal radio spectrum, its partially limb-brightened structure and its associated optical filaments which have the characteristic shock excited spectrum that is seen from SNRs. However, the classification of W 50 as a SNR may be incorrect, since the integrated energy loss from SS 433 could be sufficient to produce the shocks in the interstellar medium that constitute W 50, without the need for a supernova (SN) explosion (*e.g.*, Königl 1983). At the distance of 5 kpc that is inferred for SS 433, W 50 is $\sim 85 \times 170$ pc^2, which is considerably larger than any SNR with a comparable surface brightness (*e.g.*, Green 1984). It is not at present clear whether or not a parent SN for W 50 can be discarded completely, but the fact that W 50 is elongated east-west in alignment with the radio jets (and X-ray lobes, see below) associated with SS 433, implies that SS 433 has at least had an appreciable distorting effect on the morphology of W 50, even if it is not wholly responsible for it.

10.3.3 SS 433: X-ray and γ-ray observations

Einstein observations (Seward *et al.* 1980) of SS 433 and its surroundings revealed that most of the X-ray emission (at least in the 1-3 keV band) is from a compact source coincident with SS 433, with about a tenth of the total from "lobes" extending for about ± 40 arcmin east and west from SS 433, within W 50 (Fig. 10.8), aligned with its major axis. These "lobes" have a softer spectrum than that of emission from the central source (Watson *et al.* 1983), which, together with the strong iron line emission, favour a thermal rather than synchrotron emission mechanism. The X-ray emission is variable on time scales from hours upwards (Seaquist *et al.* 1982), with some correlation with radio variability, but only limited evidence for any 13-day periodicity. Watson *et al.* (1986) show that the strong iron X-ray line emission is Doppler shifted to different energies as expected by the 163-day periodicity in the "kinematic" model, which implies that this X-ray emission is from the jets rather than the central source. Variations in the intensity of the red- and blue-shifted iron lines also fit those expected due to obscuration by a large accretion disc, and set an upper limit of 10^9 m on the size of the X-ray emitting jets. The total X-ray luminosity is $\sim 10^{28}$ W, which is small compared with both the inferred kinetic energy input to the jets and the optical luminosity of SS 433.

Lamb *et al.* (1983) report one (or possibly two) γ-ray lines from SS 433 near 1 MeV, which, if isotropic, imply a luminosity of 2×10^{30} W, an appreciable power compared to the optical luminosity of the source, and the kinetic energy input for the jets. However, there are problems interpreting these lines, and their reality is brought into question by the failure of MacCullen *et al.* (1985) to confirm the observations.

10.3.4 SS 433: Underlying physics

Possibly the most significant parameter derived for the jets from SS 433 is the speed of 0.26 c, as it is very near (Milgrom 1979b) to the value required to Doppler shift the Lyman continuum wavelength limit to that of Lyman α, which is expected if "line-locking" is the acceleration mechanism (*e.g.*, Strittmatter & Williams 1976 for its application to quasars; Shapiro, Milgrom & Rees 1982). Material is accelerated by absorption of radiation from a luminous source that is cut off past the Lyman limit (presumably due to cool gas), until the velocity is such that the available continuum photons can no longer "line-lock" with Lyman α. This mechanism is favoured by the fact that the speed of the jets remains constant in spite of large changes in the line intensity (Margon 1981). On the other hand, there are several problems if "line-locking" is responsible for the jet production, notably the low upper limits to the mass loss rate and jet kinetic energy that can be produced, although Katz (1986) argues that these values ($\sim 2\times10^{13}$ kg s^{-1} and $\sim 6\times10^{28}$ W respectively) can be reconciled with observations (see also Shapiro, Milgrom & Rees 1986). Margon (1984) and Katz (1986) discuss in some detail the outstanding problems associated with the acceleration of the jets, their energetics, and the accretion dynamics of the SS 433 system.

The obvious outward similarity of the twin jets of SS 433 and those seen in many extragalactic sources has prompted discussion about whether the underlying physics of the collimation and acceleration of these jets is the same or not, even though the inferred masses of the central sources vary by $> 10^5$. Models of accretion onto a low mass black hole as a simulation of SS 433 are presented by Eggum, Coroniti & Katz (1985), who also discuss scaling the solutions to extragalactic jets (see also Rees 1982). Katz (1986), however, notes the differences between extragalactic jets and those in SS 433, and in particular that it is not clear how to reconcile the scaled thermal emission from SS 433's jets with observed (non-thermal) emission from extragalactic jets.

10.4 Jets in SNRs, non-thermal radio sources and X-ray binaries?

The usual classification of W 50 as a SNR has prompted several discussions of SNRs, in particular those containing compact sources, and of other Galactic non-thermal radio sources as possible analogues of W 50 and SS 433. At present, however, there are no strong candidates for any close relative to SS 433, and although the term "jet" has been applied in various contexts to several SNRs, there is little direct evidence for highly collimated features. These objects are discussed in this section, together with X-ray binaries for which indications of underlying jets have been found.

10.4.1 Non-thermal radio sources from accreting binaries?

Improved radio observations by Shaver *et al.* (1985a) and Becker & Helfand (1985) of G5.3−1.0 and G357.7−0.1, two non-thermal radio sources that have generally been catalogued as SNRs, revealed remarkable structures with bright, narrow curved "filaments", and a compact source near the edge of each source. These are features not generally seen in SNRs (although background compact sources are ever present, and some remnants do have striking radio filaments), which led Helfand & Becker (1985) to propose these objects as a new class of Galactic non-thermal radio sources caused by relativistic jets from accreting binary systems which have high space velocities. It is not, however, clear that the accreting jet model is needed to explain the striking structure of the radio emission from G5.3−1.0 and G357.7−0.1. Both sources could be close to the Galactic Centre, where the interstellar medium may be such that conventional supernovae produce remnants very different from those in the outer Galaxy: for example, the level of turbulence and the concentration of magnetic field might both be expected to be higher close to the Galactic Centre. In the case of G5.3−1.0, the probable association of a pulsar with the compact source (Manchester, D'Amico & Tuohy 1985) and a more recent radio map of the surrounding region (Caswell *et al.* 1987) imply that G5.3−1.0 is part of a composite "shell"/"filled-centre" remnant (*e.g.*, Weiler 1983), the brighter radio emission being strongly influenced by the magnetic fields and/or relativistic particles produced over a considerable time by the pulsar. (Also Shaver *et al.* (1985b) have shown the compact source near G357.7−0.1 is probably a compact H^+ region, and therefore unlikely to contain an accreting binary system as proposed by Helfand & Becker.)

Related to Helfand & Becker's model is the suggestion by Manchester (1987) that the radio structure of most SNRs can be described in terms of two overlapping loops, due to either bi-annular flow from the pre-SN star, the SN explosion itself, or from jets from an associated pulsar or X-ray binary. Although this model seems appropriatc for some remnants (*e.g.*, G320.4−1.2, Manchester & Durdin 1983; G109.1−0.1, see below), there are equally, if not more, valid conventional explanations for the structure of almost all SNRs. Furthermore, it is difficult, in general, to explain the enhanced loops in terms of jets from a central source, as the presumably relatively light beams have to penetrate a great deal of swept-up circumstellar and interstellar media.

10.4.2 "Jets" in particular SNRs

G109.1−0.1 is a SNR with a distinctive semicircular radio structure (*e.g.*, Hughes *et al.* 1984), and it has often been proposed as a close relation of W 50 and SS 433 as it contains a binary X-ray pulsar (Fahlman & Gregory 1983). Gregory & Fahlman (1983) proposed that G109.1−0.1's structure is determined by precessing twin jets from the central pulsar (cf. the generalization by Manchester discussed above). This proposition is supported by the apparent ridge of X-ray emission from the pulsar to a radio "hotspot" near the periphery of the remnant, which presumably marks the point of interaction of one of the jets with the SNR rim. There is, however, no direct evidence for highly collimated jets in G109.1−0.1: the radio "hotspot" turns out probably to be a background extragalactic source (Gregory 1986), and the X-ray ridge could simply be a chance, approximately linear enhancement. Also, the binary system in G109.1−0.1 is evidently of low mass, quite different from that in SS 433.

Kesteven *et al.* (1987) present radio maps of two SNRs, G332.4+0.1 and G315.8−0.0, which show collimated extensions – "jets" – from their shells. In each case the "jets" are evidently one-sided, in contrast to the clearly two-sided jets from SS 433 in W·50. The "jet" in G315.8−0.0 is particularly striking, being a well-defined linear extension $\sim$ 8 arcmin long from a faint loop of emission $\sim$ 12 arcmin in diameter. The "jet" in G332.4+0.1 is only well defined near the edge of the remnant shell, after which it broadens quickly into a large "plume" of faint extended emission which is considerably larger than the remnant shell. The structure of "jet" and "plume" of G332.4+0.1 hint of a "break-out" of the SNR shock into a region of lower density, a phenomenon which may also account for the extension seen to the shell of the SNR G348.7+0.3 (CTB 37A, Milne *et al.* 1979; Downes 1984). If this is the case, it is surprising that the radio emission from the "jet" in G332.4+0.1 is not polarized, and has a flat, possibly thermal spectrum (Roger *et al.* 1985) rather than a non-thermal spectrum like that of the remnant's shell.

The "jet" in the Crab Nebula was discovered by van den Bergh (1970) from deep optical plates as a faint extension to the boundary of the Crab Nebula. Gull & Fesen (1982) presented the first detailed pictures (particularly in [O^{++}]) of this remarkable feature which show the "jet" is a limb-brightened, apparently tubular extension $\sim$ 45 arcsec across, protruding $\sim$ 75 arcsec from the northern boundary of the Crab Nebula (see also Fesen & Gull (1986) for detailed optical images, and Velusamy (1984) or Wilson, Samarasinha & Hogg (1985) for radio observations). Although the term "jet" is not appropriate for this feature as it is not highly collimated, one of the models proposed to explain it (Benford 1984) requires particle jets or beams from the Crab's pulsar. Explanation of the apparent tubular morphology of the "jet" with the beam model is, however, difficult, as the pulsar is not on the axis of the "jet". Alternative models exist (a "star trail" of material lost from the Crab's massive progenitor, Blandford *et al.* (1983); or a "shadowed flow" of fast ejecta past a dense cloud in the interstellar medium surrounding the supernova, Morrison & Roberts (1985)) which explain the morphology of the "jet" more naturally.

CTB 80 is an unusual non-thermal radio source that is generally classed as a SNR, although it has a morphology quite unlike conventional remnants (*e.g.*, Mantovani *et al.* 1985). It consists of a flat radio spectrum core (Strom, Angerhofer & Dickel 1984), which contains a pulsar (Strom 1987; Kulkarni *et al.* 1987), and a surrounding "plateau" from which curved "arcs" protrude. Rather confusingly these "arcs" have been referred to as "jets", although they are not well collimated.

10.4.3 Jets in X-ray binaries

Cyg X-3, a Galactic X-ray binary with a period of $\sim$ 4.8 hours, is unusual as it undergoes occasional outbursts that include radio flares (*e.g.*, Johnston *et al.* 1986). MERLIN and VLA observations of a radio flare in 1983 (Spencer *et al.* 1986) show an originally unresolved structure developing into an expanding double, presumably indicating underlying jets. The preferred expansion rate implies (for a distance of $\sim$ 10 kpc) an expansion speed of $\sim$ 0.35 c. This value depends not only on the distance and the orientation to the plane of the sky, but also on which flare is associated with the production of the observed structures: intriguingly, a speed like that seen in SS 433 cannot be excluded. Unfortunately Cyg X-3 is highly obscured, so no optical observations are available. In contrast to SS 433, Cyg X-3 is a low-mass binary system, and is not surrounded by any non-thermal radio emission like W 50. High resolution radio observations of Sco X-1, another low-mass X-ray binary, show an expanding triple radio structure, which is interpreted as a compact core and oppositely directed jets (Geldzahler & Fomalont 1986). Sco X-1 is, however, evidently quite different to both SS 433 and Cyg X-3 as its inferred expansion speed is only $\sim$ 35 km s^{-1}. Cir X-1, an X-ray binary with a period of $\sim$ 16.5 days, is similar to Cyg X-3 in that it shows occasional outbursts, and is similar to SS 433 as it is surrounded by non-thermal diffuse radio emission (Haynes *et al.* 1986). There is, however, no evidence at present for any jets in Cir X-1.

10.5 Jets in the Galactic Centre

The subject of jets in our own Galactic Centre (GC) is fascinating from many points of view, and acts as a focus and discussion point for astronomers with many different specialities. On the one hand, the GC stands at the extreme upper end of the range of activity among objects in our Galaxy, with possible large scale versions of the energetic outflow phenomena we can see happening in young stars and stellar remnants. On the other hand, it stands at the extreme lower end of the range of activity that occurs in active galactic nuclei, radio galaxies and quasars. Thus it forms a very useful bridging point between Galactic and extragalactic astronomy in the context of jets and activity.

Much literature and several excellent reviews exist on the subject of the general environment at the GC, the range and scales of activity and the detailed reasons for belief in a central powerhouse in the form of a $10^6 M_\odot$ black hole. For orientation's sake, this will be briefly reviewed in the next section. The main aim of this contribution however, will be to gather together and critically review the specific evidence for *jets* at the GC, and then to discuss what sort of objects these really are, and whether we can indeed link them with other outflow phenomena in our own Galaxy, and/or with the properly collimated flows we call jets in radio galaxies and quasars.

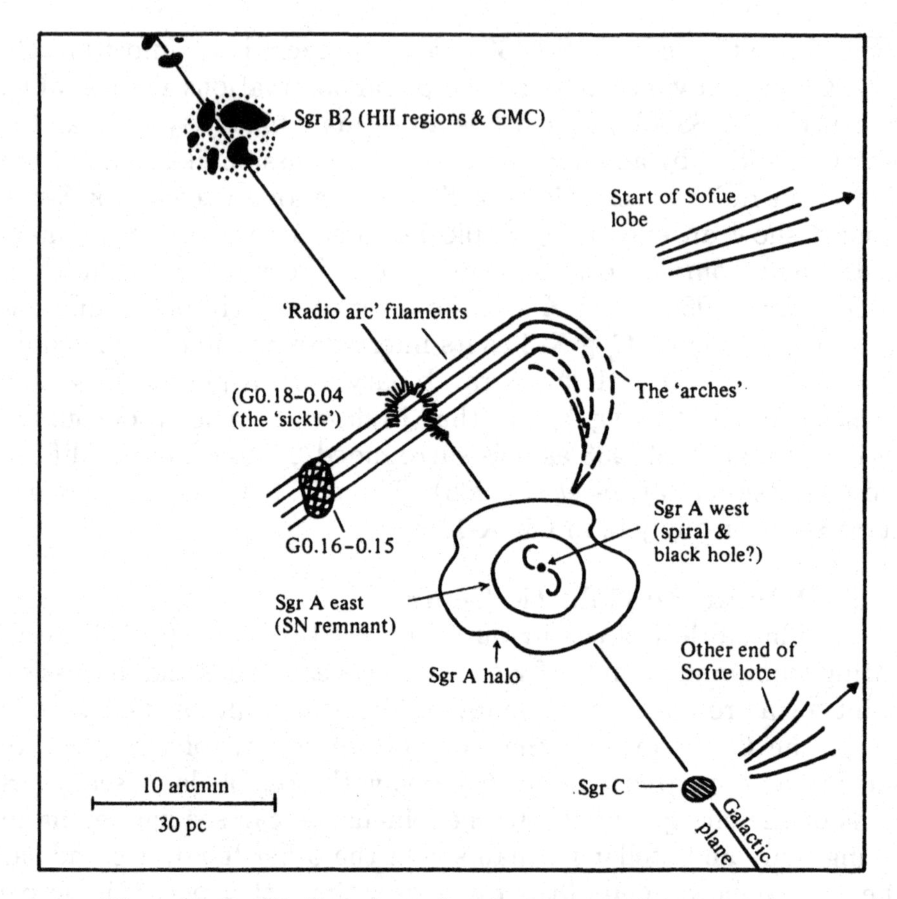

Fig. 10.9. Finding chart for the larger scale structures near the Galactic Centre.

10.5.1 A tourist's guide to the Galactic Centre

As a first step we draw attention to the "finding chart" for the Galactic Centre region shown in Fig. 10.9. The chart covers the inner $\sim$ 120 pc of the Galaxy, assuming a GC distance of 10 kpc. (This has been the distance customarily adopted, and gives a convenient conversion of $1' \sim 3$ pc, although recent evidence (see for example Feast 1987) points to a lower value and 8.5 kpc is now recommended by the IAU.) The main features which must be pointed out in a discussion of activity at the GC include the following:

10.5.1.1 The Radio Arc

This feature, running from $b \sim 0\overset{\circ}{.}1$ to $b \sim -0\overset{\circ}{.}2$ at $l \sim 0\overset{\circ}{.}15$,

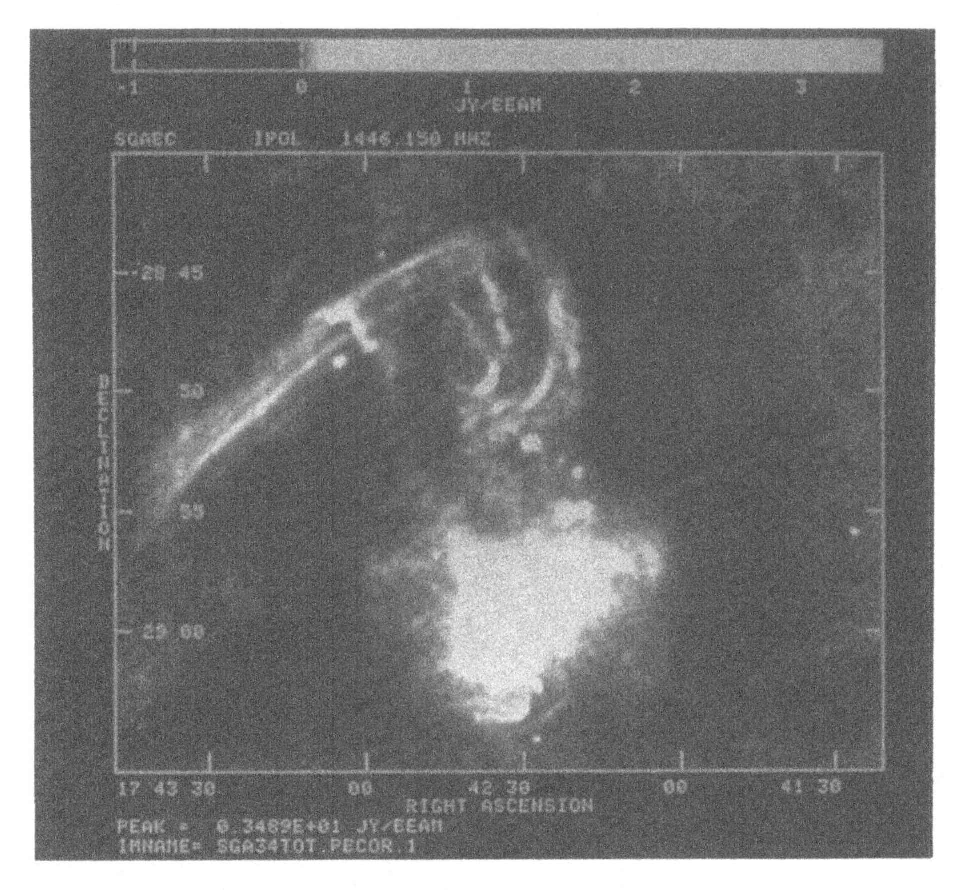

Fig. 10.10. VLA radio picture of the Arc and Sgr A at 1.4 GHz. Photo courtesy of F. Yusef-Zadeh.

was first evident in early single-dish low resolution maps of the Galactic Centre region (*e.g.*, Pauls *et al.* 1976), but its true nature was not discovered until higher resolution VLA maps were made by Yusef-Zadeh, Morris & Chance (1984). These show a remarkable structure of narrow aligned filaments, running roughly perpendicular to the Galactic plane (Fig. 10.10). There is evidence in higher resolution observations for *braiding*, and for interaction of the filaments with a thermal structure G0.18−0.04 sitting near the Galactic plane (Fig. 10.11). On larger scales, evidence for a helical structure apparently winding round the Radio Arc can be seen in Fig. 10.10. The geometry of these structures points strongly towards the magnetic field as being the controlling factor in the

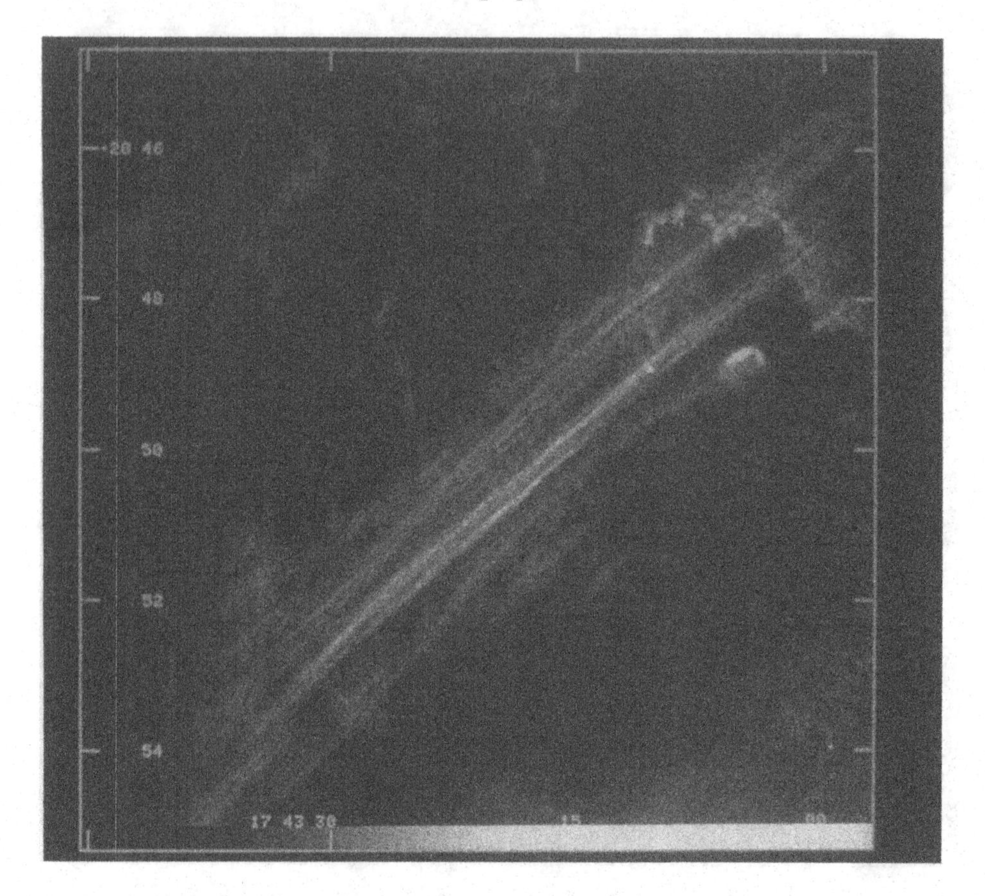

Fig. 10.11. 5 GHz VLA picture of the filaments of the Arc near the thermal structure G0.18−0.04 (visible to the NW). Photo courtesy of F. Yusef-Zadeh. (See also Yusef-Zadeh & Morris 1987b.)

dynamics (Yusef-Zadeh & Morris 1987a,c; Benford 1990; Pudritz 1990), although alternative explanations in terms of shock fronts (Bally *et al.* 1988) have also been proposed. The *polarization* structure of the Arc is also remarkable. Fig. 10.12, from Seiradakis *et al.* (1985), shows polarized emission in two lobes from areas just off the ends of the main Arc, with a central core of polarized emission in the middle, at the point where the Arc crosses the plane. Already, in this core/lobe morphology, there may appear to be a remarkable parallel with an extragalactic jet. However, the spectrum of emission from the Arc (Yusef-Zadeh 1986) together with evidence of interactions with neutral and molecular material (Lasenby, Lasenby & Yusef-Zadeh 1989) suggest strongly that this appearance is due

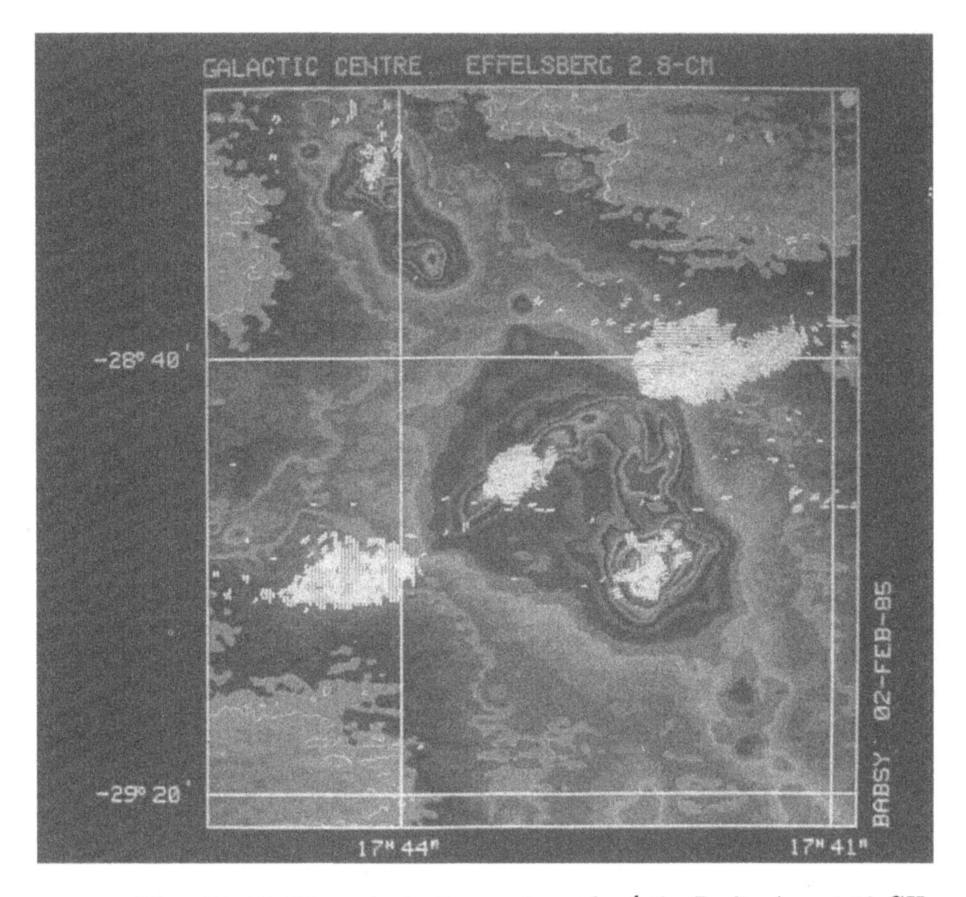

Fig. 10.12. The polarization vectors along the Radio Arc at 10 GHz superimposed on a colour coded plot of the total intensity emission at 10 GHz. Maps made with the 100 m Effelsberg telescope, from Seiradakis *et al.* (1985). Reprinted by permission from *Nature*, **317**, 697. Copyright © 1985 Macmillan Magazines Limited.

to screening, and depolarization at intervening positions, rather than any intrinsic jet-like geometry for the emission.

The total intensity emission corresponding to the polarized emission shows that the Arc has extensions to the NW and SE beyond the outline shown in Fig. 10.10. These extensions are evident in the low contrast radio image shown in Fig. 10.13 (again from Seiradakis *et al.* 1985), where symmetrical continuations to the Arc in each direction can be seen. In fact, a 5 GHz map by Altenhoff *et al.* (1979) (Fig. 10.14) shows a continuation of the Arc to the NW (at

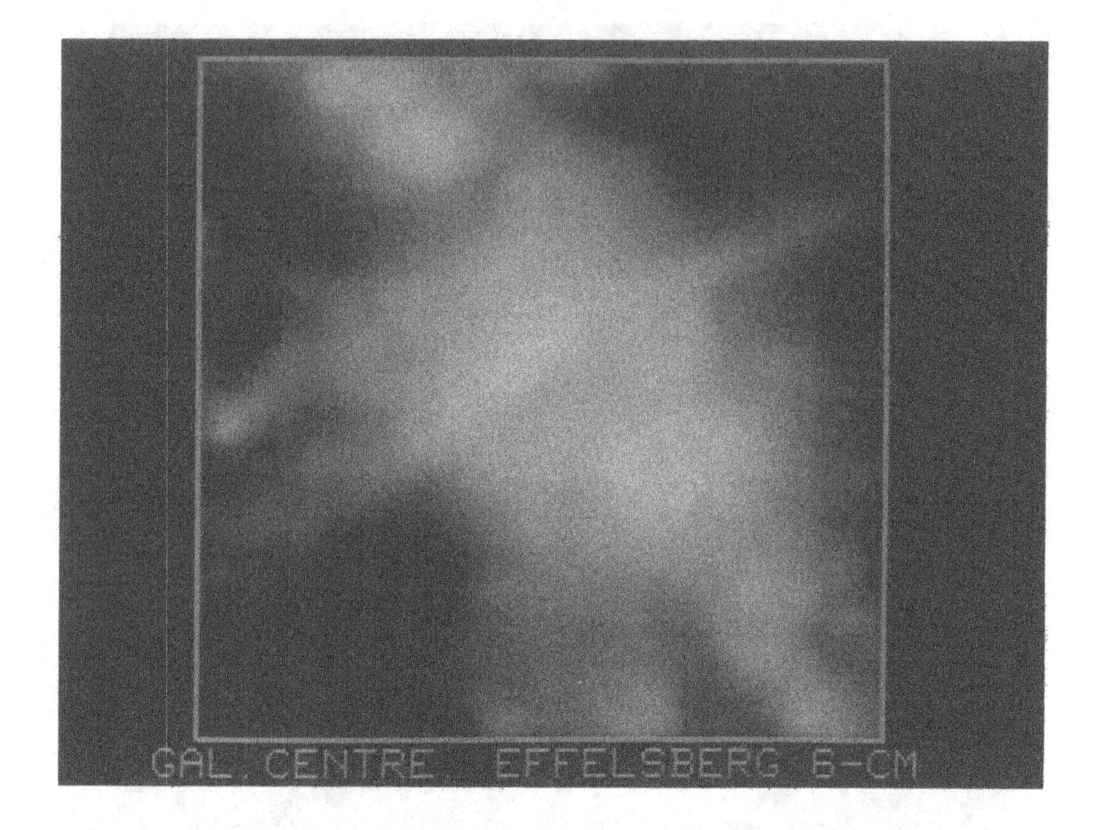

Fig. 10.13. 5 GHz radio picture of the Arc and Sgr A region showing the low surface brightness extensions to the Arc. Map made with the 100 m Effelsberg telescope, from Seiradakis *et al.* (1985). Reprinted by permission from *Nature*, **317**, 697. Copyright © 1985 Macmillan Magazines Limited.

still lower surface brightness) by a further $\sim 0\overset{\circ}{.}7$, corresponding to a total of $\sim$ 160 pc above the plane.

10.5.1.2 The Sofue Lobe

Sofue & Handa (1984), on the basis of the 10 GHz map of the region that is shown in Fig. 10.15, identified this NW extension to the Arc as being just one side of a large lobe-shaped structure, with the equivalent opposite side running from $b \sim 1°$ to $0°$ at $l \sim -0\overset{\circ}{.}6$, with possibly a connecting bridge at the top, about 200 pc above the plane. (The matching western ridge is also evident in the Altenhoff

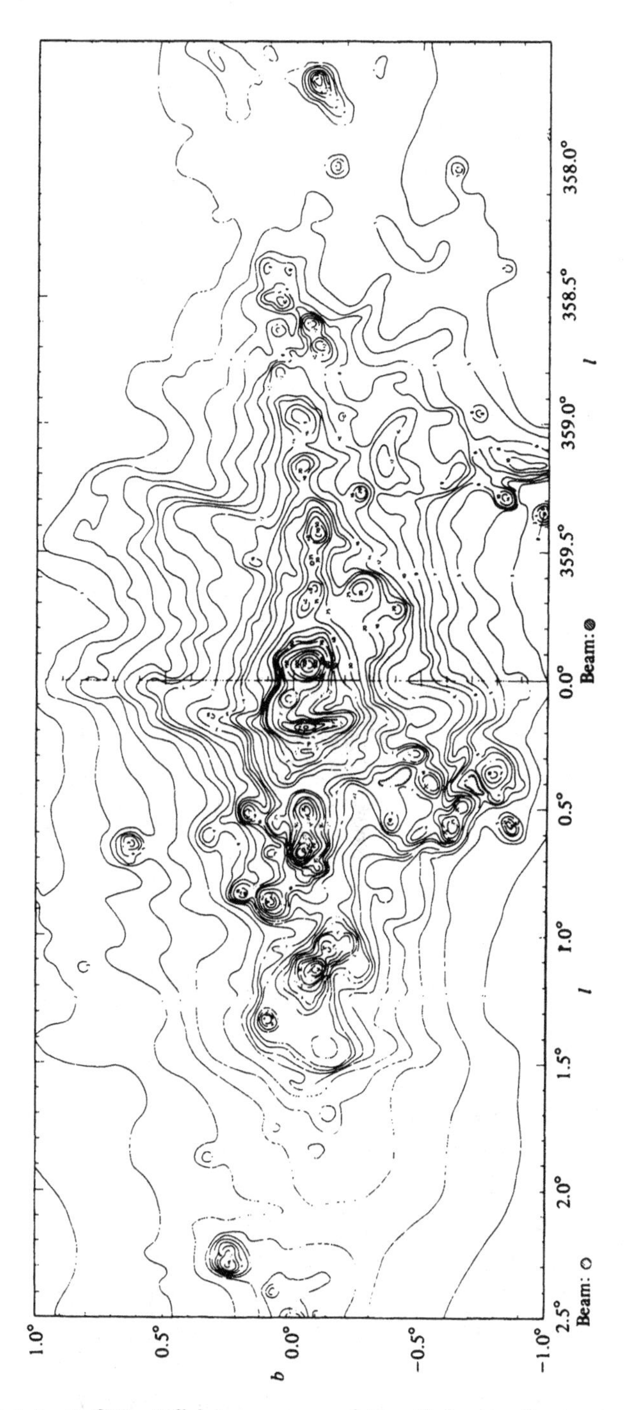

Fig. 10.14. 5 GHz Effelsberg map of the Galactic Centre region in (l, b) coordinates; from Altenhoff *et al.* (1979).

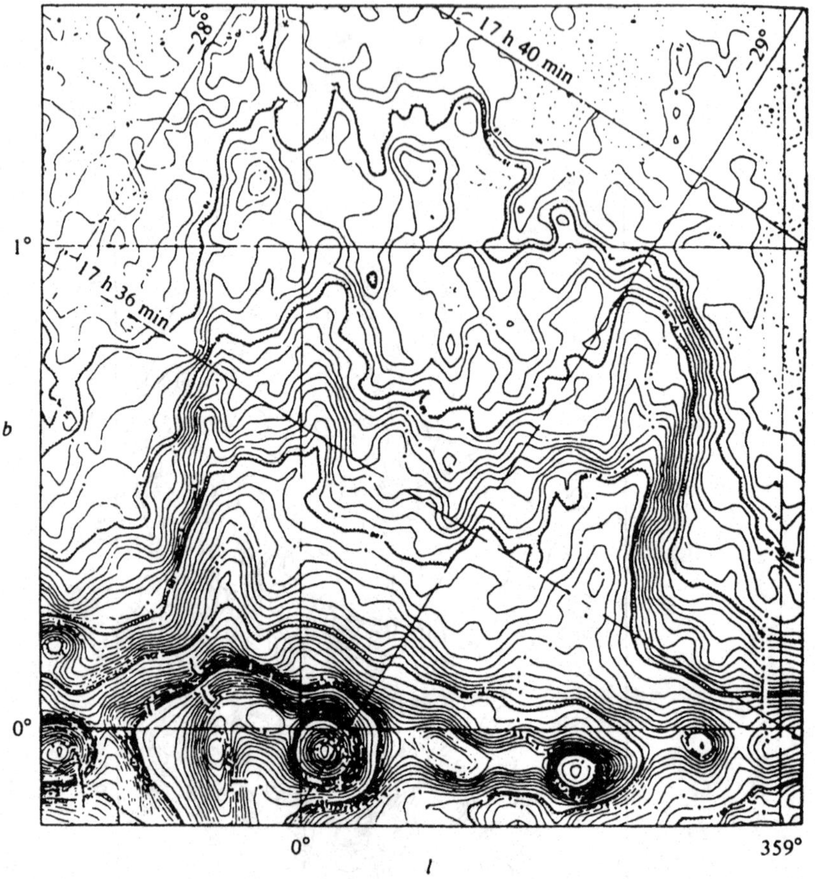

Fig. 10.15. 10.7 GHz Nobeyama map showing the "Sofue Lobe". Note the indication of a connecting bridge at the top. Reprinted by permission from *Nature*, **310**, 568. Copyright © 1984 Macmillan Magazines Limited.

et al. (1979) map, Fig. 10.14.) This was a very significant finding, suggesting an immediate analogy with the broad predominantly one-sided emission features (one-sided referring to an asymmetry about the *plane* of the galaxy) seen in some other nearby spiral galaxies (Hummel, van Gorkom & Kotanyi 1983, and see §10.5.3.1). Strengthening the suggestion that the feature at $l = -0\overset{\circ}{.}6$ is a true counterpart to the Radio Arc, was the discovery of a sharply bent filament at the point where this feature intersects the plane at Sgr C (Liszt 1985) and more recently a further filamentary structure, with every appearance of a magnetic origin, a short way to the NW

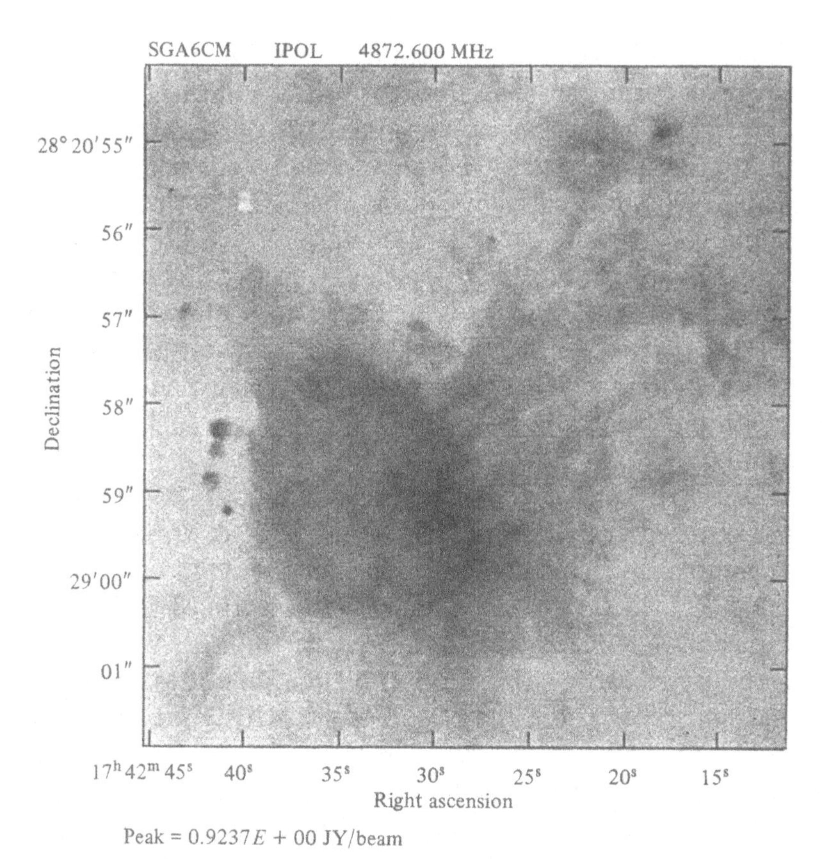

Fig. 10.16. 5 GHz VLA map of Sgr A showing the location of the ionized streamers of Sgr A West.

discovered by Yusef-Zadeh & Bally (1990). Further evidence for the reality of the lobe structure (now known as the "Sofue Lobe") is discussed in Reich *et al.* (1984) and Serabyn (1987), and we shall return to its status as a possible indicator of jet- or outflow-like activity below.

10.5.1.3 **Sgr A**

Moving in closer to the true dynamical GC, we encounter the shell and halo-like structure of Sgr A (Figs. 10.10 and 10.16). This feature, variously thought of as a supernova remnant, or possibly the result of an explosive event of a different kind directly linked with the GC, has the steep-spectrum non-thermal radio emission typical of supernova remnants elsewhere in the Galaxy, and is certainly thought to be expanding rapidly. Yusef-Zadeh *et al.* (1984) have drawn attention to the elliptical outline of the source, with the ratio of axis lengths along the plane to perpendicular to the plane of approximately 2:1. For an initially spherical explosion propagating through a *poloidal* field at the GC (*i.e.*, perpendicular to the plane, as suggested by the filaments of the Arc), one would expect the Alfvénic and bulk velocities to add coherently in a direction parallel to the plane, giving rise to an ellipticity as seen, for field strengths of $\sim 10^{-9}$ Tesla, typical of what is inferred on other grounds for the GC region (although Yusef-Zadeh & Morris (1987c) estimate that the field could be as high as 10^{-7} Tesla in some regions).

At the western end of the non-thermal Sgr A East shell, one sees superimposed the thermal structure of Sgr A West. This contains the famous spiral streamers, lying on the inside of a rotating molecular disc, which occupies Galactic radii of ~ 3 to 7 pc. (For composite see Fig. 10.17.) At the centre of the Sgr A West structure lies the point radio source Sgr A*, thought to be at the exact dynamical centre of the Galaxy. Close to this, but not exactly coincident, a star cluster is visible in the infrared. Very large velocities ($\sim 600 - 700$ km s^{-1}) have been measured in [Ne$^+$], He^0 and H^0, in the gas in this region (*e.g.*, Lacy *et al.* 1987), leading to analogies with Broad Line Regions in radio galaxies and quasars. However, an explanation in terms of mass outflow from Wolf-Rayet type stars, similar to that seen in 30 Doradus, is equally viable (Allen 1990), and, as with so much else in the very centre of the Galaxy, there is no clear evidence as yet which can tell us if something truly spectacular is occurring, or simply examples of what we see elsewhere in the Galaxy, albeit in a unique context.

The "spiral" ionized features, the rotating molecular ring and the high velocity gas are used to set constraints on the mass distribution near the centre as a function of radius, and it is from these studies (as well as a time variable e^+e^- annihilation line) that evidence has accrued for a possible black hole of mass $\sim 10^6 M_\odot$. This evidence, together with much more detailed descriptions of the inner GC region can be found in, for example, Genzel & Townes (1987) and Serabyn (1987). We note here however, that while the status of a (small) black hole looks good (though see Ozernoy (1990) for many caveats), no evidence at all has been revealed in studies of

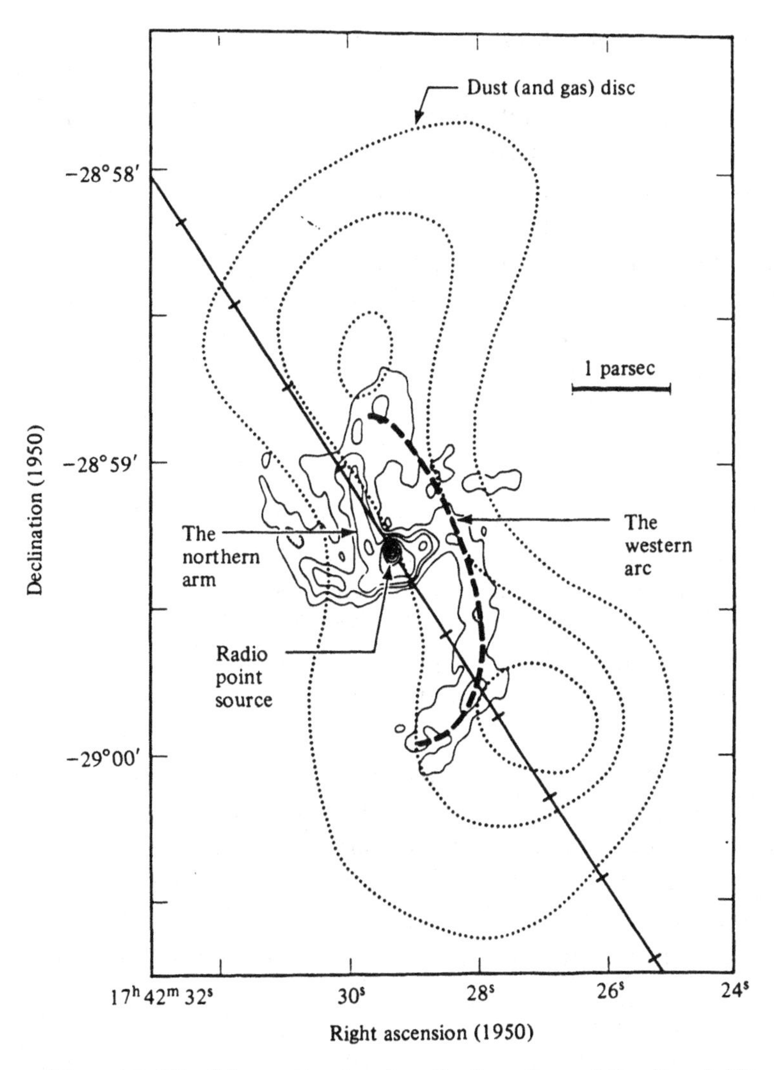

Fig. 10.17. Schematic showing the location of the Sgr A West point source with respect to the ionized streamers and the rotating disc of dust and molecular gas; from Genzel & Townes (1987).

these inner regions (say < 10 pc) for either a *jet* or a genuine accretion disc. A genuine jet is more likely to have left evidence for its presence on larger scales, and the accretion disc is presumably still to be excavated in the detailed structure of the Sgr A* point source, when that becomes available.

10.5.2 **Evidence for jets**

Out of the features discussed above, the most promising candidate for a jet-like structure at the GC is the Sofue Lobe, visible in Fig. 10.15 and discussed further in §10.5.3.1. Three other features have been tentatively identified with "jets" however, all on the basis of their low-frequency radio emission. Note that since we do not expect any jet at the GC to be currently active, it is plausible that if it is still visible it will be as a fossil jet, with an ageing electron population and highly steep spectrum emission. These low-frequency radio features will now be discussed separately, with particular emphasis on the third, a possible true kiloparsec scale jet, with the Sofue Lobe as its base.

10.5.2.1 **A low energy jet emanating from Sgr A**

An elongated ridge of emission at 160 MHz, emanating from the Sgr A East non-thermal shell, was discovered by Yusef-Zadeh *et al.* (1986) in observations made with the Culgoora circular array, and later confirmed by Kassim, LaRosa & Erickson (1986) *via* Clark Lake observations at 110.6 and 123 MHz. Two composites, the first showing the 160 MHz map in relation to the 1.4 GHz VLA map, and the second comparing all three low frequency contour maps, are displayed in Fig. 10.18(a) and (b). It is evident from the 110.6 MHz map in particular that this "jet" feature has a much steeper spectral index than the Sgr A East shell itself – Kassim *et al.* estimate $\alpha \geq 0.7$ (with $S \propto \nu^{-\alpha}$) for the jet versus $\alpha \geq 0.25$ for the central source. (The uncertainties are due mainly to uncertainties in the degree of absorption occurring in the ionized gas which is likely to be surrounding the regions and along the lines of sight.) The length of the jet (as evident in the 110.6 and 160 MHz observations) is about 30 pc and it extends more or less directly along the minor (rotation) axis of the Galaxy. The total luminosity is $\sim 2\times10^{26}$ W. Yusef-Zadeh *et al.* reject the possibility that the "jet" is in fact merely due to a *gap* in the intervening absorption, which is indeed a worry, on the basis of its positional improbability, and the lack of evidence for an intervening H^+ region at higher frequencies. The maps of Kassim *et al.* make it clear that there is a good deal of absorption occurring in the whole GC region, but even so, the appearance of features which are known from higher frequency observations (such as the non-thermal structure G0.16−0.15 on the Radio Arc and the

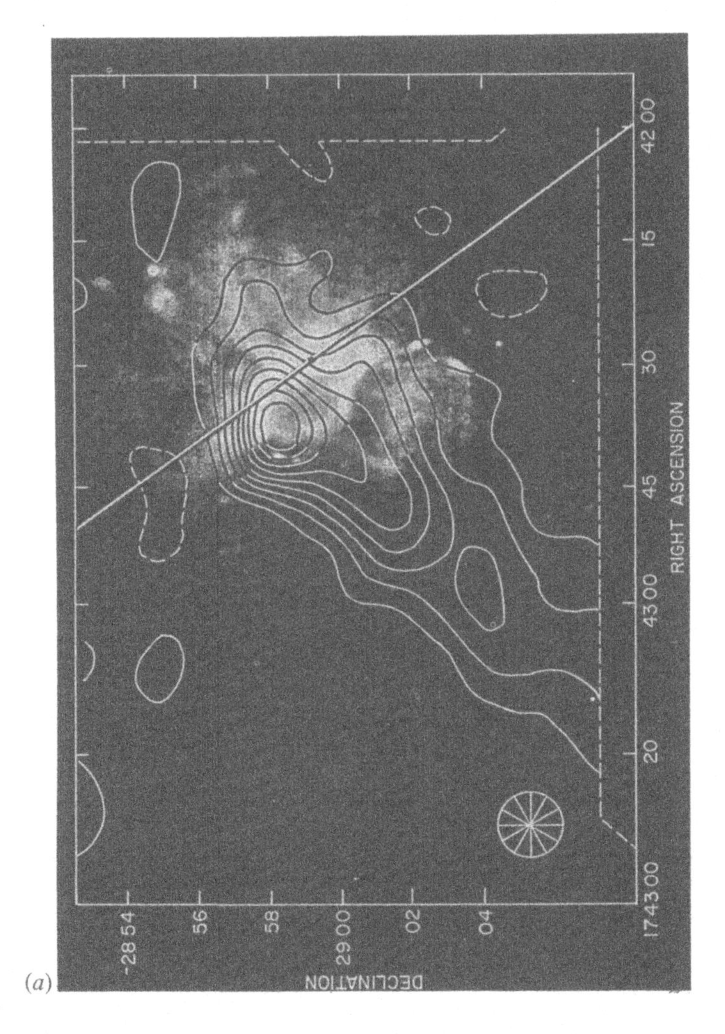

(*a*)

Fig. 10.18a. Overlay of the 160 MHz Culgoora map with the 1.4 GHz VLA map (greyscale). (See also Yusef-Zadeh *et al.* 1986.)

NW and SE lobes corresponding to the polarized emission found by Seiradakis *et al.*) do give confidence that the jet feature is real, and that it only disappears at higher frequencies due to a general merging with the background. Both sets of authors consider this feature in relation to similar one-sided emission features seen in other nearby spiral galaxies, and this will be discussed further in §10.5.4.

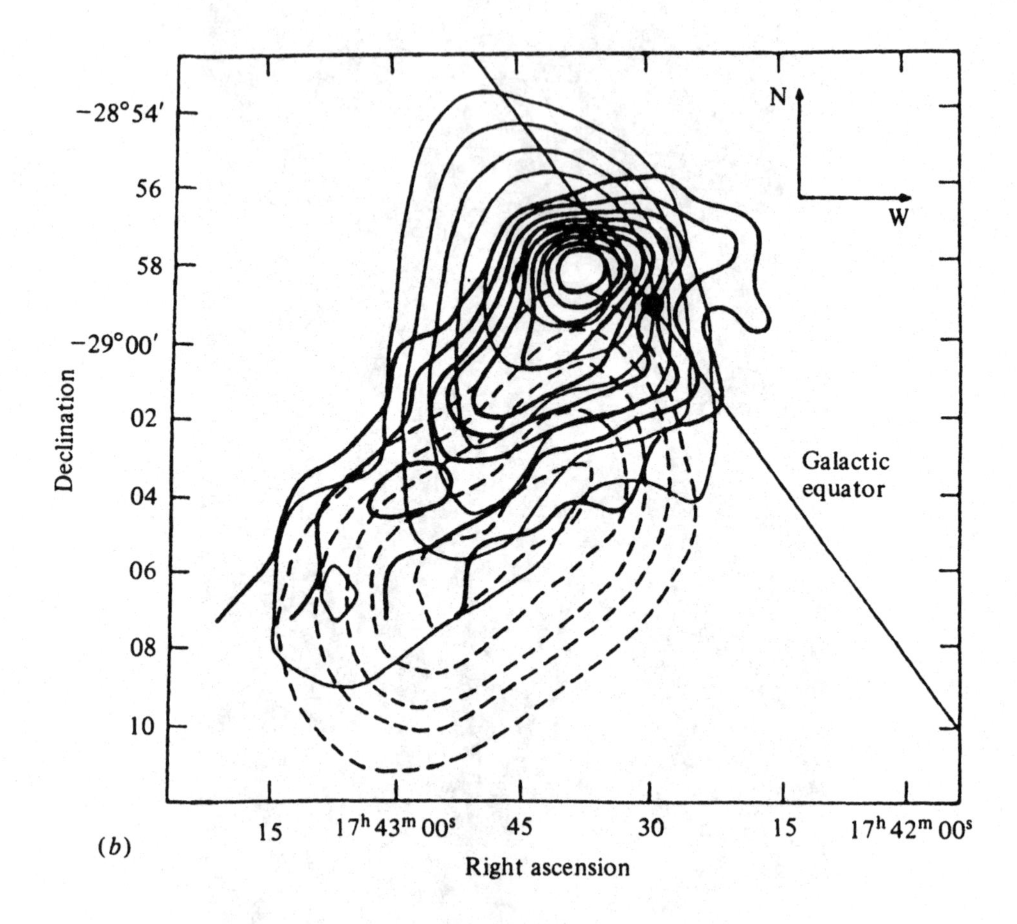

Fig. 10.18b. Composite of all three low-frequency maps showing the Sgr A "jet": thick lines 160 MHz Culgoora map; thin lines 123 MHz Clark Lake map; dashed lines 110.6 MHz Clark Lake map (from Kassim, LaRosa & Erickson 1986). Reprinted by permission from *Nature*, **322**, 522. Copyright © 1986 Macmillan Magazines Limited.

10.5.2.2 The "Northern Galactic Lobe"

In an earlier set of observations with the Clark Lake telescope, LaRosa & Kassim (1985) discovered a feature at 57.5 and 80 MHz lying near to the Galactic plane but approximately 34 arcmin (100 pc) NE from the nucleus, close to Sgr B1. They report that the very steep spectrum nature of this extended object ($\alpha \geq 1.0$) and the lack of counterparts to the source in more extended surveys of the Galactic plane and GC area, mean that it is unlikely to be a foreground Galactic object, such as a supernova remnant. Also its positional coincidence with the GC – where it is the dominant emission component at both 57.5 and 80 MHz – make it unlikely to be an extragalactic source seen through a gap in the Galactic absorption. If it is at the distance of the GC its luminosity is $\sim 1\times10^{26}$ W and LaRosa & Kassim suggest that the appropriate extragalactic counterpart would be the (generally two-sided) lobes straddling the nuclear radio source in Seyfert galaxies (*e.g.*, Wilson 1981). The reality of the feature is confirmed by the Clark Lake maps at 110.6 and 123 MHz mentioned in the preceding section, but unfortunately the 160 MHz Culgoora map (Yusef-Zadeh *et al.* 1986; Yusef-Zadeh 1986) does not cover the region. A higher resolution map of this "Northern Galactic Lobe", as Kassim *et al.* (1986) call it, would clearly be of great interest.

10.5.2.3 **A kpc scale jet?**

Sofue (1990) has recently drawn attention to a feature approximately 25° long which is visible in the GC portion of the 408 MHz all-sky survey of Haslam *et al.* (1982). This feature, shown in Fig. 10.19, has the appearance of a knotty but well collimated jet, emerging at or near the GC and running to positive Galactic latitudes along a line of approximately constant declination (near −21°) for most of its length. Sofue discusses this jet in the context of images he has made using a background filtering technique (see Sofue & Reich 1979), but in fact a narrow elongated feature of this kind is already clearly evident in the original 408 MHz survey with careful choice of contours or greyscale levels (*e.g.*, Fig. 10.19). Haslam *et al.* (1981) discuss this portion of the survey (taken with the Parkes telescope) in some detail, and conclude that the feature represents internal ridging within a nearby loop (Loop I) caused by a local supernova. A neutral hydrogen spur running approximately parallel to but offset from the feature (Heiles 1975) is adduced as additional evidence for a local origin. However, Sofue proposes that the alignment with the GC is real, and that the feature is a true jet, $\sim$ 4 kpc in projected length, and with a total luminosity of $\sim 2\times10^{28}$ W, making it certainly the largest and most energetic of the jet phenomena considered so far, and therefore (if the Sofue interpretation is correct) by far the most promising candidate out of

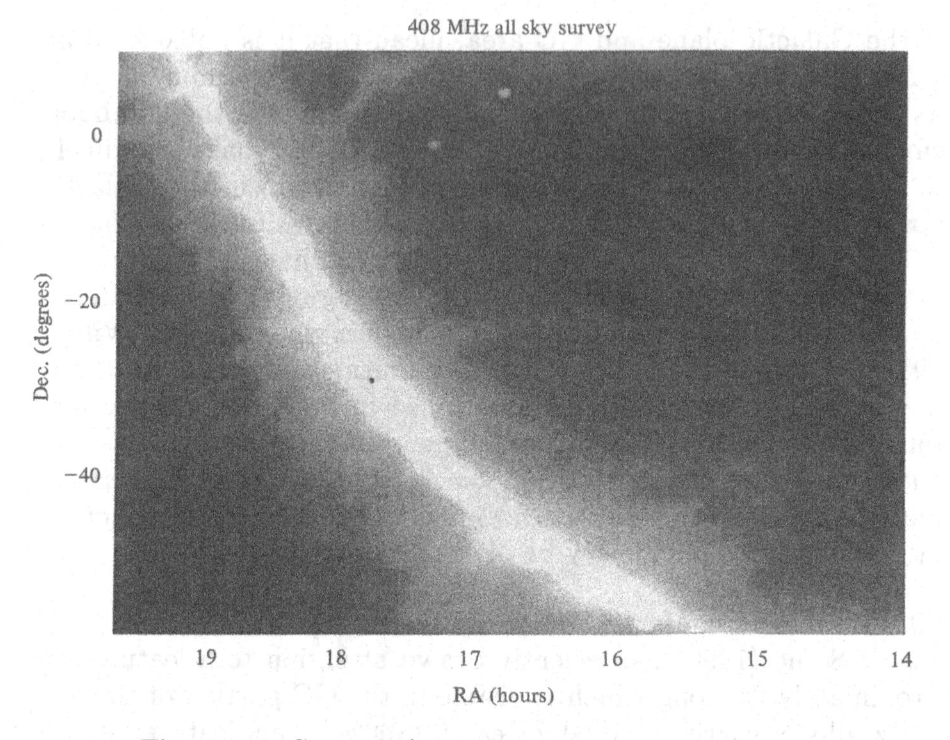

Fig. 10.19. Greyscale (logarithmic levels) of the Galactic Centre portion of the 408 MHz all-sky survey (Haslam *et al.* 1982). The dynamical centre is marked with a dot. (Figure prepared using digitized data kindly supplied by Dr John Osbourne.)

those considered in this section for comparison with extragalactic jets in active galaxies and quasars.

The status of an association with a ridge of neutral hydrogen is unclear. Kundt (1987b) has claimed that a locus, representing the path of a jet, can be drawn connecting various high velocity H^0 clouds, starting at the GC and ending up beyond the Solar radius. The inner portion of this H^0 jet (it should be stated that the jet is supposed to be composed of an e^+e^- plasma, and that the H^0 clouds arise *via* condensation along its route) lies near the Sofue feature just discussed (called by Sofue the GC jet – GCJ), but appears to curve away from the perpendicular to the plane towards positive rather than negative Galactic longitudes. Also, it must be said that the locus picked out by Kundt over this region is not at all evident to the eye in the H^0 map he is using (Burton

& Liszt 1978), whereas the GCJ and the H^0 spur in the Heiles map discussed by Haslam *et al.* (1981) are definitely real features, although whether connected, and whether any of these features are actually at the GC, remains to be definitively resolved. A further piece of evidence on the Sofue GCJ – that it is only one of several spurs visible in the 408 MHz survey spreading out perpendicular to the plane – is used to suggest opposite conclusions by Sofue and by Haslam *et al.* (1981). The latter suggest a common origin for these features in supernova explosions, their large apparent size being simply the result of a local origin, whereas Sofue claims that they are preferentially located towards the inner portion of the Galaxy, $l = -40°$ to $+40°$, and are thus intrinsically associated with the central region of the Galaxy.

A type of observation which could resolve these problems is the demonstration of a physical link between the GCJ and an object known to be at the GC. Sofue claims such a link, established *via* 1.4 GHz observations, with the "Sofue Lobe" structure discussed briefly in §10.5.1. His idea is that the GCJ is a cylindrical structure having the GCL (the GC, or Sofue, Lobe) as its base. The linear width of the jet, $\sim 2°$, near its base is wider than the GCL ($\sim 0\overset{\circ}{.}8$), but 1.4 GHz maps of the GCJ made with the Effelsberg telescope show some fragmentary structure leading from the western lobe to the beginning of the jet. However, much other small-scale structure exists also in these regions, and it is not clear as yet that a physical link has been demonstrated, although the overall picture of a unified origin for the lobe and the jet remains attractive.

10.5.3 The Sofue Lobe (GCL)

As discussed above, this pair of lobes, which contains within it features which are undoubtedly located at the GC (*e.g.*, the Radio Arc and its extensions on one side, and Sgr C at the other) forms the most promising candidate for an example of outflow activity in the GC region, being visible at high frequencies (*e.g.*, at 10 GHz) where absorption is not a problem and with a morphology very similar to the predominantly one-sided lobes seen in other spirals by Hummel *et al.* (1983). In this section, in order to establish whether it has anything to tell us specifically about *jets*, we shall briefly review the possibilities for the physical origin of this lobe, and then in §10.5.4 discuss its place, along with other examples of jet-like activity mentioned above, in the context of nuclear activity in galaxies generally.

10.5.3.1 The physical origin of the GCL

The first mechanism proposed for the GCL was that of a shock front caused by an explosion at the Galactic nucleus propagating out into an ellipsoidal disc (Sofue 1984). With disc dimensions of 230 and 100 pc, the longer dimension along the plane, and a density contrast inside/outside the disc of 1000:1, Sofue was able to obtain a shape similar to that of the GCL, and calculated a dynamical age for the lobe of $\sim 2\times10^6$ yr. The necessity for an explosion, and that we should see the lobe at a special time, is avoided in the more detailed models of Uchida & Shibata. In a series of articles (see particularly Uchida, Shibata & Sofue (1985) and Shibata & Uchida (1987) for application to the GCL) they have developed a "magnetic twist" model for the generation of outflow activity, applicable to cosmic jets, the GCL and also young stars (see §10.2.4). The feature common to these applications is the idea of a disc of rotating accreting gas, threaded by a perpendicular (poloidal) magnetic field. As the material contracts towards the centre, and rotates, it twists up the magnetic field so that it acquires a toroidal component. The $\mathbf{j} \times \mathbf{B}$ force on the gas then has an upward component in the surface layer above the plane and downward below the plane. The model thus predicts an ordered cylindrical outflow in opposing directions perpendicular to the plane, and a stationary state is achieved by relaxation of the twisted field lines back to their poloidal configuration, coupled with an increased tendency to collapse caused by the net transfer of angular momentum to the outflow lobes. Numerical simulations have verified that structures on the scale of the GCL do indeed form this way, on the required dynamical time scales, for a poloidal magnetic field typical of that required for the central region, $\sim 2\times10^{-9}$ Tesla. Note that a specific prediction of this model, namely that the magnetic field line of sight component should change sign as the plane is crossed, has been verified by Tsuboi *et al.* (1985) for the eastern lobe. They also showed that along the eastern GCL ridge the magnetic field runs approximately parallel to the ridge, as expected from the stationary state poloidal field configuration (see Fig. 10.20). However, the model clearly does not fit the facts in that it predicts a cylin-

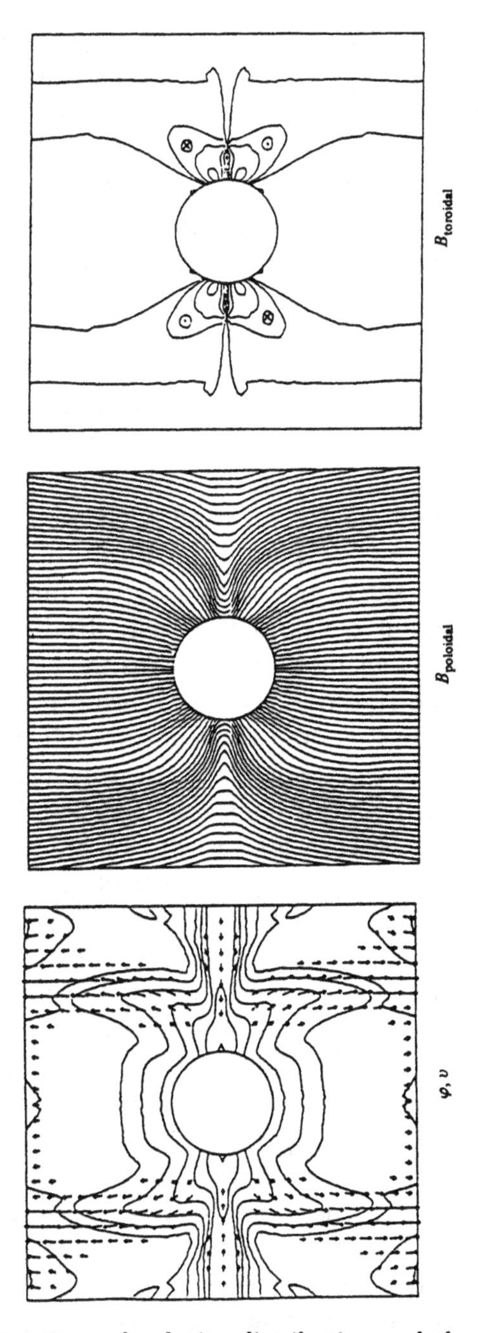

Fig. 10.20. Density and velocity distribution and the poloidal and toroidal components of the magnetic field in the models of Uchida *et al.* (1985). Reprinted by permission from *Nature*, **317**, 699. Copyright © 1985 Macmillan Magazines Limited.

drical symmetry about the rotation axis of the Galaxy, whereas the eastern lobe is closer to Sgr A West (the presumed axis) than the western lobe, and the eastern lobe has far more pronounced magnetic phenomena (the Radio Arc and the polarization lobes) than the western (the filaments at Sgr C and as discovered by Yusef-Zadeh & Bally). Also, even more strikingly, there is a very pronounced asymmetry above and below the plane, with only the fossil jet emanating from Sgr A found by Yusef-Zadeh *et al.* giving any indication for activity directed towards negative latitudes, and certainly no obvious counterpart to the GCL. (There is a possible hint of a continuation of the eastern Sofue Lobe in the Altenhoff *et al.* map, Fig. 10.14.)

A different approach towards explaining many of the asymmetric jet-like phenomena at the GC and in particular the GCL, has been considered by Sofue & Fujimoto (1987). They invoke a sporadic intrinsically one-sided jet emitted by the nucleus at random orientations at different times in the past. Radio features form at the points where this jet hits obstructions in the surrounding material; in particular the Radio Arc and Arches are supposed to be the results of such collisions, and the eastern GCL ridge is formed by the magnetically induced transport upwards (*via* an Uchida-Shibata type mechanism) of the energetic charged particles produced at this point. The western lobe would be the result of an earlier jet outburst pointing in the opposite direction. Obviously such an explanation is similar to that invoked for the complex structure seen in many extragalactic radio sources, and is similar to the ideas of Kassim *et al.* (1986) for the origin of the low-frequency radio jet, and their Northern Galactic Lobe. However, in producing structures as complex as the filaments of the Radio Arc itself, more detailed magnetohydrodynamical modelling may be needed, along the lines of the theories proposed recently by Heyvaerts, Norman & Pudritz (1988) (see also Pudritz 1990 and Benford 1990), and the connection with the larger scale ejection phenomena may then be lost. For the formation of the GCL, the ideas of Uchida & Shibata are attractive in that they do not require yet another jet to explain the observed outflow and relate in a unified way to other outflow phenomena observed to exist and known to be connected with ordered magnetic fields – *e.g.*, the generation of outflows in star formation, see §10.2.4. Added to that they have a predictive

Table 10.2 *Comparison of jet luminosity versus collimation (defined as the ratio of length to width) for the Galactic Centre "jets" and several extragalactic counterparts. Only the dominant lobe is taken in cases of double sided emission.*

Object Name	Luminosity (W)	Collimation
FJ	2.0×10^{26}	2.2
NGL	1.0×10^{26}	1.3
GCJ	2.0×10^{28}	10.0
GCL	2.0×10^{27}	1.4
NGC 3097	2.0×10^{31}	2.5
NGC 2992	1.1×10^{31}	2.2
NGC 4388	1.6×10^{30}	1.4
NGC 4438	0.8×10^{31}	2.0

element rather than being mainly descriptive, although special arrangements of material and densities will be needed to explain the observed asymmetries in terms of them.

Finally we should mention a general qualitative explanation for the GCL and the minor axis emission lobes in other spirals seen by Hummel *et al.* (1983) (NGC 2992, 3079, 4388, 4438 and 6500) and Duric *et al.* (1983) (NGC 3079). This is where large Parker instabilities amplified by relativistic electron pressure lead to the formation of "magnetic bubbles" which rise out of the plane. In this version, the nearest equivalent to the GCL elsewhere in our Galaxy could be the "jet" in the Crab Nebula, drawn attention to by Gull & Fesen (1982) (see §10.4), where again buoyancy forces in a plasma may be leading to upflow and ejection of material (Kundt 1988).

10.5.4 **Comparisons with extragalactic jets**

It will have been noticed that a unified quantitative picture for the origin of the jet-like activity at the GC is conspicuously lacking. However, it will still be useful here to draw together the information we have on the four candidate jets – the fossil jet (FJ) linked to Sgr A; the Northern Galactic Lobe (NGL); the Galactic

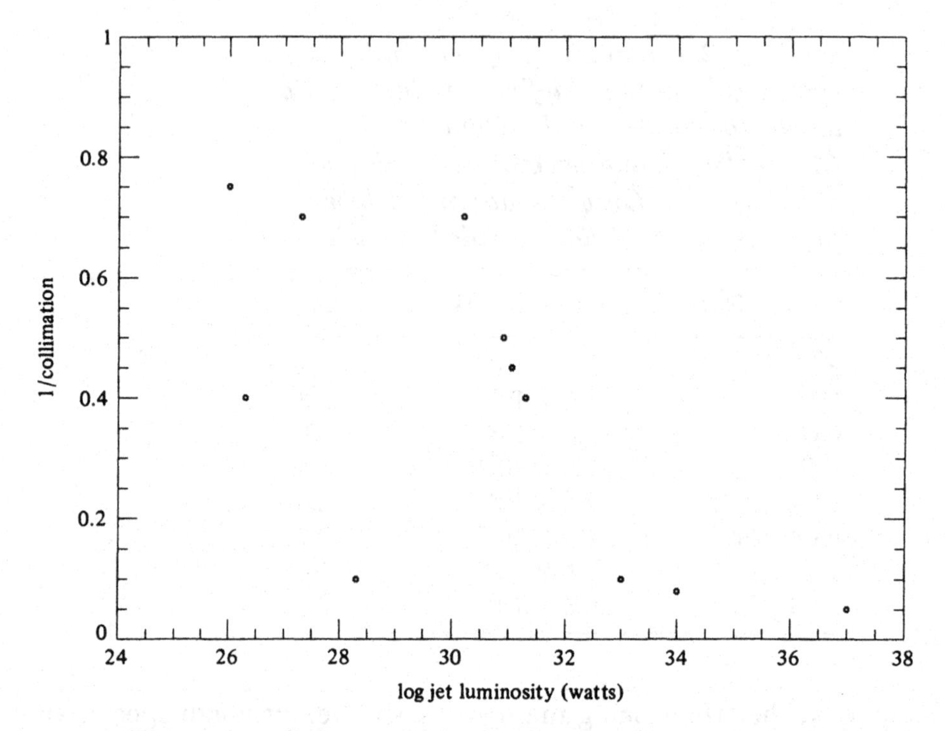

Fig. 10.21. Comparison of jet luminosity versus collimation (defined as the ratio of length to width) for the Galactic Centre "jets" and several extragalactic counterparts.

Centre Lobes (GCL) and Sofue's kpc scale "jet" (GCJ) – in order to make a systematic comparison with the properties of comparable extragalactic features. By this means, even if some of the above mechanisms seem unlikely to be applicable to, or give useful clues on extragalactic jets, we can still see if the features seen in our GC are really atypical. The directly observable critical parameters for a jet or outflow are probably luminosity and collimation, and these are displayed in Table 10.2 for the four GC candidates and some extragalactic examples.

The same numbers are plotted in Fig. 10.21. As already noticed for the case of the GCL by Sofue (1986), there is a clear correlation on which the GC "jets" fit really quite well, although at very low power. This does indeed suggest that although very disparate in morphology, and by no means clearly linked in origin, each of the GC jets we have been discussing at least has extragalactic counterparts, and that our Galaxy (and other nearby spirals) should not be considered as purely normal, but rather "normal" should be redefined so as to mean containing within it scaled down versions of all the types of energetic outflow phenomena seen elsewhere.

It is a pleasure to thank all those colleagues who have provided material in advance of publication and have allowed us to summarize it here. We also thank various colleagues in Cambridge and elsewhere for many stimulating discussions, and in particular we thank Ralph Pudritz and John Richer for their comments on the manuscript.

Symbols used in Chapter 10

Symbol	Meaning
A_v	visual extinction
B, $\mathbf{B}$, B_ϕ	magnetic field, azimuthal component of B
c	speed of light
c_S	speed of sound
E, E_{CO}	energy, energy in CO wind
g	momentum gain factor
G	universal gravitational constant
h	Planck constant
i	inclination of jet to plane of sky
$\mathbf{j}$	current density
J	rotational energy level quantum number
J	specific angular momentum
k_B	Boltzmann constant
$\mathbf{L}$, L	luminosity
L_W	mechanical luminosity of wind
$L_\odot$	solar luminosity
L_*	luminosity of driving source
m_p	proton mass
$\mathcal{M}$	Mach number
M_{core}	mass of (stellar) core
$\dot{M}_a$	mass accretion rate
$\dot{M}_{sh}$	mass outflow rate in shell
$\dot{M}_w$	mass outflow rate in wind

$M_\odot$	solar mass
n	polytropic index
$n_{\rm ext}$	external matter number density
$n_{\rm H_2}$	molecular hydrogen number density
n_j	jet number density
n_s	post jet shock number density
p, $p_{\rm CO}$	momentum, momentum in CO wind
P	pressure
r	radial variable
$r_{\rm A}$	Alfvén radius
r_0	radius at feet of field lines
R	characteristic length scale
$R_{\rm d}$	diameter of disc
$R_{\rm core}$	radius of (stellar) core
S	flux density
$t_{\rm dyn}$	dynamical time scale
T_j	temperature of jet
$T_{\rm W}$	torque associated with wind
v	vibro-rotational energy level quantum number
v	velocity
$v_{\rm crit}$	critical velocity of wind
$v_{\rm f}$	final velocity of flow
$v_{\rm max}$	maximum velocity of flow
$v_{\rm s}$	velocity of shock
$v_{\rm sh}$	velocity of shell
$v_{\rm W}$	velocity of wind
v_∞	terminal velocity of wind
v_ϕ	azimuthal component of velocity
V	volume
α	spectral index ($S \propto \nu^{-\alpha}$)
$\Delta v_{\rm FWZP}$	spectral line width
ϵ	efficiency
$\theta_{\rm max}$	semi-angle of flow
λ	wavelength
ν	frequency
$\Pi_{\rm v}$	momentum flux
$\rho_{\rm ext}$	external matter mass density
τ	optical depth
$\Phi(r)$	magnetic flux distribution

$\Omega(r)$	rotation law
∇	vector differential operator

References

Abell, G. O. & Margon, B., 1979, *Nature*, **279**, 701.
Adams, F. C. & Shu, F. H., 1986, *Ap. J.*, **308**, 836.
Adams, F. C., Lada, C. J. & Shu, F. H., 1987, *Ap. J.*, **312**, 788.
Allen, D., 1990, In *The Galactic Center, IAU Symposium 136*, ed. Morris, M. (Kluwer: Dordrecht, Netherlands), in press.
Altenhoff, W. J., Downes, D., Pauls, T. & Schraml, J., 1979, *Astr. Ap. Suppl.*, **35**, 23.
Bally, J., 1982, *Ap. J.*, **261**, 568.
Bally, J., 1986, In *Masers, Molecules and Mass Outflows in Star Forming Regions, Proceedings of* 1st *Haystack Symposium*, ed. Haschick, A. D. (Haystack Observatory: USA), p. 189.
Bally, J. & Lada, C. J., 1983, *Ap. J.*, **265**, 824.
Bally, J. & Stark, A. A., 1983, *Ap. J.*, **266**, L61.
Bally, J., Snell, R. L. & Predmore, R., 1983, *Ap. J.*, **272**, 154.
Bally, J., Stark, A. A., Wilson, R. W. & Henkel, C., 1988, *Ap. J.*, **324**, 223.
Batrla, W. & Menten, K. M., 1985, *Ap. J.*, **298**, L19.
Becker, R. H. & Helfand, D. J., 1985, *Nature*, **313**, 115.
Beckwith, S., Persson, S. E., Neugebauer, G. & Becklin, E. E., 1978, *Ap. J.*, **223**, 464.
Beckwith, S., Zuckerman, B., Skrutskie, M. F. & Dick, H. M., 1984, *Ap. J.*, **287**, 793.
Benford, G., 1984, *Ap. J.*, **282**, 154.
Benford, G., 1990, In *The Galactic Center, IAU Symposium 136*, ed. Morris, M. (Kluwer: Dordrecht, Netherlands), in press.
Bertout, C., 1987, In *Circumstellar Matter*, eds. Appenzeller, I. & Jordan, C. (Reidel: Dordrecht, Netherlands), p. 23.
Bieging, J., 1984, *Ap. J.*, **286**, 591.
Bieging, J. & Cohen, M., 1985, *Ap. J.*, **289**, L5.
Blandford, R. D. & Payne, D. G., 1982, *M. N. R. A. S.*, **199**, 883.
Blandford, R. D. & Rees, M. J., 1974, *M. N. R. A. S.*, **169**, 395.
Blandford, R. D., Kennel, C. F., McKee, C. F. & Ostriker, J. P., 1983, *Nature*, **301**, 586.
Blome, H. J. & Kundt, W., 1988, *Ap. Sp. Sci.*, **148**, 343.
Blondin, J. M., Königl, A. & Fryxell, B. A., 1989, *Ap. J. Lett.*, **337**, L37.
Bodenheimer, P., Tenorio-Tagle, G. & Yorke, H. W., 1979, *Ap. J.*, **233**, 85.
Burton, W. B. & Liszt, H. S., 1978, *Ap. J.*, **225**, 815.
Cabrit, S. & Bertout, C., 1986, *Ap. J.*, **307**, 313.
Cantó, J., Rodriguez, L. F., Barral, J. F. & Carral, P., 1981, *Ap. J.*, **244**, 102.
Caswell, J. L., Kesteven, M. J., Komesaroff, M. M., Haynes, R. F., Milne, D. K., Stewart, R. T. & Wilson, S. G., 1987, *M. N. R. A. S.*, **225**, 329.
Cohen, M., Bieging, J. H. & Schwartz, P., 1982, *Ap. J.*, **253**, 707.
Cohen, M., Emerson, J. P. & Beichman, C., 1989, *Ap. J.*, **339**, 445.

Collins, G. W. & Newsom, G. H., 1987, *Ap. J.*, **308**, 144.
Crampton, D. & Hutchings, J. B., 1981, *Ap. J.*, **251**, 604.
Davidson, J. A. & Jaffe, D. T., 1984, *Ap. J.*, **277**, L13.
Davidson, K. & McCray, R., 1980, *Ap. J.*, **241**, 1082.
DeCampli, W. M., 1981, *Ap. J.*, **244**, 124.
Dopita, M. A., Schwartz, R. D. & Evans, I., 1982, *Ap. J.*, **263**, L73.
Downes, A. J. B., 1984, *M. N. R. A. S.*, **210**, 845.
Downes, A. J. B., Pauls, T. & Salter, C. J., 1986, *M. N. R. A. S.*, **218**, 393.
Draine, B. T., 1983, *Ap. J.*, **270**, 519.
Duric, N., Seaquist, E. R., Crane, P. C., Bignell, R. C. & Davis, L. E., 1983, *Ap. J.*, **272**, L11.
Dyson, J. E., 1984, *Ap. Sp. Sci.*, **106**, 181.
Dyson, J. E., 1987, In *Circumstellar Matter*, eds. Appenzeller, I. & Jordan, C. (Reidel: Dordrecht, Netherlands), p. 159.
Dyson, J. E., Cantó, J. & Rodriguez, L. F. 1988, In *Mass Outflows from Stars and Galactic Nuclei*, eds. Bianchi, L. & Gilmozzi, R. (Kluwer: Dordrecht, Netherlands), p. 299.
Edwards, S., Strom, S. E., Snell, R. L., Jarrett, T. H., Beichman, C. A. & Strom, K. M., 1986, *Ap. J.*, **307**, L65.
Elston, R. & Baum, A. J., 1987, *Astron. J.*, **94**, 1633.
Eggum, G. E., Coroniti, F. V. & Katz, J. I., 1985, *Ap. J.*, **298**, L41.
Fahlman, G. C. & Gregory, P. C., 1983, In *Supernova Remnants and their X-ray Emission, IAU Symposium 101*, eds. Danziger, I. J. & Gorenstein, P. (Reidel: Dordrecht, Netherlands), p. 445.
Feast, M. W., 1987, In *The Galaxy, Proceedings of the NATO Summer School*, eds. Carswell, R. & Gilmore, G. (Reidel: Dordrecht, Netherlands), p. 1.
Fesen, R. A. & Gull, T. R., 1986, *Ap. J.*, **306**, 259.
Garden, R., Geballe, T. R., Gatley, I. & Nadeau, D., 1986, *M. N. R. A. S.*, **220**, 203.
Geldzahler, B. J. & Fomalont, E. B., 1986, *Ap. J.*, **311**, 805.
Geldzahler, B. J., Downes, A. J. B. & Shaffer, D. B., 1981, *Astr. Ap.*, **98**, 205.
Genzel, R. & Townes, C. H., 1987, *Ann. Rev. Astr. Ap.*, **25**, 377.
Genzel, R., Reid, M. J., Moran, J. M. & Downes, D., 1981, *Ap. J.*, **244**, 884.
Gilmore, W. & Seaquist, E. R., 1980, *Astron. J.*, **85**, 1486.
Grasdalen, G. L., Strom, E. E., Strom, K. M., Capps, R. W., Thompson, D. & Castelaz, M., 1984, *Ap. J.*, **283**, L57.
Green, D. A., 1984, *M. N. R. A. S.*, **209**, 449.
Gregory, P. C., 1986, *Can. J. Phys.*, **64**, 479.
Gregory, P. C. & Fahlman, G. C., 1983, In *Supernova Remnants and their X-ray Emission, IAU Symposium 101*, eds. Danziger, I. J. & Gorenstein, P. (Reidel: Dordrecht, Netherlands), p. 429.
Gull, T. R. & Fesen, R. A., 1982, *Ap. J.*, **260**, L75.
Gull, S. F. & Northover, K. J. E., 1973, *Nature*, **244**, 80.
Hartmann, L. & MacGregor, K. B., 1982, *Ap. J.*, **259**, 180.
Haslam, C. G. T., Klein, U., Salter, C. J., Stoffel, H., Wilson, W. E., Cleary, M. N., Cooke, D. J. & Thomasson, P., 1981, *Astr. Ap.*, **100**, 209.

Haslam, C. G. T., Salter, C. J., Stoffel, H. & Wilson, W. E., 1982, *Astr. Ap. Suppl.*, **47**, 1.

Haynes, R. F., Komesaroff, M. M., Little, A. G., Jauncey, D. L., Caswell, J. L., Milne, D. K., Kesteven, M. J., Wellington, K. J. & Preston, R. A., 1986, *Nature*, **324**, 233.

Heiles, C., 1975, *Astr. Ap. Suppl.*, **20**, 37.

Helfand, D. J. & Becker, R. H., 1985, *Nature*, **313**, 118.

Herbig, G. H. & Jones, B. F., 1981, *Astron. J.*, **86**, 1242.

Heyer, M. H., 1988, *Ap. J.*, **324**, 311.

Heyvaerts, J., Norman, C. A. & Pudritz, R. E., 1988, *Ap. J.*, **330**, 718.

Hjellming, R. M. & Johnston, K. J., 1981a, *Nature*, **290**, 100.

Hjellming, R. M. & Johnston, K. J., 1981b., *Ap. J.*, **246**, L141.

Hughes, V. A., Harten, R. H., Costain, C. H., Nelson, L. A. & Riner, M. R., 1984, *Astron. J.*, **283**, 147.

Hummel, E., van Gorkom, J. H. & Kotanyi, C. G., 1983, *Ap. J.*, **267**, L5.

Jackson, J. M., Ho, P. T. P. & Haschick, A. D., 1988, *Ap. J. Lett.*, **333**, L73.

Johnston, K. J., Geldzahler, B. J., Spencer, J. H., Waltman, E. B., Klepczynski, W. J., Josties, F. J, Angerhofer, P. E., Florkowski, D. R., McCarthy, A. D. & Matsakis, D. N., 1984, *Astron. J.*, **89**, 509.

Johnston, K. J., Spencer, J. H., Simon, R. S., Waltman, E. B., Pooley, G. G., Spencer, R. E., Swinney, R. W., Angerhofer, P. E., Florkowski, D. R., Josties, F. J., McCarthy, A. D., Matsakis, D. N., Reese, D. E. & Hjellming, R. M., 1986, *Ap. J.*, **309**, 707.

Jones, B. F. & Herbig, G. H., 1982, *Astron. J.*, **87**, 1223.

Kahn, F. D., 1976, *Astr. Ap.*, **50**, 145.

Kahn, F. D., 1987, In *Circumstellar Matter*, eds. Appenzeller, I. & Jordan, C. (Reidel: Dordrecht, Netherlands), p. 571.

Kaifu, N., Suzuki, S., Hasegawa, T., Morimoto, M., Inatani, J., Nagane, K., Miyazawa, K., Chikada, Y., Kanzawa, T. & Akabane, K., 1984, *Astr. Ap.*, **134**, 7.

Kassim, N. E., LaRosa, T. N. & Erickson, W. C., 1986, *Nature*, **322**, 522.

Katz, J. I., 1986, *Comm. Ap.*, **11**, 201.

Katz, J. I. & Piran, T., 1982, *Ap. Lett.*, **23**, 11.

Kawabe, R., Ogawa, H., Fukui, Y., Takano, T., Takaba, H., Fujimoto, Y., Sugitani, K. & Fujimoto, M., 1984, *Ap. J.*, **282**, L73.

Kenyon, S. J. & Hartmann, L., 1987, *Ap. J.*, **323**, 714.

Kesteven, M. J., Caswell, J. L., Roger, R. S., Milne, D. K., Haynes, R. F. & Wellington, K. J., 1987, In *The Origin and Evolution of Neutron Stars, IAU Symposium 125*, eds. Helfand, D. J. & Huang, J. -H. (Reidel: Dordrecht, Netherlands), p. 125.

Königl, A., 1982, *Ap. J.*, **261**, 115.

Königl, A., 1983, *M. N. R. A. S.*, **205**, 471.

Königl, A., 1986, *Can. J. Phys.*, **64**, 362.

Königl, A., 1989, *Ap. J.*, **342**, 208.

Koo, Bon-Chul, 1989, *Ap. J.*, **337**, 318.

Kuhi, L. V., 1964, *Ap. J.*, **140**, 1409.

Kulkarni, S. R., Clifton, T. R., Backer, D. C., Foster, R. S., Frutcher, A. S. & Taylor, J. H., 1987, *Nature*, **331**, 50.

Kundt, W., 1985, *Astr. Ap.*, **236**, L127.
Kundt, W., 1987a, *Ap. Sp. Sci.*, **134**, 407.
Kundt, W., 1987b, *Ap. Sp. Sci.*, **129**, 95.
Kundt, W., 1988, In *Astrophysical Jets and their Engines*, ed. Kundt, W. (Reidel: Dordrecht, Netherlands), p. 1.
Kwan, J. & Scoville, N. Z., 1976, *Ap. J.*, **210**, L39.
Kwan, J. & Tademaru, E., 1988, *Ap. J.*, **32**, L41.
Lacy, J. H., Lester, D. F., Arens, J. F., Peck, M. C. & Gaalema, S., 1987, In *The Galactic Centre, Proceedings of the Symposium in honour of Charles H. Townes*, ed. Backer, D. (American Institute of Physics).
Lada, C. J., 1985, *Ann. Rev. Astr. Ap.*, **23**, 267.
Lamb, R. C., Ling, J. C., Mahoney, W. A., Reigler, G. R., Wheaton, W. A. & Jacobson, A. S., 1983, *Nature*, **305**, 37.
LaRosa, T. N. & Kassim, N. E., 1985, *Ap. J.*, **299**, L13.
Larson, R. B., 1969, *M. N. R. A. S.*, **145**, 271.
Larson, R. B., 1984, *M. N. R. A. S.*, **206**, 197.
Lasenby, J., Lasenby, A. N. & Yusef-Zadeh, F., 1989, *Ap. J.*, **343**, 177.
Leibowitz, E. M., 1985, *M. N. R. A. S.*, **210**, 279.
Lenzen, R., 1987, In *Circumstellar Matter*, eds. Appenzeller, I. & Jordan, C. (Reidel: Dordrecht, Netherlands), p. 127.
Lester, D. F., Becklin, E. E., Genzel, R., & Wynn-Williams, C. G., 1985, *Astron. J.*, **90**, 2331.
Liszt, H., 1985, *Ap. J.*, **293**, L65.
Little, L. T., Dent, W. R. F., Heaton, B., Davies, S. R. & White, G. J., 1985, *M. N. R. A. S.*, **217**, 227.
Lizano, S., Heiles, C., Rodriguez, L. F., Koo, B., Shu, F. H., Hasegawa, T., Hayashi, S. & Mirabel, I. F., 1988, *Ap. J.*, **328**, 763.
Loushin, R., Crutcher, R. & Bieging, J., 1988, In *Molecular Clouds in the Milky Way and External Galaxies*, eds. Dickman, R. L., Snell, R. L. & Young, J. S. (Springer Verlag: Berlin).
MacCullen, C. J., Huters, A. F., Stury, P. D. & Leventhal, M., 1985, *Ap. J.*, **291**, 486.
Manchester, R. N., 1987, *Astr. Ap.*, **171**, 205.
Manchester, R. N. & Durdin, J. M., 1983, In *Supernova Remnants and their X-ray Emission, IAU Symposium 101*, eds. Danziger, I. J. & Gorenstein, P. (Reidel: Dordrecht, Netherlands), p. 421.
Manchester, R. N., D'Amico, N. & Tuohy, I. R., 1985, *M. N. R. A. S.*, **212**, 975.
Mantovani, F., Reich, W., Salter, C. J. & Tomasi, P., 1985, *Astr. Ap.*, **145**, 50.
Margon, B., 1981, *Ann. New York Acad. Sci.*, **375**, 403.
Margon, B., 1984, *Ann. Rev. Astr. Ap.*, **22**, 507.
Margon, B., Anderson, S. F., Aller, L. H., Downes, R. A. & Keyes, C. D., 1984, *Ap. J.*, **281**, 313.
Mazeh, T., Kemp, J. C., Leibowitz, E. M., Meningher, H. & Mendelson, H., 1987, *Ap. J.*, **317**, 824.
Milgrom, A., 1979a, *Astr. Ap.*, **76**, L3.
Milgrom, A., 1979b, *Astr. Ap.*, **78**, L9.

Milne, D. K., Goss, W. M., Kesteven, M. J., Haynes, R. F., Wellington, K. J., Caswell, J. L. & Skellern, D. J., 1979, *M. N. R. A. S.*, **316**, 44.
Mitchell, G. F., Allen, M., Beer, R., Dekany, R., Huntress, W. & Maillard, J., 1988, *Ap. J.*, **327**, L17.
Moriarty-Schieven, G. H. & Snell, R. L., 1988, *Ap. J.*, **332**, 364.
Moriarty-Schieven, G. H., Snell, R. L., Strom, S. E., Schloerb, F. P., Strom, K. M. & Grasdalen, G. L., 1987, *Ap. J.*, **319**, 742.
Morrison, P. & Roberts, D., 1985, *Nature*, **313**, 661.
Mouschovias, T. Ch., 1981, In *Fundamental Problems in the Theory of Stellar Evolution*, eds. Sugimoto, D., Lamb, D. Q. & Schramm, D. N. (Reidel: Dordrecht, Netherlands), p. 127.
Mundt, R., 1986, *Can. J. Phys.*, **64**, 407.
Mundt, R., 1988, In *Formation and Evolution of Low Mass Stars*, eds. Dupree, A. K. & Lago, M. T. V. T. (Kluwer: Dordrecht, Netherlands), p. 257.
Mundt, R., Brugel, E. W. & Bührke, T., 1987, *Ap. J.*, **319**, 275.
Ozernoy, L., 1990, In *The Galactic Center, IAU Symposium 136*, ed. Morris, M. (Kluwer: Dordrecht, Netherlands), in press.
Padman, R. & Richer, J. S., 1990, In *Submillimetre and Millimetre Wave Astronomy*, ed. Webster, A. S. (Kluwer: Dordrecht, Netherlands), in preparation.
Panagia, N. & Felli, M., 1975, *Astr. Ap.*, **39**, 1.
Pauls, T., Downes, D., Mezger, P. G. & Churchwell, E., 1976, *Astr. Ap.*, **46**, 407.
Persson, S. E., McGregor, P. J. & Campbell, B., 1988, *Ap. J.*, **326**, 339.
Phillips, J. P. & Beckman, J. E., 1980, *M. N. R. A. S.*, **193**, 245.
Pringle, J. E., 1989, *M. N. R. A. S.*, **236**, 107.
Pudritz, R. E., 1988, In *Galactic and Extragalactic Star Formation*, eds. Pudritz, R. E. & Fich, M. (Kluwer: Dordrecht, Netherlands), p. 135.
Pudritz, R. E., 1990, In *The Galactic Center, IAU Symposium 136*, ed. Morris, M. (Reidel: Dordrecht, Netherlands), in press.
Pudritz, R. E. & Norman, C. A., 1983, *Ap. J.*, **274**, 677.
Pudritz, R. E. & Norman, C. A., 1985, *Ap. J.*, **301**, 571.
Rees, M. J., 1982, In *Extragalactic Radio Sources, IAU Symposium 97*, eds. Heeschen, D. S. & Wade, C. M. (Reidel: Dordrecht, Netherlands), p. 211.
Reich, W., Fürst, E., Steffen, P., Reif, K. & Haslam, C. G. T., 1984, *Astr. Ap. Suppl.*, **58**, 197.
Reipurth, B., 1989, *Astr. Ap.*, **220**, 249.
Reipurth, B., Bally, J., Graham, J. A., Lane, A. P. & Zealey, W. J., 1986, *Astr. Ap.*, **164**, 51.
Richer, J. S., Hills, R. E., Padman, R., Scott, P. F. & Russell, A. P. G., 1990, In *Submillimetre and Millimetre Wave Astronomy*, ed. Webster, A. S. (Kluwer: Dordrecht, Netherlands), in preparation.
Rodriguez, L. F., Cantó, J., Torrelles, J. M. & Ho, P. T. P., 1986, *Ap. J.*, **301**, 25.
Rodriguez, L. F., 1987, In *Star Forming Regions*, eds. Peimbert, M. & Jugaku, J. (Reidel: Dordrecht, Netherlands), p. 239.
Roger, R. S., Milne, D. K., Kesteven, M. J., Haynes, R. F. & Wellington, K. J., 1985, *Nature*, **316**, 44.

Romney, J. D., Schilizzi, R. T., Fejes, I. & Spencer, R. E., 1987, *Ap. J.*, **321**, 822.
Rudolph, A. & Welch, W. J. W., 1988, *Ap. J.*, **326**, L31.
Russell, A. P. G., 1987, *Ph. D. Thesis*, University of Cambridge.
Sanders, D. B. & Willner, S. P., 1985, *Ap. J.*, **293**, L39.
Sargent, A. I. & Beckwith, S., 1987, *Ap. J.*, **323**, 294.
Sato, S., Nagata, T., Nakajima, T., Nishida, M., Tanaka, M. & Yamashita, T., 1985, *Ap. J.*, **291**, 708.
Schwartz, R. D., 1983, *Ann. Rev. Astr. Ap.*, **21**, 209.
Scoville, N. Z., 1981, In *Infrared Astronomy, IAU Symposium 96*, eds. Wynn-Williams, C. G. & Cruikshank, D. P. (Reidel: Dordrecht, Netherlands), p. 187.
Scoville, N. Z. & Kwan, J., 1976, *Ap. J.*, **206**, 718.
Seaquist, E., Gilmore, W., Johnston, K. J. & Grindlay, J., 1982, *Ap. J.*, **260**, 220.
Seiradakis, J. H., Lasenby, A. N., Yusef-Zadeh, F., Wielebinski, R. & Klein, U., 1985, *Nature*, **317**, 697.
Serabyn, E., 1987, In *Astrophysical Jets and their Engines*, ed. Kundt, W. (Reidel: Dordrecht, Netherlands), p. 47.
Seward, F., Grindlay, J., Seaquist, E. & Gilmore, W., 1980, *Nature*, **287**, 806.
Shapiro, P. R., Milgrom, M. & Rees, M. J., 1982, In *Extragalactic Radio Sources, IAU Symposium 97*, eds. Heeschen, D. S. & Wade, C. M. (Reidel: Dordrecht, Netherlands), p. 209.
Shapiro, P. R., Milgrom, M. & Rees, M. J., 1986, *Ap. J. Suppl.*, **60**, 393.
Sharp, N. A., Calvani, M. & Turolla, R., 1984, *Comm. Ap.*, **10**, 53.
Shaver, P. A., Salter, C. J., Patnaik, A. R., van Gorkom, J. H. & Hunt, G. C., 1985a, *Nature*, **313**, 113.
Shaver, P. A., Pottasch, S. R., Salter, C. J., Patnaik, A. R., van Gorkom, J. H. & Hunt, G. C., 1985b, *Astr. Ap.*, **147**, L23.
Shibata, K. & Uchida, Y., 1987, *Publ. Astron. Soc. Japan*, **39**, 559.
Shu, F. H., Adams, F. C. & Lizano, S., 1987, *Ann. Rev. Astr. Ap.*, **25**, 23.
Shu, F. H., Lizano, S., Ruden, S. P. & Najita, J., 1988, *Ap. J.*, **328**, L19.
Simon, T., Simon, M. & Joyce, R. R., 1979, *Ap. J.*, **230**, 127.
Snell, R. L. & Edwards, S., 1981, *Ap. J.*, **251**, 103.
Snell, R. L. & Schloerb, P. F., 1985, *Ap. J.*, **295**, 490.
Snell, R. L., Loren, R. B. & Plambeck, R. L., 1980, *Ap. J.*, **239**, L17.
Snell, R. L., Scoville, N. Z., Sanders, D. B. & Erickson, N. R., 1984, *Ap. J.*, **284**, 176.
Snell, R. L., Bally, J., Strom, S. E. & Strom, K. M., 1985, *Ap. J.*, **290**, 587.
Sofue, Y., 1984, *Publ. Astron. Soc. Japan*, **36**, 539.
Sofue, Y., 1986, *Can. J. Phys.*, **64**, 527.
Sofue, Y., 1990, In *The Galactic Center, IAU Symposium 136*, ed. Morris, M. (Kluwer: Dordrecht, Netherlands), in press.
Sofue, Y. & Fujimoto, M., 1987, *Ap. J.*, **319**, L73.
Sofue, Y. & Handa, T., 1984, *Nature*, **310**, 568.
Sofue, Y. & Reich, W., 1979, *Astr. Ap. Suppl.*, **38**, 25.
Spencer, R. E., 1984, *M. N. R. A. S.*, **209**, 869.
Spencer, R. E., Swinney, R. W., Johnston, K. J. & Hjellming, R. M., 1986, *Ap. J.*, **309**, 694.

Strittmatter, P. A. & Williams, R. E., 1976, *Ann. Rev. Astr. Ap.*, **14**, 307.

Strom, R. G., 1987, *Ap. J.*, **319**, L103.

Strom, R. G., Angerhofer, P. E. & Dickel, J. R., 1984, *Astr. Ap.*, **139**, 43.

Strom, S. E., Grasdalen, G. L. & Strom, K. M., 1974, *Ap. J.*, **191**, 111.

Strom, S. E., Strom, K. M., Grasdalen, G. L., Capps, R. W. & Thompson, DeAnne, 1985, *Astron. J.*, **90**, 2575.

Takano, T., Fukui, Y., Ogawa, H., Takaba, H., Kawabe, R., Fujimoto, Y., Sugitani, K. & Fujimoto, M., 1984, *Ap. J.*, **282**, L69.

Torrelles, J. M., Rodriguez, L. F., Cantó, J., Carral, P., Marcaide, J., Moran, J. M. & Ho, P. T. P., 1983, *Ap. J.*, **274**, 214.

Tsuboi, M., Inoue, M., Handa, T., Tabara, H. & Kato, T., 1985, *Publ. Astron. Soc. Japan*, **37**, 359.

Uchida, Y. & Shibata, K., 1985, *Publ. Astron. Soc. Japan*, **36**, 105.

Uchida, Y., Shibata, K. & Sofue, Y., 1985, *Nature*, **317**, 699.

Uchida, Y., Kaifu, N., Shibata, K., Hayashi, S. S. & Hasegawa, T., 1987, In *Star Forming Regions*, eds. Peimbert, M. & Jugaku, J. (Reidel: Dordrecht, Netherlands), p. 287.

Van Buren, D., 1985, *Ap. J.*, **294**, 567.

van den Bergh, S., 1970, *Ap. J.*, **160**, L27.

Velusamy, T., 1984, *Nature*, **308**, 251.

Vermeulen, R. C., Schilizzi, R. T., Icke, V., Fejes, I. & Spencer, R. E., 1987, *Nature*, **328**, 309.

Vogel, S. N., Bieging, J. H., Plambeck, R. L., Welch, W. J. & Wright, M. C. H., 1985, *Ap. J.*, **296**, 600.

Warren-Smith, R. F., Draper, P. W. & Scarrott, S. M., 1987, *M. N. R. A. S.*, **227**, 749.

Watson, M. G., Stewart, G. C., Brinkmann, W. & King, A. R., 1986, *M. N. R. A. S.*, **222**, 261.

Watson, M. G., Willingdale, R., Grindlay, J. E. & Seward, F. D., 1983, *Ap. J.*, **273**, 688.

Weaver, R., McCray, R., Castor, J., Shapiro, P. & Moore, R., 1977, *Ap. J.*, **218**, 377.

Weiler, K. W., 1983, *Observatory*, **103**, 85.

Williams, A. G. & Gull, S. F., 1984, *Nature*, **310**, 33.

Williams, A. G. & Gull, S. F., 1985, *Nature*, **313**, 34.

Wilson, A. S., 1981, In *Extragalactic Radio Astronomy*, eds. Heeschen, D. S. & Wade, C. M. (Reidel: Dordrecht, Netherlands), p. 179.

Wilson, A. S., Samarasinha, N. H. & Hogg, D. E., 1985, *Ap. J.*, **294**, L121.

Wright, A. E. & Barlow, M. J., 1975, *M. N. R. A. S.*, **170**, 41.

Yusef-Zadeh, F., 1986, *Ph. D. Thesis*, Columbia University.

Yusef-Zadeh, F. & Bally, J., 1990, In *The Galactic Center, IAU Symposium 136*, ed. Morris, M. (Kluwer: Dordrecht, Netherlands), in press.

Yusef-Zadeh, F. & Morris, M., 1987a, *Ap. J.*, **320**, 545.

Yusef-Zadeh, F. & Morris, M., 1987b, *Ap. J.*, **322**, 721.

Yusef-Zadeh, F. & Morris, M., 1987c, *Astron. J.*, **94**, 1178.

Yusef-Zadeh, F., Morris, M. & Chance, D., 1984, *Nature*, **310**, 557.

Yusef-Zadeh, F., Morris, M., Slee, O. B. & Nelson, G. J., 1986, *Ap. J.*, **300**, L47.
Zuckerman, B., Kuiper, T. B. H. & Rodriguez-Kuiper, E. N., 1976, *Ap. J.*, **209**, L137.

Index of Objects

Index of Subjects